AF182220

Progress in Botany

Volume 82

Series Editors
Ulrich Lüttge
Department of Biology
Technical University of Darmstadt
Darmstadt, Germany

Francisco M. Cánovas
Department of Molecular Biology and Biochemistry
University of Málaga
Málaga, Spain

Hans Pretzsch
Chair for Forest Growth and Yield Science
School of Life Sciences Weihenstephan
Technical University of Munich
Freising, Bayern, Germany

María-Carmen Risueño
Pollen Biotechnology of Crop Plants Group
Centro de Investigaciones Biologicas (CIB) CSIC Madrid
Madrid, Spain

Christoph Leuschner
Plant Ecology
University of Goettingen
Göttingen, Germany

Progress in Botany is devoted to all the colourful aspects of plant biology. The annual volumes consist of invited reviews spanning the fields of molecular genetics, cell biology, physiology, comparative morphology, systematics, ecology, biotechnology and vegetation science, and combine the depth of the frontiers of research with considerable breadth of view. Thus, they establish unique links in a world of increasing specialization. Progress in Botany is engaged in fostering the progression from broad information to advanced knowledge and finally deepened understanding. All chapters are thoroughly peer-reviewed.

More information about this series at http://www.springer.com/series/124

Francisco M. Cánovas • Ulrich Lüttge •
María-Carmen Risueño • Hans Pretzsch
Editors

Progress in Botany Vol. 82

Springer

Editors
Francisco M. Cánovas
Department of Molecular Biology
and Biochemistry
University of Málaga
Málaga, Spain

Ulrich Lüttge
Department of Biology
Technical University of Darmstadt
Darmstadt, Germany

María-Carmen Risueño
Pollen Biotechnology of Crop Plants Group
Centro de Investigaciones Biologicas (CIB)
CSIC Madrid
Madrid, Spain

Hans Pretzsch
Chair for Forest Growth and Yield Science
School of Life Sciences Weihenstephan
Technical University of Munich
Freising, Bayern, Germany

ISSN 0340-4773 ISSN 2197-8492 (electronic)
Progress in Botany
ISBN 978-3-030-68619-2 ISBN 978-3-030-68620-8 (eBook)
https://doi.org/10.1007/978-3-030-68620-8

© The Editor(s) (if applicable) and The Author(s), under exclusive license to Springer Nature Switzerland AG 2021
This work is subject to copyright. All rights are solely and exclusively licensed by the Publisher, whether the whole or part of the material is concerned, specifically the rights of translation, reprinting, reuse of illustrations, recitation, broadcasting, reproduction on microfilms or in any other physical way, and transmission or information storage and retrieval, electronic adaptation, computer software, or by similar or dissimilar methodology now known or hereafter developed.
The use of general descriptive names, registered names, trademarks, service marks, etc. in this publication does not imply, even in the absence of a specific statement, that such names are exempt from the relevant protective laws and regulations and therefore free for general use.
The publisher, the authors, and the editors are safe to assume that the advice and information in this book are believed to be true and accurate at the date of publication. Neither the publisher nor the authors or the editors give a warranty, expressed or implied, with respect to the material contained herein or for any errors or omissions that may have been made. The publisher remains neutral with regard to jurisdictional claims in published maps and institutional affiliations.

This Springer imprint is published by the registered company Springer Nature Switzerland AG.
The registered company address is: Gewerbestrasse 11, 6330 Cham, Switzerland

Contents

Contributors

Marcel Pascal Beier Graduate School of Agricultural Science, Tohoku University, Sendai, Japan
Graduate School of Agricultural and Life Sciences, The University of Tokyo, Tokyo, Japan

Michaela Böhm Department of Biology, Friedrich-Alexander-University, Erlangen, Germany

Heather D. Coleman Biology Department, Syracuse University, Syracuse, NY, USA

Antonio de Vicente Departamento de Microbiología, Universidad de Málaga, Málaga, Spain
Instituto de Hortofruticultura Subtropical y Mediterránea "La Mayora", Universidad deMálaga, Consejo Superior de Investigaciones Científicas (IHSM–UMA–CSIC), Málaga, Spain

Luis A. del Río Group of Antioxidants, Free Radicals and Nitric Oxide in Biotechnology, Food and Agriculture, Department of Biochemistry and Cell and Molecular Biology of Plants, Estación Experimental del Zaidín, Consejo Superior de Investigaciones Científicas (CSIC), Granada, Spain

Rebecca Ellingston Biology Department, Syracuse University, Syracuse, NY, USA

Alberto Fereres Instituto de Ciencias Agrarias – Consejo Superior de Investigaciones Científicas (ICA-CSIC), Madrid, Spain

Dolores Fernández-Ortuño Departamento de Microbiología, Universidad de Málaga, Málaga, Spain
Instituto de Hortofruticultura Subtropical y Mediterránea "La Mayora", Universidad deMálaga, Consejo Superior de Investigaciones Científicas (IHSM–UMA–CSIC), Málaga, Spain

R. Grote Karlsruhe Institute of Technology, Institute of Meteorology and Climate Research, Atmospheric Environmental Research, Garmisch-Partenkirchen, Germany

Toshihiko Hayakawa Graduate School of Agricultural Science, Tohoku University, Sendai, Japan

Riley Higgins Biology Department, Syracuse University, Syracuse, NY, USA

José I. Hormaza Department of Subtropical Fruit Crops, Instituto de Hortofruticultura Subtropical y Mediterránea "La Mayora" (IHSM–UMA–CSIC), Málaga, Spain

Keiki Ishiyama Graduate School of Agricultural Science, Tohoku University, Sendai, Japan

Jaime Jiménez Instituto de Ciencias Agrarias – Consejo Superior de Investigaciones Científicas (ICA-CSIC), Madrid, Spain
Department of Entomology and Nematology, University of Florida, Gainesville, FL, USA

Joohyun Kang Division of Integrative Bioscience and Biotechnology, POSTECH, Pohang, Republic of Korea
Institute of Plant and Microbial Biology, University Zurich, Zurich, Switzerland

Soichi Kojima Graduate School of Agricultural Science, Tohoku University, Sendai, Japan

Christian Körner Department of Environmental Sciences, Institute of Botany, University of Basel, Basel, Switzerland

G. Heinrich Krause Institute of Plant Biochemistry, Heinrich Heine University Düsseldorf, Düsseldorf, Germany

Georg Kreimer Department of Biology, Friedrich-Alexander-University, Erlangen, Germany

Youngsook Lee Division of Integrative Bioscience and Biotechnology, POSTECH, Pohang, Republic of Korea

Jorge Lora Department of Subtropical Fruit Crops, Instituto de Hortofruticultura Subtropical y Mediterránea "La Mayora" (IHSM–UMA–CSIC), Málaga, Spain

Ulrich Lüttge Department of Biology, Technical University of Darmstadt, Darmstadt, Germany

Pierre Marraccini CIRAD, UMR IPME, Montpellier, France
IPME, University Montpellier, CIRAD, IRD, Montpellier, France
Agricultural Genetics Institute, LMI RICE2, Hanoi, Vietnam

Jesús Martínez-Cruz Departamento de Microbiología, Universidad de Málaga, Málaga, Spain
Instituto de Hortofruticultura Subtropical y Mediterránea "La Mayora", Universidad deMálaga, Consejo Superior de Investigaciones Científicas (IHSM–UMA–CSIC), Málaga, Spain

Enrico Martinoia Division of Integrative Bioscience and Biotechnology, POSTECH, Pohang, Republic of Korea
Institute of Plant and Microbial Biology, University Zurich, Zurich, Switzerland

Aránzazu Moreno Instituto de Ciencias Agrarias – Consejo Superior de Investigaciones Científicas (ICA-CSIC), Madrid, Spain

A. Moser-Reischl Chair for Forest Growth and Yield Science, Technical University Munich, Freising, Germany

Yerim Park Biology Department, Syracuse University, Syracuse, NY, USA

S. Pauleit Chair of Strategic Landscape Planning and Management, Technical University of Munich, Freising, Germany

Alejandro Pérez-García Departamento de Microbiología, Universidad de Málaga, Málaga, Spain
Instituto de Hortofruticultura Subtropical y Mediterránea "La Mayora", Universidad deMálaga, Consejo Superior de Investigaciones Científicas (IHSM–UMA–CSIC), Málaga, Spain

Álvaro Polonio Departamento de Microbiología, Universidad de Málaga, Málaga, Spain
Instituto de Hortofruticultura Subtropical y Mediterránea "La Mayora", Universidad deMálaga, Consejo Superior de Investigaciones Científicas (IHSM–UMA–CSIC), Málaga, Spain

H. Pretzsch Chair for Forest Growth and Yield Science, Technical University Munich, Freising, Germany

M. A. Rahman Chair of Strategic Landscape Planning and Management, Technical University of Munich, Freising, Germany

T. Rötzer Chair for Forest Growth and Yield Science, Technical University Munich, Freising, Germany

Bethanie Viele Biology Department, Syracuse University, Syracuse, NY, USA

Dan Wang Biology Department, Syracuse University, Syracuse, NY, USA

Klaus Winter Smithsonian Tropical Research Institute, Balboa, Ancon, Republic of Panama

Curriculum Vitae

Christian Körner was born in 1949, grew up in Salzburg, Austria (1949) and enrolled at the University of Innsbruck (Austria). In 1973, he received a Master's diploma qualifying him for a teacher's career in natural history. Yet, in the same year, he started teaching undergraduates in botany in Innsbruck and received a PhD in 1977 for his thesis on plant–water relations of alpine plants. In 1982, he became a senior lecturer ('habilitation' for botany), and in 1989, he was appointed full professor for botany at the University of Basel, Switzerland (retired in 2014).

Academic memberships and awards: Austrian Academy of Sciences, German National Academy of Sciences Leopoldina, Honorary Membership of the Ecological Society of America, Marsh Award of the British Ecological Society, Dr. h.c. Illia University Tbilisi and University of Innsbruck.

Academic services: 25 years editorial board and editor in chief for Oecologia, editorial boards of Science magazine, Trends in Ecology and Evolution, Global Change Biology, and others. Chair of Global Mountain Biodiversity Assessment (GMBA), task leader in GCTE/IGBP, Chair of Proclim, the Swiss climate change platform, and of the Forum Biodiversity, both at the Swiss Academy of Sciences. Various academic functions at the University of Basel, with that as an ombudsperson on-going.

Publications: c. 400, several edited books, textbook on Alpine Plant Life and Alpine Treeline, and the plant ecology part of the major university text 'Strasburger's Plant Sciences' around 60,000 citations in Google scholar.

Major research fields: alpine ecology, forest ecology, influences of temperature, water, and CO_2 on plants and ecosystems.

For more details see: https://duw.unibas.ch/en/koerner/

Tools Shape Paradigms of Plant-Environment Interactions

Christian Körner

Contents

Abstract The direction of science is often driven by contemporary theory, and theory emerges from consolidated empirical knowledge. What we know emerges from what we explore, and we explore what we have technical tools for. I feel that

C. Körner (✉)
Department of Environmental Sciences, Institute of Botany, University of Basel, Basel, Switzerland
e-mail: ch.koerner@unibas.ch

© Springer Nature Switzerland AG 2020

Progress in Botany (2021) 82: 1–42, https://doi.org/10.1007/124_2020_41,
Published online: 29 December 2020

technical opportunities contributed strongly towards what is held as a contemporary, widely accepted theory. However, the presumed causality may become reverted, if one accounts for those less explored questions, for which tools are missing. Here, I will reflect on decades of research experience in empirical plant sciences, mainly plant water relations, plant carbon relations and biogeography, during which some mainstream paradigms became challenged. Scientific theory passes through waves and cycles and is even linked to fashion. Insight that seemed established at one time may become outdated by novel concepts facilitated by novel methods, and as time progresses, old concepts may find a revival. In the following chapter, I will illustrate such shifts in awareness and misleading paradigms that were driven by the contemporary availability of methods rather than stringent logics. Examples include plant responses to drought stress; the drivers of plant growth in general, as well as in the context of rising atmospheric CO_2 concentrations; and how physiological plant ecology can contribute to resolving biogeographical questions such as range limits of plant species and plant life forms. My résumé is that explanations of plant responses to the environment are predominantly below ground and require an understanding of developmental and meristematic processes, whereas available tools often lead to attempts at above-ground answers based on primary metabolism (e.g. photosynthesis). Further, well-understood processes at the organ (leaf) level are losing relevance at the community or ecosystem level, where much less understood mechanisms come into action (e.g. stand density control). While the availability of certain convenient methods can open new research arenas, it may also narrow the scope and may direct theory development towards easily measurable parameters and processes.

Keywords CO_2, Drought, Elevation, Growth, Hydraulics, Limitation, Low temperature, Meristems, Nutrients, Photosynthesis, Productivity, Stomata, Water potential, Xylem

1 Prologue

My parents took us, my brother and me, to wild places, such as unspoiled lakeshores or Mediterranean sand dunes where we camped for 3 weeks away from civilization. This was in the 1950s. I had my own garden plot when I was 7, but we never could afford a car with the post-war income my father earned as a community engineer in Salzburg (Austria) and my mother as a freelance professional photographer. I decided to become a biologist at 10. At that age, I entered the 8-year science-oriented gymnasium, with subject-specific teachers, some of whom stayed with class for all 8 years till final exams. One of them was my biology teacher. For a 10-year-old, it was almost unbelievable that talking about plants and animals can be a profession. No question, this was going to become my job. After graduation and founding a family in Innsbruck, my school teacher career became very brief, however. Walter Larcher, my later doctor father, felt I should better teach undergraduates at the University, and he offered me a job which also paid me through my PhD years from 1973 to 1977, a rather fortunate situation, because in those days, doctoral students were not paid and project money was scarce, a situation explained so well by Erwin Beck in his reflections (Beck 2016).

My graduation project in 1972 had already paved the path towards empirical plant biology. I was given the task to develop and test a so-called leaf diffusion porometer. Today an indispensable instrument to assess plant-water relations (specifically, the degree of stomatal opening in leaves), there was no trustworthy technical solution on the market in the early 1970s. The mechanical, electronic and aerodynamic challenges were substantial. It so happened that W. Larcher had just employed the young physicist Alexander Cernusca to establish environmental physics at Botany. In the post-war years, his father built the first automatic, biometeorological station for the federal high-elevation forest research group that was based in Innsbruck as well (Aulitzky 1961). They worked on treeline questions, a field I wrote a book on, many years later (Körner 2012). Growing up, academically, door to door with a physicist with such interests and skills (e.g. Cernusca 1976), helped me in developing this porometer but also exposed me to micrometeorology and physical solutions to plant science questions, which shaped my thinking till today. With a brand-new design of an electronic humidity sensor made in Switzerland (the country that should later become my academic home), my instrument not only worked very well (Körner and Cernusca 1976) but also made it a hot seller by the then German company Walz. They produced a semi-automatic version and sold 14 instruments in 2 years, before LI-COR stepped in with their null-balance porometer around 1978. I was proud that LI-COR adopted my absolute calibration routine with capillary evaporimeters, the readings of which revealed the amount of vapour that entered the chamber per unit of time and caused the humidity sensor to produce a signal as if a leaf were releasing vapour to the chamber. With these experiences, I was set to become an *experimental plant biologist* or *physiological ecologist*.

In the late 1960s, Botany at Innsbruck had joined the International Biological Program, IBP, that aimed at assessing the biological productivity of terrestrial ecosystems and explaining its main drivers. The Innsbruck team led by W. Larcher belonged to the so-called 'Tundra-Biome' module, contributing to alpine heath productivity and its climatic drivers and carbon, nutrient and water relations (Larcher 1977). I became the 'water man' and learned to appreciate the added value of multi-thematic approaches and international comparisons and cooperation. So, my second academic heritage from those years was the *comparative approach* in field-oriented plant sciences.

I was very fortunate that I could meet in person, both pioneers of comparative physiological ecology, Otto Stocker (Darmstadt; Lüttge 1979; see Stocker 1956, 1976 as examples) and Arthur Pisek (Innsbruck; Larcher 1975). I believe my entire work was in their tradition that was so elegantly developed into the modern era by Walter Larcher, a student of Pisek (see Larcher 2003). I still remember the occasion when I explained my porometer to the ca. 80-year-old Arthur Pisek who had already played with the idea to measure directly what he was exploring indirectly (Pisek et al. 1970). His way of a comparative, ecophysiological characterization of as broad as possible a spectrum of species belonging to a specific habitat type was novel but also timeless and guided much of my alpine research (examples in Pisek and Cartellieri 1934, Pisek 1960, Pisek et al. 1973). When I decided to work with W. Larcher, it was for this conceptual framework that always aimed at the 'big picture', the overarching pattern. Species always differ, and these differences matter and need to be accounted for. In a modern phrase, this would be addressed as *functional biodiversity research*, the third legacy from that period.

And then, in 1989, it happened that I was offered a full professorship at the University of Basel, Switzerland. My inauguration lecture at Botany in 1990 was entitled 'The Art of Asking Small Questions', meaning that very elaborate methods may consume so much time and energy that the work becomes restricted to one species or even one genotype. Hence, to cover diversity and obtain a big sample and eventually arrive at a 'big picture', the challenge is rather to ask cleverly designed 'small questions' tied to simple but meaningful traits that can be explored in many species and locations (discussion in Körner 2018a). Leaving Austria, I brought with me the history of 100 years of alpine research in Innsbruck, mostly published in German, which, I felt, deserves wider visibility, complemented by more recent data and insights I took from travelling many other mountain regions during the 1980s. In 1995, I decided to wrap this up in what became *Alpine Plant Life* (Körner 1999a, 2003a, 2021). I started my professional life in Basel with a sense for multidisciplinary work, a reasonable foundation in environmental physics, and an open mind for new methods and alternative explanations of plant responses to the environment. The following four sections summarize ideas I gained over almost half a century of empirical work with plants (see Körner 2018a).

2 Plant-Water Relations Revisited

As water moves from the soil to the atmosphere through a plant, it passes through a series of resistors. In analogy to Ohm's law, the drivers, resistors and resulting fluxes can be identified and measured (including the variable stomatal diffusive resistance as defined by Gaastra 1959; see also Cowan 1965). My porometer permitted to quantify and monitor the diffusive resistance of the leaf epidermis (or the reciprocal, the leaf diffusive conductance, g) – a tool and new concepts for leaf gas exchange that emerged in these days that shaped my early approaches to plant-water relations.

2.1 Stomatal Conductance Correlates with Photosynthesis

In the 1970s, it was still possible to know almost everything about a subject. Within a very short period of time, porometers had produced hundreds of data sets from

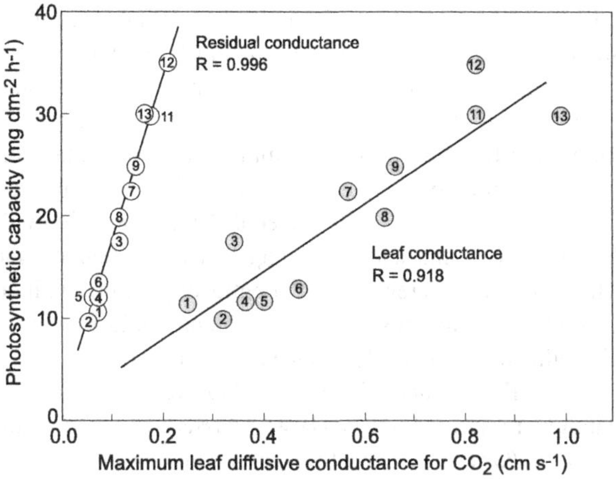

Fig. 1 Maximum leaf diffusive conductance (g) and maximum rate of net photosynthesis (A) of leaves of C3 plants correlate linearly across the major plant functional types across the globe (Körner et al. 1979). *1* succulents, *2* evergreen conifers, *3* deciduous woody plants, *4* herbs from shaded habitats, *5* evergreen broadleaved wood plants, *6* desert and steppe shrubs, *7* deciduous orchard trees, *8* wild-growing graminoids, *9* cultivated C3 grasses, *10* cultivated C4 grasses (not shown; g similar to 11, but A twice as high), *11* herbaceous crop plants, *12* wild herbs from open habitats, *13* plants from aquatic habitats and swamps

around the globe. How do they compare? What are the patterns across the plant kingdom? While IBP was clearly after 'big global patterns', my stomata world was ready for a search for big patterns as well. Indeed, maximum stomatal conductance turned out to correlate linearly with the maximum rate of photosynthesis (then called photosynthetic capacity) across plant life forms (Fig. 1; Körner et al. 1979). So, the conductance for gas diffusion turned out to be tightly associated with the capacity of the mesophyll to absorb CO_2. The regression shown in Fig. 1 was strongly driven by plant functional groups with low and high extremes of g_{max}. Years later, a closer look at the central range of the regression, where almost all tree taxa are nested, revealed a surprisingly narrow range of maximum leaf diffusive conductance, averaging at 218 ± 24 mmol m^{-2} s^{-1} across all 151 taxa examined from different climatic regions (expressed in the molar units that became more popular since then; Körner 1994). While still allowing for some interspecific variation, such a unifying number is rare in plant ecology. What made the variation to condense was the application of a common projected leaf area reference, even when leaves had a strong three-dimensional shape such as in some conifer needles.

The 1979 *Photosynthetica* paper took more than a year to become printed after acceptance. By pure coincidence, the mechanism of this interrelationship shown in Fig. 1 was explained by a group in Canberra at the Australian National University in a paper by Wong et al. (1979) that appeared at the same time. Both studies illustrate that plants optimize the gain in C for the unavoidable loss of water, solving the eternal compromise between 'hunger and thirst' in plants. Known for authorities like

Ralph Slatyer, Barry Osmond, Ian Cowan and Graham Farquhar, this was the place to go for a postdoc. The time in Canberra became most influential in several respects.

I started in Canberra, by testing stomatal response theories along an elevation gradient in the Australian Snowy Mountains. Although the idea that stomata should respond to evaporative forcing (ambient humidity) in a way that minimizes water loss for a given amount of carbon captured was around for several years (Cowan and Troughton 1971, Lange et al. 1971, Schulze et al. 1972), the field evidence was still scarce. *Eucalyptus* in Australia and *Nothofagus* in New Zealand should be ideal objects to test this hypothesis in regions with contrasting humidity. Indeed, both taxa exhibited a clear-cut stomatal response to ambient humidity that could *not* be explained by drought stress (critically reduced leaf water potential as assessed with a pressure chamber; Körner and Cochrane 1985, Körner et al. 1986). Once a threshold of vapour pressure deficit (vpd) was passed, g declined almost linearly. For *Eucalyptus pauciflora* in a dry atmosphere, the vpd threshold for stomatal control of transpiration was twice as high (c. 15 hPa) as for *Nothofagus menziesii* in a humid environment (c. 8 hPa). So, stomata clearly control vapour flux even in well-watered trees, and there was adaptation to the regional climate. Later it was shown that such vpd responses also relate to both photosynthetic capacity and xylem hydraulics (Franks and Farquhar 1999).

2.2 *Transpiration Drives Shoot Water Potential*

A second observation during this field work with *Eucalyptus pauciflora* in the Snowy Mountains made me rethink the contemporary plant-water relations concepts. Counter to widely held assumptions, leaf water potentials became reduced as the rate of transpiration rose, as long as soil moisture was not severely limiting (pushing leaf water potential beyond a critical threshold). The popular view was (and still often is) that a declining leaf water potential should affect stomata and these should reduce flux. In reality, vpd and g jointly determine transpiration, with the inevitable consequence that the leaf water potential must fall across the given hydraulic resistances as long as the water tension remains subcritical for stomatal action (for most trees less than −1.6 to −1.8 MPa). Except when the water potential falls below such thresholds (when deep roots fail to reach adequate soil moisture), leaf water potential is a result of transpiration and not the other way round (Fig. 2; for an early account see Landsberg et al. 1975). At any point in the hydraulic capillary continuum from the root cortex to the leaf mesophyll, the drop in water potential is the consequence of a flux across resistors, with the atmospheric evaporative forcing 'pulling' the inelastic water capillaries across all these resistors. This hierarchy of causalities is still not widely appreciated.

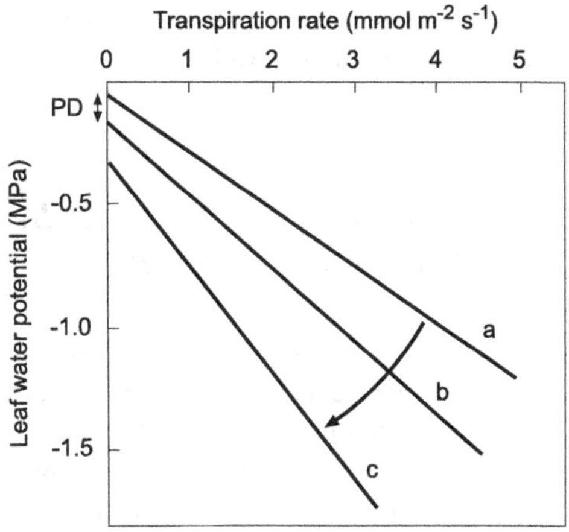

Fig. 2 Unless soils are severely dehydrated, the rate of transpiration determines the reduction of shoot water potential (not the other way round). A conceptual diagram based on works by Körner and Cochrane 1985, published in a standard textbook. PD pre-dawn leaf water potential (Körner 2013a)

2.3 A Water/Carbon Relations Proxy: Stable Carbon Isotope Discrimination

For me as the 'water man' from Innsbruck, the time in Canberra (1980/1981 and 1989) opened an arena of new topics: stable carbon isotopes in plants (thanks to Graham Farquhar) and plant nutrient relations and plant carbohydrate storage (thanks to Suan Chin Wong). They had a new mass-spectrometer lab there, the CSIRO Department of Forestry next door had an automated nutrient analysis facility, and Chin Wong developed a new enzymatic assay for non-structural carbohydrates in plants (see Sects. 3.2.1 and 4.2). With new tools and new paradigms in sight, my attention gradually moved away from stomata and water potential. Having worked in the Tyrolean Alps, the Great Caucasus, the Andes, the Australian Alps and the mountains of New Zealand and Papua New Guinea, lots of samples waited to become ground and analysed for global patterns related to elevation, complementing measured stomatal and mesophyll properties.

And the global patterns did emerge: provided a large enough sample of species was collected and elevation-related drought was avoided (drought is not an intrinsic feature of mountains), stable carbon isotope discrimination declines with elevation globally. Hence, the photosynthetic efficiency in capturing CO_2 increases relative to constraints by stomatal diffusion (Körner et al. 1988, 1991). This is a clear-cut proof that atmospheric pressure and, thus, the partial pressures of CO_2 and O_2 were the main drivers, confirmed many years later from exploring the Basel herbarium for a large set of congeneric samples from contrasting elevations and latitudes (Zhu et al. 2010). So, high-elevation plants cope extremely well with reduced CO_2 partial pressure, as photosynthetic gas exchange studies confirmed in the high Alps (Körner and Diemer 1987, see Sect. 3). An increased photosynthetic efficiency goes hand in hand with higher leaf nutrient concentrations in high-elevation plants (Körner 1989). With an on average thicker leaf mesophyll, a higher stomatal density is required (when compared in congeneric species from different elevations; Körner et al.

1989a, b). Thus, water relations and carbon relations remain connected at leaf level as one moves to high elevations.

2.4 *When Leaf-Level Processes Become Overruled by Stand-Level Adjustments*

It is widely held that plant life on land became vigorous, once stomata were 'invented'. Their control of water loss seems like the ultimate solution for plants to cope with periodic water shortage. Yet, from an investment and amortization perspective, this could mean shutting down an expensive machinery (leaves) for much of the time when water falls short. From a 'management perspective', it would be more efficient to install less of that machinery (foliage) per unit land area but keep it running for most of the time. Most physiological plant ecologists do not like this sort of reasoning because there is little to measure with fancy equipment. Hence, the fundamental role of leaf area per unit land area (LAI) or plant spacing (e.g. Whitehead et al. 1984), leaf area per total plant mass (LAR), and the role of rooting depth (e.g. Jackson et al. 1996; our own works in Aleppo pine in Sarris et al. 2013) tend to become neglected. Reduced stand density and LAI and higher investment in roots and species replacement towards specialist taxa are by far the most important long-term biological responses to water shortage (Körner 2013a), with stomata meeting a sort of fire brigade function for rapid intervention. Researchers have repeatedly been surprised that plants keep transpiring and even growing under severe drought as a result of such stand-level adjustments. A good example is *Festuca orthophylla*, the dominant species in the Bolivian Altiplano, at 4,250 m elevation with annual precipitation of c. 300 mm confined to a few months, with hot days and freezing nights during most of the year. We found that these tussocks keep tillering and expanding leaves throughout the long dry season (Monteiro et al. 2011; Monteiro and Körner 2013). The explanation is plant density control (Fig. 3). The root spheres of these tall grass tussocks cover land six times larger than the tussock itself (or clusters of fragmented tussocks), and they do not overlap among neighbours, with the consequence that the water available per unit of tussock area exceeds 1,500 mm, sufficient to make a year-long living. Note the sandy substrate (and, thus, a lack of a capillary continuum) prevents deep moisture from becoming exhausted by evaporation from bare soil. An entirely different strategy is adopted by perennial forbs scattered among these tussocks: tiny leaf rosettes, firmly attached to or even sunken into the ground, resting on huge below-ground structures such as tap roots (Patty et al. 2010). So there are different solutions to the same problem with spacing and plant architecture as the dominant factors.

Plant-water relations came a bit out of fashion in the late 1980s till the early 2000s, with most questions seemingly resolved. As a consequence, most young researchers nowadays had never heard about pressure-volume curves (Körner and Cochrane 1985). Simple diffusion porometers almost disappeared from the market

Fig. 3 Wide plant spacing or other ways of reducing leaf area per unit land area is the most important long-term adjustment to drought. Once land cover is tuned to moisture availability, stomatal responses play a minor role. The photograph shows the exploration of the size of the root sphere of tall grass tussocks in the Bolivian Altiplano at 4,250 m asl

and gave place to more abundant and sophisticated CO_2 gas exchange equipment that delivered leaf diffusive conductance as a by-product (see the carbon chapter for the consequences). The use of pressure chambers became largely reduced to student courses. Only a few faculties globally had staff that was able to teach basics in plant-water relations, and only a hand full of labs maintained research in this field, largely on xylem hydraulics.

2.5 *Hydraulic Failure in Trees: Cause or By-Product of Tree Mortality?*

Big drought waves in the south-western USA and clever instrumentation suddenly changed the mood, and plant-water relations came back to front stage via plant hydraulics. About 90% of all papers that employ the phrase 'hydraulic failure' appeared in the last decade (e.g. McDowell 2011, Anderegg et al. 2012, Choat et al. 2012, Urli et al. 2013). Trees are considered dying, because their stem conduits could not meet the demand by foliage. Attracted by those aspects of the soil-plant-atmosphere continuum that can be easily measured (xylem conductivity, fraction of embolized xylem), other aspects that were considered key to plant survival under drought in the early half of the last century received less attention (e.g. tissue dehydration tolerance). While trees that have lost their conduits' integrity by xylem embolism are likely to enter problems when it comes to recovery as moisture returns, the currently often implied chain of causalities is quite incomplete (Körner 2019). Trees build stems largely to compete with their neighbours (and, in doing so, manage to resist an hurricane), using the same structures for stability that can also serve water transport. Given the dominant mechanical demand, circumstantial and direct evidence obtained since nearly 50 years ago indicates that the conduit systems' capacity is way oversized. Not surprisingly, most of the conductive rings are taken out of function after a few years, resulting in a narrow, conducting sapwood. In ring-porous oaks, all rings but the last one are deactivated.

Why would evolution select for such narrow, conducting sapwood fractions if conductive capacity was likely to become an existential limitation under drought? People have done 'cruel' things to small stems, cutting them in half, even crosswise (e.g. Mackay and Weatherley 1973), with no effect on leaf transpiration. I got involved in the latest of such attempts, although the biggest in scale ever: we cut 30–40-m-tall trunks in situ in half and observed canopy responses from the gondola of a crane (see Fig. 7): no effect on transpiring leaves under bright summer weather when fluxes peak (Dietrich et al. 2018). While one can question the dummy nature of a cut compared to widespread embolism, it seems obvious that drought is the most unlikely of all conditions during which the capacity of a piping system can become exhausted, when that system can supply crowns at maximum transpiration with only part of it being intact. We found flux rates to double in intact xylem near the edge of the cut (indicating spare conductive capacity), like water flow in a creek narrowed by the walls of a lock. Once stomata are closed in response to a severe drought, the flux will approach zero. As I explained in this recent comment (Körner 2019), xylem embolism is better viewed as a by-product, not the cause of mortality, when roots fail to supply water, once the soil-root capillary continuum became interrupted. That debate will continue, but inevitably, the focus has to turn to roots at one end and dehydration tolerance at the other end (leaves and shoots) within the soil-plant-atmosphere continuum.

How does water get through roots? An interesting question as posed by Steudle and Peterson (1998). And how does water get to the root, when the soil passes through critical dehydration? And why did evolution retain tracheids, after the far more efficient vessels had been 'invented'? And why are some trees dying from severe drought with hardly any embolism and others do well after drought with a lot of embolism (Johnson et al. 2018)? The likely explanation is that all this has little to do with hydraulic failure of the stem xylem as such but with the roots' failure to obtain water. When drought tolerance was discussed in the early part of the last century, this was a question of how much tissue dehydration is tolerable. The stem's xylem as such was never considered the bottleneck when people started to quantify the segments of the soil-plant-atmosphere continuum (SPAC; see, e.g. Jarvis 1975, Jackson et al. 2000). A xylem that becomes embolized because the roots fail to replenish crown water loss represents a water reservoir to the tree (with tension relaxed in the broken capillary system), a finite source of apoplastic water for attached parenchyma, from the paratracheal one to that in the mesophyll. How long does that reservoir last under minimum transpiration in order to survive

(Čermák et al. 2007)? How much can a tree reduce such minimum losses compared to its apoplastic stores? What is the role of minor conduits and tracheids for facilitating recovery, when most vessels became dysfunctional? And what is the role of roots in protecting trees from critical dehydration? In my view these are the decisive questions. Stem hydraulics received so much attention in the drought mortality debate because their conductivity had been quantified, and not because they explain mortality. Tools shape paradigms. There is no question that water flux in xylem is still full of unknowns and is a most fascinating research area (e.g. Meinzer et al. 1999, Tyree 2003, Holbrook and Zwieniecki 2005, Mayr and Sperry 2010), but roots are the weakest link between soil and foliage (Jackson et al. 2000), though much harder to study.

3 Plant Carbon Relations in a Source-Sink Context

Since De Saussure (1804) discovered that plants 'eat air', it fascinated plant scientists but also the broad public how they do it, and it soon became obvious that the process of CO_2 assimilation is central to life on earth. And so it was to my early career.

3.1 Photosynthesis Under Low CO_2 Partial Pressure

If anywhere, I felt, CO_2 assimilation should limit plant growth in 'thin', high-altitude air and in cold places. Thus, high-elevation photosynthesis needs to be understood in a broad range of species. With water-cooled triple-walled glass cylinders placed in situ over the dominant alpine sedge species in the Alps (Fig. 4), it became clear that the rate of net photosynthesis at 2,300 m elevation is largely light limited, with temperature playing a minor role because of perfect thermal acclimation (Körner 1982). The reference for this judgement was a fully illuminated, horizontal leaf's capacity to absorb CO_2 at optimum temperature. Clouds, suboptimal leaf angles and mutual shading in the canopy almost halved potential seasonal assimilation and, thus, presumably also growth. This work was entirely driven by my then assumption that growth in alpine plants (or plants in general) is photosynthesis driven and thus carbon limited. Later works proved this assumption to be wrong.

With a substantial Austrian Science Foundation grant, I could buy the then most advanced technology to study leaf photosynthesis and its response functions to light, temperature and CO_2 concentration in situ. Testing high alpine plants' photosynthetic response to elevated CO_2 was particularly tempting given the intrinsically low partial pressure of CO_2 at high elevation and the ongoing atmospheric CO_2 enrichment (Körner 1992). This battery and mass-flow controller-operated system worked fast enough, so that it was possible to study replicated response functions of net photosynthesis in 12 pairs of congeneric species at 2,600 and 600 m elevation during

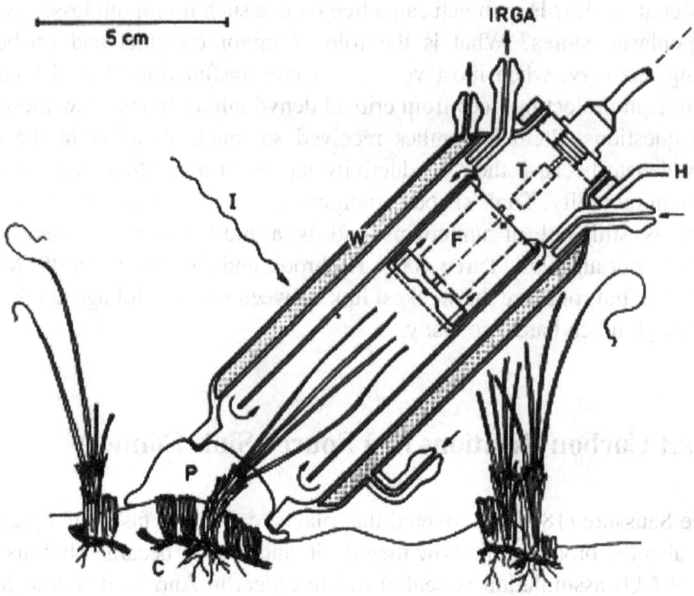

Fig. 4 In situ studies of alpine plant photosynthesis with a three-walled air-conditioned glass chamber mounted to *Carex curvula* tillers (Körner 1982). *F* fan, *W* water jacket, *H* holder, *T* temperature and humidity sensor, *P* plastic silicon rubber, *C* rhizome system

the short alpine summer. The outcome was expected to explain why alpine plants are so small and grow so little compared to their low elevation relatives and how much these high-elevation plants might profit from future atmospheric CO_2 enrichment.

Alpine species turned out to absorb CO_2 far more efficiently than congeneric low-elevation species (in line with the carbon isotope data from around the globe, see Sect. 2). By this efficiency, they not only neutralize the 20% diminished partial pressure of CO_2 at the 2,600 m test site near Innsbruck compared to the valley floor (600 m) but also exhibited higher maximum rates of assimilation (Körner and Diemer 1987). Hence, there was no evidence that these plants faced unit leaf area-based carbon constraints of growth any greater than low-elevation species do. Elevating CO_2 supply caused rates of photosynthesis to strongly increase. When artificially grown under elevated CO_2 in full sunlight, these alpine species retained this high carbon capturing capacity (Körner and Diemer 1994). However, time was not yet mature to question the carbon limitation paradigm that drove all this leaf gas exchange research.

3.2 Effects of Elevated CO_2 on Plants and Ecosystems

Through my move to Basel, the CO_2 enrichment experiments with alpine plants discussed above became interrupted. Meetings with the small but growing research community that had an interest in exploring plant responses to elevated CO_2 at IIASA (Laxenburg near Vienna), Woods Hole (USA) and Lunteren (Netherlands) made it clear that we do need realistic ecosystem-level experiments in this field. The conference dinner of the Lunteren meeting in the artificial jungle at Het Heijderbos was a key experience: during the train ride back to Basel, I designed an experiment with tropical model ecosystems that capitalized on the brand-new experimental greenhouse at the Institute of Botany in Basel.

3.2.1 Tropical Forests

We built four highly diverse, humid tropical plant communities composed of 15 species each, with understory, shrub, tree and liana taxa, planted as small clonal propagules, provided by the Basel tropical greenhouse staff. These complex but exactly replicated communities in a split-plot design were grown on a low-fertility, leaf litter compost-sand mixture. The communities were rapidly filling the 17 m^3 air-conditioned containments that either received elevated or ambient CO_2. Hot and humid conditions caused plants to grow so vigorously that the taller life forms

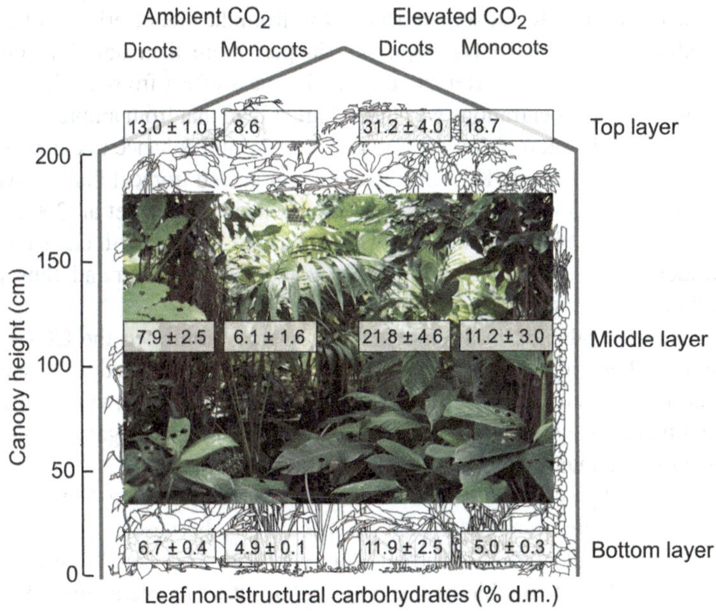

Fig. 5 Model jungles composed of 15 humid tropical species belonging to different plant functional groups and exposed to contrasting CO_2 supply reveal no CO_2 'fertilization' effect on biomass production, but enhanced soil CO_2 release and nutrient leaching. Elevated CO_2 concentrations caused non-structural carbohydrate concentrations in leaves (shown here) to rise dramatically (except for monocots in the understory at 1–2% sunlight), while leaf area index levelled off at 7 (Körner and Arnone 1992)

pushed the roof after 3 months (Fig. 5). The expected CO_2 fertilization effect on biomass production was not found. In what became my first paper in *Science* magazine (Körner and Arnone 1992), we showed massive starch accumulation in the upper canopy, higher root mass, much higher soil CO_2 release and increased nutrient leaching through a lysimeter drainage to the basement of the building. Counter to expectation, LAI levelled off slightly lower under elevated CO_2. In retrospective, this makes sense, given the overabundance of photoassimilates.

These works found a continuation at the Smithsonian Tropical Research Institute (STRI) in Panama under a Mellon Foundation grant in cooperation with Klaus Winter. We grew seedlings of tropical trees and shrubs in situ under elevated and ambient CO_2 concentrations in very deep forest shade on Barro Colorado Island in order to test the hypothesis that low light is the most likely condition to induce a CO_2 fertilization effect. Why? Because elevated CO_2 shifts the light compensation point of photosynthesis, and these plants grew right at the light compensation point, so that the CO_2 concentration became decisive for whether they can grow at all (Fig. 4.1 in Lüttge 2008). And indeed, providing those seedlings with extra CO_2 induced a very strong growth stimulation at otherwise very small absolute growth rates (Würth et al. 1998a). Such deep-shade responses to elevated CO_2 may decide upon later forest structure. Should liana seedlings take a similar advantage, the likelihood of lianas reaching the canopy top would rise. And lianas grown from Yucatan seed sources in deep shade on original forest soil we shipped in from the site of origin to Basel did indeed exhibit exceptional CO_2 responses (in part more than doubling biomass; Granados and Körner 2002). Remarkably the biomass effect from pre-industrial to then ambient CO_2 concentrations was much bigger than that from ambient to similar CO_2 increases in the future. More aggressive lianas would enhance tree turnover and reduce forest carbon stocks (Körner 2004a, b), and indeed tropical lianas have been observed to become more aggressive in recent decades (Phillips et al. 2002, 2004). Thus, a biodiversity effect exerted by a CO_2-enhanced liana growth can shorten the mean residence time of C in the biggest biological carbon reservoir and, thus, reduce its contribution to the global biomass C stock.

Experiments conducted in a temperate zone forest confirmed the CO_2 x shade interaction. When light is reduced to very low levels, elevated CO_2 causes a significant relative stimulation of growth of seedlings (Hättenschwiler and Körner 2000) and lianas (Zotz et al. 2006) irrespective of nutrition, yet again, at otherwise very low absolute rates of growth.

How will a fully sunlit tropical forest canopy respond to rising CO_2 concentrations? Still today, this remains a field of speculation, because it is very hard to estimate or model sink control of carbon capture in late successional forests. Scaling from tree or plot level to net biome responses entails other problems related to stand dynamics (e.g. Philipps et al. 2004; Brienen et al. 2015), which determine the carbon stock: faster growth commonly reduces the mean residence time of carbon, that is, tree longevity (Körner 2017, Büntgen et al. 2019). Whatever the long-term carbon sink capacity of a tropical forest is, one would expect that individual leaves in the canopy, artificially exposed to elevated CO_2, would never show any C-overflow symptoms (that means an accumulation of non-structural carbohydrates, NSC).

Fig. 6 Providing 12 cm^2 leaf sections at the top of a humid tropical forest in Panama with extra CO_2 derived from soil: isotopic labelling confined to exposed leaf tissue, carbohydrate accumulation despite a near to infinite carbon sink of such tall trees compared to a small leaf section and no stomatal response to elevated CO_2 (Körner and Würth 1996; Würth et al. 1998b)

Huge trees would always permit dissipating extra C obtained by a single leaf. It seemed worth testing this hypothesis in order to obtain at least a hint at top canopy C relations in the light of rising CO_2.

At the Smithsonian crane site near Panama City, we trapped CO_2 released from the soil under 4 m^2 of black plastic foil on the ground. Inevitably that CO_2 carries a stable carbon isotope signal ca. 20 permille more negative than in free-air CO_2. We then pumped this isotopically labelled CO_2 to a manifold box in the 40–50 m tall forest canopy. From the crane's gondola, we attached small transparent plastic cups to the underside of big leaves and connected these to the manifold box so that the cups became continuously flushed with extra CO_2 from soil (Fig. 6). After a week only, leaf tissue attached to the cup was packed with extra NSC compared to controls (Würth et al. 1998b), very similar to what we observed in our greenhouse jungles in Basel. Leaf sections under the cup carried a clear carbon isotope signal reflecting the soil-derived CO_2. These data demonstrate that NSC accumulates in foliage in response to elevated CO_2 irrespective of source-sink relationships. Neither NSC nor the ^{13}C signal was dissipated within a given leaf blade but remained confined to the exposed leaf section. Another surprising observation was that stomatal conductance was not reduced in leaf sections exposed to elevated CO_2, challenging a common hypothesis. With this rather simple and cheap experimental system, we could demonstrate across different tropical tree species that NSC accumulation is an intrinsic response to elevated CO_2, irrespective of sink capacity. While we do need entire forest CO_2 enrichment experiments, such small-scale studies can illuminate basic questions for this largest of all biological C reservoirs on earth (for a review see Körner 2009).

3.2.2 Alpine Vegetation

Established at my new post in Basel, I received a substantial grant by the Swiss Science Foundation for the first in situ CO_2 enrichment experiment in alpine grassland, using a special design of open top chambers with an excentric air release collar (1992–1995). After four seasons in a high-CO_2 environment at 2,500 m asl, either with or without mineral fertilizer addition, the productivity of this grassland remained unresponsive to elevated CO_2, but a relatively low dose of NPK fertilizer (40 kg $ha^{-1}a^{-1}$) alone almost doubled above-ground biomass (Schäppi and Körner 1996; Körner et al. 1997).

At that point, the carbon limitation hypothesis was falsified for fully illuminated vegetation, both in a vigorously growing humid tropical model jungle and a slow growing alpine test system. The results for the alpine one were confirmed later in a free-air CO_2 enrichment (FACE) experiment, with glacier forefield vegetation, in which elevated CO_2 even exerted negative effects on plant biomass (Inauen et al. 2012), which we attributed to competitive inhibition of nutrient availability by an exudate-driven microbial community. In a recent experiment, we tested the gradual effect of shading from no shade to half of sunlight for three seasons in the same type of late successional grassland, which had no effect on bulk above-ground biomass (but at maximum shade below-ground biomass became measurably reduced after 3 years), underlining that photosynthesis was not constraining above-ground biomass production (Möhl et al. 2020). These results suggest that alpine vegetation is not carbon limited as I was assuming when studying high-elevation gas exchange in my early career (see Sect. 3.1). Whether the similar results obtained with the tropical model ecosystem reflect real-world responses awaits to be explored in large-scale in situ experiments.

3.2.3 Temperate Forests

How could it be that providing photosynthesis with more CO_2 stimulated its rate per unit leaf area instantaneously but had no effect on growth and biomass accumulation? Were experimental conditions prohibitive for growth in these specific cases? And if so, where did all the extra carbon go? Whatever the explanation, it was clear that larger-scale and more realistic experiments were needed, performed in the field, without interfering with the nutrient cycle, no soil disturbance, no enclosures and preferentially with forest trees, given so many hopes are tied to their ability to capture more carbon in a CO_2-rich future. A site was to be found with as many as possible tree species with adult individuals in a small area that could be surveyed from a huge building crane at the level of tree crowns 30–40 m above ground. After spending 1–2 tons per day of industrial CO_2, cleaned to food quality and released in the forest canopy for eight seasons, using a free-air CO_2 release system, the growth response of the exposed deciduous trees was zero (Körner et al. 2005, Bader et al. 2013). There was a clear isotopic fingerprint of the added CO_2 in tree rings,

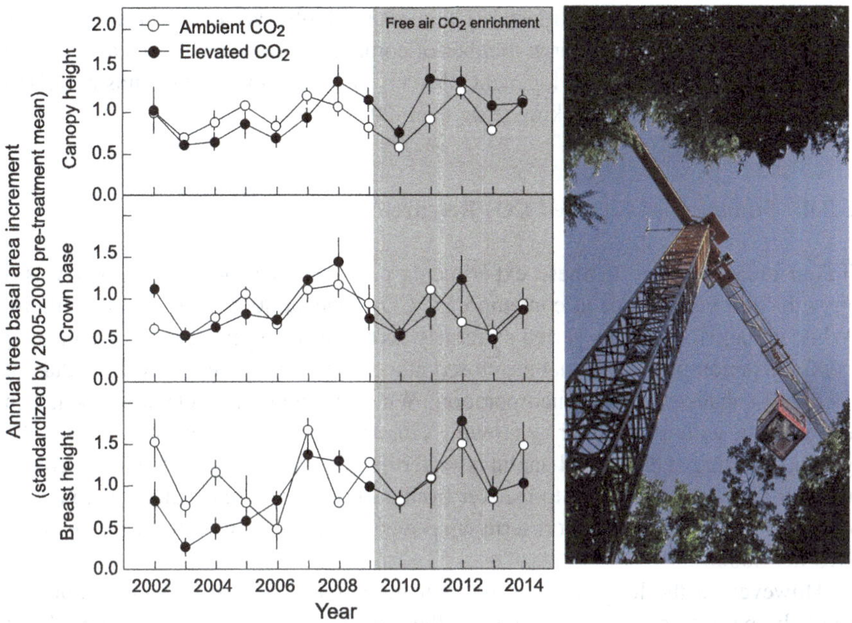

Fig. 7 Mature, 36-m-tall *Picea abies* trees take no growth advantage from five seasons of canopy-scale CO_2 enrichment. The diagram shows standardized tree ring responses for three heights(Klein et al. 2016b). The canopy crane operated for 20 years near Basel facilitated a large number of research works in mature forest trees

confirming efficient CO_2 enrichment (Keel et al. 2006), and per unit leaf area photosynthesis remained stimulated (though somewhat downregulated) by year 8 (Bader et al. 2010).

We repeated the same treatment for 5 years in 36-m-tall spruce trees, with the same result (Klein et al. 2016a; Fig. 7). Other research groups working with test forests under steady state (closed canopy, constant leaf and root turnover) arrived at the same answer (boreal forest, Sigurdsson et al. 2013; temperate forest, Norby et al. 2010; eucalypt forest, Ellsworth et al. 2017), all lining up with the observation that evergreen oaks growing for 30 years around natural CO_2 springs in Tuscany took advantage from elevated CO_2 when they were young, after coppicing, but the signal disappeared at canopy closure (Hättenschwiler et al. 1997). The fate of extra C was suspected to dissipate through a myriad of metabolic leaks, most likely associated with phloem transport and processes in the rhizosphere, each too small to produce a measurable signal, but summing up to a substantial C release (Mildner et al. 2014, Klein et al. 2016b). Very recently that view had been supported by data collected at the Australian Eucalyptus FACE experiment (Jiang et al. 2020).

An exciting by-product of the experiment with tall spruce trees at the Swiss Canopy Crane site was the observation that the stable carbon isotope signal applied to the tree crowns with CO_2 derived from fossil fuel could be traced to roots and stems of neighbouring beech and other tree species, indicating substantial belowground interspecific carbon transfer (Klein et al. 2016b). We could show that these mature forest trees share a large number of common ectomycorrhiza taxa, of which the recently discovered, phylogenetically very old Basidiomycota genus *Sebacina* plays a key role for C trafficking (Rog et al. 2020).

3.2.4 Summary of Elevated CO_2 Research

Taken together, none of these experiments provided evidence that carbon was a growth-limiting resource at contemporary CO_2 concentrations and in full sunlight. While climatic conditions varied extremely across these experiments, the remaining common denominator was that the closed nutrient cycle in these test systems did not facilitate enhanced carbon incorporation. When the nutrient cycle was opened in experiments with trees, either by growing saplings in potting mixtures, disturbing the ground, starting with spaced saplings in fertile soils, applying horticultural conditions or allowing the system to recover from a shoot to root imbalance due to recent disturbances, this opens a short-term window for a so-called CO_2 fertilization effect (Körner 2006b).

However, in the long run, soil nutrient limitation is almost inevitable, but it is unlikely that nitrogen supply alone sets the ultimate limit for several reasons. Most importantly, the nitrogen cycle is, in principle, open, with the size of the cycling N-pool adjusting over the years to the availability of other nutrients (e.g. P depending on weathering but, for instance, also Mo if one considers N_2 fixation; Barron et al. 2009; Wurzburger et al. 2012). Not surprisingly growth and productivity of tropical forests were not found N limited in these and related works. In addition, the forest test site in Switzerland that showed no growth stimulation by elevated CO_2 received substantial rates of anthropogenic N deposition since the 1970s with peaks close to 50 kg N ha^{-1} a^{-1} in the test region and rates coming down during the last decades to 20–25 kg N ha^{-1} a^{-1}, with P, Mg or K more likely candidates for the next limiting resource, similar to what has been concluded for the Australian Eucalyptus FACE, cited above. Accordingly, N/P ratios in tree foliage widened critically in central Europe in recent years (Braun et al. 2010). So, the issue is far more complex and cannot be reduced to N limitation. The only experiment so far that tested growth responses to elevated CO_2 on two different soil types on which the test species *Fagus sylvatica* and *Picea abies* co-occur naturally ended with a surprise: the patterns observed on calcareous soil was the opposite to the pattern on the siliceous soil (Spinnler et al. 2002). Admittedly, these saplings grew on dug-out soils in large open-top chambers, facilitating vigorous growth and, thus, a CO_2-fertilization effect. Yet, had we employed only one of these soil types, we had written two entirely different publications (*Fagus* 'winning' over *Picea* on calcareous soil and vice versa on the acidic soil). Quite clearly, soils rule tree responses to elevated CO_2, and responses of young saplings cannot be extrapolated to adult trees.

In summary, these results suggest that we are living in a CO_2-saturated world, with a growth stimulation by CO_2 via photosynthesis restricted to fertile locations and, perhaps, to fresh clearings and young plantations. It is not even clear whether naturally formed forest gaps during the earliest infilling period will facilitate a positive CO_2 effect, because the root spheres of the surrounding trees will consume any extra nutrients. The only other situation under which a rising CO_2 concentration is likely to be effective on plant growth under natural conditions is deep shade.

In contrast to CO_2 and N_2, both theoretically infinitely available, only depending in rates of fixation, all other minerals required to make a living (18 other than C, H, O, and N for plants; Elser et al. 2010) are finite per unit land area (disregarding minute dust deposition). These other mineral nutrients are derived from weathering, and their plant-available fraction in soils is heavily competed for by plants and microorganism ever since. Could a CO_2-rich atmosphere affect soil nutrient acquisition and release more nutrients from the soil matrix (Körner and Arnone 1992; Schleppi et al. 2012; Hungate et al. 2013)? And if so, how long could that be sustained? Open questions. The most plausible explanation why vegetation that arrived at steady-state canopy and root dynamics did not show enhanced growth in response to elevated CO_2 concentrations is the finite mineral nutrient capital per unit land area. When soils were disturbed (and thus activated) or when CO_2 was applied to initially well-spaced plantlets (space as a surrogate for extra nutrients; e.g. Talhelm et al. 2014), elevated CO_2 concentration was found to stimulate young tree growth (Körner 2006b). So, the rate of nutrient addition (sensu Ingestad 1982) defines the rate at which a plant can incorporate C. From bacteria to elephants, from daisy to giant sequoia, it requires chemical elements other than C in a proportional quantity to build an organism. Why should there be left-over soil mineral resources that could support enhanced CO_2 capture in a CO_2-enriched world?

3.3 When Elevated CO_2 Acts via Plant-Water Relations

CO_2 enrichment was found to reduce stomatal conductance in many (though not all) species (first observed by Darwin 1898, the son of Charles Darwin). This effect could slow water consumption by plants and, thus, prolong periods of high soil moisture and nutrient availability. In seminatural, periodically dry grassland, this moisture-saving effect is the most likely explanation for a CO_2-driven stimulation of productivity (long-grass prairie, Owensby et al. 1994; temperate calcareous grassland, Niklaus et al. 1998; Niklaus and Körner 2004; grassland in Tasmania, Hovenden et al. 2014; desert annual grassland, Grünzweig and Körner 2001; for a summary and short-grass prairie data, see Morgan et al. 2004). The fundamental problem with interpreting such results is twofold: First, the exposed grassland patches are islands in an otherwise untreated landscape (Leuzinger et al. 2015). Would all vegetation respond that way, the atmosphere would receive less water vapour, and thus, evaporative forcing would rise. Second, reduced transpiration causes leaves to warm, which in turn stimulates transpiration. The island effect puts a question mark on any water-driven CO_2 effect on growth. In his PhD works, Matthias Volk showed that such CO_2 effects on growth can be produced under ambient CO_2 by simulating the moisture dynamics otherwise produced by elevated CO_2 (Volk et al. 2000). Thus, projections based on water-driven CO_2 effects rest on assumptions regarding the future climate (precipitation, humidity). And such climatic effects depend on the patterns (spacing) of rainfall events in both grassland and forests (Bachman et al. 2010; Leuzinger and Körner 2010). Thus secondly, it is very important to separate potential photosynthesis-driven and stomata-driven CO_2 effects. Finally, all projects that involved more than one species reported species-specific effects; hence, species differ in their responses to elevated CO_2 in all climatic zones explored but also to available soil moisture (different rooting depths), strongly arguing against mono-specific test systems.

4 Growth Controls Photosynthesis

So, what is it that controls plant growth? When I went to school, it was self-evident that growth follows photosynthesis. The more a plant photosynthesizes, the faster it will grow, and the more biomass will accumulate. In a way that assumption is still correct, except for the implied hierarchy of the causalities. The uptake of carbon and the investment of carbon cannot be decoupled in the long run, so a measure of either one reflects the other. This is the place were manipulative experiments (see Sect. 3) and plant nutrient stoichiometry come into play. Since the role of plant nutrition is so intimately connected to carbon relations, it was discussed in Sect. 3.2.4. Here, I will focus on water and temperature constraints.

4.1 Water Shortage Affects Meristems, Long Before It Affects Stomata and Photosynthesis

The experiences from CO_2 enrichment research can be extended to other growth determining factors. Water shortage was long believed to affect carbon uptake via stomata. Under drought stress, stomata may close and, thus, limit CO_2 assimilation, a textbook classic, not supported by evidence. For half a century, it is known that for building a cell, it requires a machinery that synthesizes cell wall components and a finely tuned osmotic system that controls cell size enlargement. As discovered and explained by John Boyer (see Fig. 9 in his 2017 contribution to this series of essays), this delicate process is far more sensitive to water shortage than is leaf gas exchange through stomata. Drought stress affects cell production (meristematic activity) long before it affects leaf gas exchange (Muller et al. 2011). This explains why drought-stressed plants accumulate non-structural carbohydrates (mostly starch) rather than suffering from assimilate shortage (Körner 2003b). At the scale of a tropical forest represented by 15 adult tree species in Panama, my PhD student Mirjam Würth evidenced that periods with ample rain enhanced growth and reduced carbohydrate stores, and dry periods caused carbohydrates to accumulate from tip to toe (Fig. 8),

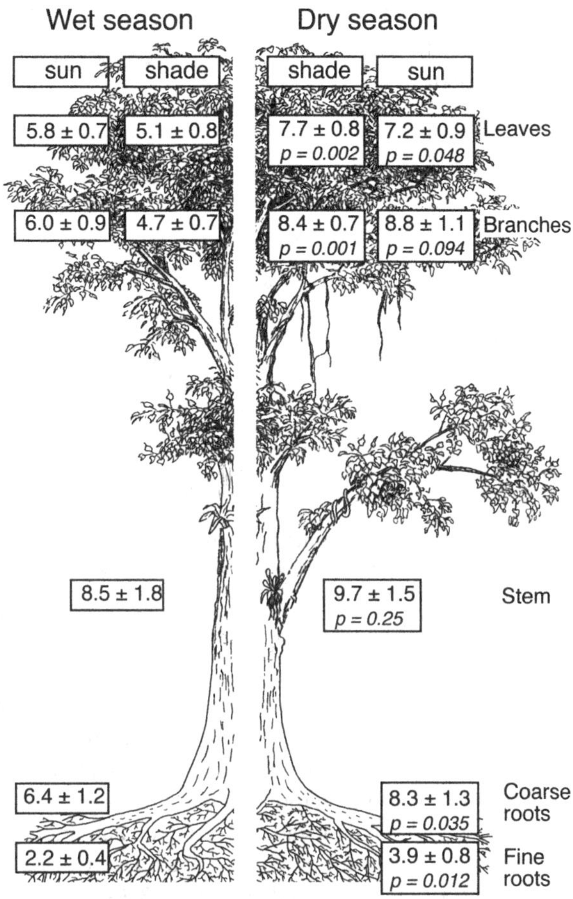

Fig. 8 Non-structural carbohydrate reserves in 15 adult tropical tree species in Panama during the rainy and dry season. Note the higher concentrations in all organs during the dry season. Canopy samples were taken with the Smithsonian canopy crane (Würth et al. 2005)

the opposite of the classical paradigm that assumes that drought causes plants to starve. Drought prevents plants from investing photoassimilates into structures.

4.2 *Low Temperature Limits Growth Before It Limits Photosynthesis*

A similar story applies to low temperature (for a review see Körner 2006a). Since low temperature can reduce the rate of net photosynthesis, it is commonly held that plants suffer from carbon shortage when it gets cold and, thus, grow slow. The causality is the opposite. While cold-adapted plants photosynthesize till they freeze (with c. 30% of full capacity at 0°C; dating back to works already summarized by Pisek et al. 1973), plants do not grow at 0°C (Nagelmüller et al. 2017). In fact, there is hardly any growth below 5°C (Alvarez-Uria and Körner 2007; Rossi et al. 2007;

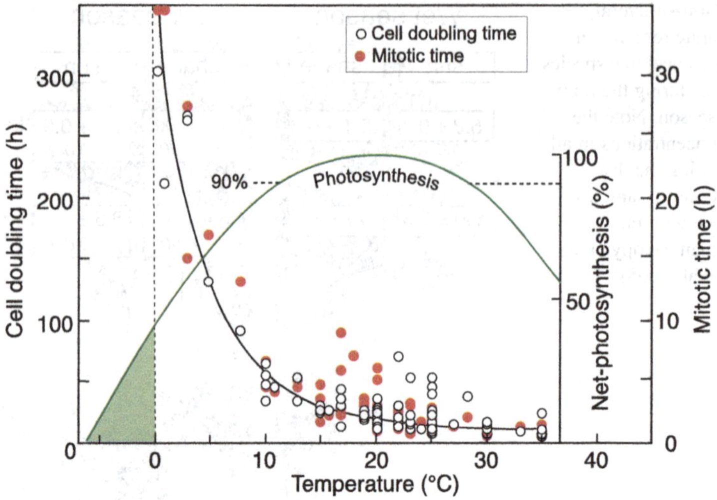

Fig. 9 A scheme of the temperature response of net photosynthesis in cold-adapted plants (source activity, mean response for many alpine species) and the corresponding cell duplication rate or mitotic rate (sink activity, data from the literature from various sources). Note that low temperatures constrain cell production far more than CO_2 assimilation (simplified from Körner 2003a)

Schenker et al. 2014; Nagelmüller et al. 2016), a temperature at which net photosynthesis of leaves reaches 50–60% of maximum rates (Fig. 9; Körner 2003b, Körner 2015). This means low temperatures affect cell production (meristematic activity) long before carbon capture declines. Not surprisingly, trees at the low-temperature, high-elevation treeline contain more non-structural carbohydrates than trees at lower elevation, as Günter Hoch evidenced in his PhD (Hoch and Körner 2003) and confirmed later for 13 locations around the globe (Hoch and Körner 2012).

Nutrient availability, water shortage and low temperature all affect growth (meristems) long before these limitations affect photosynthesis (Körner 2015). This leads to a paradigm of plant growth in which demand for C controls CO_2 assimilation, instead of assimilate supply controlling investment. Trying to understand alpine plant life (Körner 2003a, 2021), alpine treelines (Körner 2012) and the consequences of elevated CO_2 on plant growth and productivity, these observations are drawing a picture in which C sink activity controls much of the terrestrial biosphere's C cycle. The rate of photosynthesis (source activity) is tracking growth and, thus, supplies assimilates on demand set by meristems, which operate under direct nutrient, moisture and temperature control (Körner 2003a, b, 2015). Textbooks deserve substantial adjustments in this respect! A research focus away from gas exchange towards meristem functioning is needed. When we ask ourselves what is really basic in plant growth control, most scientists, including myself, would have mentioned photosynthesis and respiration some 30 years ago, with the insight that building a cell undergoes more basic constraints in a real-world context still not a very popular view.

4.3 Why Growth and Productivity Rather than Temperature Drive Respiration

Here, I add a brief comment on mitochondrial respiration. Most commonly considered a burden to a plant, a similar change in paradigm is required as for the role of photosynthesis. Plants do not respire for 'fun'; hence, they do not execute the inevitable temperature response captured by an Arrhenius plot, but they can be expected (in the long run) to respire in response to a demand for ATP, NAD(P)H and reactive oxygen species (ROS) and other energy equivalents (simply addressed as 'ATP demand' here). A major ATP demand is created by growth ('growth respiration'), the energy required to build structures and reserves. Active nutrient uptake induces another demand for ATP, as does maintaining the cellular machinery. As a consequence, dark respiration is best considered the result of plant vigour (Reich et al. 1998, O'Leary et al. 2019), whatever drives it. When plants are grown under conditions in which carbon is the only limiting resource (the majority of examples taken from laboratory experiments conducted with plants receiving ample water and nutrients and grow under optimal temperature conditions), carbohydrate supply sets the rate of growth and thus respiration. In a natural setting, where nutrients, soil moisture or temperature are directly constraining meristem activity, respiration is driven by sink activity (demand control). At ecosystem scale and over long periods of time, respiration is not driven by temperature, as many models still assume, but by the productivity of the ecosystem. Should a higher temperature (or whatever other environmental stimulation) facilitate a higher productivity, there is more biomass to be built and more substrate to be recycled (Raich and Nadelhoffer 1989). So in essence, net primary production (NPP) is controlling respiration and not the other way round. Riccarda Caprez illustrated this in her PhD project in the Swiss Alps by comparing NPP and respiration across an elevation gradient (Caprez et al. 2012).

4.4 Why Are Bonsais Small?

I am closing this section with a few words on bonsais. They are forced to slow growth and dwarfed stature by withholding resources, perhaps a model that also helps understanding small plant size in harsh mountain environments (but see Sect. 5.2). What is it that ultimately causes bonsais to be so small? Do they produce smaller cells or a smaller number of cells in response to the severe shortage of resources? I set out exploring this, when bonsais became available at the Innsbruck Botanical Garden for which the mother trees were known. So, together with my ever so helpful and dedicated assistant Susanna Riedl, we looked into the anatomy of bonsais and their mother tree, and the project profited greatly from discussions with Pete John, a developmental biologist I learned to know in Canberra. In short, the 30–50 times smaller leaves of bonsais (compared to leaves of the mother trees) have cells as large or even larger than the leaves of their mother trees, but they have smaller numbers of cells (Körner et al. 1989a). From this we learn (perhaps a triviality in retrospective) that it is the cell cycle and associated cell differentiation that control growth and plant size, and not the size of cells, a most conservative trait, for several hydraulic, mechanic and genomic reasons.

5 Applying Physiological Plant Ecology to Plant Biogeography

As we aim at understanding plant performance from the first principle, it is not an easy task to define a starting point in terms of both the processes considered basic and the phenomena to be explained. Should we aim at explaining growth, stress survival, plant reproduction or plant distribution? Aren't the four interrelated? Growth and stress resistance coming first, successful reproduction second and range filling, as a consequence, last? When I started my career, carbon assimilation was considered central in explaining plant performance. A common outcome of such physiological ecology studies was that plants are amazingly well adjusted to the environment they inhabit. If they wouldn't, we would not find plants for study. Yet, at one point, a species reaches its natural range limit. Finding out what the ultimate constraints are at the edge of their fundamental niche is the most fruitful application of physiological ecology. Once resolved, such range limits can be predicted.

5.1 Explaining the Low Temperature Range Limit of Tree Species

When the European Research Council (ERC) granted me an advanced grant during my last 5 years as professor of the University of Basel, my team was expected to answer where, why and how dominant European tree species reach their low temperature range limit (species of beech, oak, ash, cherry and others; Körner et al. 2016). With three PhD students and three postdoctoral researchers involved, we did arrive at functional explanations for these range limits – to my knowledge among the few in any plant species, which sounds like a paradox, given these are the most fundamental questions plant ecologists could ask. This project combined biogeography, micrometeorology, reproductive biology, stress physiology, developmental biology (phenology), a study of growth and related physiological traits as well as exploring genotypic versus phenotypic components of adjustments (Fig. 10). Each of these research fields alone could not have expected to come up with a conclusive answer.

We identified the undisturbed, low temperature range limits of seven dominant deciduous species, both in the Alps and in polar regions, with highest elevation and

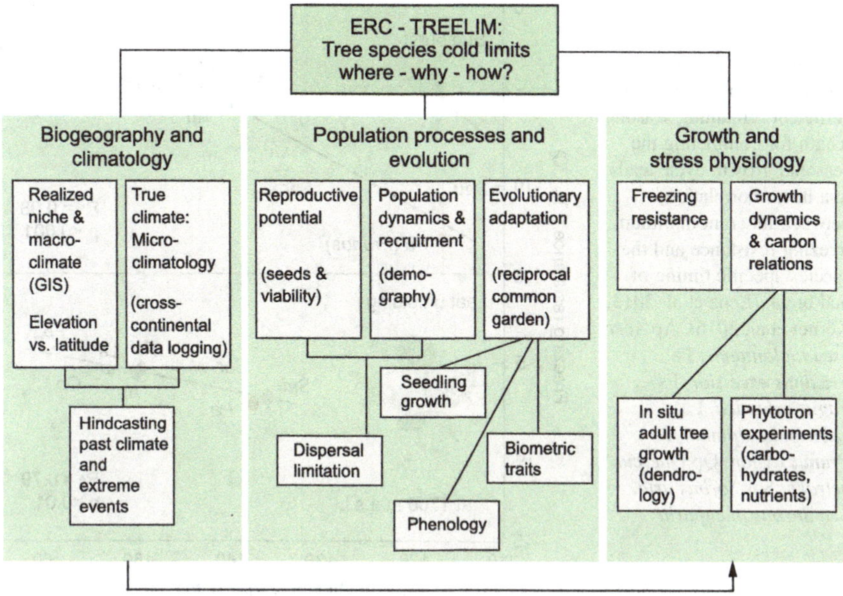

Fig. 10 Research fields to be covered in an attempt at explaining the low temperature edge of the fundamental niche of deciduous tree species in Europe (Körner et al. 2016)

latitude limits as proxies of the edge of the fundamental niche (Randin et al. 2013). We assessed the on-site temperature regime by installing hundred data loggers in the Alps and Sweden, the tops of tree crowns included, linked the data to long-term weather records (low temperature extremes in particular) and, thus, hindcasted the most likely return rates of extremes (Kollas et al. 2014a, b). We explored viability of seeds collected at the range limit and found neither seed nor viability limitation (Kollas et al. 2012). The safety margins against freezing damage were found narrowest at flushing in spring (Lenz et al. 2013) and were closely correlated with the species-specific flushing date (Fig. 11; Lenz et al. 2013, 2016), which was controlled by genotypic requirements of heat sums and, depending on species, by photoperiod (Basler and Körner 2012, 2014). There was no indication that any of these range limits were associated with insufficient growth or carbohydrate storage (Lenz et al. 2014). When transplanted, even beyond the current adult range limit, species largely retained part of their phenology controls and revealed less phenotypic plasticity the higher the elevational origin of the seed family (Vitasse et al. 2013, 2014). With the same conceptual approach, we (research partners and students of the Botany Institute of the Chinese Academy of Sciences in Kunming, China) are currently exploring the range limit of broadleaved evergreen oaks at 4,270 m elevation in the Baima Snow Mountains of N-Yunnan only 100 m below the local conifer treeline. Again we find a fine-tuned flushing phenology that prevents freezing damage but critically constrains the remaining season length (Yang et al. 2020).

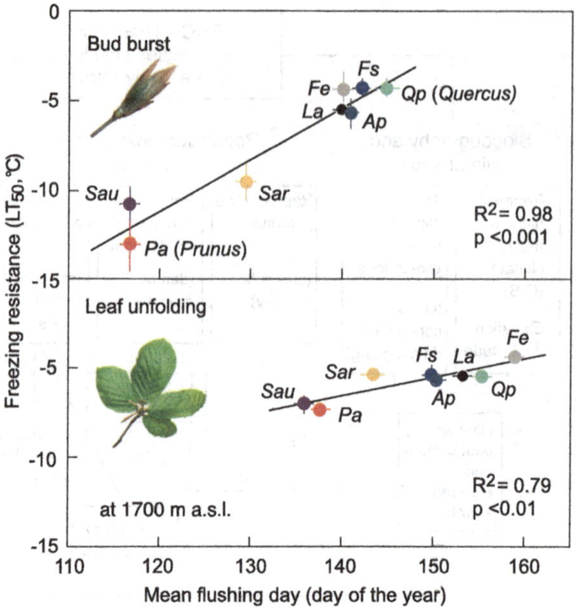

Fig. 11 The species-specific trade-off between late flushing in order to escape freezing damage and sufficient remaining season length for completing the seasonal growth cycle leads to a linear correlation between inherent minimum freezing resistance and the species-specific timing of bud break (Lenz et al. 2013, Körner et al. 2016). Ap *Acer pseudoplantanus*, Fe *Fraxinus excelsior*, Fs *Fagus sylvatica*, La *Laburnum alpinum*, Pa *Prunus avium*, Qp *Quercus petraea*, Sar *Sorbus aria*, Sau *Sorbus aucuparia*

We concluded that the low temperature range limits of these tree species (including the oaks in China) are not related to their carbon relations (water availability is not an issue in these cases), but reflect the finely tuned interplay between three factors: the timing of spring flushing (phenology), freezing tolerance at and after flushing and, as a trade-off of the flushing date, the duration of the remaining season and, thus, the life history-associated requirements for shoot and fruit maturation in order to survive the next winter. Through these works, phenology – often just regarded a biological thermometer – revealed a decisive role in setting a species range limit through the adjustment of time of bud break with regard to the inherent freezing tolerance of a species. The results underline the importance of photoperiod control (in interaction with concurrent temperatures) for securing plants against the curiosities of spring weather by preventing them from simply tracking rising spring temperatures. These findings also explain the declining temperature sensitivity of spring phenology as warm climatic conditions occur earlier in the year in the course of climatic warming. Photoperiod also controls the timely transition to autumnal senescence of foliage and hardening of perennial tissues in order to arrive at sufficient freezing tolerance, before the advent of low temperature extremes (Körner and Basler 2010).

5.2 *The Alpine Treeline: A Unique Biogeographic Delineation*

The fact that growth is far more constrained by low temperature than is CO_2 assimilation is the currently best explanation of the climatic treeline globally. The treeline very closely follows a ca. 6°C isotherm of seasonal mean temperature across latitudes (Körner and Paulsen 2004, Paulsen and Körner 2014), an idea first coined by Alexander von Humboldt (Körner and Spehn 2019). The reason why trees reach a thermal limit beyond which alpine taxa flourish is related to aerodynamics. By their architecture (erect stature), trees can hardly decouple from ambient air temperature (Körner 2007a, 2012; Fig. 12), while low-stature alpine vegetation 'engineers' its own microclimate by pronounced aerodynamic constraints to heat exchange close to the ground, arriving at far warmer conditions than reported by a weather station (Scherrer and Körner 2009, 2011; Körner and Hiltbrunner 2018). It is the thermal limitation of meristematic activity and neither CO_2 uptake, stress tolerance nor moisture or nutrient availability that control the treeline, although the latter two can pose substantial additional growth limitations, but hardly affect the elevational position of the treeline (for drought effects see Hoch and Körner 2005). So, bioclimatology and the biology of the growth process itself hold the best explanation for this most prominent biogeographic delineation that permitted quantifying the global land area above the treeline, the alpine life zone that accounts for 3.6 Mio km^2 or 2.6% of the land outside Antarctica (Körner et al. 2011, 2017). The position of the high elevation and arctic treeline isotherm can now be predicted for any point on earth, with the main technical limitation, the quality of climatic data obtained from climate data bases (Paulsen and Körner 2014). For unknown reasons, the model fails in Mediterranean mountains; a regional lack of suitable tree taxa may be one explanation. The great advantage of that tool is that the potential treeline position

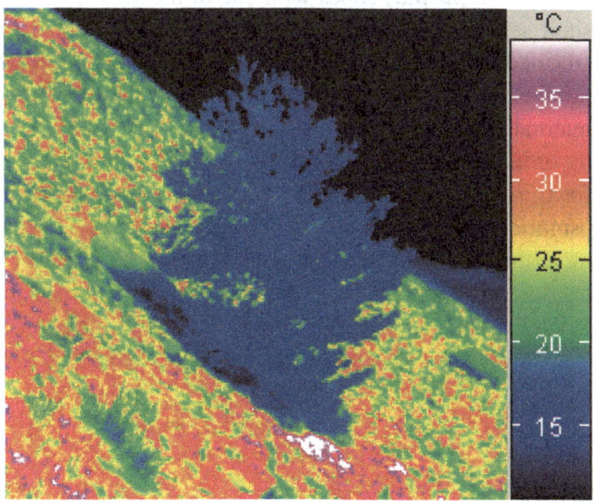

Fig. 12 A thermal image illustrating that trees operate much closer to ambient air temperature than low-stature alpine grassland. This is the aerodynamic reason why there is a treeline. Note, this is not a photograph, but an assemblage of 76,800 point-temperature records converted to a color scale (adapted from Körner 2007a, b)

can even be predicted for places, where forests have been destructed (Noroozi and Körner 2018). So, the treeline is a potential bioclimatic life form boundary, irrespective of whether trees have been cut or burned locally or became eradicated by avalanches or storms and, thus, are absent from treeline (Körner 2020a).

5.3 Alpine Plant Life

Although trying to understand alpine plant life was a central task of my entire career (Körner 1999a, 2003a), we started too late to ask similar questions as discussed above; hence the range limits of alpine taxa still await exploration. Embedded in traditional thinking, research confirmed that these plants are not carbon limited, and freezing damage is an issue during the growing season only, but safety margins are wide and habitat specific (Larcher et al. 2010; Taschler and Neuner 2004), except for inflorescences (Ladinig et al. 2013), thus pushing selection towards a clonal life strategy. In the temperate zone, growth of alpine plants is primarily constrained by the length of the growing season. To complete their seasonal life cycle, they employ large below-ground reserves, rapid flushing, adjusted flowering phenology and photoperiod-controlled senescence before the onset of late-season freezing. Productivity of closed alpine vegetation expressed per day of growing season and unit land area is not inferior to any other humid environment, so the duration of the period suitable for growth is the main explanation for low biomass and productivity (Körner 2021). The small size of high alpine plant species does not reflect a climate- or nutrient-enforced dwarfing, but these plants are 'small by design' and, thus, capitalize on the aerodynamically sheltered and warm climate near the ground (Körner et al. 1989b; 'solar heating'; Körner and Hiltbrunner 2018).

Only recently, we could illustrate the way in which a 'classical' snow bed species (*Soldanella pusilla*) copes with one of the most extreme types of habitat in terms of short season length (Körner et al. 2019): bi-seasonal ('evergreen') foliage, closely attached to the ground (thus trapping heat once out of snow) and growing an entire inflorescence (including the stalk) under metres of snow at 0°C, initiated by an internal clock in January, to be ready at snow melt. In addition, this species is filling its tissues up to almost half of dry matter with storage reserves with the biggest fraction represented by the tetrasaccharide stachyose, thus avoiding a too high osmotic pressure while otherwise being readily available for metabolism. These reserves suffice for 2 years should snow not disappear in a year. Presumed lignification constraints during growth at 0°C are confined to non-essential tissues (sclerenchyma) but, surprisingly, not to vessels. While plants cannot grow below 0°C and growth becomes extremely slow at temperatures below 5°C (see Sect. 4), photosynthesis would reach high rates at such low temperatures once plants become released from snow (Sect. 3).

5.4 Experiments by Nature

Counter to expectation, alpine plant species richness is higher than one would expect from alpine land area, despite the environmental constraints, and it is far higher than biodiversity in the arctic life zone although the arctic land area is roughly twice as large as the alpine one (Chapin and Körner 1994, Körner et al. 2011). Topographic diversity and microclimatic diversity strongly contribute to this phenomenon (Körner 2002, 2004a, b, 2008). Mountains in general are hot spots of biodiversity, given the contrasting climates that become compressed over a narrow geographical range. These quasi-'experiments by nature' offer great opportunities for research (Körner 2000, 2007b, 2020b, 2021). Yet, in contrast to the Arctic, the research community that could assess and explain mountain diversity is so fragmented as mountains are. So, in the aftermath of the Rio conference, the Global Mountain Biodiversity Assessment (GMBA) network was founded in 2000, and I had the pleasure to run it together with Eva Spehn till my retirement and edit several volumes that assembled researchers and data from different geographical regions (Körner and Spehn 2002, Spehn et al. 2006). These initiatives led to the mountain chapter of the Millennium Ecosystem Assessment report (Körner and Ohsawa 2005), a résumé of the state and risks of the world's mountain systems.

Counter to common belief, I consider mountain biota the least vulnerable compared to other biomes with respect to climatic warming, because nowhere else do organisms find alternative microclimates at such close proximity than on mountains. In addition to their high diversity and often retained naturalness, this is one extra reason to protect mountain ecosystems and monitor their long-term fate. This is a field of research I became strongly involved recently at several sites in the Austrian, Italian and Swiss Alps (Körner 2018b). These highly standardized, multidisciplinary monitoring activities took great advantage from experiences gained at the ideal facilities at the Furkapass research station (www.alpfor.ch), 300 m above treeline in the Swiss Alps (see Sect. 6). Together with Erika Hiltbrunner, who runs this station, we assembled 47 experts to assess the organismic inventory of this alpine biome and arrived at 2,100 named taxa, with many new discoveries in what many people consider a barren and biologically near-to-empty terrain (Hiltbrunner and Körner 2018). The mosaics of different life conditions in the treeless alpine belt permit studying plant responses to climatic contrasts over a few metres that would otherwise occur over kilometres of elevation or more than thousand kilometres of latitude (Fig. 13). From such works at sharp snow-melt gradients, it became clear that monitoring long-term changes in these slowly responding and patchy alpine biota requires a permanent plot approach, with contrasting groups of organismic taxa and abiotic conditions assessed repeatedly over time following a strict protocol.

Fig. 13 The diversity of microhabitats (in this example by contrasting snow duration) in alpine landscapes permits species with different climatic preferences to co-exist at close proximity – an ideal setting to test climate change effects (Furkapass, Swiss Alps, 2,450 m asl)

These 'experiments by nature' permit identifying species range limits and their potential shifts within a few metres.

5.5 *The Cold Edge of Plant Life*

The ultimate edge of angiosperm life as it gets very cold is neither found in the Arctic or Antarctic, nor at the highest locations were angiosperms have been found so far (around 6,000–6,300 m in the Himalayas; Körner 2003a; Klimes and Dolezal 2010), but – as far as it is known until now, in Switzerland. Old mountaineer reports about plant life at the summit of the Dom, which caused me to have a closer look. With one of my three daughters, her husband and a mountain guide, we climbed the 4,545 m summit in 2008; got enveloped in fog once we reached the top, just in time to bury data loggers in – what a surprise – massive moss cushions on a rock at the summit, amidst a world of ice and snow; and descended. Well instructed, the guide went back next summer, could recover one of the data loggers and identified five, more than hand size, cushions of *Saxifraga oppositifolia* a few metres downslope at 4,505 m on a little rock terrace that we could not reach for the dense fog at our first visit. The temperatures recorded up there, in what were comparatively good summers, identify this location as the coldest place on earth where an angiosperm was found so far (Körner 2011). And these plants were not alone: several species of fungi, including mycorrhizal fungi, Collembola and mites, occupied these 'islands in the sky' (Oehl and Körner 2014) – the cold edge of higher plant life.

6 Some Lessons Learned from Almost Half a Century of Research and Teaching

Maybe there are two main lessons that outweigh any other: One is the immense diversity of solutions plants have evolved for the same 'problem'. Not even the harshest locations on earth have selected plants with one unique solution for life. Second, available methods and mainstream reasoning can bias our research efforts as well as the emerging paradigms. These were the messages of what I considered a humble little paper that made it to one of the 100 considered most influential papers published by the British Ecological Society journals (Körner 1991; see Ülo Niinemets in Grubb and Whittaker 2013). Accounting for biological variation as well as selecting questions, rather than the availability of methods, requires a theory-based identification of traits that matter. Yet, many traits considered functional (sensu 'important') are often selected for convenience (Körner 2018a). My impression is that the recent functional traits euphoria is at a decline and plant functional groups (except for the most trivial ones such as trees and grasses) have not met expectations (for an early account, see Körner 1993). I expect response traits (dynamic traits) to become more successful (see the discussion in Körner 2018a), and the keystone species concept also appears to be more promising. After all, species matter.

Embedded in these overarching lessons, there are more subtle ones. One of these is perhaps that one has to go through certain practical experiences for later departing to new grounds with confidence. Had I not gone through all the details of plant gas exchange work myself, I am not sure if I had given up with the carbon centric view at the plant world. Another lesson is the often neglected dominance of structure over processes. It seems to be more rewarding, particularly to young researchers, to view a digital display, rather than to use a spade or a root sieve, with the latter ones often holding the answer (Fig. 3). A third lesson is that theory and potentially also mathematical models should play a bigger role in designing and/or interpreting experimental research. Had modellers applied humble logics early on, they should have spotted the black boxes experimentalists bypassed for the lack of tools to tackle these unresolved questions. As an example, it should have been clear upfront that soil mineral nutrients are finite and plant life is governed by nutrient stoichiometry so that, inevitably, the nutrient cycle drives the carbon cycle and not the other way round (one of the reasons why carbon cycle models fail; e.g. Huntzinger et al. 2017). It is hard to believe that people with a strong theoretical backing permitted experimentalists to confuse fluxes of carbon (e.g. rates of growth) with pool sizes of carbon (storage) as still happens almost weekly in the international literature (Körner 2017, Büntgen et al. 2019). It is far more difficult to study, explain and model mean residence time of carbon in an ecosystem (which sets the pool size) or stochastic outbursts of carbon in case of disturbance, than to quantify the influx of carbon and its biochemistry (Körner 2003c). In a steady state, input-output models cannot arrive at carbon pool size, just as in an overflowing bathtub; the flux across the tub is not telling with regard to the tub's volume (Körner 2017). For a review of such concepts that developed in our Basel Botany lab see Leuzinger and Hättenschwiler (2013).

Another set of messages I took, particularly from working in mountains, is that what we believe is harsh by physical experience is also harsh or 'stressing' for native plants (Körner 1999b, 2003d). Over decades, people have commented on the severe limitations of arctic and alpine plants by low temperature, and now we are told that warming puts these plants at risk. Either one or both of these notions must be wrong. In reality, cold environments are dominated by low-stature plants, with the actual temperatures experienced, controlled by aerodynamic restrictions of heat exchange and entirely misrepresented by weather stations. Mosaics of seasonal thermal life conditions can vary by more than twice the worst IPCC warming scenarios over a few metres distance in such environments (Scherrer and Körner 2009, 2011). Even within a single inflorescence, temperatures may increase by 10 K from the edge of petals to the ovaries under full sun (Dietrich and Körner 2014). It is very unfortunate that biologists do not receive even the most basic training in meteorology and micrometeorology. Further, life under snow has been poorly examined but may hold surprising answers as was shown for the *Soldanella* case in Sect. 5.3.

Among the practical lessons, one certainly is the imperative of hands-on biology in the field (Körner 2020b). A greenhouse can help at times, and a well-controlled phytotron may be indispensable to clarify certain issues, but these can never substitute field data. Quite often logistics, work conditions or the desire to 'control' things lend towards escaping to artificial environments. While natural climatic conditions might be approximated by sophisticated technology, that of soil conditions cannot. This is where well-equipped field stations can play a central role. Over the last two decades, the Alpine Research and Education station Furka (www.alpfor.ch), at the Swiss Furkapass at 2,440 m asl, facilitated around 11 PhD and 36 MSc projects, all profiting from the fact that alpine nature is at the doorstep (Fig. 14).

Over my 48 years of academic teaching, I may have given full-term courses to 5,000 students, and many others attended excursions and graduate schools. I guess, I guided some 30 students to a doctoral degree and perhaps around 100 to graduation (MSc). I do not have any statistics on this. Over the last 15 years, I also gave courses in Georgia (Illia University, Tbilisi), in China (Kunming Institute of Botany of the Chinese Academy of Sciences), in Chile (Katalapi graduate summer school near Puerto Mont), in Brazil (University of Campinas, near Sao Paulo), twice in Svalbard (UNIS, Longyearbyen) and annually in the Swiss Alps at the Furka station. As with scientific research, which is not just collecting data but condensing knowledge, the task of academic teaching, in my view, is to deliver a message and concepts, a well-thought distillate of the immense body of knowledge, very different from a review. This was also my philosophy when I wrote the textbooks *Alpine Plant Life* (Georgian and Chinese translations), *Alpine Treelines* (Chinese translation) and the four plant ecology chapters of *Strasburger's Plant Sciences* (translated from German to Russian, Italian and English): drawing lines of reasoning, showing overarching patterns, coining paradigms or questioning them. Although immensely time-consuming, it is very educating to write a textbook.

Any lessons from my academic teaching? Why keep lecturing at times of Wikipedia and electronic textbooks? I think the most important added value in

Fig. 14 When the field is the lab, it requires appropriate infrastructure. The research station ALPFOR at Furkapass, Swiss Alps, 2,430 m asl permits field studies almost at the doorstep in the core of the alpine belt of the Alps

face-to-face teaching is that facts can be loaded with life. Another one is bridging across disciplines. A third one for classroom teaching is *less is more*. I warn young academic teachers against bringing more information to the classroom than their own brain can keep (except for some complex diagrams). And I encourage them to show feelings about the subject they teach (a friendly comment by Hättenschwiler and Arnone 2013). This is considered emotional. Yes it is. Learning is also tied to emotions, just as creative science is not so far apart from arts, as many think.

Acknowledgements I thank Ulrich Lüttge for inviting me to write this chapter and commenting an earlier draft. I also gratefully acknowledge comments by Erika Hiltbrunner and Günter Hoch. Thanks to Susanna Riedl for helping with the artwork. What I summarized here could not have been achieved without the hard work and dedication by a large number of graduate students and academic staff at the University of Basel and in partner institutions. I heartily thank all of them.

References

Alvarez-Uria P, Körner C (2007) Low temperature limits of root growth in deciduous and evergreen temperate tree species. Funct Ecol 21:211–218

Anderegg WRL, Berry JA, Smith DD, Sperry JS, Anderegg LDL, Field CB (2012) The roles of hydraulic and carbon stress in a widespread climate-induced forest die-off. Proc Natl Acad Sci U S A 109:233–237

Aulitzky H (1961) Die Bodentemperaturen in der Kampfzone oberhalb der Waldgrenze und im subalpinen Zirben-Lärchenwald. Mitt forstl Versuchsw Österr 59:153–208
Bachman S, Heisler-White JL, Pendall E, Williams DG, Morgan JA, Newcomb J (2010) Elevated carbon dioxide alters impacts of precipitation pulses on ecosystem photosynthesis and respiration in a semi-arid grassland. Oecologia 162:791–802
Bader MKF, Siegwolf R, Körner C (2010) Sustained enhancement of photosynthesis in mature deciduous forest trees after 8 years of free air CO_2 enrichment. Planta 232:1115–1125
Bader MKF, Leuzinger S, Keel SG, Siegwolf RTW, Hagedorn F, Schleppi P, Körner C (2013) Central European hardwood trees in a high-CO_2 future: synthesis of an 8-year forest canopy CO_2 enrichment project. J Ecol 101:1509–1519
Barron AR, Wurzburger N, Bellenger JP, Wright SJ, Kraepiel AML, Hedin LO (2009) Molybdenum limitation of asymbiotic nitrogen fixation in tropical forest soils. Nat Geosci 2:42–45
Basler D, Körner C (2012) Photoperiod sensitivity of bud burst in 14 temperate forest tree species. Agric For Meteorol 165:73–81
Basler D, Körner C (2014) Photoperiod and temperature responses of bud swelling and bud burst in four temperate forest tree species. Tree Physiol 34:377–388
Beck E (2016) Torn between nature and lab: a dying breed of plant scientists? Prog Bot. https://doi.org/10.1007/124_2016_2
Boyer JS (2017) Plant water relations: a whirlwind of change. Prog Bot. https://doi.org/10.1007/124_2017_3
Braun S, Thomas VFD, Quiring R, Flückiger W (2010) Does nitrogen deposition increase forest production? The role of phosphorus. Environ Pollut 158:2043–2052
Brienen RJW, Phillips OL, Feldpausch TR et al (2015) Long-term decline of the Amazon carbon sink. Nature 519:344–348
Büntgen U, Krusic PJ, Piermattei A, Coomes DA, Esper J, Myglan VS, Kirdyanov AV, Camarero JJ, Crivellaro A, Körner C (2019) Limited capacity of tree growth to mitigate the global greenhouse effect under predicted warming. Nat Commun 10:2171. https://doi.org/10.1038/s41467-019-10174-4
Caprez R, Niklaus PA, Körner C (2012) Forest soil respiration reflects plant productivity across a temperature gradient in the Alps. Oecologia 170:1143–1154
Čermák J, Kučera J, Bauerle WL, Phillips N, Hinckley TM (2007) Tree water storage and its diurnal dynamics related to sap flow and changes in stem volume in old-growth Douglas-fir trees. Tree Physiol 27:181–198
Cernusca A (1976) Energy exchange within individual layers of a meadow. Oecologia 23:141–149
Chapin FS III, Körner C (1994) Arctic and alpine biodiversity: patterns, causes and ecosystem consequences. Trends Ecol Evol 9:45–47
Choat B, Jansen S, Brodribb TJ, Cochard H, Delzon S, Bhaskar R, Bucci SJ, Feild TS, Gleason SM, Hacke UG, Jacobsen AL, Lens F, Maherali H, Martinez-Vilalta J, Mayr S, Mencuccini M, Mitchell PJ, Nardini A, Pittermann J, Pratt RB, Sperry JS, Westoby M, Wright IJ, Zanne AE (2012) Global convergence in the vulnerability of forests to drought. Nature 491:752–755
Cowan IR (1965) Transport of water in the soil-plant-atmosphere system. J Appl Ecol 2:221–239
Cowan IR, Troughton JH (1971) The relative role of stomata in transpiration and assimilation. Planta 97:325–336
Darwin F (1898) Observations on stomata. Phil Trans R Soc London 190:531–621
De Saussure T (1804) Recherches chimiques sur la vegetation. Nyon, Paris
Dietrich L, Körner C (2014) Thermal imaging reveals massive heat accumulation in flowers across a broad spectrum of alpine taxa. Alp Bot 124:27–35
Dietrich L, Hoch G, Kahmen A, Körner C (2018) Losing half of the conductive area hardly impacts the water status of mature trees. Sci Rep 8:15006
Ellsworth DS, Anderson IC, Crous KY, Cooke J, Drake JE, Gherlenda AN, Gimeno TE, Macdonald CA, Medlyn BE, Powell JR, Tjoelker MG, Reich PB (2017) Elevated CO2 does not increase eucalypt forest productivity on a low-phosphorus soil. Nat Clim Chang 7:279–283

Elser JJ, Fagan WF, Kerkhoff AJ, Swenson NG, Enquist BJ (2010) Biological stoichiometry of plant production: metabolism, scaling and ecological response to global change. New Phytol 186:593–608
Franks PJ, Farquhar GD (1999) A relationship between humidity response, growth form and photosynthetic operating point in C3 plants. Plant Cell Environ 22:1337–1349
Gaastra P (1959) Photosynthesis of crop plants as influenced by light, carbon dioxide, temperature, and stomatal diffusion resistance. Meded Landbouwhogesch Wageningen 59(13):1–68
Granados J, Körner C (2002) In deep shade, elevated CO2 increases the vigor of tropical climbing plants. Glob Chang Biol 8:1109–1117
Grubb P, Whittaker J (2013) 100 influential papers published in 100 years of the British Ecological Society journals. Comment 25 by U Niinemets. British Ecological Society, London
Grünzweig JM, Körner C (2001) Growth, water and nitrogen relations in grassland model ecosystems of the semi-arid Negev of Israel exposed to elevated CO_2. Oecologia 128:251–262. https://doi.org/10.1038/nature13281
Hättenschwiler S, Arnone JA (2013) A tribute to Christian Körner for his 25 years of service on the Oecologia editorial board. Oecologia 171:605–611
Hättenschwiler S, Körner C (2000) Tree seedling responses to in situ CO2-enrichment differ among species and depend on understorey light availability. Glob Chang Biol 6:213–226
Hättenschwiler S, Miglietta F, Raschi A, Körner C (1997) Thirty years of in situ tree growth under elevated CO_2: a model for future forest responses? Glob Chang Biol 3:436–471
Hiltbrunner E, Körner C (2018) Hotspot Furka. Biologische Vielfalt im Gebirge. Alpine Forschungs- und Ausbildungsstation Furka and University of Basel, ISBN: 978-3-033-06701-1
Hoch G, Körner C (2003) The carbon charging of pines at the climatic treeline: a global comparison. Oecologia 135:10–21
Hoch G, Körner C (2005) Growth, demography and carbon relations of *Polylepis* trees at the world's highest treeline. Funct Ecol 19:941–951
Hoch G, Körner C (2012) Global patterns of mobile carbon stores in trees at the high-elevation tree line. Glob Ecol Biogeogr 21:861–871
Holbrook NM, Zwieniecki MA (2005) Vascular transport in plants. Elsevier, Amsterdam
Hovenden MJ, Newton PCD, Wills KE (2014) Seasonal not annual rainfall determines grassland biomass response to carbon dioxide. Nature 511:583–586
Hungate BA, Dijkstra P, Wu ZT, Duval BD, Day FP, Johnson DW, Megonigal JP, Brown ALP, Garland JL (2013) Cumulative response of ecosystem carbon and nitrogen stocks to chronic CO_2 exposure in a subtropical oak woodland. New Phytol 200:753–766
Huntzinger DN, Michalak AM, Schwalm C et al (2017) Uncertainty in the response of terrestrial carbon sink to environmental drivers undermines carbon-climate feedback predictions. Sci Rep 7:4765. https://doi.org/10.1038/s41598-017-03818-2
Inauen N, Körner C, Hiltbrunner E (2012) No growth stimulation by CO_2 enrichment in alpine glacier forefield plants. Glob Chang Biol 18:985–999
Ingestad T (1982) Relative addition rate and external concentration driving variables used in plant nutrition research. Plant Cell Environ 5:443–453
Jackson RB, Canadell J, Ehleringer JR, Mooney HA, Sala OE, Schulze ED (1996) A global analysis of root distributions for terrestrial biomes. Oecologia 108:389–411
Jackson RB, Sperry JS, Dawson TE (2000) Root water uptake and transport: using physiological processes in global predictions. Trends Plant Sci 5:482–488
Jarvis PG (1975) Water transfer in plants. In: De Vries DA (ed) Heat and mass transfer in the environment of vegetation. Scripta, Washington, pp 369–394
Jiang M, Medlyn BE, Drake JE et al (2020) The fate of carbon in a mature forest under carbon dioxide enrichment. Nature 580:227–231
Johnson DM, Domec JC, Berry ZC et al (2018) Co-occurring woody species have diverse hydraulic strategies and mortality rates during extreme drought. Plant Cell Environ 41:576–588
Keel SG, Siegwolf RTW, Körner C (2006) Canopy CO_2 enrichment permits tracing the fate of recently assimilated carbon in a mature deciduous forest. New Phytol 172:319–329

Klein T, Bader MKF, Leuzinger S, Mildner M, Schleppi P, Siegwolf RTW, Körner C (2016a) Growth and carbon relations of mature Picea abies trees under 5 years of free-air CO_2 enrichment. J Ecol 104:1720–1733
Klein T, Siegwolf RTW, Körner C (2016b) Belowground carbon trade among tall trees in a temperate forest. Science 352:342–344
Klimes L, Dolezal J (2010) An experimental assessment of the upper elevational limit of flowering plants in the western Himalayas. Ecography 33:590–596
Kollas C, Vitasse Y, Randin CF, Hoch G, Körner C (2012) Unrestricted quality of seeds in Europesn borad-leaved tree species growing at the cold boundary of their distribution. Ann Bot 109:473–480
Kollas C, Körner C, Randin CF (2014a) Spring frost and growing season length co- control the cold range limits of broad-leaved trees. J Biogeogr 41:773–783
Kollas C, Randin CF, Vitasse Y, Körner C (2014b) How accurately can minimum temperatures at the cold limits of tree species be extrapolated from weather station data? Agric For Meteorol 184:257–266
Körner C (1982) CO_2 exchange in the alpine sedge Carex curvula as influenced by canopy structure, light and temperature. Oecologia 53:98–104
Körner C (1989) The nutritional status of plants from high altitudes. A worldwide comparison. Oecologia 81:379–391
Körner C (1991) Some often overlooked plant characteristics as determinants of plant growth: a reconsideration. Funct Ecol 5:162–173
Körner C (1992) Response of alpine vegetation to global climate change. Catena Suppl 22:85–96
Körner C (1993) Scaling from species to vegetation: the usefulness of functional groups. In: Schulze ED, Mooney HA (eds) Biodiversity and ecosystem function, Ecological studies, vol 99. Springer, Berlin, pp 117–140
Körner C (1994) Leaf diffusive conductances in the major vegetation types of the globe. In: Schulze ED, Caldwell MM (eds) Ecophysiology of photosynthesis, Ecological studies, vol 100. Springer, Berlin, pp 463–490
Körner C (1999a) Alpine plant life, 1st edn. Springer, Berlin
Körner C (1999b) Alpine plants: stressed or adapted? In: Press MC, Scholes JD, Barker MG (eds) Physiological plant ecology, The 39th symposium of the British Ecological Society held at the University of York 7–9 September 1998. Blackwell, Oxford, pp 297–311
Körner C (2000) Why are there global gradients in species richness? Mountains might hold the answer. Trends Ecol Evol 15:513–514
Körner C (2002) Alpine ecosystems. In: Encyclopedia of life sciences, vol 1. Nature, London, pp 392–393
Körner C (2003a) Alpine plant life, 2nd edn. Springer, Berlin
Körner C (2003b) Carbon limitation in trees. J Ecol 91:4–17
Körner C (2003c) Slow in, rapid out – carbon flux studies and Kyoto targets. Science 300:1242–1243
Körner C (2003d) Limitation and stress – always or never? J Veg Sci 14:141–143
Körner C (2004a) Mountain biodiversity, its causes and function. Ambio Spec Rep 13:11–17
Körner C (2004b) Through enhanced tree dynamics carbon dioxide enrichment may cause tropical forests to lose carbon. Philos Trans R Soc Lond Ser B Biol Sci 359:493–498
Körner C (2006a) Significance of temperature in plant life. In: Morison JIL, Morecroft MD (eds) Plant growth and climate change. Blackwell Publishing Ltd, Oxford, pp 48–69
Körner C (2006b) Plant CO_2 responses: an issue of definition, time and resource supply. New Phytol 172:393–411
Körner C (2007a) Climatic treelines: conventions, global patterns, causes. Erdkunde 61:315–324
Körner C (2007b) The use of "altitude" in ecological research. Trends Ecol Evol 22:569–574
Körner C (2008) Alpine ecosystems and the high-elevation treeline. In: Jorgensen SE, Fath BD (eds) Ecosystems, Encyclopedia of ecology, vol 1. Elsevier, Oxford, pp 138–144

Körner C (2009) Responses of humid tropical trees to rising CO_2. Annu Rev Ecol Evol Syst 40:61–79
Körner C (2011) Coldest places on earth with angiosperm plant life. Alp Bot 121:11–22
Körner C (2012) Alpine treelines. Springer, Basel
Körner C (2013a) Plant-environment interactions. In: Bresinsky A, Körner C, Kadereit JW, Neuhaus G, Sonnewald U (eds) Strasburger's plant sciences. Springer, Berlin
Körner C (2013b) Growth controls photosynthesis – mostly. Nova Acta Leopold 391:273–283
Körner C (2015) Paradigm shift in plant growth control. Curr Opin Plant Biol 25:107–114
Körner C (2017) A matter of tree longevity. Science 355:130–131
Körner C (2018a) Concepts in empirical plant ecology. Plant Ecol Divers 11:405–428
Körner C (2018b) Comparative, long-term ecosystem monitoring across the Alps: Austrian Hohe Tauern National Park, South-Tyrol and Swiss Central Alps. In: Bauch K (ed) Proceedings symposium for research in protected areas. Salzburger Nationalparkfonds/Austrian Academy of Sciences, Mittersill/Vienna, pp 331–337
Körner C (2019) No need for pipes when the well is dry - a comment on hydraulic failure in trees. Tree Physiol 39:695–700
Körner C (2020a) Climatic controls of the global high elevation treelines. In: Encyclopedia of the world's biomes. Elsevier, pp 275–281. https://doi.org/10.1016/B978-0-12-409548-9.11998-0
Körner C (2020b) Experiments by nature – strength in realism. In: Burt TP, Thompson DBA (eds) Curious about nature: a passion for fieldwork. Cambridge University Press, Cambridge. Chap. 27
Körner C (2021) Alpine plant life, 3rd edn. Springer, Berlin
Körner C, Arnone JA III (1992) Responses to elevated carbon dioxide in artificial tropical ecosystems. Science 257:1672–1675
Körner C, Basler D (2010) Phenology under global warming. Science 327:1461–1462
Körner C, Cernusca A (1976) A semi-automatic diffusion porometer and its performance under alpine field conditions. Photosynthetica 10:172–181
Körner C, Cochrane PM (1985) Stomatal responses and water relations of *Eucalyptus pauciflora* in summer along an elevational gradient. Oecologia 66:443–455
Körner C, Diemer M (1987) In situ photosynthetic responses to light, temperature and carbon dioxide in herbaceous plants from low and high altitude. Funct Ecol 1:179–194
Körner C, Diemer M (1994) Evidence that plants from high altitudes retain their greater photosynthetic efficiency under elevated CO_2. Funct Ecol 8:58–68
Körner C, Hiltbrunner E (2018) The 90 ways to describe plant temperature. Perspect Plant Ecol Evol Syst 30:16–21
Körner C, Ohsawa M (2005) Mountain systems. In: Hassan R, Scholes R, Ash N (eds) Ecosystems and human Well-being: current state and trends, vol 1. Island Press, Washington, pp 681–716
Körner C, Paulsen J (2004) A world-wide study of high altitude treeline temperatures. J Biogeogr 31:713–732
Körner C, Spehn EM (2002) Mountain biodiversity, a global assessment. Parthenon, Boca Raton
Körner C, Spehn E (2019) A Humboldtian view of mountains. Science 365(6458):1061–1061
Körner C, Würth M (1996) A simple method for testing leaf responses of tall tropical forest trees to elevated CO_2. Oecologia 107:421–425
Körner C, Scheel JA, Bauer H (1979) Maximum leaf diffusive conductance in vascular plants. Photosynthetica 13:45–82
Körner C, Bannister P, Mark AF (1986) Altitudinal variation in stomatal conductance, nitrogen content and leaf anatomy in different plant life forms in New Zealand. Oecologia 69:577–588
Körner C, Farquhar GD, Roksandic Z (1988) A global survey of carbon isotope discrimination in plants from high altitude. Oecologia 74:623–632
Körner C, Pelaez Menendez-Riedl S, John PCL (1989a) Why are Bonsai plants small? A consideration of cell size. Aust J Plant Physiol 16:443–448
Körner C, Neumayer M, Pelaez Menendez-Riedl S, Smeets-Scheel A (1989b) Functional morphology of mountain plants. Flora 182:353–383

Körner C, Farquhar GD, Wong SC (1991) Carbon isotope discrimination by plants follows latitudinal and altitudinal trends. Oecologia 88:30–40

Körner C, Diemer M, Schäppi B, Niklaus P, Arnone J (1997) The responses of alpine grassland to four seasons of CO_2 enrichment: a synthesis. Acta Oecol 18:165–175

Körner C, Asshoff R, Bignucolo O, Hättenschwiler S, Keel SG, Pelaez-Riedl S, Pepin S, Siegwolf RTW, Zotz G (2005) Carbon flux and growth in mature deciduous forest trees exposed to elevated CO_2. Science 309:1360–1362

Körner C, Paulsen J, Spehn EM (2011) A definition of mountains and their bioclimatic belts for global comparison of biodiversity data. Alp Bot 121:73–78

Körner C, Basler D, Hoch G, Kollas C, Lenz A, Randin CF, Vitasse Y, Zimmermann NE (2016) Where, why and how? Explaining the low-temperature range limits of temperate tree species. J Ecol 104:1076–1088

Körner C, Jetz W, Paulsen J, Payne D, Rudmann-Maurer K, Spehn EM (2017) A global inventory of mountains for bio-geographical applications. Alp Bot 127:1–15

Körner C, Riedl S, Keplinger T, Richter A, Wiesenbauer J, Schweingruber F, Hiltbrunner E (2019) Life at 0°C: the biology of the alpine snowbed plant Soldanella pusilla. Alp Bot. https://doi.org/10.1007/s00035-019-00220-8

Ladinig U, Hacker J, Neuner G, Wagner J (2013) How endangered is sexual reproduction of high-mountain plants by summer frosts? Frost resistance, frequency of frost events and risk assessment. Oecologia 171:743–760

Landsberg JJ, Beadle CL, Biscoe PV, Butler DR, Davidson B, Incoll LD, James GB, Jarvis PG, Martin PJ, Neilson RE, Powell DBB, Slack EM, Thorpe MR, Turner NC, Warrit B, Watts WR (1975) Diurnal energy, water and CO2 exchanges in an apple (*Malus pumila*) orchard. J Appl Ecol 12:659–684

Lange OL, Lösch R, Schulze ED, Kappen L (1971) Responses of stomata to changes in humidity. Planta 100:76–86

Larcher W (1975) Arthur Pisek 1894-1975. Ber Dt Bot Ges 88:497–502

Larcher W (1977) Ergebnisse des IBP-Projekts "Zwergstrauchheide Patscherkofel". In: Sitzungsber Oesterr Akad Wiss, Mathem-naturwiss Kl, Abt I, vol 186, pp 301–371

Larcher W (2003) Physiological plant ecology, 4th edn. Springer, Berlin

Larcher W, Kainmüller C, Wagner J (2010) Survival types of high mountain plants under extreme temperatures. Flora 205:3–18

Lenz A, Hoch G, Vitasse Y, Körner C (2013) European deciduous trees exhibit similar safety margins against damage by spring freeze events along elevational gradients. New Phytol 200:1166–1175

Lenz A, Vitasse Y, Hoch G, Körner C (2014) Growth and carbon relations of temperate deciduous tree species at their upper elevation range limit. J Ecol 102:1537–1548

Lenz A, Hoch G, Körner C, Vitasse Y (2016) Convergence of leaf-out towards minimum risk of freezing damage in temperate trees. Funct Ecol. https://doi.org/10.1111/1365-2435.12623

Leuzinger S, Hättenschwiler S (2013) Beyond global change: lessons from 25 years of CO2 research. Oecologia 171:639–651

Leuzinger S, Körner C (2010) Rainfall distribution is the main driver of runoff under future CO_2 concentration in a temperate deciduous forest. Glob Chang Biol 16:246–254

Leuzinger S, Fatichi S, Cusens J, Körner C, Niklaus PA (2015) The 'island effect' in terrestrial global change experiments: a problem with no solution? AoB Plants 7:plv092

Lüttge U (1979) Otto Stocker zum 90.Gebuststag. Ber Dt Bot Ges 92:1–6

Lüttge U (2008) Physiological ecology of tropical plants. Springer, Berlin

Mackay JFG, Weatherley PE (1973) The effects of transverse cuts through the stems of transpiring woody plants on water transport and stress in the leaves. J Exp Bot 24:15–28

Mayr S, Sperry JS (2010) Freeze-thaw-induced embolism in Pinus contorta: centrifuge experiments validate the 'thaw-expansion hypothesis' but conflict with ultrasonic emission data. New Phytol 185:1016–1024

McDowell NG (2011) Mechanisms linking drought, hydraulics, carbon metabolism, and vegetation mortality. Plant Physiol 155:1051–1059

Meinzer FC, Goldstein G, Franco AC, Bustamante M, Igler E, Jackson P, Caldas L, Rundel PW (1999) Atmospheric and hydraulic limitations on transpiration in Brazilian cerrado woody species. Funct Ecol 13:273–282

Mildner M, Bader MKF, Leuzinger S, Siegwolf RTW, Körner C (2014) Long-term C^{13} labeling provides evidence for temporal and spatial carbon allocation patterns in mature Picea abies. Oecologia 175:747–762

Möhl P, Hiltbrunner E, Körner C (2020) Halving sunlight reveals no carbon limitation of aboveground biomass production in alpine grassland. Glob Chang Biol. https://doi.org/10.1111/gcb.14949

Monteiro JAF, Körner C (2013) Leaf turnover and herbivory in the tall tussock grass Festuca orthophylla in the Andean Altiplano. Alp Bot 123:13–20

Monteiro JAF, Hiltbrunner E, Körner C (2011) Functional morphology and microclimate of Festuca orthophylla, the dominant tall tussock grass in the Andean Altiplanto. Flora 206:387–396

Morgan JA, Pataki DE, Körner C, Clark H, Del Grosso SJ, Grünzweig JM, Knapp AK, Mosier AR, Newton PCD, Niklaus PA, Nippert JB, Nowak RS, Parton WJ, Polley HW, Shaw MR (2004) Water relations in grassland and desert ecosystems exposed to elevated atmospheric CO_2. Oecologia 140:11–25

Muller B, Pantin F, Genard M, Turc O, Freixes S, Piques M, Gibon Y (2011) Water deficits uncouple growth from photosynthesis, increase C content, and modify the relationships between C and growth in sink organs. J Exp Bot 62:1715–1729

Nagelmüller S, Hiltbrunner E, Körner C (2016) Critically low soil temperatures for root growth and root morphology in three alpine plant species. Alp Bot 126:11–21

Nagelmüller S, Hiltbrunner E, Körner C (2017) Low temperature limits for root growth in alpine species are set by cell differentiation. AoB Plants 9:plx054. https://doi.org/10.1093/aobpla/plx054

Niklaus PA, Körner C (2004) Synthesis of a six-year study of calcareous grassland responses to in situ CO_2 enrichment. Ecol Monogr 74:491–511

Niklaus PA, Spinnler D, Körner C (1998) Soil moisture dynamics of calcareous grassland under elevated CO2. Oecologia 117:201–208

Norby RJ, Warren JM, Iversen CM, Medlyn BE, McMurtrie RE (2010) CO_2 enhancement of forest productivity constrained by limited nitrogen availability. Proc Natl Acad Sci U S A 107:19368–19373

Noroozi J, Körner C (2018) A bioclimatic characterization of high elevation habitats in the Alborz mountains of Iran. Alp Bot 128:1–11

O'Leary BM et al (2019) Core principles which explain variation in respiration across biological scales. New Phytol 222:670–686

Oehl F, Körner C (2014) Multiple mycorrhization at the coldest place known for angiosperm plant life. Alp Bot 124:193–198

Owensby CE, Auen LM, Coyne PI (1994) Biomass production in a nitrogen-fertilized, tallgrass prairie ecosystem exposed to ambient and elevated levels of CO_2. Plant Soil 165:105–113

Patty L, Halloy SRP, Hiltbrunner E, Körner C (2010) Biomass allocation in herbaceous plants under grazing impact in the high semi-arid Andes. Flora 205:695–703

Paulsen J, Körner C (2014) A climate-based model to predict potential treeline position around the globe. Alp Bot 124:1–12

Philipps OL, Baker TR, Arroyo L, Higuchi N, Killeen TJ, Laurance WF, Lewis SL, Lloyd J, Vinceti B et al (2004) Pattern and process in Amazon tree turnover 1976–2001. Phil Trans R Soc Lond B 359:381–407

Phillips OL, Vasquez Martinez R, Arroyo L, Baker TR, Killeen T, Lewis SL, Malhi Y, Monteagudo Mendoza A, Neill D, Nunez Vargas P, Alexiades M, Ceron C, Di Fiore A, Erwin T, Jardim A,

Palacios W, Saldias M, Vincenti B (2002) Increasing dominance of large lianas in Amazonian forests. Nature 418:770–774
Pisek A (1960) Pflanzen der Arktis und des Hochgebirges. In: Ruhland W (ed) Handbuch der Pflanzenphysiologie, vol 5. Springer, Berlin, pp 377–413
Pisek A, Cartellieri E (1934) Zur Kenntnis des Wasserhaushalts der Pflanzen. III Alpine Zwergsträucher. In: Fitting H (ed) Jahrbücher für Wissenschaftliche Botanik, vol 79. Gebr Bornträger, Leipzig, pp 131–190
Pisek A, Knapp H, Ditterstorfer J (1970) Maximal opening width and morphology of stomata, with dates of their size and number. Flora 159:459–479
Pisek A, Larcher W, Vegis A, Napp-Zinn K (1973) The normal temperature range. In: Precht H, Christophersen J, Hensel H, Larcher W (eds) Temperature and life. Springer, Berlin, pp 102–194
Raich JW, Nadelhoffer KJ (1989) Belowground carbon allocation in forest ecosystems: global trends. Ecology 70:1346–1354
Randin CF, Paulsen J, Vitasse Y, Kollas C, Wohlgemuth T, Zimmermann NE, Körner C (2013) Do the elevational limits of deciduous tree species match their thermal latitudinal limits? Glob Ecol Biogeogr 22:913–923
Reich PB, Walters MB, Ellsworth DS, Vose JM, Volin JC, Gresham C, Bowman WD (1998) Relationships of leaf dark respiration to leaf nitrogen, specific leaf area and leaf life-span: a test across biomes and functional groups. Oecologia 114:471–482
Rog I, Rosenstock NP, Körner C, Klein T (2020) Share the wealth trees with greater ectomycorrhizal species overlap share more carbon. Mol Ecol 29:2321–2333. https://doi.org/10.1111/mec.15351
Rossi S, Deslauriers A, Anfodillo T, Carraro V (2007) Evidence of threshold temperatures for xylogenesis in conifers at high altitudes. Oecologia 152:1–12
Sarris D, Siegwolf R, Körner C (2013) Inter- and intra-annual stable carbon and oxygen isotope signals in response to drought in Mediterranean pines. Agric For Meteorol 168:59–68
Schäppi B, Körner C (1996) Growth responses of an alpine grassland to elevated CO_2. Oecologia 105:43–52
Schenker G, Lenz A, Körner C, Hoch G (2014) Physiological minimum temperatures for root growth in seven common European broad-leaved tree species. Tree Physiol 34:302–313
Scherrer D, Körner C (2009) Infra-red thermometry of alpine landscapes challenges climatic warming projections. Glob Chang Biol 16:2602–2613
Scherrer D, Körner C (2011) Topographically controlled thermal-habitat differentiation buffers alpine plant diversity against climate warming. J Biogeogr 38:406–416
Schleppi P, Bucher-Wallin I, Hagedorn F, Körner C (2012) Increased nitrate availability in the soil of a mixed mature temperate forest subjected to elevated CO2 concentration (canopy FACE). Glob Chang Biol 18:757–768
Schulze ED, Lange OL, Buschbom U, Kappen L, Evenari M (1972) Stomatal responses to changes in humidity in plants growing in the desert. Planta 108:259–270
Sigurdsson BD, Medhurst JL, Wallin G, Eggertsson O, Linder S (2013) Growth of mature boreal Norway spruce was not affected by elevated CO_2 and/or air temperature unless nutrient availability was improved. Tree Physiol 33:1192–1205
Spehn EM, Liberman M, Körner C (2006) Land use change and mountain diversity. CRC Press, Boca Raton
Spinnler D, Egli P, Körner C (2002) Four-year growth dynamics of beech-spruce model ecosystems under CO_2 enrichment on two different forest soils. Trees 16:423–436
Steudle E, Peterson CA (1998) How does water get through roots? J Exp Bot 49:775–788
Stocker O (1956) Die Dürreresistenz (Drought resistance). In: Encyclopedia of plant physiology, vol III. Springer, Berlin, pp 696–731
Stocker O (1976) The water-photosynthesis syndrome and the geographical plant distribution in the Saharan Desert. In: Lange OL, Kappen L, Schulze E-D (eds) Water and plant life, Ecological studies, vol 19. Springer, Berlin, pp 506–521

Talhelm AF, Pregitzer KS, Kubiske ME, Zak DR, Campany CE, Burton AJ, Dickson RE, Hendrey GR, Isebrands JG, Lewin KF, Nagy J, Karnosky DF (2014) Elevated carbon dioxide and ozone alter productivity and ecosystem carbon content in northern temperate forests. Glob Change Biol 20:2492–2504

Taschler D, Neuner G (2004) Summer frost resistance and freezing patterns measured in situ in leaves of major alpine plant growth forms in relation to their upper distribution boundary. Plant Cell Environ 27:737–746

Tyree MT (2003) The ascent of water. Nature 423:923

Urli M, Porte AJ, Cochard H, Guengant Y, Burlett R, Delzon S (2013) Xylem embolism threshold for catastrophic hydraulic failure in angiosperm trees. Tree Physiol 33:672–683

Vitasse Y, Hoch G, Randin CF, Lenz A, Kollas C, Scheepens JF, Körner C (2013) Elevational adaptation and plasticity in seedling phenology of temperate deciduous tree species. Oecologia 171:663–678

Vitasse Y, Lenz A, Kollas C, Randin CF, Hoch G, Körner C (2014) Genetic vs. non-genetic responses of leaf morphology and growth to elevation in temperate tree species. Funct Ecol 28:243–252

Volk M, Niklaus PA, Körner C (2000) Soil moisture effects determine CO_2 responses of grassland species. Oecologia 125:380–388

Whitehead D, Jarvis PG, Waring RH (1984) Stomatal conductance, transpiration, and resistance to water uptake in a *Pinus sylvestris* spacing experiment. Can J For Res 14:692–700

Wong SC, Cowan IR, Farquhar GD (1979) Stomatal conductance correlates with photosynthetic capacity. Nature 282:424–426

Würth MKR, Winter K, Körner C (1998a) In situ responses to elevated CO_2 in tropical forest understorey plants. Funct Ecol 12:886–895

Würth MKR, Winter K, Körner C (1998b) Leaf carbohydrate responses to CO_2 enrichment at the top of a tropical forest. Oecologia 116:18–25

Würth MKR, Pelaez-Riedl S, Wright SJ, Körner C (2005) Non-structural carbohydrate pools in a tropical forest. Oecologia 143:11–24

Wurzburger N, Bellenger JP, Kraepiel AML, Hedin LO (2012) Molybdenum and phosphorus interact to constrain asymbiotic nitrogen fixation in tropical forests. PLoS One 7:e33710

Yang Y, Sun H, Körner C (2020) Explaining the exceptional 4270 m high elevation limit of an evergreen oak in the south-eastern Himalayas. Tree Phys, in press. https://doi.org/10.1093/treephys/tpaa070

Zhu Y, Siegwolf RTW, Durka W, Körner C (2010) Phylogenetically balanced evidence for structural and carbon isotope responses in plants along elevational gradients. Oecologia 162:853–863

Zotz G, Cueni N, Körner C (2006) In situ growth stimulation of a temperate zone liana (*Hedera helix*) in elevated CO_2. Funct Ecol 20:763–769

Gene Expression in Coffee

Pierre Marraccini

Contents

P. Marraccini (✉)
CIRAD, UMR IPME, Montpellier, France

IPME, University Montpellier, CIRAD, IRD, Montpellier, France

Agricultural Genetics Institute, LMI RICE2, Hanoi, Vietnam
e-mail: marraccini@cirad.fr

© Springer Nature Switzerland AG 2020
Progress in Botany (2021) 82: 43–112, https://doi.org/10.1007/124_2020_42,
Published online: 28 December 2020

Abstract Coffee is cultivated in more than 70 countries of the intertropical belt where it has important economic, social and environmental impacts. As for many other crops, the development of molecular biology technics allowed to launch research projects for coffee analyzing gene expression. In the 90s decade, the first expression studies were performed by Northern-blot or PCR, and focused on genes coding enzymes of the main compounds (e.g., storage proteins, sugars, complex polysaccharides, caffeine and chlorogenic acids) found in green beans. Few years after, the development of 454 pyrosequencing technics generated expressed sequence tags (ESTs) obviously from beans but also from other organs (e.g., leaves and roots) of the two main cultivated coffee species, *Coffea arabica* and *C. canephora*. Together with the use of real-time quantitative PCR, these ESTs significantly raised the number of coffee gene expression studies leading to the identification of (1) key genes of biochemical pathways, (2) candidate genes involved in biotic and abiotic stresses as well as (3) molecular markers essential to assess the genetic diversity of the *Coffea* genus, for example. The development of more recent Illumina sequencing technology now allows large-scale transcriptome analysis in coffee plants and opens the way to analyze the effects on gene expression of complex biological processes like genotype and environment interactions, heterosis and gene regulation in polypoid context like in *C. arabica*. The aim of the present review is to make an extensive list of coffee genes studied and also to perform an inventory of large-scale sequencing (RNAseq) projects already done or on-going.

1 Introduction: Once Upon a Time – The Story of Gene Expression in Coffee Plants

Despite the economic importance of coffee in international market, the knowledge about coffee molecular biology, and particularly regarding gene cloning and expression, can be considered as relatively recent. The first coffee genes described in the literature correspond to the complementary DNA (cDNA) sequences of α-galactosidase (Zhu and Goldstein 1994) and metallothionein I-like protein (Moisyadi and Stiles 1995), the first being a *Short Communication* in *Gene* and the second, a *Plant Gene Register* in *Plant Physiology*. Both articles only reported the cloning of these cDNAs without analyzing the expression of corresponding genes in coffee tissues.

This was the situation when I just arrived in Nestlé-Tours Research and Development Centre to initiate a project aiming to identify genes involved in coffee cup quality. Based on all the researches describing the importance of storage proteins (particularly in cereals) in the quality of final products, our interest was logically focused first to characterize these proteins in coffee fruits. Then in 1999, we reported the first article describing the expression of *csp1* (*coffee storage protein*) gene coding the 11S proteins accumulated during bean development (Rogers et al. 1999a). At that time, gene expression studies were always performed by Northern blot experiments requiring both high quantities of total RNAs and the preliminary cloning of studied genes in order to synthetize their corresponding radio-labelled DNA probes. This situation persisted until the beginning of the 2000s, and in 2004, there were only 1,570 nucleotide sequences and 115 proteins from coffee deposited in GenBank/EMBL databases.

Few years after, with the development of high-throughput sequencing techniques, the first coffee EST ("expressed sequence tag") sequencing projects were realized, and in 2016 there were 35,153; 25,574; and 25,574 unigenes available in public databases for *Coffea arabica*, *Coffea canephora*, and *Coffea eugenioides*, respectively. Then, the development of real-time quantitative PCR (RT-qPCR) technology significantly accelerated the number of coffee gene expression studies. The access to these ESTs also permitted to set up a 15 K microarray ("PUCECAFE") DNA chip which was used to perform the first large-scale expression analyses aiming to understand transcription networks in flowers, mature beans, and leaves of *C. canephora*, *C. eugenioides*, and *C. arabica* (Privat et al. 2011). The same chip was also used to analyze the leaf expression of homeologous genes in response to changing temperature between *C. arabica* and its two ancestral parents, *C. canephora* and *C. eugenioides* (Bardil et al. 2011).

Soon after came the next-generation Illumina RNA sequencing (RNAseq) method enabling to perform expression analyses of thousands of genes by in silico approaches. The first article using such techniques was published by Combes et al. (2013) who studied the transcriptome in leaves of *C. arabica* submitted to warm and cold conditions suitable to *C. canephora* and *C. eugenioides*, respectively. Since this work, numerous other RNAseq studies were published, and many others are actually ongoing. Using all these data, it is now possible to generate reference transcriptomes which should help us to identify candidate genes (CGs) correlated with agronomic and quality traits in coffee.

2 Coffee Gene Expression

2.1 Reference Genes for qPCR Experiments

Since the development of EST sequencing projects (for reviews, see Lashermes et al. 2008; de Kochko et al. 2010, 2017; and Tran et al. 2016), RT-qPCR experiments, using either SYBR Green fluorochrome or specific TaqMan probes, are nowadays used in routine to study coffee gene expression. In order to quantify the expression levels, these experiments require the use of endogenous reference genes (as internal controls) which must be previously validated for particular tissues (Bustin 2002; Bustin et al. 2009). In that sense, several articles were published to identify the best reference genes to be used in different coffee tissues and growth conditions.

The first were published in 2009 showing that *GAPDH* (coding the glyceraldehyde 3-phosphate dehydrogenase) and *UBQ10* (coding ubiquitin) were stable reference genes for normalization of qPCR experiments in different tissues of *C. arabica*, particularly in leaves and roots under drought stress (Barsalobres-Cavallari et al. 2009; Cruz et al. 2009). These two genes are also the most suitable for data normalization when analyzing multiple or single stresses in leaves of *C. arabica* and *C. canephora* (Goulao et al. 2012). In another study, Fernandes-Brum et al. (2017a) showed that the most stable reference genes were AP47 (coding the clathrin adaptor protein medium subunit), *UBQ*, (ubiquitin 60S), *RPL39* (ribosomal protein L39), and *EF1α* (elongation factor 1-alpha) in all tissues of *C. arabica*, while *GAPDH* and *UBQ*, together with *ADH2* (class III alcohol dehydrogenase) and *ACT* (β-actin), were the most stable for all tissues of *C. canephora*.

When analyzing the caffeine biosynthetic pathway, Sreedharan et al. (2018) showed that *GAPDH* and *UBQ* were the reference genes presenting the lowest variability in leaves and developing endosperm of *C. canephora* between control samples and treatments with salicylic acid (SA), methyl jasmonate (MeJA), light exposure, and PEG, which permitted the quantification of xanthosine methyltransferase (NMT) coding genes. In fact, *UBQ* was commonly used as a reference gene to normalize expression studies during bean development (Salmona et al. 2008; Joët et al. 2009, 2010, 2014; Cotta et al. 2014; Dussert et al. 2018) as well as in other coffee tissues, such as leaves and flower buds (Marraccini et al. 2011, 2012; Vieira et al. 2013; Mofatto et al. 2016). Even though several studies reported that *RPL39* was not the most accurate reference (Cruz et al. 2009; de Carvalho et al. 2013), this gene was also used as a reference to compare expression profiles of several genes in developing beans and also in different organs such as leaves, stems, branches, roots, and flowers (Lepelley et al. 2007, 2012a, b; Pré et al. 2008; Privat et al. 2008; Simkin et al. 2006, 2008; Bottcher et al. 2011).

On the other hand, *GAPDH* and *UBQ* appeared to be the less stable reference genes for transcript normalization in *C. arabica* hypocotyls inoculated with *Colletotrichum kahawae* (causing the coffee berry disease (CBD)), for which the use of *IDE* (coding insulin degrading enzyme) and *β-Tub9* (coding *β*-tubulin) (Figueiredo et al. 2013) as references is recommended. In another study, de Carvalho et al. (2013) showed that *GAPDH* together with *MDH* (coding malate dehydrogenase) and *EF1α* can be used as reference genes in leaves and roots of *C. arabica* subjected to N-starvation and heat stress, while *UBQ10* was the most suitable reference for salt stress treatments. Using RefFinder, a web-based tool integrating geNorm, NormFinder, and BestKeeper programs (Xie et al. 2012), Martins et al. (2017) showed that *MDH* (malate dehydrogenase) presented the highest mRNA stability to study leaf gene expression in both *C. arabica* and *C. canephora* species subjected to single or multiple abiotic stresses such as elevated temperature and CO_2 concentration ($[CO_2]$). In another work, Freitas et al. (2017) showed that the *24S* (ribosomal protein 24S) and *PP2A* (protein phosphatase 2A) genes were the most suitable references to study expression in embryogenic and non-embryogenic calli, embryogenic cell suspensions, and somatic embryos at different developmental stages in *C. arabica*.

2.2 *Gene Expression in Coffee Species*

At the time of writing this review (I apologize if I forgot mentioning some studies), the number of genes for which expression studies have been carried out individually was around 700. Most of these studies were performed by RT-qPCR using specific primer pairs designed against coffee ESTs generated by sequencing projects. In a chronological order, the first project was the Nestlé and Cornell initiative which generated around 63,000 ESTs from six cDNA libraries from fruits and leaves (at different developmental stages) of *C. canephora* clones of the Indonesian Coffee and Cocoa Research Institute (ICCRI) (Lin et al. 2005). Next was the IRD project which led more than 10,400 ESTs also from fruits and leaves of *C. canephora* (Poncet et al. 2006). Finally, the "Brazilian Coffee Genome" Project (BCGP), coordinated by the UNICAMP [University of Campinas] and the Embrapa [Empresa Brasileira de Pesquisa Agropecuária]), produced more than 200,000 ESTs (Vieira et al. 2006; Mondego et al. 2011) from *C. arabica* ($\approx$187,000), from *C. canephora* ($\approx$15,500), and also from *C. racemosa* ($\approx$10,500). In order to identify the maximum of genes, this project used 43 cDNA libraries; most of them were built from transcripts extracted from fruits and leaves at different developmental stages but also from different plant organs (flowers, roots) and tissues (calli, cell suspensions, etc.) subjected to various biotic (e.g., roots infected with nematodes, stems infected with *Xylella* spp., leaves infected with miner *Leucoptera coffeella* and rust fungus *Hemileia vastatrix*) and abiotic (e.g., suspension cells treated with NaCl and chemicals such as acibenzolar-S-methyl and brassinosteroids) stresses.

As reported in Tables 1, 2, and 3, most of these expression studies were performed in *C. arabica* ($n \approx 550$ genes) and *C. canephora* ($n \approx 100$ genes), with a repartition reflecting quite well the importance of *C. arabica* (59%) and *C. canephora* (41%) species in the worldwide coffee production (ICO 2020). These expression studies were more limited in other coffee species such as in *C. racemosa* ($n = 25$), *C. eugenioides* ($n = 18$), and *C. liberica* ($n = 5$). For 70 genes, expression analyses were performed on both *C. arabica* and *C. canephora* species. However, a limited number of studies (described in Sect. 2.3) analyzed gene expression simultaneously in *C. arabica*, *C. canephora*, and *C. eugenioides* using specific primers and qPCR for each homeolog in each species. Several articles also reported in silico gene expression profiles which were not confirmed by RT-qPCR (Table 4).

2.3 *Coffee Gene Expression in* C. arabica*: A Tricky Case*

Before discussing gene expression in coffee, it is important to remember that *C. arabica* ($2n = 4\times = 44$) is an allotetraploid coffee species derived from a natural hybridization event between the two diploid ($2n = 2\times = 22$) species *C. canephora* and *C. eugenioides* (Lashermes et al. 1999) which occurred approximately

Table 1 List of coffee genes studied at the transcriptional level

Function	Gene name	Gene numbers	Expression	Tissues	Species	References
Bean development						
11S storage protein	CaCSP1	Y16976	N	BD	Ca	Marraccini et al. (1999)
	CaCSP1	Cc03_g05570	Q	BD/VT	Ca/Cc	Simkin et al. (2006); Dussert et al. (2018)
	CaCSP1	Cc03_g05570	sQ	BD/VT	Ca	De Castro and Marraccini (2006)
Aspartic proteinase	CaAP2	JU319520	Q	BD	Ca	Abreu et al. (2012)
Cysteine proteinase	CcCP1	JF950589	Q	BD/BG/VT	Ca/Cc	Abreu et al. (2012); Lepelley et al. (2012a)
	CcCP4	JF950590/JF950591	Q	BD/BG/VT	Ca/Cc	Abreu et al. (2012); Lepelley et al. (2012a)
	CaCP23	JU319517	Q	BD/BG/VT	Ca/Cc	Abreu et al. (2012)
Cysteine proteinase inhibitor	CcCPI1	JF950585	Q	BD/BG/VT	Cc	Lepelley et al. (2012a)
	CcCPI2	JF950586	Q	BD/BG/VT	Cc	Lepelley et al. (2012a)
	CcCPI3	JF950587	Q	BD/BG/VT	Cc	Lepelley et al. (2012a)
	CcCPI4	JF950588	Q	BD/BG/VT	Cc	Lepelley et al. (2012a)
Expansin	CaEXPA1	GQ434001	N	BD/VT	Ca	Budzinski et al. (2010)
	CaEXPA2	GQ434002	N	BD/VT	Ca	Budzinski et al. (2010)
	CaEXPA3	GQ434003	N	BD/VT	Ca	Budzinski et al. (2010)
Pectin methylesterase	CaPME4	JN863081	N	BD/VT	Ca	Cação et al. (2012)
Late embryogenic abundant proteins	CcLEA1	DQ333961	sQ	BD/VT	Ca/Cc	Hinniger et al. (2006)
Eukaryotic initiation factor 1	CaSUI1	AJ519839	N	VT	Ca	Gaborit et al. (2003)
	CaSUI1	AJ519839	sQ	VT	Ca	De Castro and Marraccini (2006)
Isocitrate lyase	ICL	XM_027208879	sQ	Bd	Ca	Selmar et al. (2006)
	ICL	XM_027208879	sQ	Bd	Ca	Bytof et al. (2007)
	ICL	XM_027208879	sQ	Bd	Ca	Kramer et al. (2010)

Catalase	CAT3	Cc10_g00580	Q	GS/Bd	Ca	Santos et al. (2013)
Peroxidase	PER3	Cc05_g04990	Q	GS/Bd	Ca	Santos et al. (2013)
Isoflavone reductase-like protein	CaIRL	FJ972200	N	R/L/ BD	Ca	Brandalise et al. (2009)
Prolyl oligopeptidase	CaPOP	JN572042/ JN572043/ JN572044	N/Q	BD/L	Ca	Singh et al. (2011)
Sugars/polysaccharides						
Endo-β-mannanase	ManA	AJ293305	N	GS	Ca	Marraccini et al. (2001)
	ManB	AJ278996	N	GS	Ca	Marraccini et al. (2001)
Mannan synthase	CcManS1	EU568115/ Cc06_g04240	Q	BD	Ca/Cc	Pré et al. (2008); Joët et al. (2014); Dussert et al. (2018)
	CcManS2	EU716311	Q	BD	Cc	Pré et al. (2008); Joët et al. (2014)
Galactomannan galactosyltransferase	CaGMGT1	EU568117	Q	BD	Ca/Cc	Pré et al. (2008); Joët et al. (2014)
	CcGMGT2	EU716313	Q	BD	Cc	Pré et al. (2008); Joët et al. (2014)
Glycosyltransferase	CcXT1	EU760961	Q	BD	Cc	Pré et al. (2008); Joët et al. (2014)
	CcGT1	EU716312	Q	BD	Cc	Pré et al. (2008); Joët et al. (2014)
α-Galactosidase	CaGAL1	AJ877911	N/Q	BD	Ca	Marraccini et al. (2005); Joët et al. (2014)
	CcGAL1	AJ877912	N/Q	BD	Ca	Marraccini et al. (2005); Joët et al. (2014)
β-Galactosidase	CaβGal	HQ283330	sQ	BD	Ca	Figueiredo et al. (2011)
Sucrose synthase 1	CaSUS1	AM087674	N/Q	BD/VT	Ca	Geromel et al. (2006, 2008b); Joët et al. (2014)
	CrSUS1	AM087674	N/Q	BD	Cr	Geromel et al. (2008a)
	SS2	DQ834312	Q	BD/L	Ca/Cc	Privat et al. (2008)
Sucrose synthase 2	CaSUS2	AM087675	N	BD/VT	Ca	Geromel et al. (2006, 2008b)
	CrSUS2	AM087675	N	BD	Cr	Geromel et al. (2008a)
	SS1	DQ826510	Q	BD/L	Ca/Cc	Privat et al. (2008)

(continued)

Table 1 (continued)

Function	Gene name	Gene numbers	Expression	Tissues	Species	References
Invertase	CcINV1	DQ834314	Q	BD/L	Ca/Cc	Privat et al. (2008)
	CcINV2	DQ834315	Q	BD/L	Ca/Cc	Privat et al. (2008)
	CcINV3	DQ834316	Q	BD/L	Ca/Cc	Privat et al. (2008)
	CcINV4	DQ842235	Q	BD/L	Ca/Cc	Privat et al. (2008)
Invertase inhibitor	CcInvI1	DQ834317	Q	L	Cc	Privat et al. (2008)
	CcInvI2	DQ834318	Q	L	Cc	Privat et al. (2008)
	CcInvI3	DQ834319	Q	L	Cc	Privat et al. (2008)
	CcInvI4	DQ834320	Q	L	Cc	Privat et al. (2008)
Sucrose phosphate synthase	CcSPS1	DQ834321	Q	L	Cc	Privat et al. (2008)
	CcSPS2	DQ842234	Q	L	Cc	Privat et al. (2008)
Sucrose phosphatase	CcSP1	DQ834313	Q	L	Cc	Privat et al. (2008)
1-Aminocyclopropane-1-carboxylate oxidase (ACO)	CaACO	KC686714	N	BD/VT	Ca	Pereira et al. (2005)
Polyols						
Mannose-6-phosphate reductase	CaM6PR	GT648734	Q	L	Ca	Freire et al. (2013)
	CcM6PR	GT648734	N/Q	L	Cc	Marraccini et al. (2012)
	CaM6PR	GW488867	N/Q	L	Cc	de Carvalho et al. (2014)
	CcM6PR	GT649509	N/Q	L	Cc	de Carvalho et al. (2014)
Phosphomannose isomerase	CaPMI	GT709210	N/Q	L	Cc	de Carvalho et al. (2014)
	CcPMI	DV688525	N/Q	L	Cc	de Carvalho et al. (2014)
Mannitol dehydrogenase	CaMTD	GW445924	N/Q	L	Cc	de Carvalho et al. (2014)
	CcMTD	GT652950	N/Q	L	Cc	de Carvalho et al. (2014)
Galactinol synthase	CaGolS1	GQ497218	N	L/VT	Ca	dos Santos et al. (2011)
	CaGolS2	GQ497220/ Cc03_g00450	N/Q	BD/L/ VT	Ca	dos Santos et al. (2011); Ivamoto et al. (2017a)
	CaGolS3	GQ497219/ Cc02_g35350	N/Q	BD/L/ VT	Ca	dos Santos et al. (2011); Ivamoto et al. (2017a)
	CaGolS4	Cc11_g15250	Q	BD/L	Ca	Ivamoto et al. (2017a)
	CcGolS1	GQ497218	N/Q	L	Cc	dos Santos et al. (2015)

Raffinose synthase	CaRS1	Cc05_g15530	Q	BD/L	Ca	Ivamoto et al. (2017a)
Isoprenoid (MVA) biosynthesis						
3-Hydroxy-3-methyglutaryl-CoA reductase	CaHMGR1	HQ540670	N	BD/L/F	Ca	Tiski et al. (2011)
	CaHMGR1	HQ540671	N	BD/L/F	Ca	Tiski et al. (2011)
Cytochrome P450s						
Triterpene biosynthesis	CaCYP72A15	Cc05_g08890	Q	L	Ca	Ivamoto et al. (2017b)
Jasmonic acid catabolism	CaCYP94B1	Cc01_g18610	Q	L	Ca	Ivamoto et al. (2017b)
Monoterpenoid biosynthesis	CaCYP76C4	Cc02_g36410	Q	L	Ca	Ivamoto et al. (2017b)
Lipoxygenase pathway	CaCYP74A1	Cc10_g03570	Q	L	Ca	Ivamoto et al. (2017b)
Homoterpene biosynthesis	CaCYP82C2	Cc04_g10600	Q	L	Ca	Ivamoto et al. (2017b)
Ent-kaurene oxidase	CaCYP701A3	Cc10_g03710	Q	L	Ca	Ivamoto et al. (2017b)
Monoterpenes hydroxylation	CaCYP71A25	Cc04_g11300	Q	L	Ca	Ivamoto et al. (2017b)
PAL pathway						
Phenylalanine ammonia lyase	CcPAL1	AAN32866	Q	BD/VT	Cc	Lepelley et al. (2012b)
	CcPAL2	AEO94540	Q	BD/VT	Cc	Lepelley et al. (2012b)
	CcPAL3	EO94541	Q	BD/VT	Cc	Lepelley et al. (2012b)
Chlorogenic acid synthesis						
Hydroxycinnamoyl-CoA quinate hydroxycinnamoyl transferase	CcHQT	EF153931	Q	BD/VT	Cc	Lepelley et al. (2007, 2012b)
Hydroxycinnamoyl-CoA shikimate/ quinate hydroxycinnamoyl transferase	CcHCT	EF137954	Q	BD/VT	Cc	Lepelley et al. (2007)
p-Coumarate 3-hydroxylase	CcC3H1	EF153932	Q	BD/VT	Cc	Lepelley et al. (2007)
Caffeoyl-CoA 3-O methyltransferase	CcCCoAOMT1	EF153933	Q	BD/VT	Cc	Lepelley et al. (2007)
Carotenoid biosynthetic pathway						
Phytoene synthase	CcPSY	DQ157164	Q	BD/VT	Ca/Cc	Simkin et al. (2008, 2010)
Phytoene desaturase	CcPDS	DQ357179	Q	BD/VT	Ca/Cc	Simkin et al. (2008, 2010)
z-Carotene desaturase	CcZDS	DQ357180	Q	BD/VT	Ca/Cc	Simkin et al. (2008, 2010)
Plastid terminal oxidase	CcPTOX	DQ233245	Q	BD/VT	Ca/Cc	Simkin et al. (2008, 2010)

(continued)

Table 1 (continued)

Function	Gene name	Gene numbers	Expression	Tissues	Species	References
Lycopene e-cyclase	CcLCY-E	DQ357178	Q	BD/VT	Ca/Cc	Simkin et al. (2008, 2010)
b-Carotene hydroxylase	CaCRTR-B	DQ157169	Q	BD/VT	Ca/Cc	Simkin et al. (2008, 2010)
Zeaxanthin epoxidase	CcZEP	DQ357177	Q	BD/VT	Ca/Cc	Simkin et al. (2008, 2010)
Violaxanthin de-epoxidase	CaVDE	DQ234768	Q	BD/VT	Ca/Cc	Simkin et al. (2008, 2010)
Carotenoid cleavage dioxygenase 1	CaCCD1	DQ157170	Q	BD/VT	Ca/Cc	Simkin et al. (2008, 2010)
Fibrillin family	CcFIB1	DQ157168	Q	BD/VT	Ca/Cc	Simkin et al. (2008, 2010)
9-Cis-epoxycarotenoid dioxygenase	CaNCED3	DQ157167	Q	BD/VT	Ca/Cc	Simkin et al. (2008, 2010)
	CcNCED3	DQ157167	Q	R/L	Cc	Costa et al. (2015); Thioune et al. (2017)
Trigonelline						
Trigonelline synthase	CTgS1	AB054842	sQ	BD/L/F	Ca	Mizuno et al. (2014)
	CTgS2	AB054843	sQ	BD/L/F	Ca	Mizuno et al. (2014)
Lipids						
Oleosin	CaOLE-1	AY928084/ Cc02_g04750	sQ	BD/VT	Ca/Cc	Simkin et al. (2006); Dussert et al. (2018)
	CcOLE-2	AY841272	sQ	BD/VT	Ca/Cc	Simkin et al. (2006); Dussert et al. (2018)
	CcOLE-3	AY841273	sQ	BD/VT	Ca/Cc	Simkin et al. (2006); Dussert et al. (2018)
	CcOLE-4	AY841274	sQ	BD/VT	Ca/Cc	Simkin et al. (2006); Dussert et al. (2018)
	CcOLE-5	AY841275	sQ	BD/VT	Ca/Cc	Simkin et al. (2006); Dussert et al. (2018)
Steroleosin	CcSTO-1	AY841276	Q	BD/VT	Ca/Cc	Simkin et al. (2006); Dussert et al. (2018)
Non-specific lipid transfer protein (nsLTP)	CaLTP1*	HG008739	Q	BD/PB	Ca	Cotta et al. (2014); Mofatto et al. (2016)
	CaLTP2*	HG008740	Q	BD/PB	Ca	Cotta et al. (2014); Mofatto et al. (2016)
	CaLTP3*	HG008741	Q	BD/PB	Ca	Cotta et al. (2014); Mofatto et al. (2016)
	CcLTP3	HG323822	Q	BD	Cc	Cotta et al. (2014)
Linoleate 13S-lipoxygenase 3-1, chloroplastic	LOX3	Cc00_g30760	Q	L	Ca/Cc	Scotti-Campos et al. (2019)
Linoleate 9S-lipoxygenase 5	LOX5A	Cc02_g33320	Q	L	Ca/Cc	Scotti-Campos et al. (2019)

Linoleate 9S-lipoxygenase 5, chloroplastic	LOX5B	Cc03_g03580	Q	L	Ca/Cc	Scotti-Campos et al. (2019)
ω-3 fatty acid desaturase, chloroplastic	FAD3	Cc02_g06400	Q	L	Ca/Cc	Scotti-Campos et al. (2019)
Cell protection and TAFs						
Dehydrin	CcDH1	DQ323987/ DQ323988	sQ/Q	BD/L	Cc	Hinniger et al. (2006); Thioune et al. (2017)
	CcDH2	DQ323989	sQ/Q	BD	Cc	Hinniger et al. (2006)
	CcDH3	DQ333960	sQ/Q	BD	Cc	Hinniger et al. (2006)
	CaDH3	DQ333961	sQ/Q	Bd	Ca	Kramer et al. (2010)
	CaDH1	JP709195/ JP709196	sQ/Q	R/L/CS	Ca	Santos and Mazzafera (2012)
	CaDH2	DQ323989	sQ/Q	R/L	Ca	Santos and Mazzafera (2012)
	CaDH3	DQ333960/ JP709199	sQ/Q	R/L/CS	Ca	Santos and Mazzafera (2012)
Metallothionein-like protein 1	CaMT21	Cc06_g02650	Q	L	Ca	Barbosa et al. (2017)
Metallothionein I	CaMT1	CF588839	Q	L	Ca/Cc/ Cd	Fortunato et al. (2010)
Glutathione reductase	CaGRed	GT020496	Q	L	Ca/Cc/ Cd	Fortunato et al. (2010)
Dehydroascorbate reductase	CaDHAR	GT731606	Q	L	Ca/Cc/ Cd	Fortunato et al. (2010)
Chitinase class III	Cachi3-1	CF588617	Q	L	Ca/Cc/ Cd	Fortunato et al. (2010)
Chitinase class III	Cachi3-2	CF589150	Q	L	Ca/Cc/ Cd	Fortunato et al. (2010)
Chitinase class IV	Cachi4-1	CF588708	Q	L	Ca/Cc/ Cd	Fortunato et al. (2010)
Class III peroxidase	CaPRX	JN705803	Q	VT	Ca/Cc	Severino et al. (2012)

(continued)

Table 1 (continued)

Function	Gene name	Gene numbers	Expression	Tissues	Species	References
Endoplasmic reticulum heat-shock protein 90	CcHSP90	KU049664	Q	L	Cc	Thioune et al. (2017)
MYB transcription factor	CcMb102	KT698109	Q	L	Cc	Thioune et al. (2017)
Arabidopsis transcription factor ABF2	CcABF2	KU049665	Q	L	Cc	Thioune et al. (2017)
Arabidopsis NON-YELLOW COLOR ING gene	CcNYE1	KU049666	Q	L	Cc	Thioune et al. (2017)
Arabidopsis thaliana ACTIVATING FACTOR 1	CcATAF1	KU049666	Q	L	Cc	Thioune et al. (2017)
Dehydration-responsive-element-binding protein 1D	CcDREB1D	Cc02_g03430	Q	L	Cc	Marraccini et al. (2012); Vieira et al. (2013); Thioune et al. (2017)
	CaDREB1D	Cc02_g03430	Q	L/R	Ca/Cc	Alves et al. (2017, 2018)
Photosynthesis						
Rubisco small subunit	CaRBCS1	AJ419826	N	L	Ca	Marraccini et al. (2003)
	CcRBCS1*	FR728242	Q	L	Ca/Cc/Ce	Marraccini et al. (2011)
Chlorophyll a-/b-binding protein CP24	CaCP24	GT010356	Q	L	Ca/Cc/Cd	Batista-Santos et al. (2011)
Chlorophyll a-/b-binding 22 kDa protein	CaCP22	HQ130481	Q	L	Ca/Cc/Cd	Batista-Santos et al. (2011)
Photosystem II 10 kDa polypeptide precursor	CaPII10a	CF588865	Q	L	Ca/Cc/Cd	Batista-Santos et al. (2011)
Cytochrome f	CaCytf (petA)	GW469919	Q	L	Ca/Cc/Cd	Batista-Santos et al. (2011)
Photosystem I subunit	CaPI (psaB)	EF044213	Q	L	Ca/Cc/Cd	Batista-Santos et al. (2011)

Caffeine						
3,7-Dimethylxanthine N-methyltransferase	CaMTL1	AB039725	sQ	R/St/L	Ca	Ogawa et al. (2001); Kumar and Giridhar (2015); Kumar et al. (2017)
	CaMTL2	AB048792	sQ	R/St/L	Ca	Ogawa et al. (2001); Kumar and Giridhar (2015); Kumar et al. (2017)
	CaMTL3	AB048793	sQ	R/St/L	Ca	Ogawa et al. (2001); Kumar and Giridhar (2015); Kumar et al. (2017)
7-Methylxanthine N-methyltransferase	CaMXMT1	AB048794	sQ	R/St/L	Ca	Ogawa et al. (2001); Kumar and Giridhar (2015); Kumar et al. (2017)
3,7-Dimethylxanthine N-methyltransferase	CaMTL1	AB039725	sQ	L/BD	Ca	Uefuji et al. (2003)
	CaXMT1	AB048793	sQ	L/BD	Ca	Uefuji et al. (2003)
7-Methylxanthine N-methyltransferase	CaMXMT1	AB048794	sQ	L/BD	Ca	Uefuji et al. (2003)
	CaMXMT2	AB084126	sQ	L/BD	Ca	Uefuji et al. (2003)
3,7-Dimethylxanthine N-methyltransferase	CaDXMT1	AB084125	sQ	L/BD	Ca	Uefuji et al. (2003)
7-Methylxanthine N-methyltransferase	CCS1 (CtCS6)	AB086414	sQ	BD/VT	Ca	Mizuno et al. (2003a, b); Koshiro et al. (2006)
3,7-Dimethylxanthine N-methyltransferase	CtCS7	AB086415	sQ	BD/VT	Ca	Mizuno et al. (2003a, b); Koshiro et al. (2006)
7-Methylxanthine N-methyltransferase	CTS2	AB054841	N/sQ	BD/VT	Ca	Mizuno et al. (2003a, b); Koshiro et al. (2006)
7-Methylxanthosine	CmXRS1	AB034699	sQ	BD/VT	Ca	Mizuno et al. (2003a, b); Koshiro et al. (2006)
Methionine synthase	MS	AF220054	sQ	BD/VT	Ca	Mizuno et al. (2003a, b); Koshiro et al. (2006)
Xanthosine methyltransferase	CcXMT1	JX978509	Q	BD/L	Ca/Cc	Perrois et al. (2015)
7-Methylxanthine N-methyltransferase	CcMXMT1	JX978507	Q	BD/L	Ca/Cc	Perrois et al. (2015)
3,7-Dimethylxanthine N-methyltransferase	CcDXMT	JX978506	Q	BD/L	Ca/Cc	Perrois et al. (2015)

(continued)

Table 1 (continued)

Function	Gene name	Gene numbers	Expression	Tissues	Species	References
Xanthosine methyltransferase	CaXMT1	JX978514	Q	BD/L	Ca/Cc	Perrois et al. (2015)
	CaXMT2	JX978515	Q	BD/L	Ca/Cc	Perrois et al. (2015)
7-Methylxanthine N-methyltransferase	CaMXMT1	JX978511	Q	BD/L	Ca/Cc	Perrois et al. (2015)
	CaMXMT2	JX978512	Q	BD/L	Ca/Cc	Perrois et al. (2015)
3,7-Dimethylxanthine N-methyltransferase	CaDXMT2	KJ577792	Q	BD/L	Ca/Cc	Perrois et al. (2015)
7-Methylxanthosine synthase	CmXRS1	AB034699	sQ/Q	BD	Ca	Maluf et al. (2009)
Theobromine synthase	CTS2	AB054841	sQ/Q	BD	Ca	Maluf et al. (2009)
Caffeine synthase	CCS1	AB086414	sQ/Q	BD	Ca	Maluf et al. (2009)
Flowering						
AGAMOUS	CaC03	GU332281	Q	F/R/L	Ca	de Oliveira et al. (2010)
APETALA3	CaC12	GU332287	Q	F/R/L	Ca	de Oliveira et al. (2010)
SEPALLATA3	CaC14	GU265820	Q	F/R/L	Ca	de Oliveira et al. (2010)
FLOWERING LOCUS C	CaFLC	HQ845334	Q	F/R/B/L	Ca	Barreto et al. (2012); Vieira et al. (2019)
FRIGIDA-like	CaFRL4	HQ845335	Q	F/R/B/L	Ca	Barreto et al. (2018)
	CcFRL-1*	Cc01_g15840	Q	F/L/BD/SE	Ca	Vieira et al. (2019)
	CcFRL-2*	Cc03_g03790	Q	F/L/BD/SE	Ca	Vieira et al. (2019)
	CcFRL-3*	Cc04_g05540	Q	F/L/BD/SE	Ca	Vieira et al. (2019)
	CcFRL-4*	Cc05_g14640	Q	F/L/BD/SE	Ca	Vieira et al. (2019)
	CcFRL-5*	Cc00_g14390	Q	F/L/BD/SE	Ca	Vieira et al. (2019)

Somatic embryogenesis (SE)						
BABY BOOM-like gene	BBM1	Cc09_g04020	Q	SE(DS)	Ca/Cc	Nic-Can et al. (2013); Silva et al. (2015); Torres et al. (2015); Pinto et al. (2019)
AP2/ERF-like transcription factor	CaERF-like	AY522505	Q	SE(DS)	Ca/Cc	Nic-Can et al. (2013); Silva et al. (2015); Torres et al. (2015); Pinto et al. (2019)
LEAFY COTYLEDON1	CcLEC1	Cc09_g00330	sQ/Q	SE(DS)	Cc	Nic-Can et al. (2013)
WUSCHEL-RELATED HOMEOBOX4	CcWOX4	Cc10_g04700	sQ/Q	SE(DS)	Cc	Nic-Can et al. (2013)
Somatic embryogenesis receptor-like kinase 1	SERK1	Cc10_g06160	Q	SE (DS)/ CS/R/L	Ca/Cc	Silva et al. (2014); Torres et al. (2015); Pérez-Pascual et al. (2018)
Cyclin-dependent kinase type A	CaCDKA	AJ496622	sQ	SE	Ca	Valadez-González et al. (2007)
Auxin response factor (ARF)	CcARF5	Cc10_g01900	Q	SE(DS)	Cc	Quintana-Escobar et al. (2019)
	CcARF6	Cc09_g08740	Q	SE(DS)	Cc	Quintana-Escobar et al. (2019)
	CcARF9	Cc08_g16330	Q	SE(DS)	Cc	Quintana-Escobar et al. (2019)
	CcARF18	Cc06_g03950	Q	SE(DS)	Cc	Quintana-Escobar et al. (2019)
Auxin/indole-3-acetic acid regulator	CcAux/IAA7	Cc03_g04670	Q	SE(DS)	Cc	Quintana-Escobar et al. (2019)
	CcAux/IAA12	Cc01_g17790	Q	SE(DS)	Cc	Quintana-Escobar et al. (2019)
Gretchen Hagen 3 protein	CaGH3.9	Cc01_g20620	Q	SE(DS)	Cc	Pinto et al. (2019)
	CaGH3.13	Cc05_g05640	Q	SE(DS)	Cc	Pinto et al. (2019)
	CaGH3.15	Cc05_g12940	Q	SE(DS)	Cc	Pinto et al. (2019)
	CaGH3.16	Cc07_g06610	Q	SE(DS)	Cc	Pinto et al. (2019)
Water transport						
Aquaporin	CaPIP2;1	LM654169	sQ/Q	R/L	Ca	dos Santos and Mazzafera (2013); Miniussi et al. (2015)
	CaPIP2;2	LM654170	sQ/Q	R/L	Ca	dos Santos and Mazzafera (2013); Miniussi et al. (2015)
	CaPIP1;1	LM654171	Q	R/L	Ca	Miniussi et al. (2015)

(continued)

Table 1 (continued)

Function	Gene name	Gene numbers	Expression	Tissues	Species	References
	CaPIP1;2	LM654172	sQ/Q	R/L	Ca	dos Santos and Mazzafera (2013); Miniussi et al. (2015)
	CaTIP4;1	LM654173	Q	R/L	Ca	Miniussi et al. (2015)
	CaTIP2;1	LM654174	Q	R/L	Ca	Miniussi et al. (2015)
	CaTIP1;1	LM654175	Q	R/L	Ca	Miniussi et al. (2015)
	CaTIP1;2	LM654176	sQ/Q	R/L	Ca	dos Santos and Mazzafera (2013); Miniussi et al. (2015)
	CaTIP1;3	LM654177	Q	R/L	Ca	Miniussi et al. (2015)
Phytate and ferritin biosynthesis						
Myo-inositol 1-phosphate synthase	CaMIPS1	GU108583	Q	BD	Ca	Nobile et al. (2010)
Phosphatidyl inostitol kinase 1	CaIPK1	EZ421795	Q	BD	Ca	Nobile et al. (2010)
Phosphatidyl inostitol kinase 2	CaIPK2	EZ421796	Q	BD	Ca	Nobile et al. (2010)
Ferritin	CaFER1	GQ913984	Q	BD	Ca	Nobile et al. (2010); Bottcher et al. (2011)
	CaFER2	GU001880	Q	BD	Ca	Nobile et al. (2010); Bottcher et al. (2011)
	CaFER3	EZ421798	Q	BD	Ca	Nobile et al. (2010)
Self-incompatibility RNases						
Putative S-RNase (AR1a allele)	Ca1a	FN547919	sQ	P/S/L	Ca/Cc	Asquini et al. (2011)
Putative S-RNase (CA1a allele)	Cc1a	FN547910	sQ	P/S/L	Ca/Cc	Asquini et al. (2011)
Putative S-RNase (CA1b allele)	Cc1b	FN547911	sQ	P/S/L	Ca/Cc	Asquini et al. (2011)
Putative S-RNase (CA1c allele)	Cc1c	FN547912	sQ	P/S/L	Ca/Cc	Asquini et al. (2011)
Putative S-RNase (CA1d allele)	Cc1d	FN547913	sQ	P/S/L	Ca/Cc	Asquini et al. (2011)
Biotic stress (CLR)						
Protein unknown function	CaR111	CF589193	Q	L	Ca	Ganesh et al. (2006)
Non-race-specific disease resistance	CaNDR1	CO773976	Q	L	Ca	Ganesh et al. (2006); Couttolenc-Brenis et al. (2020)
WRKY transcription factor	CaWRKY1	CO773974	Q	L	Ca	Ganesh et al. (2006)
	CaWRKY1a*	DQ335599	Q	L	Ca/Cc/Ce	Petitot et al. (2008)
	CaWRKY1b*	DQ335598	Q	L	Ca/Cc/Ce	Petitot et al. (2008)

R gene nucleotide-binding site leucine-rich repeat	CaNBS-LRR	GT030058	Q	L	Ca	Diola et al. (2013); Couttolenc-Brenis et al. (2020)
Biotic stress (CLM)						
Class III chitinase PR-8	CaPR8	CK484623	MA	L/VT	Ca	Mondego et al. (2005)
Probable microsomal signal peptidase complex SPC25	CaSPC25	CK484624	MA	YL/GB	Ca	Mondego et al. (2005)
Photosystem I	CaPSAH	CK484626	MA	L/F/GB	Ca	Mondego et al. (2005)
Putative calcium exchanger	CaCAX9	CK484625	MA	L/GB	Ca	Mondego et al. (2005)
BEL1-related homeotic protein 29	CaBEL	CK484627	MA	F/BD	Ca	Mondego et al. (2005)
Biotic stress (CBD)						
Receptor-like kinase	RLK	CF589181**	Q	H	Ca	Figueiredo et al. (2013)
Pathogenesis-related protein 10	PR10	CF589103**	Q	H	Ca	Figueiredo et al. (2013)
Abiotic stress						
Gene responding to dehydration stress	CcRD22	nd	Q	L	Cc	Menezes-Silva et al. (2017)
	CcRD29B	nd	Q	L	Cc	Menezes-Silva et al. (2017)
	CcRAB18	nd	Q	L	Cc	Menezes-Silva et al. (2017)
Putative cytosolic ascorbate peroxidase (cAPX)	APXc	JQ013438	Q	L	Ca/Cc	Ramalho et al. (2018b)
Membrane-bound ascorbate peroxidase (mAPX)	APXm	Q013439	Q	L	Ca/Cc	Ramalho et al. (2018b)
Stromatic ascorbate peroxidase (sAPX)	APXt+s	JQ013441	Q	L	Ca/Cc	Ramalho et al. (2018b)
Putative class III peroxidase (POX4)	PX4	JQ013435	Q	L	Ca/Cc	Ramalho et al. (2018b)
Violaxanthin de-epoxidase	VDE2	DQ234768	Q	L	Ca/Cc	Ramalho et al. (2018b)
Ascorbate peroxidase	APX	Q42564***	Q	R	Ca	Bazzo et al. (2013)
Superoxide dismutase	SOD	O81235***	Q	R	Ca	Bazzo et al. (2013)
Catalase	CAT	Q42547***	Q	R	Ca	Bazzo et al. (2013)
Citrate synthase	CS	Q9SJH7***	Q	R	Ca	Bazzo et al. (2013)

(continued)

Table 1 (continued)

Function	Gene name	Gene numbers	Expression	Tissues	Species	References
Malate dehydrogenase	MDH	Q9ZP06***	Q	R	Ca	Bazzo et al. (2013)
Germin-like protein	GLP	Q9LEA7***	Q	R	Ca	Bazzo et al. (2013)
Mg transporter	MGT1	Q9S9N4***	Q	R	Ca	Bazzo et al. (2013)
Phospholipase C1	PLC1	Cc02g06510	Q	CS	Ca	González-Mendoza et al. (2020)
Phospholipase C2	PLC2	Cc06g01320	Q	CS	Ca	González-Mendoza et al. (2020)
Phospholipase C3	PLC3	Cc02g06530	Q	CS	Ca	González-Mendoza et al. (2020)
Phospholipase C4	PLC4	Cc01g14270	Q	CS	Ca	González-Mendoza et al. (2020)
N-transport and metabolism						
Nitrate and ammonium transporters	CaAMTa	GW473095/ Cc03_g06810	sQ/Q	R	Ca	dos Santos et al. (2017, 2019)
	CaAMTb	GW483639/ Cc01_g14140	sQ/Q	R	Ca	dos Santos et al. (2017, 2019)
	CaAMTc	GT683246/ Cc07_g19360	sQ/Q	R	Ca	dos Santos et al. (2017, 2019)
	CaNRTa	GW479551/ Cc02_g36020	sQ/Q	R	Ca	dos Santos et al. (2017, 2019)
	CaNRTb	GW442751/ Cc06_g08580	sQ/Q	R	Ca	dos Santos et al. (2017, 2019)
	CaNRTc	GT693501/ Cc04_g15770	sQ/Q	R	Ca	dos Santos et al. (2017, 2019)
Cytosolic glutamine synthetase	CaGS1	GW485208/ Cc07_g13290	Q	L	Ca	Baba et al. (2020)
Plastid glutamine synthetase	CaGS2	GR998899/ Cc07_g13290	Q	L	Ca	Baba et al. (2020)
Nitrate reductase	CaNR	GT687366/ Cc00_g12040	Q	L	Ca	Baba et al. (2020)
Asparagine synthetase	CaAS	GW450683/ Cc01_g14420	Q	L	Ca	Baba et al. (2020)

Circadian clock						
Zeitlupe protein family	ZTL	Cc06_g13030	Q	L	Ca/Ce/Cc	Bertrand et al. (2015)
ZTL protein (proline-rich protein 5)	PRR5	Cc02_g00820	Q	L	Ca/Ce/Cc	Bertrand et al. (2015)
Timing of CAB expression 1	TOC1	Cc04_g14990	Q	L	Ca/Ce/Cc	Bertrand et al. (2015); Breitler et al. (2020)
Circadian clock regulator – late elongated hypocotyl	CcLHY	Cc02_g39990	Q	L	Ca/Ce/Cc	Bertrand et al. (2015); Toniutti et al. (2019b); Breitler et al. (2020)
Circadian clock regulator – Gigantea	CcGIGANTEA	Cc10_g15270	Q	L	Ca/Ce/Cc	Bertrand et al. (2015); Toniutti et al. (2019b); Breitler et al. (2020)
Circadian clock regulator – Gigantea	CcLUX-ARRYTHMO	Cc06_g20160	Q	L	Ca	Toniutti et al. (2019b); Breitler et al. (2020)
Starch degradation	CcGWD1	Cc11_g15490	Q	L	Ca	Toniutti et al. (2019b); Breitler et al. (2020)
Chlorophyll biosynthesis	CcPOR1A	Cc05_g12370	Q	L	Ca	Toniutti et al. (2019b); Breitler et al. (2020)
Chlorophyll biosynthesis	CcPOR1B	Cc05_g06850	Q	L	Ca	Toniutti et al. (2019b)
Starch degradation	CcISA3	Cc10_g06640	Q	L	Ca	Toniutti et al. (2019b)

The genes mentioned in this list were described in articles describing expression studies of a limited ($n < 10$) number of genes. Coffee species: *Ca*, *Coffea arabica*; *Cc*, *C. canephora*; *Cd*, *C. dewevrei*; *Ce*, *C. eugenioides*; *Cr*, *C. racemosa*. Gene numbers: the numbers correspond to GenBank accession numbers (https://www.ncbi.nlm.nih.gov) and to the gene names of *C. canephora* reference genome (http://coffee-genome.org/) or to the SOL Genomics Network (**: https://solgenomics.net/). Expression techniques used: *N* Northern blot, *Q* real-time quantitative PCR (RT-qPCR), *sQ* semiquantitative RT-PCR. Tissues: *BD* bean under development (different stages), *Bd* bean under drying process, *CS* cell suspension, *F* flowers, *GB* green beans, *GS* germinating seeds, *H* hypocotyl, *L* leaves, *P* pistil, *PB* plagiotropic buds, *R* roots, *S* stamen, *St* stem, *SE* somatic embryogenesis, *SE(DS)* somatic embryogenesis (at different stages), *VT* various tissues, *YF* young flowers. Expression studies performed for more genes or in the frame of RNAseq projects are described in Tables 2 and 3, respectively. (*): expression analysis *CaCe* and *CaCc* homeologs in *C. arabica* cv. Catuaí. (***): in the absence of nucleic accession numbers, UNIPROT codes are given (https://www.uniprot.org/). *CLM* coffee leaf miner, *CLR* coffee leaf rust, *CBB*, coffee berry borer

Table 2 List of coffee genes studied at the transcriptional level

Topic	*N*	Techniques	Tissues	Species	References
Coffee fruit development	111	Q	BD	Ca	Salmona et al. (2008)
	137	Q	BD	Ca	Joët et al. (2009, 2012)
	26	Q	BD	Ca	Joët et al. (2014)
	28	sQ	BD	Ca	Gaspari-Pezzopane et al. (2012)
	10	Q	BD	Ca	Ságio et al. (2014)
Genetic resources	10	Q	B/L	Ce	Yuyama et al. (2016)
Flowering (MADS box)	18	Q	F	Ca	de Oliveira et al. (2014)
Somatic embryogenesis	17	Q	L	Cc	Pérez-Pascual et al. (2018)
	19	Q	SE (DS)	Ca	de Oliveira et al. (2019)
DREB-like genes	31	Q	L/R	Ca/Cc	Torres et al. (2019)
Abiotic stress (drought)	49	N(8)/Q(41)	L	Cc	Marraccini et al. (2012)
	35	Q	L	Cc	Vieira et al. (2013)
	48	Q	L	Ca	Nguyen Dinh et al. (2016)[a]
Abiotic stress (cold)	19	Q	L	Cc	Dong et al. (2019b)
Abiotic stress (heat stress/high CO2)	12	Q	L	Cc	Martins et al. (2016)
Abiotic stress (T°C)	23	Q	BD	Ca	Joët et al. (2014)
Biotic stress (CLM)	23	Q	L	Ca/Cr	Cardoso et al. (2014)
Biotic stress (CBD)	14	Q	H	Ca	Diniz et al. (2017)
Biotic stress (CLR/NEM/JA)	18	Q	L	Ca	Ramiro et al. (2010)
	21	Q	L	Ca	Diola et al. (2013)
Photosynthesis	8	Q	L	Ca	Avila et al. (2020)

The genes mentioned in this list were described in articles reporting expression studies of a number of genes ≥8. The legend is identical to that of Table 1. The reader needs to access to the articles to know what genes were studied

[a]RT-qPCR study performed to analyze tRNA splicing and gene expression of chloroplast genes

10,000–50,000 years ago (Cenci et al. 2012). Consequently, the transcriptome of *C. arabica* is a mixture of transcripts expressed from homeologous genes harbored by its two sub-genomes, respectively, namely, *CaCc* (also referred as C^a) for *C. canephora* sub-genome and *CaCe* (also referred as E^a) for *C. eugenioides* sub-genome.

In the first attempt to analyze gene expression contributions of each sub-genome in *C. arabica*, Vidal et al. (2010) used qPCR coupled with allele-specific combination TaqMAMA-based method (Li et al. 2004) and developed a pipeline to find SNP (single nucleotide polymorphism) haplotypes of *CaCc* and *CaCe* homeologs in the ESTs of the BCGP. Of the 2069 contigs studied, these authors observed a biased expression for 22% of them, with 10% overexpressing *CaCc* homeologs and 12% overexpressing *CaCe* homeologs, therefore showing that the two sub-genomes do

Table 3 List of high-throughput expression studies

	Species	Tissues	n° EST*/ uni.	cDNA lib.	RNA tech.	NCBI	N	Expression	References
Genetic resources									
	Cc	BD/L	±50000*/ 13,175	5	454	DV663352- DV713545	No	No	Lin et al. (2005)
	Cc	BD/L	5814/4606	2	454	EE191792- EE200565	No	No	Poncet et al. (2006)
	Ca	BD/L/ F/VT	32961*/ 10,799	3	454	nd	No	No	Montoya et al. (2007)
	Ca/Cc/ Ce	L	2092	1	454	DQ655733- DQ655790	No	No	Aggarwal et al. (2007)
	Ca/Cc/ Cr	VT	214964*/ 32,155	43	454	GT640310- GT640366 GT669291- GT734396 GW427076- GW492625 GT645618- GT658452	No	No	Vieira et al. (2006); Mondego et al. (2011)
	Ce	L/F	36,935	2	Illumina	SRP052722	10	Q	Yuyama et al. (2016)
	Ca	L/F/BD	65,364	7	Illumina	PRJNA339585	4	Q	Ivamoto et al. (2017a)
	Ca/Ce	L	56,216	5	Illumina	nd	No	No	Combes et al. (2013)
	Ca	R	34,654	9	Illumina	ERP017352	12	Q	dos Santos et al. (2019)
	Ca	BG	14,005	3	Illumina	PRJNA305756	7	Q	da Silva et al. (2019)
	Ce/Cc	L	14,206	17	Illumina	PRJEB7565	No	No	Combes et al. (2015)
	Ca/Cc/ Ce	L	nd	4	Illumina	PRJEB5543	No	No	Lashermes et al. (2016)
	Ca/Cc/ Ce	L	15,522	12	Microarray	GSE24682	No	No	Bardil et al. (2011)

(continued)

Table 3 (continued)

	Species	Tissues	n° EST*/ uni.	cDNA lib.	RNA tech.	NCBI	N	Expression	References
	Ca/Cc/ Ce	L	15,522	12	Microarray	GSE24754- GSE24682	111	Q	Privat et al. (2011)
	Ca	BD	15,522	1	Microarray	GSE107949	6	Q	Dussert et al. (2018)
	Cc	R	25,574	8	Illumina	nd	3	Q	Costa (2014); Costa et al. (2015)
Somatic embryogenesis									
	Cc	SE(DS)	nd	12	Illumina	GSE128888	6	Q	Quintana-Escobar et al. (2019)
Biotic stress									
Coffee berry borer (CBB)	Ca/Cl	B	6048/5952	2	454	nd	5	Q	Idárraga et al. (2012)
Coffee leaf rust (CLR)	Ca	L	527	2	SSH/454	CF588584- CF589197	10	sQ	Fernandez et al. (2004)
	Ca	L	13,951	1	454	nd	No	No	Fernandez et al. (2012)
	Ca	L	43,159	10	Illumina	PRJNA353233- 353185-353182	13	Q	Florez et al. (2017)
	Ca	L	4,895	23	Illumina	PRJNA448416	No	No	Echeverría-Beirute et al. (2019)
Coffee leaf miner (CLM)	Ca	L	±1500*	6	SSH/ macroarray	CK484622- CK484627 CV998038- CV998046 CV998048-CV99050 CX068758- CX068760	5	N	Mondego et al. (2005)
	Ca	L	±22000*	6	Microarray	nd	18	Q	Cardoso et al. (2014)

Abiotic stress									
SAR chemical inducer	Ca	L/R	1587/138	2	454	AM232089-AM232226	8	Q	De Nardi et al. (2006)
Drought	Ca	PB	41,512	4	454	PRJNA282394	38	Q	Mofatto et al. (2016)
Salt	Ca	L	19,581	2	Illumina	nd	No	No	Haile and Kang (2018)
Cold	Cc	L/BD	nd	7	Illumina	PRJNA561881	38 (L)	Q	Dong et al. (2019a)
Others									
Circadian clock	Ca	L	nd	3	Illumina	nd	7	Q	Toniutti et al. (2019b)
Self-incompatibility RNases	15 species	L/F/BD/R/P	61	15 species	454	JN035305-JN035366	3	sQ	Nowak et al. (2011)

The legend is identical to that described for Table 1. *SAR* systemic acquired resistance induced by benzo(1,2,3)thiadiazole-7-carbothioic acid-S-methyl ester (BTH). Coffee species: *Cl Coffea liberica*. Tissues: *PB* plagiotropic buds. The number of DNA sequences generated for each project corresponds to ESTs (*) or unigenes/contigs. The sequencing techniques (RNA tech.) used were 454-pyrosequencing or Illumina for RNAseq or SSH (suppression subtractive hybridization) and microarray. More information about RNAseq projects already achieved or still ongoing are available at https://www.ncbi.nlm.nih.gov/bioproject using the (NCBI) numbers. *N* for each project, the number of genes for which expression was analyzed by N (Northern blot), Q (RT-qPCR), or sQ (semiquantitative RT-PCR), is indicated. *nd* information not given or found. *no* gene expression studies not performed

Table 4 Expression studies performed in silico (without checking gene expression by RT-qPCR)

Topic	Genes	References
Genetic resources	Several genes	Mondego et al. (2011)
	Several genes	Combes et al. (2012)
Abiotic stress	Several genes expressed under drought	Vinecky et al. (2012)
	Several genes expressed under drought	Marraccini et al. (2012)*
	Drought memory genes	de Freitas Guedes et al. (2018)
Biotic stress	NBS-LRR and others	Alvarenga et al. (2010)
	Genes of SA, JA, and ET pathway	Diniz et al. (2017)
Bean development	LEA and other genes (bean and other tissues)	Dussert et al. (2018)**
Flowering development	MADS box	de Oliveira et al. (2010)
Photosynthesis	Photosynthetic genes	Bang and Huyen (2015)
Caffeine transport (purine permease)	Purine permeases	Kakegawa et al. (2019)
Diterpene biosynthesis	Several genes	Sant'Ana et al. (2018)

(*) and (**): studies also cited in Tables 2 and 3, respectively

not contribute equally to the transcriptome of *C. arabica*. By analyzing gene ontology (GO), these authors also proposed that the *CaCe* sub-genome expressed genes of proteins involved in basal biological processes (such as those related to photosynthesis, carbohydrate metabolic processes, aerobic respiration, and phosphorylation). On the other hand, the *CaCc* sub-genome contributed to adjust Arabica expression (e.g., to biotic and abiotic stresses) through the expression of genes of regulatory proteins such as those related to hormone stimuli (mainly auxin), GTP signal transduction, translation, and ribosome biogenesis proteasome activity.

The 15 K "PUCECAFE" microarray (Privat et al. 2011) was also used to perform genome-wide expression study in order to analyze the effects of warm and cold temperatures on leaf gene expression of *C. arabica* and those of its two ancestral parents (*C. canephora* and *C. eugenioides*) (Bardil et al. 2011). Even though this global gene expression analysis did not allow determining the relative contributions of homeologs to the *C. arabica* leaf transcriptome, it revealed the existence of transcription profile divergences between the allopolyploid and its parental species that were greatly affected by growth temperature. Two other "in silico" analyses that studied the effects of warm vs. cold temperature in *C. arabica* were performed. The first one used SNP ratio quantification to monitor the relative expression of 13 homeologous gene pairs in five organs (cotyledons, young leaves, leaves, stems, and roots) in addition of warm/cold temperatures (Combes et al. 2012). No case of gene silencing or organ-specific silencing was detected, but 10 out of 13 sampled genes showed biased expression: 4 genes toward *CaCe*, 4 genes toward *CaCc*, and 2 genes toward *CaCe* or *CaCc* depending on the organ considered. In the second study, the effects of warm/cold temperatures on *C. arabica* leaf transcriptome were analyzed by RNA sequencing (Combes et al. 2013). The relative homeologous gene expression, assessed in 9,959 and 10,628 pairs of homeologs in warm and cold growing conditions, respectively, revealed that 65% of these genes had an equivalent expression level, while the rest (35%) showed biased homeologous expression. Although the warm and cold conditions were suitable for *C. canephora* or *C. eugenioides* parental species, respectively, neither sub-genome appeared preferentially expressed to compose the final transcriptome of *C. arabica*.

Because *CaCc* and *CaCe* sub-genomes of *C. arabica* have low sequence divergence (with an average difference for genes of only 1.3%) (Cenci et al. 2012), we can conclude that all the studies analyzing gene expression in *C. arabica* by "wet lab" approaches (e.g., Northern blot experiments for the most ancient and even RT-qPCR using primer pairs probably designed in highly conserved cDNA regions) quantify the transcripts expressed by both *CaCc* and *CaCe* sub-genomes.

However, few studies succeed in discriminating specifically the expression of *CaCc* and *CaCe* homeologs in *C. arabica*. All of them (described below) used the presence of SNPs or the small insertions and deletions (INDELs), for example, present in the 3′ and 5′ untranslated regions (UTRs), to design *CaCc* and *CaCe* primer pairs which permitted to identify homeologous differential expression (HDE) by qPCR. The first one concerned the expression of the *CaWRKY1a* (*CaCc*) and *CaWRKY1b* (*CaCe*) genes in *C. arabica* (Petitot et al. 2008, 2013) coding transcription factors known to be associated with plant defense responses to biotic and abiotic stresses (reviewed in Ülker and Somssich 2004; Eulgem 2006). In this species, both homeologs were concomitantly expressed in leaves and roots under all treatments (salicylic acid and infection by leaf rust [*H. vastatrix*] and root-knot nematode (RKN) *Meloidogyne exigua*), suggesting that they undergo the same transcriptional control.

A different situation was observed in *C. arabica* for the *RBCS1* gene with the predominant expression of the homeolog *CaCe* (over the *CaCc* homeolog) in the leaves of non-introgressed ("pure") cultivars such as Typica, Bourbon, and Catuaí (Marraccini et al. 2011), suggesting that specific suppression of *RBCS1 CaCc* expression occurred during the evolutionary processes that generated the *C. arabica* species. This situation fits with the concept of genome dominance (or genome expression dominance) for which the total expression of homeologs of a given gene in an allopolyploid is statistically the same as only one of the parents (Grover et al. 2012). However, *RBCS1 CaCe* and *CaCc* homeologs were co-expressed (with the same order of magnitude) in the leaves of *C. arabica* Timor hybrid HT832/2 used to create the IAPAR59; Tupi and Obabã cultivars of *C. arabica*, for example; as well as in Icatú which comes from a cross between *C. canephora* and *C. arabica* Bourbon. For all these "introgressed" Arabica cultivars, *CaCc* expression was always higher than *CaCe*. The existence of a bias in favor of *CaCc* homeologs suggests that one (or several) genetic factor of *C. canephora* species was introgressed in *C. arabica* together with the HdT (hybrid of Timor, a spontaneous hybrid between *C. arabica and C. canephora*) genes conferring resistance to leaf rust and activated (or unrepressed) the *CaCc* sub-genome.

In a work analyzing the effects of abiotic stress on the expression of genes of the mannitol biosynthesis pathway, de Carvalho et al. (2014) reported that the *CaCc* homeologs of *CaM6PR* (coding mannose-6-phosphate reductase), *CaPMI* (coding phosphomannose isomerase), and *CaMTD* (coding the NAD+-dependent mannitol dehydrogenase, oxidizing mannitol to produce mannose) were also highly expressed in leaves of *C. arabica* IAPAR59 subjected to drought, high salinity, and heat-shock stress.

HDE was also observed when analyzing expression of *nsLTP* (encoding non-specific lipid transfer proteins) genes in the separated tissue of developing beans (Cotta et al. 2014). More precisely, transcripts of *CaLTP3* (*CaCc*) homeolog were detected at different stages of pericarp development, while *CaLTP1/2* (*CaCe*) homeologs were weakly expressed in this tissue. However, both *CaLTP* homeologs were highly expressed during the first stages of endosperm development. In another study, we also reported the high expression of *CaCc* and *CaCe* homeologs of *CaLTP* genes in the plagiotropic buds of the drought-tolerant cultivar "IAPAR59" subjected to water limitation but not in those of the drought-susceptible cultivar "Rubi" (Mofatto et al. 2016). This could be related to the thicker cuticle observed on the abaxial leaf surface in IAPAR59 compared to Rubi.

In a more recent study, Vieira et al. (2019) analyzed the expression of five *FRI GIDA-like* (*FRL*) genes in flowers, beans, and somatic embryos of *C. arabica*. As previously reported (Combes et al. 2013), gene silencing was not detected for *CaFRL* genes, both *CaCc* and *CaCe* homeologs being expressed in all tissues analyzed. However, HDE was observed, for example, during early stages of flower development with a bias toward the expression of *CaCc* homeolog of *CaFRL2*, while a bias toward a *CaCc* homeolog *CaFRL4* was noticed in the latter stages of endosperm development. However, for this latter gene, a bias toward the overexpression of *CaCe* homeolog was observed in somatic embryos. This homeostasis of gene expression observed in the allopolyploid *C. arabica* could explain why this species had a greater phenotypic plasticity compared to its *C. canephora* parent (Bardil et al. 2011; Bertrand et al. 2015).

3 Gene Expression in Coffee Tissues

3.1 Beans

Several thousands of bean cDNAs were generated in the frame of the first coffee EST sequencing projects. For example, the Nestlé and Cornell project used three fruit libraries of *C. canephora* realized at early (whole cherries, 18–22 WAP), middle (endosperm and perisperm, 30 WAP [weeks after pollination]), and late (endosperm and perisperm, 42–46 WAP) stages of fruit development, leading to 9,843; 10,077; and 9,096 ESTs, respectively (Lin et al. 2005). On the other hand, the IRD and C ENICAFE sequencing projects also generated, respectively, more than 5,800 ESTs from *C. canephora* and 9,500 ESTs from *C. arabica* but without mentioning the fruit

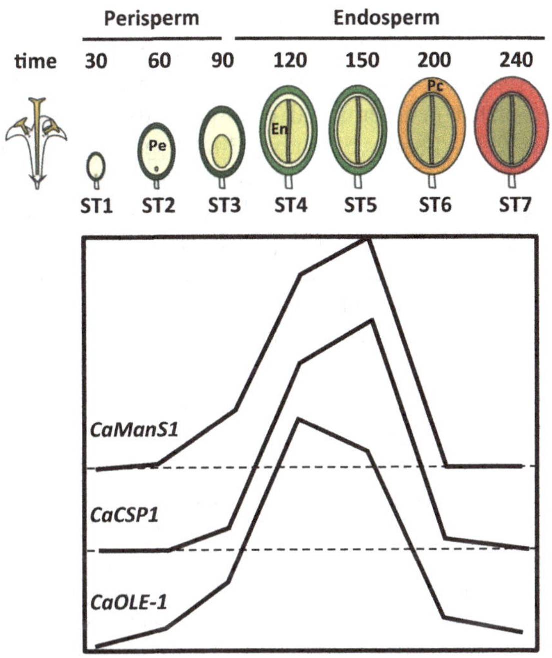

Fig. 1 Schematic representation of the seven developmental stages and tissue changes occurring during fruit development of *C. arabica*. The time is indicated in days after flowering (DAF). Tissues: *Pe* perisperm, *En* endosperm, *Pc* pericarp. RT-qPCR gene expression profiles of *CaCSP1*, *CaOLE-1*, and *CaManS1* (coding for 11S globulin, oleosin, and mannan synthase, respectively) are chosen to illustrate accumulation of storage proteins, triacylglycerols, and cell wall polysaccharides. Adapted from Dussert et al. (2018)

developmental stage (Poncet et al. 2006; Montoya et al. 2007), while the BCGP project produced 14,779 ESTs from 2 fruit libraries (FR1 and FR2) of *C. arabica* and 15,162 from 2 libraries (FR4 and FV2) of *C. racemosa* (Vieira et al. 2006; Mondego et al. 2011).

Regarding the 700 genes reported in Tables 1, 2, and 3, most expression studies were performed in developing coffee beans in which it is not a surprise if we consider that the analysis of its transcriptome is absolutely required to understand the basis of genetic and environmental variations in coffee quality. The time between anthesis and full ripening varies between *C. arabica* (from 6 to 8 months) and *C. canephora* (from 9 to 11 months), and it is usually referred to as days (or weeks) after anthesis (DAA), flowering (DAF), or pollination (DAP) (De Castro and Marraccini 2006). The different stages of developing coffee cherries are mainly defined on its size and also in accordance to the changes of exocarp (pulp) color occurring during the latest maturation steps (Pezzopane et al. 2003; Morais et al. 2008; Gaspari-Pezzopane et al. 2012; Vieira et al. 2019).

Considering the bean and its own tissues, it is now very well known that some important changes occur during its development. Soon after fecundation and up to mid-development (e.g., 90–120 DAF for *C. arabica*), the bean is mainly constituted of perisperm (maternal) which is thereafter progressively replaced by the endosperm which hardens as it ripens during the maturation phase (Fig. 1). For a practical point of view, most of the gene expression studies performed during bean development (referred to as BD in Tables 1, 2, and 3) analyzed the bean as a whole without extracting RNA from separated perisperm and endosperm issues. If it is true to consider that perisperm represents the main tissue in the earliest stages of development (up to 90 DAF), this is no more the case after, when it is reduced to the fine silver skin membrane surrounding the bean. Several works analyzed expression in separated perisperm and endosperm tissues like those studying expansins and HMGRs (human 3-hydroxy-3-methyglutaryl-CoA reductase) (Budzinski et al. 2010) or enzymes of the mevalonic acid (MVA) pathway involved in the biosynthesis of cafestol and kahweol diterpenes (Tiski et al. 2011).

In 2008, Salmona et al. performed a transcriptomic approach combining targeted cDNA arrays, containing 266 selected candidate gene sequences and RT-qPCR on a large subset of 111 genes to decipher the transcriptional networks during the *C. arabica* bean development. This study was the first dividing coffee bean development in seven stages (ST1 0–60 DAF, small fruit with aqueous perisperm; ST2 60–90 DAF, perisperm surrounding a very small liquid endosperm; ST3 90–120 DAF, aqueous endosperm growing and replacing the perisperm; ST4 120–150 DAF, soft milky endosperm; ST5 150–210 DAF, hard white endosperm with green pericarp; ST6 210–240 DAF, ripening cherries with pericarp turning to yellow; ST7 > 240 DAF, mature cherries with red pericarp) (Fig. 1). Few years later, the same research group completed this study by combining gene expression and metabolite profiles (analyzed by high-performance liquid chromatography) in order to identify the key metabolic pathways of coffee bean development (Joët et al. 2009, 2010, 2012).

Regarding sucrose metabolism, Geromel et al. (2006, 2008b) reported high expression of *CaSUS1*, coding the sucrose synthase isoform 1, at the earlier stages of endosperm development (ST4), and high expression of *CaSUS2* (sucrose synthase isoform 2) at the later stages of endosperm development (ST6–7) but also in the perisperm at 205 DAF (Joët et al. 2009). Even restricted at a fine membrane surrounding the endosperm, the high *SUS2* expression detected at that time in the perisperm could contribute to the peak of sucrose detected at the latest development stages in both pericarp and endosperm tissues (Rogers et al. 1999b).

Together with other studies, the genes involved in the most important biochemical pathways were now studied like those involved in sucrose (Geromel et al. 2006, 2008b; Privat et al. 2008; Joët et al. 2014), raffinose (dos Santos et al. 2011, 2015; Ivamoto et al. 2017a) metabolism, polysaccharide synthesis such as galactomannans (Marraccini et al. 2005; Pré et al. 2008; Joët et al. 2014; Dussert et al. 2018), lipid synthesis and transport (Simkin et al. 2006; Cotta et al. 2014; Dussert et al. 2018), caffeine (Ogawa et al. 2001; Uefuji et al. 2003; Mizuno et al. 2003a, b; Koshiro et al. 2006; Perrois et al. 2015; Maluf et al. 2009; Kumar and Giridhar 2015; Kumar et al. 2017), chlorogenic acids (CGAs) (Lepelley et al. 2007, 2012b), carotenoids (Simkin et al. 2010), trigonellines (Mizuno et al. 2014), storage proteins (Marraccini et al. 1999; Simkin et al. 2006; Dussert et al. 2018), and dehydrins and LEAs (Hinniger et al. 2006) (Table 1). Altogether, these studies revealed the existence of several phases during coffee bean development. The first one (perisperm-specific) is characterized by the synthesis of CGA occurring early in the perisperm and accumulation of chitinases, as also confirmed by 2D gel electrophoresis and protein sequencing (De Castro and Marraccini 2006; Alves et al. 2016). More recently, Ivamoto et al.

(2017a) performed the first large-scale transcriptome analysis of *C. arabica* beans during initial (from 30 to 150 DAF) developmental stages, showing the predominant expression of genes of catalytic protein, kinases, cytochrome P450, and binding site domains in the perisperm, for example. The second phase (between ST3 and ST6) is characterized by the activation of cell wall polysaccharide (mainly galactomannans and arabinogalactans) biosynthetic machinery and the synthesis of storage proteins (Marraccini et al. 1999; Pré et al. 2008; Joët et al. 2014; Dussert et al. 2018) (Fig. 1). The third phase concerns the metabolic rerouting of CGA characterized by the *HCT1* expression peak during the latest stages of seed development and the synthesis, storage, and exports of fatty acids requiring oleosins and LTPs (lipid transfer proteins). Finally, the last (endosperm-specific) stage is characterized by the sucrose synthesis and accumulation and dehydration of beans. These steps were recently confirmed by the recent long-read sequencing full-length (LRS) coffee bean transcriptome (Cheng et al. 2018). In that case, the last steps of coffee bean development were characterized by the drastic drop of chitinase transcripts and the great upregulation of genes coding late embryogenesis abundant (LEA) proteins, heat-shock proteins (HSPs), and ROS (reactive oxygen species) scavenging (e.g., superoxide dismutases, catalases, glutathione reductases, glutaredoxins, and glutathione peroxidases) and antioxidant (e.g., dehydroascorbate reductases, glutathione reductases, monodehydroascorbate reductases, and thioredoxins) enzymes, for example (Dussert et al. 2018).

The regulation of gene expression during coffee bean development should implicate specific transcription factors (TFs). In a recent study, Dong et al. (2019a) identified 63 *NAC-like* genes in the reference genome of *C. canephora*, coding TFs well-known to play important functions in plant development and stress regulations (Puranik et al. 2012). After FPKM (Fragments Per Kilobase of transcript per Million mapped reads) treatment of RNAseq data generated at different stages of fruit development, these authors identified 54 *CcNAC* genes with DEG (differentially expressed gene) profiles during the bean development which were verified by qPCR for 10 of them. This led to classify the *CcNAC* genes with continuous upregulated expression as positive regulator of bean development, while those showing downregulated expression were considered as negatively correlated with bean development.

In addition to the gene expression studies performed during coffee bean development, several works also analyzed gene expression in beans during drying (Bytof et al. 2007; Kramer et al. 2010; Santos et al. 2013; Selmar et al. 2006) and germination (da Silva et al. 2019; Lepelley et al. 2012a; Marraccini et al. 2001; Santos et al. 2013) processes.

3.2 *Leaves*

In the frame of the Nestlé/Cornell (Lin et al. 2005) and IRD (Poncet et al. 2006) sequencing projects, 8,942 and 4,606 ESTs were generated from *C. canephora* leaves, respectively, while 12,024 ESTs were also sequenced from *C. arabica* leaves by CENICAFE (Montoya et al. 2007). On the other hand, the BCGP produced 26,931 ESTs from 4 leaf libraries (LV4, LV5, young leaves from orthotropic branches, and LV8, LV9, mature leaves from plagiotropic branches) of *C. arabica*, as well as 5,567 ESTs of *C. arabica* leaves infected with leaf miner and leaf rust (RM1 library), and 13,111 ESTs from 2 leaf libraries (SH1 and SH3) of *C. canephora* plants grown under water deficit (Vieira et al. 2006; Mondego et al. 2011; Vinecky et al. 2012). In this project, leaf ESTs were also generated in the SS1 (960 ESTs), SH2 (7,368 ESTs), and AR1-LP1 (5,664 ESTs) cDNA libraries from tissue pools of *C. arabica* plantlets well-watered, drought-stressed, and treated with arachidonic acid, respectively. Since these studies, numerous projects aiming to study the effects of biotic and abiotic stresses in leaves by RNAseq were performed (see Sects. 4 and 5).

In coffee, leaves are important organs not only as source organs performing photosynthesis and sugar biosynthesis (Campa et al. 2004) but also because they synthesize many other biochemical compounds such as caffeine (Frischknecht et al. 1986; Ashihara et al. 1996; Zheng and Ashihara 2004; Ashihara 2006), chlorogenic acids (CGAs) (Ky et al. 2001; Bertrand et al. 2003; Campa et al. 2017), and trigonelline (Zheng et al. 2004; Zheng and Ashihara 2004) which are further exported to beans and involved in the final cup quality (Leroy et al. 2006).

From the data of Tables 1, 2, and 3, leaf expression studies were reported for more than 400 genes. The first published concerned the three methyltransferases of the caffeine pathway encoded by the XMT (xanthosine N-methyltransferase), MXMT (7-methylxanthine-N-methyltransferase or theobromine synthase), and DXMT (3,7-dimethylxanthine-N-methyltranferase or caffeine synthase) genes (Ogawa et al. 2001; Uefuji et al. 2003; Mizuno et al. 2003a, b). These studies, initially performed by semiquantitative PCR, were further completed by RT-qPCR to better specify the expression of *CaXMT1*, *CaMXMT1*, and *CaDXMT2* genes (belonging to the *C. canephora* sub-genome) and *CaXMT2*, *CaMXMT2*, and *CaDXMT1* (belonging to the *C. eugenioides* sub-genome) in young and mature leaves of *C. arabica* and *C. canephora* (Perrois et al. 2015).

Numerous other studies also detailed the leaf expression profiles of genes of photosynthesis (Marraccini et al. 2003, 2011), sugar metabolism (Privat et al. 2008), and the biosynthetic pathways of carotenoids (Simkin et al. 2008), trigonelline (Mizuno et al. 2014), CGAs (Lepelley et al. 2007, 2012b), and diterpenes (Ivamoto et al. 2017b), for example.

3.3 Roots

More than 12,000 root ESTs were produced in the frame of the BCGP from 4 libraries (RT3, roots; NS1, root infected by nematodes; RT5, roots treated with acibenzolar-S-methyl – a systemic acquired resistance [SAR] inducer; and RT8, roots stressed with aluminum) of *C. arabica* (Vieira et al. 2006; Mondego et al.

2011). In 2006, 1,587 ESTs were produced from embryonic roots of two *C. arabica* cultivars (De Nardi et al. 2006). Among them, 1,506 sequences were used to set up a cDNA microarray which led to the identification of 139 genes differentially expressed in response to induced SAR. In the frame of PhD thesis of T.S. Costa (2014), 25,574 cDNA sequences were generated from roots of drought-susceptible and drought-tolerant clones of *C. canephora* Conilon submitted to water limitation. Even though these data were not deposited in public databases, this study permitted to identify several genes with upregulated expression under drought (see Sect. 5.1). In a more recent RNAseq study, dos Santos et al. (2019) obtained 34,654 assembled contigs from N-starved roots of *C. arabica* and identified three *AMT* (coding specific transporters of ammonium) and three *NRT* (coding nitrate transporters) for which in silico gene expression profiles (dos Santos et al. 2017) were validated by RT-qPCR (dos Santos et al. 2019). Expression profiles in roots were also reported for genes of sugar (Geromel et al. 2006) and caffeine (Ogawa et al. 2001) biosynthetic pathways.

3.4 Flowers

Compared to fruits, leaves, and roots, the studies analyzing gene expression in flowers are very limited. In terms of genetic resources, the BCGP generated 23,036 ESTs from 3 cDNA libraries (FB1, FB2, and FB4) of flowers in different developmental stages and 14,779 ESTs from 2 libraries (FR1 and FR2) corresponding to a mixture of transcripts extracted from flower buds and fruits at different developmental stages (Vieira et al. 2006; Mondego et al. 2011). The CE NICAFE research group also reported the production of 8,707 EST sequences from flowers of *C. arabica* (cv. Caturra), but these data were neither released in public databases. In a recent RNAseq study, Ivamoto et al. (2017a) identified several genes that were exclusively expressed in flowers such as those coding a FASCICLIN-like arabinogalactan protein precursor (FLA3, a protein with InterPro FAS1 Domain IPR000782) and a pectin esterase inhibitor (InterPro Domain IPR006501).

The studies of Asquini et al. (2011) and Nowak et al. (2011), aiming to characterize S-RNase genes and to analyze their expression in pistils (at pre- and post-anthesis stages) and stamens of *C. arabica* and *C. canephora* flowers, were also worth noting.

Other studies characterized the genes of *C. arabica* coding MADS-box TFs (involved in the floral organ identity) and also checked the expression of *FLOWERING LOCUS C* (*FLC*), *AGAMOUS*, *APETALA3*, and *SEPALLATA3* (de Oliveira et al. 2010, 2014). In a more recent study, Vieira et al. (2019) analyzed the expression of five *FRIGIDA-like* (*FRL*) genes, coding key proteins that regulate flowering by activating FLC (Wang et al. 2006). In that case, these authors used the qPCR TaqMAMA-based method (Li et al. 2004) to identify the expression of *CaCc* and *CaCe* homeologs of *FRL* genes in *C. arabica* flowers at different development stages (see also Sects. 2.3 and 3.5). Altogether, these results should help us to understand the genetic determinisms controlling the gametophytic self-incompatibility system of *C. canephora* (Berthaud 1980; Lashermes et al. 1996; Moraes et al. 2018) and coffee male sterility (Mazzafera et al. 1990; Toniutti et al. 2019a).

3.5 Somatic Embryogenesis

In coffee, the somatic embryogenesis (SE) is important particularly to propagate elite clones of *C. canephora* and F1 hybrids of *C. arabica* that could not be spread by seeds (Etienne et al. 2018; Bertrand et al. 2019; Georget et al. 2019). This is the reason why several laboratories are working to identify the genes controlling the main phases and key developmental switches of coffee SE. This also explains the important number (12) of cDNA libraries from suspension cells, calli (primary, embryogenic, and non-embryogenic), and embryos performed in the frame of the BCGP, which generated more than 65,000 ESTs (Vieira et al. 2006; Mondego et al. 2011).

Among these genes, it was reported that the expression of *CcLEC1* (LEAFY C OTYLEDON 1, a key regulator for embryogenesis) and *CcBBM1* (BABY BOOM 1, a AP2/ERF TF associated with cell proliferation) was only observed after SE induction in *C. canephora*, whereas *CcWOX4* (WUSCHEL-RELATED HOMEO BOX4, a plant regulator of embryogenic patterning and stem cell maintenance) expression decreased during embryo maturation (Nic-Can et al. 2013). The expression of *BBM* and *SERK1* (somatic embryogenesis receptor-like kinase 1, a positive regulator of SE activating the YUCCA [flavin-containing monooxygenase]-dependent auxin biosynthesis) genes could also constitute a good parameter for evaluating the development and quality of *C. arabica* (Silva et al. 2014, 2015; Torres et al. 2015) and *C. canephora* (Pérez-Pascual et al. 2018) embryogenic cell suspensions. The fact that expression of *FLC* and *FRL* (especially that of *CaFRL-3*, *CaFRL-4*, and *CaFRL-5*) genes, initially reported as regulators of flowering development, was also observed in both zygotic and somatic embryos of *C. arabica* (Vieira et al. 2019) clearly indicates that both embryogenesis processes share common developmental pathways.

In order to better understand the transcriptomic changes occurring during SE process, Quintana-Escobar et al. (2019) recently performed the first RNAseq study analyzing different stages of SE induction in *C. canephora*. Among the genes differentially expressed, these authors identified eight *ARF* (auxin response factors) as well as seven Aux/IAA (auxin/indole-3-acetic acid regulators) and confirmed that *CcARF18* and *CcARF5* genes were highly expressed after 21 days of the SE induction. In another recent study, Pinto et al. (2019) characterized 17 *GH3* genes from *C. canephora* (encoding the Gretchen Hagen 3 already reported to be key proteins controlling somatic embryogenesis induction through auxin) and analyzed their expression profiles in cells with contrasting embryogenic potential in *C. arabica*, showing that *CaGH3.15* was correlated with *CaBBM*, a *C. arabica* ortholog of a major somatic embryogenesis regulator (Silva et al. 2015). Altogether, these genes could be useful as markers to follow the SE stage converting somatic to embryogenic cells.

4 Coffee Gene Expression in Response to Biotic Stress

Recent modeling studies have delivered warnings on the threat of climate change (CC) by increasing attacks by pests and pathogens (Avelino et al. 2004, 2015; Ghini et al. 2008, 2011, 2015; Jaramillo et al. 2011; Kutywayo et al. 2013; Magrach and Ghazoul 2015). For both *C. canephora* and *C. arabica*, the main pests and diseases are (1) the leaf rust caused by the fungus *H. vastatrix*, (2) the leaf miner *Leucoptera coffeella* (Guérin-Mèneville), (3) the root attacks caused by nematodes, (4) the fruit damages caused by the borer *Hypothenemus hampei*, and (5) the coffee berry disease (CBD) caused by the hemibiotrophic fungus *Colletotrichum kahawae* which is a major constraint of *C. arabica* coffee production in Africa (van der Vossen and Walyaro 2009).

Regarding the coffee genetic diversity, most of *C. canephora* are resistant to coffee leaf rust (CLR), while "pure" (non-introgressed) *C. arabica* are susceptible. However, Catimor and Sarchimor cultivars of *C. arabica* introgressed with the HdT are considered as totally or partially resistant to CLR (Eskes and Leroy 2004). Natural resistances to coffee berry borer (CBB) and coffee leaf miner (CLM) are rather limited in both *C. canephora* and *C. arabica* species. However, natural resistance to the CLM can be found in several wild coffee diploid species, such as in *C. racemosa* (Guerreiro-Filho et al. 1999; Guerreiro-Filho 2006), and has been introgressed into *C. arabica* to generate new cultivars (e.g., Siriema) resistant to CLR (Matiello et al. 2015). Regarding nematodes, a large genetic diversity exists particularly in diploid species (e.g., *C. canephora*, *C. liberica*, and *C. congensis*) but less in *C. arabica*, regarding the variation in resistance particularly to the root-knot *Meloidogyne* spp. from high susceptibility to near immunity as it is the case of the clone 14 of *C. canephora* Conilon (Lima et al. 2014, 2015). Information about genetic resistance to coffee berry borer (CBB) is very limited for both *C. arabica* and *C. canephora* species. However, Romero and Cortina (2004, 2007) reported a reduction of CBB growth rate when *H. hampei* is fed with *C. liberica* fruits. In another study, Sera et al. (2010) showed that *C. kapakata*, *Psilanthus bengalensis*, *C. eugenioides*, as well as genotypes introgressed with *C. eugenioides* were CBB resistant. In that case, the CBB^R of *C. eugenioides* and *C. kapakata* was observed at the pericarp level (but not in the bean), while *P. bengalensis* presented CBB^R in both tissues. In addition to be CLR^R, some *C. arabica* coming from HdT, but also the F1 hybrid cultivar Ruiru 11, were also reported as CBD^R (Omondi et al. 2004, Walyaro 1983; Van der Vossen 1985). This genetic diversity observed in the *Coffea* genus regarding these different abiotic stresses could be used to identify the genes controlling these resistances and to initiate new breeding programs aiming to create new hybrids better resistant to pests and diseases.

On the other hand, the BCGP produced more than 5,000 ESTs of *C. arabica* from RM1 (leaves infected with CLM and CLR) and NS1 (roots infected with nematodes) (Vieira et al. 2006; Mondego et al. 2011. In a recent study, genes coding for the LOX (lipoxygenase), AOS (allene oxide synthase), AOC (allene oxide cyclase), and OPR (12-oxo-phytodienoic acid reductase) enzymes involved in the production of jasmonic acid (JA), one of the key plant hormones involved in plant defense against insect pests, were identified in *C. canephora* by bioinformatic approaches (Bharathi and Sreenath 2017) but without confirming gene expression of this pathway in infested coffee plants.

4.1 Coffee Leaf Rust (CLR)

In 2004, Fernandez et al. used suppression subtractive hybridization (SSH) method and semiquantitative RT-PCR to identify *C. arabica* L. genes involved in the specific hypersensitive reaction (HR) upon infection by *H. vastatrix*. Among the genes showing HR upregulation were those coding for receptor kinases, AP2 domain and WRKY TFs, cytochromes P450, heat-shock 70 proteins, several glucosyltransferases, and NDR1, for example. Other studies showed that SA and MeJA treatments markedly upregulated the expression of *CaNDR1* (coding a non-race-specific disease resistance protein well-known to be involved in resistance signalization pathway in *Arabidopsis thaliana*) and *CaWRKY1* genes, suggesting a key role of their corresponding proteins in the molecular resistance responses of coffee to *H. vastatrix* (Ganesh et al. 2006; Cacas et al. 2011; Petitot et al. 2008, 2013). This was confirmed by Ramiro et al. (2010) who showed that in addition to *CaWRKY1*, expression of *CaWRKY3*, *CaWRKY17*, *CaWRKY19/20/21*, and *CaWRKY22* genes was also highly upregulated upon CLR. Although a significant correlation was also observed between *WRKY* expression profiles after MeJA and rust treatments, expression of coffee genes involved in JA biosynthesis, including allene oxide synthase (*CaAOS*) and lipoxygenase (*Ca9-LOX* and *Ca13-LOX*), did not support the involvement of JA in the early coffee resistance responses to CLR.

The first valuable EST dataset from *C. arabica* CIFC 147/1 (CLR resistant) infected by leaf rust was produced by Fernandez et al. (2012) who identified 205,089 ESTs and 13,951 contigs from coffee together with 57,332 ESTs and 6,763 contigs from *H. vastatrix*. Among the most abundant coffee genes expressed in rust-infected leaves were those coding for several pathogenesis-related (thaumatin-like) proteins and enzymes of carbohydrate, amino acid, and lipid transport/metabolism. Florez et al. (2017) also used the *C. arabica* cultivars Caturra (CLR susceptible) and HdT CIFC 832/1 (CLR resistant) to generate 43,159 contigs which were assembled using as a reference the genome of *C. canephora* (Denoeud et al. 2014). Among DEG profiles identified by RT-qPCR were genes coding for a putative disease resistance protein RGA1, putative disease resistance response (dirigent-like protein) family protein, and Premnaspiridione oxygenase with higher expression at early stage of rust infection in the resistant cultivar plant than in the susceptible genotype. In addition, expression of several TFs (putative basic helix-loop-helix bHLH DNA-binding superfamily protein and ethylene-responsive transcription factor 1B) was detected earlier in HdT than in Caturra, suggesting that they may be involved in the defense mechanisms of the CLR^R cultivar. In a more recent study, Echeverría-Beirute et al. (2019) performed RNAseq approach to study the effects of CLR and fruit thinning in leaves of susceptible cultivars red Catuaí (Caturra x Mundo Novo) and F1 hybrid H3 (Caturra x Ethiopian 531) of *C. arabica*. Using regression and prediction statistical models, these authors identified 460 DEGs between the inbred and the F1 hybrid. Among them, the expression of *PR* (*pathogenesis-related*) genes was upregulated in Catuaí, while those coding proteins involved in homoeostasis increased in the F1 hybrid. Even though these results were not confirmed by RT-qPCR, they validate the hypothesis of lower impact of CLR in F1 hybrids (Echeverria-Beirute et al. 2018) due to their physiological status, which itself depends on their genetic background, plant vigor, agronomic conditions, and environmental factors (Toniutti et al. 2017, 2019b).

4.2 Coffee Leaf Miner (CLM)

Although the defense mechanisms to leaf miner are not well understood, previous genetic analyses suggested that this resistance was dominant and controlled by a limited number of genes (Guerreiro-Filho et al. 1999). The first attempt to identify these genes was performed by SSH method coupled with the screening of DNA macroarrays to study gene expression in the leaves of the CLM-susceptible (CLM^S: red Catuaí) and CLM-resistant cultivar (CLM^R corresponding to a backcross of [*C. racemosa* x *C. arabica* x *C. arabica*]) infested by *L. coffeella* (Mondego et al. 2005). From the 1,500 ESTs spotted on the array, upregulated expression upon CLM infestation was observed for several ESTs coding proteins previously reported to be related to plant defense and biotic stress and similar to the phospholipase D, the lipoxygenase LOX3, the late embryogenesis abundant protein 1 (LEA1), the acid phosphatase vegetative storage protein (VSP), and the lipid transfer protein/trypsin inhibitor/seed storage domain, for example. For *CaPR8* (class III chitinase), *CaSPC25* (signal peptidase complex subunit), *CaPSAH* (photosystem I), *CaCAX9* (a putative calcium exchanger), and *CaBEL* (BEL1-related homeotic protein 29) genes, their upregulated expression upon CLM infestation suggested that they play a key role in coffee defense mechanisms against *L. coffeella*.

In a more recent study, Cardoso et al. (2014) used a 135 K microarray (NimbleGen) based on the 33,000 genes identified in the frame of the BCGP, to identify DEG genes in CLM^S and CLM^R cultivars of *C. arabica* at three stages (T0, non-infected/control; T1, egg hatching, and T2, egg eclosion) of interaction with *L. coffeella*. Even though previous studies reported that caffeine has no effect on leaf miner survival rates (Guerreiro-Filho and Mazzafera 2000; Magalhães et al. 2010), high upregulated expression of a putative *caffeine synthase* gene was reported at both T0 and T2 in CLM^R leaves compared to CLM^S ones. In the same study, expression profiles of genes involved in plant response pathways to herbivory attacks (e.g., linoleic acid cycle, phenylpropanoid synthesis, and apoptosis), as well as JA (e.g., coding lipoxygenase and enoyl-CoA hydratase) and flavonoids (e.g., coding chalcone synthase and flavanone 3-hydroxylase-like) biosynthesis, were also upregulated in CLM^R plants even in the absence (at T0) of leaf miner infestation, indicating that defense was already built up in these plants prior to infection, as a priming mechanism.

4.3 Nematodes (NEM)

Despite the important damages caused by nematodes, there are a limited number of studies analyzing the coffee gene responses to these pathogens. When studying *WRKY* genes coding transcription factors regulating plant responses to biotic stresses, Ramiro et al. (2010) reported that expression of *CaWRKY6*, *CaWRKY11*, *CaWRKY12*, *CaWRKY13/14*, *CaWRKY15*, and *CaWRKY17* genes was upregulated in roots of *C. arabica* cv. IAPAR59 infected by the RKN *Meloidogyne exigua*. In another work, Severino et al. (2012) reported upregulated expression of *CaPRX* (encoding a putative class III peroxidase) in roots inoculated with RKN *M. paranaensis* but with significant difference between susceptible (*C. arabica* cv. Catuaí) and resistant (*C. canephora* cv. Robusta) plants. The nematode-resistant (NEM^R) clone 14 of *C. canephora* Conilon (Lima et al. 2014, 2015) was also used to investigate gene expression in roots at regular days after infestation (4, 8, 12, 20, 32, and 45 DAI) by the root-knot *M. paranaensis* (Lima 2015). The RNAseq data (not yet publicly available) showed higher expression levels of several *PR* (pathogenesis-related) genes, such as those coding class III chitinase and NBS-LRR proteins, in infected roots of NEM^R clone 14 than in those of NEM^S clone 22. In addition, the peak of *NBS-LRR* transcripts was detected at 8 and 20 DAI for the clones 14 and 22, respectively, suggesting earlier expression of this gene in NEM^R than in NEM^S coffee clones (Valeriano et al. 2019). RT-qPCR experiments also showed that expression of *CcCPI1* (coding a cysteine proteinase inhibitor) was higher in roots of clone 14 than in those of 22, with or without nematode infestation, suggesting that the this protein, also highly expressed in coffee beans under development and germination (Lepelley et al. 2012a), could also play a key role in controlling nematode development. In that sense, CPIs have already been reported to inhibit proteinases in the digestive tracts, therefore reducing the destructive effects of herbivorous insects (Benchabane et al. 2010; Schluter et al. 2010), and to increase tolerance to nematodes as well as to fungal and bacterial pathogens in transgenic plants (Urwin et al. 2003; Martinez et al. 2005).

4.4 Coffee Berry Borer (CBB)

Considering that *C. arabica* fruits are more susceptible to CBB than those of *C. liberica*, Idárraga et al. (2012) constructed cDNA libraries from fruits for these two species infested with *H. hampei* and generated 3,634 singletons and 1,454 contigs. In silico analyses revealed that infested *C. arabica* berries displayed a higher number of DEG genes coding proteins involved in general stress responses, while genes coding proteins involved in insect defense were overexpressed in *C. liberica*. For some of these genes, expression profiles in infested cherries were checked by RT-qPCR. Interestingly, expression levels of genes coding a hevein-like protein, an isoprene synthase, a SA carboxyl methyltransferase, and a patatin-like protein appeared much more upregulated in *C. liberica* than in *C. arabica*. The upregulation of these genes was already reported in other plants in response to insect herbivory and JA treatments (Kiba et al. 2003; Reymond et al. 2000; Falco et al. 2001), suggesting that they could be involved in the partial resistance to CBB in *C. liberica*.

4.5 Coffee Berry Disease (CBD)

Cytological and biochemical studies revealed that coffee resistance to *C. kahawae* is characterized by restricted fungal growth associated with several host responses, such as hypersensitive-like cell death (HR), callose deposition, accumulation of phenolic compounds, lignification of host cell walls, and increased activity of oxidative and peroxidase enzymes (Silva et al. 2006; Gichuru 1997, 2007; Loureiro et al. 2012).

The first study analyzing gene expression in response to *C. kahawae* was performed by Figueiredo et al. (2013) in hypocotyls of *C. arabica* cultivars Catimor 88 (HdT derivative CBD^{R}) and Caturra CIFC 19/1 (CBD^{S}). These authors showed that expression levels of *RLK* (coding a receptor-like kinase) and *PR10* (coding a pathogenesis-related protein 10) genes were higher in Catimor than in CBD-infected Caturra. Interestingly, upregulated expression of these two genes was also reported during coffee infection with *H. vastatrix* (Fernandez et al. 2004). In order to understand the molecular mechanisms involved in coffee resistance to *C. kahawae*, Diniz et al. (2017) evaluate the expression of genes involved in SA, JA, and ethylene (ET) pathways in the same cultivars. From the 14 genes studied by RT-qPCR, these authors showed the involvement of JA and ET phytohormones rather than SA in this pathosystem. Regarding the ET pathway, the strong activation of *ERF1* gene (coding for ET receptor) at the beginning of the necrotrophic phase suggests the involvement of ethylene in tissue senescence.

4.6 Gene Expression in Response to Other Pests and Diseases

Of the two commercially cultivated coffee species, *C. arabica* and *C. canephora* are considered as susceptible and resistant, respectively, to the insect pest *Xylotrechus quadripes* known as coffee white stem borer (CWSB). Using SSH approach, Bharathi et al. (2017) identified 265 unigenes overexpressed in *C. canephora* bark tissues upon CWSB larval infestation, many of them coding putative pectin-degrading enzymes like a pectate lyase (*Cc07_g00190*[1]), three polygalacturonases (*Cc03_g15700, Cc03_g15740*, and *Cc03_g15840*), and a pectinacetylesterase (*Cc08_g04630*). By RT-qPCR, these authors also showed that the expression of *Cc07_g00190* was strongly induced at 72 h after CWSB infestation. The possible role of this pectinolytic enzyme in the production of oligogalacturonides was proposed, which could act as elicitors involved in defense responses of *C. canephora* to CWSB (Bharathi and Sreenath 2017).

5 Coffee Gene Expression in Response to Abiotic Stress

Several models predicted that CC will have strong negative impacts on both *C. canephora* and *C. arabica* species at environmental, economic, and social levels (Assad et al. 2004; Bunn et al. 2015a, b; Ovalle-Rivera et al. 2015; Davis et al. 2012, 2019; Moat et al. 2017, 2019). Drought and high air temperatures are undoubtedly the major threats to coffee production, forecasted by potential climate changes (IPCC 2013). Drought is a limiting factor that affects flowering and yield of coffee (DaMatta and Ramalho 2006), as well as bean development and biochemical composition and consequently the final cup quality (Silva et al. 2005; Vinecky et al. 2017). Increased [CO_2] in air is also a key factor for coffee plant acclimation to high temperature; strengthening the photosynthetic pathway, metabolism, and antioxidant protection; and modifying gene transcription and mineral balance (Ramalho et al. 2013; Martins et al. 2014, 2016; Ghini et al. 2015; Rodrigues et al. 2016). In this context, understanding the genetic determinism of coffee's adaptation to abiotic stress has become essential for creating new varieties (Cheserek and Gichimu 2012).

5.1 Drought

The first study analyzing the effects of drought stress was performed by Simkin et al. (2008), who reported the gene expression profiles of the carotenoid biosynthesis pathway in leaf, branch, and flower tissues of *C. arabica* subjected to water

[1]Gene names found in the Coffee Genome Hub (http://coffee-genome.org/)

withdrawal. In this work, it was shown that the transcript levels of *PTOX*, *CRTR-B*, *NCED3*, *CCD1*, and *FIB1* increased under drought, suggesting that drought favored the synthesis of xanthophylls implicated in the adaptation of plastids to changing environmental conditions by preventing photooxidative damage of the photosynthetic apparatus. On the other hand, drought was reported to decrease the *RBCS1* gene expression in both *C. arabica* and *C. canephora* species (Marraccini et al. 2011, 2012). However, this reduction was not accompanied by a decrease of RBCS1 protein in the leaves of *C. canephora* under water withdrawal. In the same work, it was also shown that the transcriptional contribution of each *RBCS1* homeolog may be affected by drought in *C. arabica* cultivars (Marraccini et al. 2011). In *C. canephora*, and whatever the clone studied, drought was also shown to downregulate the leaf expression of many genes related to photosynthesis such as *CcCAB1* (coding chlorophyll a-/b-binding proteins), *CcCA1* (coding for the carbonic anhydrase supplying CO_2 for Rubisco), as well as expression of *CcPSBO*, *CcPSBP*, and *CcPSBQ* genes coding proteins of the PSII oxygen-evolving complex (Marraccini et al. 2012; Vieira et al. 2013).

On the other hand, drought stress significantly upregulated the expression of genes coding proteins involved in maintenance, reinforcement, and protection during the dehydration-rehydration process such as dehydrins and glycin-rich and heat-shock proteins in *C. canephora* (Marraccini et al. 2012; Vieira et al. 2013) and *C. arabica* (Santos and Mazzafera 2012; Mofatto et al. 2016). Drought stress was also shown to increase expression of some *PIP* (plasma membrane intrinsic proteins) genes in the leaves and roots of different coffee species, suggesting the involvement of these aquaporins in controlling the water status in coffee plants (dos Santos and Mazzafera 2013; Miniussi et al. 2015).

In coffee, like in many other plants, drought stress was also reported to affect the metabolic pathways involved in the synthesis of many solutes such as sugars of the raffinose family oligosaccharides (RFOs) (e.g., trehalose, raffinose, and stachyose), already described to be involved in osmoprotection against abiotic stresses in plants (Kerepesi and Galiba 2000). The upregulated expression of *CaGolS2* and *CaGolS3* genes coding galactinol synthases explained the increase of raffinose and stachyose contents also observed in leaves of *C. arabica* cv. IAPAR59 plants submitted to severe water deficit (dos Santos et al. 2011). In *C. canephora* Conilon, water limitation also increased *CcGolS1* gene expression in leaves of the drought-tolerant (D^T) clone 14 but decreased the expression of the same gene in leaves of the drought-susceptible (D^S) clone 109A (dos Santos et al. 2015). Drought was also shown to upregulate the expression of *M6PR* gene coding the mannose-6-phosphate reductase in leaves of both *C. canephora* (Marraccini et al. 2012) and *C. arabica* (Freire et al. 2013). In *C. arabica* cv. IAPAR59, the increased expression of *CaPMI* (mannitol synthesis) and decreased *CaMTD* (controlling mannitol degradation) expression under drought were correlated with high mannitol levels detected in leaves under drought conditions (de Carvalho et al. 2014).

Drought also increased the expression of regulatory genes *CcRD29*, *CcRD26*, and *CcDREB1D* coding a RD29-like protein, a NAC-RD26-like TF, and an AP2/ERF DREB-like TF, respectively, in D^T (14, 73, and 120) and D^S (22) clones of

C. canephora Conilon (Marraccini et al. 2012; Vieira et al. 2013). Even though these studies highlighted the existence of different mechanisms among the D^T clones of *C. canephora* regarding water deficit, they also showed that *CcDREB1D* expression was always higher in leaves of D^T clones (particularly in clone 14) than in those of D^S clone 22 under water withdrawal (Fig. 2). Upregulated expression of the *CcDREB1D* was also reported in leaves of *C. canephora* and *C. arabica* subjected to low relative humidity (Thioune et al. 2017; Alves et al. 2018). A study of *CcDREB1D* promoter regions in the D^T clone 14 and D^S clone 22 revealed the existence of several haplotypes diverging by several SNPs and insertions/deletions (Alves et al. 2017). A functional analysis of these promoters in transgenic plants of *C. arabica* var. Caturra showed that haplotype HP16 (found in the D^T clone 14) was able to drive the expression of the *uidA* reporter gene under water deficit in leaf mesophyll and guard cells more strongly and earlier than the HP15 (present in both clones) and HP17 (only present in D^S clone 22) haplotypes (Alves et al. 2017). In a more recent work aiming to study the expression of *DREB*-like genes regarding various abiotic stresses (Torres et al. 2019), drought (mimicked by water limitation) was shown to upregulate expression of *CcDREB1B*, *CcRAP2.4*, *CcERF027*, *CcDREB1D*, and *CcTINY* mainly in leaves of *C. canephora* D^T clones, while drought (mimicked by low humidity) upregulated the expression of *CaERF053*, *CaRAP2.4*, *CaERF017*, *CaERF027*, *CaDREB1D*, and *CaDREB2A.1* in leaves of *C. arabica*. On the other hand, expression of *CcDREB2F*, *CcERF016*, and *CcRAP2.4* genes was greatly upregulated under drought specifically in the roots of D^S clone 22 (Fig. 2), which could help this clone to compensate its low efficiency in controlling stomatal closure and high reduction of net CO_2 assimilation (*A*) observed upon drought acclimation (Marraccini et al. 2012).

M.G. Cotta (2017) also analyzed the expression profiles of genes coding the PYR/PYL/RCAR-SnRK2-PP2C proteins known to be involved in the first steps of ABA perception and signal transduction in plants (Klingler et al. 2010), in leaves, and in roots of D^T (14, 73, and 120) and D^S (22) clones of *C. canephora* subjected to drought. In leaves, drought downregulated the expression of *CcPYR1*, *CcPYL2*, and *CcPYL4* genes (coding ABA receptors) and upregulated the expression of *CcAHG2* and *CcHAB* (coding PP2C phosphatases functioning as negative regulators of ABA pathway) in D^T clones. However, expression of *SnRK2* genes (coding protein kinases functioning as positive regulators of this pathway) was poorly affected by drought conditions. On the other hand, drought upregulated the expression of *PP2C* (e.g., *CcABI1*, *CcABI2*, *CcAHG3*) and *SnRK2* (e.g., *SnRK2.2*, *SnRK2.6*, and *SnRK2.7*) genes mainly in roots of *C. canephora* D^T clone 120. *CcPYL8b* was the gene most expressed in drought-stressed roots, particularly in D^T clones 73 and 120, while expression of *CaPYL8a* was upregulated by drought mainly in leaves of *C. arabica* D^T accession (Santos et al. 2019).

In *C. canephora*, Menezes-Silva et al. (2017) reported that coffee plants exposed to multiple drought events tended to display a higher expression of the *RD29B* and *RD22* genes which could be involved in acclimation to repeated drought events. Recently, de Freitas Guedes et al. (2018) performed an RNAseq study to analyze the effects of multiple drought stress on gene expression in leaves of the D^T clone

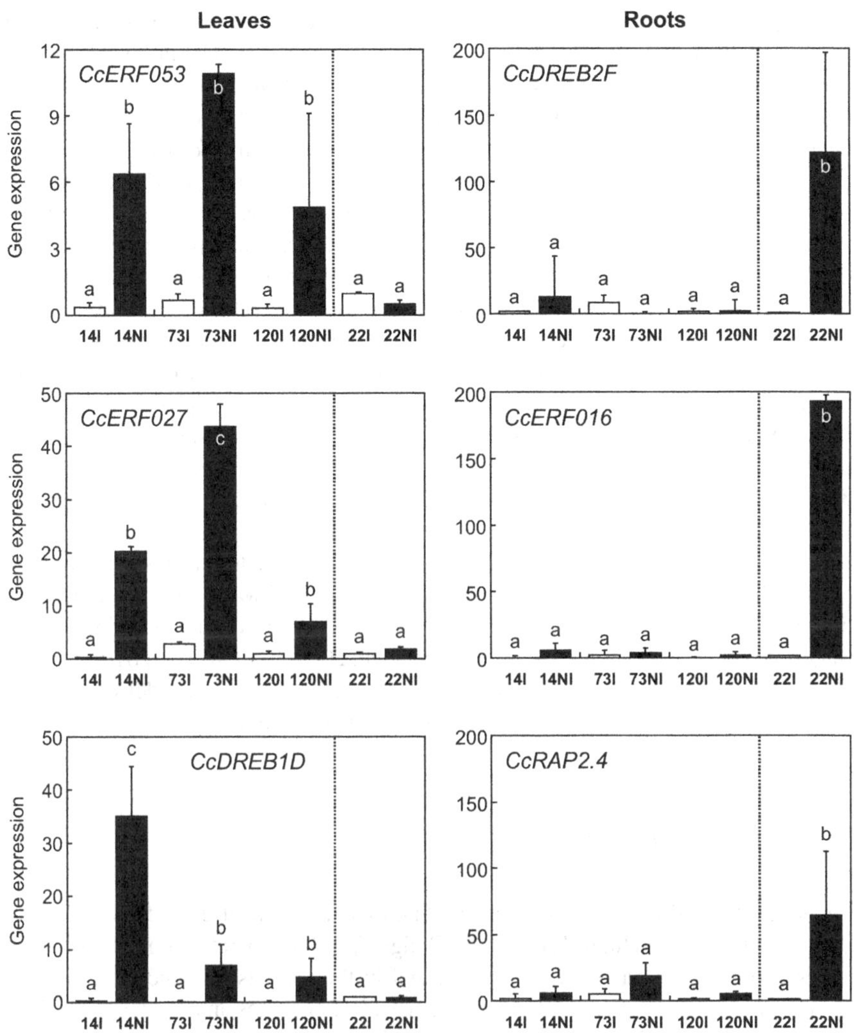

Fig. 2 Gene expression profiles of *DREB*-like genes in leaves and roots of D^T (14, 73, and 120) and D^S (22) clones of *C. canephora* Conilon subjected (*NI* not irrigated, black isobars) or not (*I* irrigated, white isobars) to water limitation. The D^T and D^S clones are separated by a vertically dotted line. Gene names are indicated in the histograms. Expression values corresponding to the mean of three biological and technical replications (±SD) are expressed in fold change relative to the expression level of the sample 22I as the reference sample (relative expression = 1). Transcript abundances were normalized using the expression of the *CcUBQ10* (Barsalobres-Cavallari et al. 2009) as the endogenous control. Treatments sharing the same letter are not significantly different. Data adapted from Torres et al. (2019)

120 and D^S clone 109 of *C. canephora*. Among the 22,764 genes generated, these authors identified 49 genes in the D^T clone (e.g., coding a MYB-like proteins or for defense-related proteins containing LRR and kinase domains), which could be involved in stress "memory."

As previously mentioned, Costa (2014) analyzed the expression profiles of several genes in roots of D^S and D^T clones of *C. canephora* Conilon submitted to water limitation. Among the identified DEGs, it is worth noting that upregulated expression was specifically observed under drought in roots of the D^T clone 14 for the *CcMJE1* (coding a protein involved in MeJA metabolism), *CcNCED3* (encoding a rate-limiting protein involved in the synthesis of abscisic acid), *CcPAP1* (coding a putative protein containing the acid phosphatase domain TIGR01675 characterizing vegetative storage proteins (VSPs)), *CcPRX1* (coding for a putative peroxidase), and *CcclXIP* (coding a chitinase-like xylanase inhibitor protein), as well as *CcM6PR*, *CcGOLS3b*, and *CcLTP4* (involved in RFOs and lipid biosynthesis pathways) genes. More recently, Vasconcelos et al. (2011) reported that the protein expressed from the *CaclXIP* cDNA (originally identified as a class III chitinase encoding gene from *C. arabica*) functioned as a chitinase-like xylanase inhibitor protein (clXIP) of fungal xylanases. Altogether, these responses suggest the existence of cross talk between abiotic and biotic pathways in roots of D^T clone 14 which could explain its drought tolerance and resistance to several species of RKN of *Meloidogyne* spp. (see Sect. 4.3).

It is also worth noting that expression of many genes cited in this section (e.g., coding dehydrins, enzymes of carotenoid and RFO pathways, and other proteins involved in stabilization of membranes and proteins) was also studied during the last stages of coffee bean development (Hinniger et al. 2006; Simkin et al. 2010; Ivamoto et al. 2017a; Dussert et al. 2018), characterized by the intense dehydration of endosperm (De Castro and Marraccini 2006; Eira et al. 2006).

5.2 *High Temperature*

The study of Bardil et al. (2011) was the first to analyze the effects of low (LT, day 26°C/night 22°C) and high (HT, day 30°C/night 26°C) temperature on homeologous genes expressed in leaves of *C. arabica* and in those of its two ancestral parents, *C. canephora* and *C. eugenioides*. Among the 15 K unigenes analyzed, around 50% appeared differentially expressed (with 25% upregulated) at low temperature between *C. arabica*, *C. canephora*, and *C. eugenioides*. Similar proportions were found at high temperature when comparing the transcriptome of *C. arabica* vs. *C. eugenioides* and *C. canephora* vs. *C. eugenioides*. However, only 8.9% of transcriptome divergence was observed when comparing *C. arabica* vs. *C. canephora*. In terms of expression patterns observed in *C. arabica*, the number of genes with "*C. canephora*-like dominance" increased from 8–14% under LT (in the Java and T18141 cultivars) to 21–26% under HT conditions. In that case, it was worth noting that transcription profiles of T18141 (a cultivar recently introgressed with *C. canephora* genome) were more similar to that of *C. canephora* than that of the "pure" (non-introgressed recently) Java cultivar. Altogether, these results indicate that *C. arabica* mainly expressed genes from its *CaCc* sub-genome under hot temperatures.

In another work, Bertrand et al. (2015) analyzed gene expression profiles in leaves of *C. arabica*, *C. eugenioides*, and *C. canephora* (cv. Nemaya) exposed to four thermal regimes (TRs: 18–14, 23–19, 28–24, and 33–29°C). Under hot temperatures, upregulated expression in *C. arabica* was observed for several genes like *Cc10_g00570* coding a catalase (CAT3) (when compared to *C. canephora*) and *Cc06_g11950* coding a photosystem II subunit X (when compared to *C. eugenioides*). On the other hand, expression profiles of *Cc05_g04680* coding a L-ascorbate oxidase homolog and those of photosynthetic genes coding light-harvesting complex (LHCII: *Cc04_g16410*) and chlorophyll a–b-binding protein (CAB: *Cc10_g00140*, *Cc05_g12720*, *Cc09_g09020*, *Cc05_g09650*, and *Cc09_g09030*), or for respiration-like genes *Cc10_g00410*, *Cc02_g25840*, and *Cc07_g00550* (coding a chloroplast glyceraldehyde-3-phosphate dehydrogenase, a chloroplast ribose-phosphate pyrophosphokinase, and a Rubisco methyltransferase, respectively), were strongly downregulated in *C. arabica* compared with its two parents.

In leaves of *C. arabica*, heat-shock conditions also upregulated the expression of *CaGolS1*, *CaGolS2*, *CaGolS3*, *CaPMI*, *CaMTD*, and *CaERF014* and downregulated expression of *CaM6PR* (dos Santos et al. 2011; de Carvalho et al. 2014; Torres et al. 2019). The interactions of high temperature and high [CO_2] on expression profiles of gene coding protective and antioxidant proteins were also studied by Martins et al. (2016) and Scotti-Campos et al. (2019) (see Sect. 5.4 below).

5.3 Cold Stress

The first studies to analyze the effects of cold stress on coffee gene expression were realized by Fortunato et al. (2010) and Batista-Santos et al. (2011) who subjected several cultivars and hybrids of *C. canephora*, *C. arabica*, and *C. dewevrei* to gradual cold treatments. These authors showed that upregulation of *CaGRed* and *CaDHAR* genes (coding a glutathione reductase (GR) and dehydroascorbate reductase, respectively) and of *CaCP22*, *CaPI*, and *CaCytf* (coding proteins involved in PSII, PSI, and Cytb6/f complex, respectively) could explain the ability of Icatu (*C. arabica* × *C. canephora*) cultivar to better support cold stress by reinforcing its antioxidative capabilities and maintaining efficient thylakoid functioning.

In their analysis of gene expression profiles in leaves of *C. arabica*, *C. eugenioides*, and *C. canephora* (cv. Nemaya) exposed to different thermal regimes, Bertrand et al. (2015) also reported upregulated expression profiles under cold stress in *C. arabica* for *Cc07_g15610* gene coding a L-ascorbate oxidase, for genes involved in respiration (e.g., *Cc02_g08980*, *Cc00_g15710*, and *Cc02_g06960* coding a phosphoenolpyruvate carboxylase kinase, a ribulose bisphosphate carboxylase small chain, and a sedoheptulose-1,7-bisphosphatase, respectively), and also for genes of photosynthesis (e.g., *Cc02_g28520*, *Cc05_g15930*, *Cc07_g10820*, and *Cc06_g19130* coding a ferredoxin-nitrite reductase, a photosystem II 10 kDa polypeptide, a ferredoxin-NADP reductase, and a ferredoxin-dependent glutamate synthase, respectively). On the other hand, overexpression of *Cc05_g10250*, *Cc00_g35890*, and *Cc05_g10310* genes (coding polyphenol oxidases) was seen in *C. canephora* under low temperatures. For the *LHY* (late elongated hypocotyl, *Cc02_g39990*) gene involved in circadian cycle, RT-qPCR experiments confirmed in silico data, showing the highest expression under low than high temperatures particularly in leaves of *C. canephora*.

More recently, two studies investigated the effects of cold stress on the leaf gene expression in *C. canephora*. In the first one, Dong et al. (2019a) performed gene expression analyses in leaves of *C. canephora* plants subjected to cold stress (C1 (7 days at day 13°C/night 8°C) followed by C2 (3 days at day 4°C/night 4°C)) but also in fruits at different stages of development. For the 38 *CcNAC* genes analyzed by qPCR in cold-stressed leaves, expression was (1) upregulated upon C1 and C2 treatments for 4 genes, (2) downregulated upon C1 (but not C2) for 10 genes, (3) upregulated upon C2 (but not C1) for 7 genes, and (4) downregulated upon both cold treatments for 17 genes. In the second work, the same authors characterized 49 *CcWRKY* genes from the reference genome of *C. canephora* and analyzed their expression profiles by qPCR for 45 of them in cold-stressed leaves as reported in the previous study (Dong et al. 2019b). This led to identify 14 *CcWRKY* genes with expression induced during the cold acclimation stage (upon C1 and C2 treatments), 17 genes upregulated by cold treatment (C2 but not C1), and 12 downregulated by both cold stress treatments. Among the 14,513 putative target genes of *CcWRKY* identified in *C. canephora* by a genome-wide analysis, 235 were categorized into response to the cold process, including carbohydrate metabolic, lipid metabolic, and photosynthesis process-related genes. Like in many other plants, these observations clearly highlight the vital regulatory role played by WRKY TFs in various developmental and physiological processes (such as seed development) but also in a range of abiotic stress (like cold, heat, drought, as well as salinity) and biotic stress (Rushton et al. 2010).

In a more recent work, Ramalho et al. (2018a) analyzed the impacts of single and combined exposure to drought and cold stress in *C. arabica* cv. Icatu, *C. canephora* cv. Apoatã, and the hybrid *C. arabica* cv. Obatã. At the physiological level, the Icatu cultivar showed a lower impact upon exposure to cold and drought stress, characterized by a reduced lipoperoxidation under stress interaction, for example. At the molecular level, simultaneous exposure of Icatu to both stresses increases the expression of genes coding ascorbate peroxidase (APX) involved in H_2O_2 removal (e.g., *APXc* [cytosolic] and *APXt+s* [stromatic]) and consequently total APX enzymatic activity. To a lesser extent, this situation was also observed in *C. canephora*, while Obatã was the less responsive genotype considering the studied genes.

5.4 CO_2 Concentration

The research group of J.C. Ramalho (Lisbon University, Portugal) published several articles studying the effects of elevated [CO_2] on coffee. They demonstrated that elevated [CO_2] mitigated the impact of heat on coffee physiology (Rodrigues et al. 2016) and also contributed to preserve the bean quality (Ramalho et al. 2018b). In a study aiming to analyze the interactions of elevated [CO_2] and high temperature on protective response mechanisms in coffee, Martins et al. (2016) showed that the maintenance (or increase) of the pools of several protective molecules (e.g., neoxanthin, lutein, carotenes, α-tocopherol, heat-shock proteins HSP70, and raffinose), activities of antioxidant enzymes (e.g., superoxide dismutase, APX, GR, and catalase [CAT]), and the upregulated expression of *ELIP* (coding chloroplast early light-induced protein) and *Chap20* (coding chloroplast 20 kDa chaperonin) genes were correlated with heat tolerance (up to day 37°C/night 30°C) at 380 and 700 μL CO_2 L^{-1} for both *C. arabica* L. cvs. Icatu and IPR108 and *C. canephora* cv. Conilon clone 153. These authors also showed that upregulated expression of genes related to protective (*ELIPS*, *HSP70*, *Chap20*, and *Chap60*) and antioxidant (*CAT*, *APXc*, *APXt+s*) proteins was largely driven by temperature, while enhanced [CO_2] promoted a greater upregulation of these genes mainly in *C. canephora* CL153 and *C. arabica* Icatu. In the more recent study analyzing the expression of genes related to lipid metabolism under elevated [CO_2], heat, and their interaction, Scotti-Campos et al. (2019) showed that the strong remodeling (unsaturation degree) of membrane lipids observed during the heat shock (from day 37°C/night 30°C to day 42°C/night 34°C) of plants grown under high [CO_2], coordinated with *FAD3* (coding for fatty acid desaturase) downregulation in *C. arabica* and upregulation of lipoxygenase-coding genes *LOX5A* (in CL153 and Icatu) and *LOX5B* (in Icatu), could contribute to long-term acclimation of coffee chloroplast membranes to climate changes.

5.5 Salt Stress

In leaves of *C. arabica* cv. IAPAR59, upregulated expression of galactinol synthase genes *CaGolS2* and *CaGolS3* was observed after irrigation with 150 mM NaCl (dos Santos et al. 2011). In the same cultivar, salt stress upregulated the expression of *CaM6PR* and *CaPMI* genes and markedly downregulated that of *CaMTD* (de Carvalho et al. 2014). In parallel, leaf mannitol contents increased gradually to reach a peak after 12 days of salt stress imposition. However, this content was lower than in leaves of plants under water deprivation, indicating that coffee plants have different responses to drought and salinity.

The effects of salt stress in leaves were recently studied by RNAseq in leaves of *C. arabica* seedlings irrigated with normal water (control, ECw [electrical conductivity] = 0.2 $dS.m^{-1}$) or with deep sea water (salt treatment, ECw = 2.3 $dS.m^{-1}$) (Haile and Kang 2018). From the 19,581 genes aligned on the reference genome of

C. canephora, in silico analyses identified 611 genes presenting significant DEG profiles between the control and salt treatment. Among the most expressed upregulated genes were *Cc00_g13890*, *Cc04_g05080*, and *Cc08_g11060*, coding for WRKY TFs; *Cc06_g01240* coding a putative trihelix TF GT-3a already reported in controlling the developmental process and response to abiotic and biotic stress (Park et al. 2004; Wang et al. 2016); and *Cc10_g04710* coding the putative ethylene-responsive (ERF011) TF. On the other hand, salt stress also downregulated the expression of *Cc05_g16570* (coding a putative MYB family transcription factor APL), *Cc02_g17440* and *Cc07_g03240* (both coding putative bHLH TFs), and *Cc02_g10740* and *Cc06_g21410* (coding putative transcription elongation factor SPT of RNA polymerase II). However, the DEG expression profiles of these TF-encoding genes were not verified in vivo by qPCR experiments.

5.6 Wounding

WRKY and *NDR* genes were previously reported as playing key roles in the molecular resistance responses of coffee to *H. vastatrix* (see Sects. 2.3 and 4.1). In the first study, Ganesh et al. (2006) reported upregulated expression of *CaNDR1*, *CaWRKY1* (see Sect. 4.1), and *CaR111* (coding a putative protein of unknown function) genes in leaves of *C. arabica* wounded by performing transversal cuts with scissors. Few years after, Petitot et al. (2008, 2013) showed that expression of both *CaWRKY1a* (*CaCc*) and *CaWRKY1b* (*CaCe*) homeologs was upregulated in wounded leaves of *C. arabica* (see Sect. 7). In parallel, wounding also markedly upregulated expression of *CaWRKY1a* and *CaWRKY1b* genes in leaves of *C. canephora* and *C. eugenioides*, respectively, confirming that both genes were functional. In addition to *CaWRKY1*, Ramiro et al. (2010) also showed that *CaWRKY19/20/21* genes, as well as *CaWRKY15* and *CaWRKY17*, were also highly induced by wounding. In another work, Brandalise et al. (2009) showed that expression of *CaIRL*, coding an isoflavone reductase-like protein, was induced in leaves of *C. arabica* submitted to a mechanical injury, leading to further study the promoter of this gene (see Sect. 8).

6 Gene Expression in F1 Hybrids of *C. arabica*

In the context, the creation of new coffee varieties better adapted to biotic and abiotic stresses to low levels of inputs and to CC is now one of the challenges of several coffee research institutes (van der Vossen et al. 2015; Bertrand et al. 2019).

In *C. arabica*, it is possible to create and select in a relatively short time (e.g., around 8 years against 25 years for conventional breeding programs) new F1 hybrid varieties with increased production (e.g., under agroforestry) and also improved aromatic quality without increasing fertilizer quantities (Bertrand et al. 2006, 2011), by crossing pure commercial line varieties with phylogenetically distant plants

Table 5 List of experiments (and related RNAseq analyses) planned in the frame of the BREED CAFS project (see www.breedcafs.eu)

Trial number	Place	Condition	Experiment
1	Univ. of Lisbon	Phytotron	Drought/[CO_2]
2	Univ. of Lisbon	Phytotron	Heterosis/[CO_2]
3	Cirad	Phytotron	Nitrogen/light/[CO_2]
4	Cirad	Phytotron	Heterosis/T°C
5	Univ. of Copenhagen	Greenhouse	Heterosis/shade
6	Cirad	Greenhouse	Heterosis/N-depletion
7	Cirad	Phytotron	Shade/nitrogen/T°C/rust
8	Cirad	Greenhouse	Drought/heterosis
9	Cirad	Greenhouse	Circadian clock
10	Cirad	Greenhouse	Heterosis/circadian clock
11	Cirad	Phytotron	Grafting/T°C
12	Cirad	Rhizoscope	Grafting/nitrogen
13	Costa Rica	Field	Grafting/elevation/shade
14	Nicaragua	Nicaragua	Heterosis/shade
15	Nicaragua	Nicaragua	25 best cultivars
16	Nicaragua	Nursery	Heterosis/T°C
17	Vietnam	Field	Drought/AFS
18	Salvador	Field	F1 hybrids/parents
19	Nicaragua	Field	Maternal effect
20	Nicaragua	Nursery	Genomic selection

corresponding to wild individuals from Ethiopia and Sudan, for example (Van der Vossen et al. 2015). The objective of the H2020 BREEDCAFS[2] (BREEDing Coffee for Agroforestry Systems) project, supported by the EU (2017–2021), is to identify robust markers (allelic, molecular, epigenetic) that could be used as early predictors to speed up future *C. arabica* breeding programs aiming to create new F1 hybrids with increased resistance and greater resilience to climate change in agroforestry systems (Bertrand et al. 2019). This project intends to compare the leaf transcriptomic profiles in F1 hybrids and cultivated varieties (and/or hybrids to their two parents) upon different abiotic stresses either performed in phytotrons and greenhouses (e.g., in order to test the effects of temperature, light, drought, CO_2, and N_2) or in field trials (or in networks of "demoplots" in farms). The numerous RNAseq studies planned to be perform within the framework of this project (Table 5) should also help us to better understand why the pure line varieties are less adapted to environmental constraints than F1 hybrids. For example, Toniutti et al. (2019b) showed that hybrid vigor (heterosis) could be explained by the modification of leaf expression profiles of several genes involved in the circadian clock (e.g., *LHY* and *GIGANTEA*), the chlorophyll synthesis (e.g., *POR1A* and

[2] www.breedcafs.eu.

PORIB), and starch degradation (e.g., *CcGWD1* and *CcISA3*) in leaves of the *C. arabica* F1 hybrid GPFA124 compared to those of the inbred Caturra line. In the same work, upregulated expression of chloroplast genes in the *C. arabica* GPFA124 was also reported (see Sect. 7).

7 Expression of Chloroplast Genes

The chloroplast genome of *C. arabica* consists of 155,189 base pairs encoding 130 genes with 18 intron-containing genes (Samson et al. 2007). In a pioneer work, Dinh et al. (2016) analyzed the effects of drought, cold, or combined drought and heat stresses on intron splicing and expression patterns of 48 chloroplast genes from *C. arabica*. By RT-qPCR, these authors showed that the transcript levels of chloroplast mRNAs were globally decreased in seedlings submitted to drought or cold treatments. For example, expression of *rbcL* (coding the large subunit of Rubisco) and *psaA* and *psaB* (coding photosystem I proteins) was significantly reduced in *C. arabica* under cold stress conditions but not under drought. Regarding intron-containing genes, it was also shown that the splicing efficiencies of *trnG*, *trnK*, and *trnA* genes increased upon drought, combined drought and heat, or cold stress treatments, while these efficiencies decreased for *trnL* under these stresses. On the other hand, the splicing efficiencies of mRNA genes *rps16*, *atpF*, *petB*, and *rpl2* were decreased upon drought but increased upon cold stress treatment.

Overexpression of *CaPsbB* gene (coding the photosystem II CP47 chlorophyll apoproteins) was also reported, either by in silico (Vieira et al. 2006; Mondego et al. 2011; Vinecky et al. 2012) or by in vivo (Mofatto et al. 2016) analyses, in leaves of drought-stressed coffee plants but also in those infected by *H. vastatrix* (Fernandez et al. 2012).

In addition to the circadian genes (see Sect. 6), Toniutti et al. (2019b) also reported increased photosynthetic electron transport efficiency in the *C. arabica* hybrid GPFA124 probably explained by higher expression of chloroplast genes *CaPsbA* and *CaPsbD* (coding the D1 and D2 proteins of PSII, respectively); *CaPetA*, *CaPetD*, and *CaPetB* (coding proteins of the cytB6/f complex); and *CaPsaA*, *CaPsaB*, and *CaPsaJ* (coding proteins of PSI), in this hybrid compared to the *C. arabica* cv. Caturra.

8 Coffee Promoters

The expression studies previously detailed also led to the identification of coffee promoters (De Almeida et al. 2008). For several of them, they were functionally characterized using the *uidA* (coding the β-glucuronidase) as the reporter gene by transgenic approaches either in *Nicotiana tabacum* or in *Coffea* sp. The first promoter was cloned from the *CaCSP1* gene of *C. arabica* coding for the 11S seed storage protein and was shown to function as a bean (endosperm)-specific promoter in transgenic tobacco plants (Marraccini et al. 1999). A similar result was also observed for the shorter and medium promoter fragments of the *CaLTP* gene coding non-specific lipid transfer proteins (Cotta et al. 2014). Leaf-specific expression was also reported for *CaRBCS1* and *CcMXMT1* coffee promoters in transgenic tobacco (Marraccini et al. 2003; Satyanarayana et al. 2005). The *SERK1* (somatic embryogenesis receptor-like kinase 1) promoter from *C. canephora* was also shown to drive the *uidA* expression in different embryo structures such as globular, heart, torpedo, and cotyledonal embryos present at 60 days after embryogenic induction (Jiménez-Guillen et al. 2018). Regarding abiotic stress, Brandalise et al. (2009) showed that the promoter of *CaIRL* was induced by wounded leaves of *N. tabacum*. In 2016, Nobres et al. analyzed the promoter function of the *CaHB12* from *C. arabica*, a gene coding member of the homeodomain-leucine zipper I subfamily (HD-Zip) and conferring greater tolerance to drought stress when overexpressed in *Arabidopsis* (Alves-Ferreira et al. 2012). The study of transgenic *A. thaliana* plants bearing *pCaHB12::GUS* constructs showed that this promoter was expressed in leaves during drought and in roots after polyethylene glycol or mannitol treatments. On the other hand, the different haplotypes of the *CcDREB1D* promoter from *C. canephora* were shown to be upregulated by different abiotic stresses in the leaves of *C. arabica* (see Sect. 5.1) and *N. tabacum* transgenic plants (Alves et al. 2017, 2018; de Aquino et al. 2018). Regarding biotic stress, Petitot et al. (2013) analyzed the promoter activities of *CaWRKY1a* (named pW1a) and *CaWRKY1b* (named pW1b) homeologous genes, previously identified to be induced by CLR infestation in the *C. arabica* leaves (see Sects. 2.3, 4.1, and 5.6), in transient assays of *N. benthamiana* leaves, and in stable transgenic plants of *C. arabica*. These authors also showed increased activities of both promoters in leaves of tobacco treated with SA or in those of coffee infected with CLR, as well as increased activities of pW1a upon wounding. The other coffee promoters already described in the literature but without being tested in transgenic plants are cited in Table 6.

9 Coffee Small RNA (sRNA)

Using small (20 ± 26 nt) homologous sequences, small RNAs (sRNA) are known to play important roles by silencing pathways at the transcriptional or translational levels. Plant sRNAs are classified as (1) microRNAs (miRNAs) which are derived from self-complementary hairpin structures and (2) small interfering RNAs (siRNAs) which are derived from double-stranded RNA (dsRNA) or hairpin precursors (Borges and Martienssen 2015). The core mechanism of sRNA production requires the endonuclease activity of DICER-LIKE 1 (DCL1) and ARGONAUTE (AGO) proteins as effectors of silencing, while siRNA biogenesis involves action of RNA-dependent RNA polymerase (RDR), Pol IV, and Pol V. With the release of the *C. canephora* genome (Denoeud et al. 2014), sRNAs were now identified.

Table 6 List of coffee promoters already described in the literature

Gene	Function	Tissue specificity	Validation	References
CaCSP1	11S storage protein	Endosperm	*Nt*	Marraccini et al. (1999)
		Nd	nd	Acuña et al. (1999)
CaRBCS1	Rubisco small subunit	Leaf	*Nt*	Marraccini et al. (2003)
CaSUI1	Translation initiation factor SUI1	Nd	nd	Gaborit et al. (2003)
CcMXMT1	*N*-methyltransferase	Leaf	*Nt*	Satyanarayana et al. (2005)
CcOLE-1	Oleosin	Nd	nd	Simkin et al. (2006)
CcDH2a	Dehydrin	Nd	nd	Hinniger et al. (2006)
CaIRL	Isoflavone reductase-like protein	Leaf (wounding)	*Nt*	Brandalise et al. (2009)
CaWRKY1	Transcription factor	Leaf (wounding, CLR, SA)	*Nb/Ca*	Petitot et al. (2013)
CaLTP	Non-specific lipid transfer proteins	Endosperm	*Nt*	Cotta et al. (2014)
CaHB12	Homeodomain-leucine zipper I subfamily	Leaf, root (drought, PEG, mannitol)	*At*	Nobres et al. (2016)
CcDREB1D	Dehydration-responsive element-binding TF	Leaf	*Ca*	Alves et al. (2017, 2018)
		Leaf (dehydration, HS, cold)	*Nt*	de Aquino et al. (2018)
CcSERK1	Somatic embryogenesis receptor-like kinase 1	SE(DS)	*Cc*	Jiménez-Guillen et al. (2018)

CLR coffee leaf rust, *SE(DS)* somatic embryogenesis (at different stages), *HS* heat shock. The gene names, with their function and tissue-specific expression, are indicated. The transgenic plants used to validate the promoters are also indicated (*At, Arabidopsis thaliana*; *Ca, Coffea arabica*; *Cc, Coffea canephora*; *Nb, Nicotiana benthamiana*; *Nt, Nicotiana tabacum*). *nd* not determined

One of the first attempts to study coffee miRNAs was performed by Nellikunnumal and Chandrashekar (2012) who identified 18 miRNAs, belonging to 12 families, from *C. canephora* ESTs by computational approaches. By RT-PCR, these authors showed that expression was detected for seven families (viz., mir156, mir169, mir172, mir319, mir393, mir395, and mir396) in *C. canephora* leaves. By the same computational approach, Rebijith et al. (2013), Loss-Morais et al. (2014), and Devi et al. (2016) also identified miRNAs in *C. arabica* and *C. canephora*, showing that the majority of their potential targets corresponded to mRNA coding proteins involved in transcriptional regulation and signal transduction pathways. In another study, Akter et al. (2014) identified a potential miRNA (named mir393) from *C. arabica* ESTs and also showed that this sequence had as potential targets several genes coding transcription factors (e.g., bHLH7 and WRKY TFs) or proteins involved in auxin signaling pathway and plant defense responses (e.g., auxin signaling F-box 2 and auxin transporter protein 1). Using a specific pipeline to search for miRNA homologs on expressed sequence tag (EST) and genome survey sequence (GSS) coffee databases, Chaves et al. (2015) identified 36 microRNAs and a total of 616 and 362 potential target genes for *C. arabica* and *C. canephora*, respectively. Using a stem-loop RT-PCR assay, these authors also detected a higher amount of miRNAs (miRNAs 171, 172, 390, and 167) in leaves of *C. arabica* than in those of *C. canephora*, suggesting a possible role of sRNA in regulating *C. arabica* transcriptome.

Fernandes-Brum et al. (2017b) identified 11 AGO proteins, nine DCL-like proteins, eight RDR proteins, and 48 other proteins implicated in the sRNA pathways. These authors also identified (1) 235 miRNA precursors producing 317 mature miRNAs belonging to 113 MIR families and (2) 2239 putative *C. canephora* miRNA targets in different pathways. In another study, Bibi et al. (2017) also identified potential miRNAs potentially targeting 150 genes coding transcription factors but also proteins involved in multiple biological and metabolic processes, hypothetical proteins, signal transduction, transporters, growth and development, stress-related processes, structural constituents, and disease-related processes, for example.

In the study analyzing coffee memory to multiple drought exposures, de Freitas Guedes et al. (2018) also reported upregulated expression of mir398 and mir408 by the drought cycles in *C. canephora*. In addition to drought, these genes were also reported to be regulated in other plants by ABA, heat, UV, and also biotic stress events (Zhu et al. 2011; Khraiwesh et al. 2012; Guan et al. 2013). Interestingly, transgenic chickpea plants overexpressing mir408 were shown to be tolerant to several stresses including drought (Hajyzadeh et al. 2015; Ma et al. 2015). In the recent study, dos Santos et al. (2019) analyzed the transcriptome in N-starved roots of *C. arabica* and also identified 86 microRNA families targeting 253 genes. RT-qPCR assays showed that expression profiles of mir169, mir171, mir167, mir393, and mir858 were upregulated in roots after N-starvation, while mircar1 was downregulated after prolonged N-restriction. Altogether, these results highlight the role that might play sRNA in modulating the expression of genes involved in the adaptive responses of coffee plants to environmental factors.

10 Conclusions

Like many other crops, gene identification and characterization are of fundamental biological interest in coffee to understand the transcription networks involved in important agronomic traits and further to identify SNPs that can serve as markers of specific phenotypes to better drive future breeding programs. In that way, the high number of large-scale expression analyses, together with the recent access to long-read sequencing of transcripts (Cheng et al. 2017), to reference transcriptomes (Yuyama et al. 2016; Cheng et al. 2018), and to reference genomes of

C. canephora (Denoeud et al. 2014) and *C. arabica* (de Kochko et al. 2015, 2017; Gaitan et al. 2015; Morgante et al. 2015; Yepes et al. 2016), now opens the way to identify SNPs associated with bean biochemical compound content (Tran et al. 2018) and adaptation to environmental factors (de Aquino et al. 2019) and to initiate marker-assisted selection (Alkimim et al. 2017) and genome-wide association studies (Andrade 2018; Sant'Ana et al. (2018); Carneiro et al. 2019). Together with the help of CRISPR/Cas9 technology (Breitler et al. 2018), it is now possible to greatly shorten the time required to create new coffee varieties with improved agronomic traits under CC.

Acknowledgments I would like to thank B. Bertrand and H. Etienne (UMR IPME, research team "CoffeeAdapt" of UMR IPME, https://umr-ipme.ird.fr/equipes/equipe-coffeeadapt) for their critical reading of the manuscript. This work was supported by the European Union, Program Horizon 2020, call H2020-SFS-2016-2, action RIA, project BREEDCAFS (http://www.breedcafs.eu/), and proposal 727934.

References

Abreu HMC, Nobile PM, Shimizu MM, Yamamoto PY, Silva EA, Colombo CA, Mazzafera P (2012) Influence of air temperature on proteinase activity and beverage quality in *Coffea arabica*. Braz J Bot 35:357–376

Acuña R, Bassüner R, Beillinson V, Cortina H, Cadena-Gómez G, Montes V, Nielsen N (1999) Coffee seeds contain 11S storage proteins. Physiol Plant 105:122–131

Aggarwal RK, Hendre PS, Varshney RK, Bhat PR, Krishnakumar V, Singh L (2007) Identification, characterization and utilization of EST-derived genic microsatellite markers for genome analyses of coffee and related species. Theor Appl Genet 114:359–372

Akter A, Islam MM, Mondal SI, Mahmud Z, Jewel NA, Ferdous S, Amin MR, Rahman MM (2014) Computational identification of miRNA and targets from expressed sequence tags of coffee (*Coffea arabica*). Saudi J Biol Sci 21:3–12

Alkimim ER, Caixeta ET, Sousa TV, Pereira AA, de Oliveira ACB, Zambolim L, Sakiyama NS (2017) Marker-assisted selection provides arabica coffee with genes from other Coffea species targeting on multiple resistance to rust and coffee berry disease. Mol Breeding 37:6

Alvarenga SM, Caixeta ET, Hufnagel B, Thiebaut F, Maciel-Zambolim E, Zambolim L, Sakiyama NS (2010) *In silico* identification of coffee genome expressed sequences potentially associated with resistance to diseases. Genet Mol Biol 33:795–806

Alves LC, de Magalhães DM, Labate MTV, Guidetti-Gonzalez S, Labate CA, Domingues DS, Sera T, Vieira LGE, Pereira LFP (2016) Differentially accumulated proteins in *Coffea arabica* seeds during perisperm tissue development and their relationship to coffee grain size. J Agric Food Chem 64:1635–1647

Alves GSC, Torres LF, Déchamp E, Breitler J-C, Joët T, Gatineau F, Andrade AC, Bertrand B, Marraccini P, Etienne H (2017) Differential fine-tuning of gene expression regulation in coffee leaves by *CcDREB1D* promoter haplotypes under water deficit. J Exp Bot 68:3017–3031

Alves GSC, Torres LF, de Aquino SO, Reichel T, Freire LP, Vieira NG, Vinecky F, This D, Pot D, Etienne H, Paiva LV, Marraccini P, Andrade AC (2018) Nucleotide diversity of the coding and promoter regions of *DREB1D*, a candidate gene for drought tolerance in *Coffea* species. Trop Plant Biol 11:31–48

Alves-Ferreira M,Waltenberg FPC, Pinto ERC, Grossi-de-Sá FM (2012) Use of the coffee homeobox gene *CaHB12* to produce transgenic plants with greater tolerance to water scarcity and salt stress. WO061911:A2
Andrade AC (2018) Developments in molecular breeding techniques in Robusta coffee. In: Lashermes P (ed) Achieving sustainable cultivation of coffee: breeding and quality traits. Burleigh Dodds Science Publishing, Cambridge
Ashihara H (2006) Metabolism of alkaloids in coffee plants. Braz J Plant Physiol 18:1–8
Ashihara H, Monteiro AM, Gillies FM, Crozier A (1996) Biosynthesis of caffeine in leaves of coffee. Plant Physiol 111:747–753
Asquini E, Gerdol M, Gasperini D, Igic B, Graziosi G, Pallavicini A (2011) S-RNase-like sequences in styles of *Coffea* (Rubiaceae). Evidence for S-RNase based gametophytic self-incompatibility? Trop Plant Biol 4:237–249
Assad ED, Pinto HS, Zullo J Jr, AMH A (2004) Impacto das mudanças climáticas no zoneamento agroclimático do café no Brasil. Pesqui Agropecu Bras 39:1057–1064
Avelino J, Willocquet L, Savary S (2004) Effects of crop management patterns on coffee rust epidemics. Plant Pathol 53:541–547
Avelino J, Cristancho M, Georgiou S, Imbach P, Aguilar L, Bornemann G, Läderach P, Anzueto F, Hruska A, Morales C (2015) The coffee rust crises in Colombia and Central America (2008–2013): impacts, plausible causes and proposed solutions. Food Secur 7:303–321
Avila RT, Martins SCV, Sanglard LMVP, dos Santos MS, Menezes-Silva PE, Detman KC, Sanglard ML, Cardoso AA, Morais LE, Vital CE, Araújo WL, Nunes-Nesi A, DaMatta FM (2020) Starch accumulation does not lead to feedback photosynthetic downregulation in girdled coffee branches under varying source-to-sink ratios. Trees 34:1–16
Baba VY, Braghini MT, dos Santos TB, de Carvalho K, Soares JDM, Ivamoto-Suzuki ST, Maluf MP, Padilha L, Paccola-Meirelles LD, Pereira LF, Domingues DS (2020) Transcriptional patterns of *Coffea arabica* L. nitrate reductase, glutamine and asparagine synthetase genes are modulated under nitrogen suppression and coffee leaf rust. Peer J 8:e8320
Bang CF, Huyen TTT (2015) *In silico* identification, classification and expression analysis of genes encoding putative light-harvesting chlorophyll A/B-binding proteins in coffee (*Coffea canephora* L.). J Agric Technol 11:2547–2561
Barbosa BCF, Silva SC, de Oliveira RR, Chalfun-Junior A (2017) Zinc supply impacts on the relative expression of a metallothionein-like gene in *Coffea arabica* plants. Plant Soil 411:179–191
Bardil A, De Almeida JD, Combes MC, Lashermes P, Bertrand B (2011) Genomic expression dominance in the natural allopolyploid *Coffea arabica* is massively affected by growth temperature. New Phytol 192:760–774
Barreto HG, Lazzari F, Ságio SA, Chalfun-Junior A, Paiva LV (2012) *In silico* and quantitative analyses of the putative *FLC-like* homologue in coffee (*Coffea arabica* L.). Plant Mol Biol Rep 30:29–35
Barreto HG, Daude MM, Lima AA, Chalfun-Junior A (2018) Expression analysis of the coffee (*Coffea arabica* L.) *FRIGIDA4-like* gene (*CaFRL4*). Rev Desafios 5:204–213
Barsalobres-Cavallari CF, Severino FE, Maluf MP, Maia IG (2009) Identification of suitable internal control genes for expression studies in *Coffea arabica* under different experimental conditions. BMC Mol Biol 10:1
Batista-Santos P, Lidon FC, Fortunato A, Leitão AE, Lopes E, Partelli F, Ribeiro AI, Ramalho JC (2011) The impact of cold on photosynthesis in genotypes of *Coffea* spp. – photosystem sensitivity, photoprotective mechanisms and gene expression. J Plant Physiol 168:792–806
Bazzo BR, Eiras AL, DeLaat DM, Siqueira WJ, Mondego JMC, Colombo CA (2013) Gene expression analysis suggests temporal differential response to aluminum in *Coffea arabica* cultivars. Trop Plant Biol 6:191–198
Benchabane M, Schluter U, Vorster J, Goulet MC, Michaud D (2010) Plant cystatins. Biochimie 92:1657–1666

Berthaud J (1980) L'incompatibilité chez *Coffea canephora*: méthode de test et déterminisme génétique. Café Cacao Thé 24:267–274

Bertrand C, Noirot M, Doulbeau S, de Kochko A, Hamon S, Campa C (2003) Chlorogenic acid content swap during fruit maturation in *Coffea pseudozanguebariae* qualitative comparison with leaves. Plant Sci 165:1355–1361

Bertrand B, Vaast P, Alpizar E, Etienne H, Davrieux F, Charmetant P (2006) Comparison of bean biochemical composition and beverage quality of arabica hybrids involving Sudanese-Ethiopian origins with traditional varieties at various elevations in Central America. Tree Physiol 26:1239–1248

Bertrand B, Alpizar E, Lara L, Santacreo R, Hidalgo M, Quijano JM, Montagnon C, Georget F, Etienne H (2011) Performance of *Coffea arabica* F1 hybrids in agroforestry and full-sun cropping systems in comparison with American pure line cultivars. Euphytica 181:147–158

Bertrand B, Bardil A, Baraille H, Dussert S, Doulbeau S, Dubois E, Severac D, Dereeper A, Etienne H (2015) The greater phenotypic homeostasis of the allopolyploid *Coffea arabica* improved the transcriptional homeostasis over that of both diploid parents. Plant Cell Physiol 56:2035–2051

Bertrand B, Breitler J-C, Georget F, Penot E, Bordeaux M, Marraccini P, Léran S, Campa C, Bonato O, Villain L, Etienne H (2019) New varieties for innovative agroforestry coffee systems. In: Côte F-X, Poirier-Magona E, Perret S, Roudier P, Rapidel B, Thirion M-C (eds) The agroecological transition of agricultural systems in the global south. Versailles, Quae, pp 161–176

Bharathi K, Sreenath HL (2017) Identification and analysis of jasmonate pathway genes in *Coffea canephora* (Robusta coffee) by *in silico* approach. Pharmacogn Mag 13:196–200

Bharathi K, Santosh P, Sreenath HL (2017) Transcripts of pectin-degrading enzymes and isolation of complete cDNA sequence of a pectate lyase gene induced by coffee white stem borer (*Xylotrechus quadripes*) in the bark tissue of *Coffea canephora* (robusta coffee). 3 Biotech 7:45

Bibi F, Barozai MYK, Din M (2017) Bioinformatics profiling and characterization of potential microRNAs and their targets in the genus *Coffea*. Turk J Agric For 41:191–200

Borges F, Martienssen RA (2015) The expanding world of small RNAs in plants. Nat Rev Mol Cell Biol 16:727–741

Bottcher A, Nobile PM, Martins PF, Conte FF, Azevedo RA, Mazzafera P (2011) A role for ferritin in the antioxidant system in coffee cell cultures. Biometals 24:225–237

Brandalise M, Severino FE, Maluf MP, Maia IG (2009) The promoter of a gene encoding an isoflavone reductase-like protein in coffee (*Coffea arabica*) drives a stress-responsive expression in leaves. Plant Cell Rep 28:1699–1708

Breitler J-C, Déchamp E, Campa C, Rodrigues LAZ, Guyot R, Marraccini P, Etienne H (2018) CRISPR/Cas9-mediated efficient targeted mutagenesis has the potential to accelerate the domestication of *Coffea canephora*. Plant Cell Tissue Organ Cult 134:383–394

Breitler J-C, Djerrab D, Leran S, Toniutti L, Guittin C, Severac D, Pratlong M, Dereeper A, Etienne H, Bertrand B (2020) Full moonlight-induced circadian clock entrainment in *Coffea arabica*. BMC Plant Biol 20:24

Budzinski IGF, Santos TB, Sera T, Pot D, Vieira LGE, Pereira LFP (2010) Expression patterns of three α-expansin isoforms in *Coffea arabica* during fruit development. Plant Biol 13:462–471

Bunn C, Läderach P, Ovalle Rivera O, Kirschke D (2015a) A bitter cup: climate change profile of global production of Arabica and Robusta coffee. Clim Chang 129:89–101

Bunn C, Läderach P, Pérez Jimenez JG, Montagnon C, Schilling T (2015b) Multiclass classification of agro-ecological zones for Arabica coffee: an improved understanding of the impacts of climate change. PLoS One 10:e0140490

Bustin SA (2002) Quantification of mRNA using real-time reverse transcription PCR (RT-PCR): trends and problems. J Mol Endocrinol 29:23–39

Bustin SA, Benes V, Garson JA, Hellemans J, Huggett J, Kubista M, Mueller R, Nolan T, Pfaffl MW, Shipley GL, Vandesompele J, Wittwer CT (2009) The MIQE guidelines: minimum information for publication of quantitative real-time PCR experiments. Clin Chem 55:611–622

Bytof G, Knopp SE, Kramer D, Breitenstein B, Bergervoet JHW, Groot SPC, Selmar D (2007) Transient occurrence of seed germination processes during coffee post-harvest treatment. Ann Bot 100:61–66

Cação SMB, Leite TF, Budzinski IGF, dos Santos TB, Scholz MBS, Carpentieri-Pipolo V, Domingues DS, Vieira LGE, Pereira LFP (2012) Gene expression and enzymatic activity of pectin methylesterase during fruit development and ripening in *Coffea arabica* L. Genet Mol Res 11:3186–3197

Cacas J-L, Petitot A-S, Bernier L, Estevan J, Conejero G, Mongrand S, Fernandez D (2011) Identification and characterization of the non-race specific disease resistance 1 (NDR1) orthologous protein in coffee. BMC Plant Biol 11:144

Campa C, Ballester JF, Doulbeau S, Dussert S, Hamon S, Noirot M (2004) Trigonelline and sucrose diversity in wild Coffea species. Food Chem 88:39–44

Campa C, Urban L, Mondolot L, Fabre D, Roques S, Lizzi Y, Aarrouf J, Doulbeau S, Breitler J-C, Letrez C, Toniutti L, Bertrand B, La Fisca P, Bidel LPR, Etienne H (2017) Juvenile coffee leaves acclimated to low light are unable to cope with a moderate light increase. Front Plant Sci 8:1126

Cardoso DC, Martinati JC, Giachetto PF, Vidal RO, Carazzolle MF, Padilha L, Guerreiro-Filho O, Maluf MP (2014) Large-scale analysis of differential gene expression in coffee genotypes resistant and susceptible to leaf miner-toward the identification of candidate genes for marker assisted-selection. BMC Genomics 15:66

Carneiro FA, de Aquino SO, Mattos NG, Valeriano JC, Carneiro WWJ, Rodrigues GC, Carvalho MAF, Veiga AD, Grattapaglia D, Silva-Júnior OB, Marraccini P, Junior ICV, Balestre M, Andrade AC (2019) Genome wide selection in *Coffea canephora* breeding. In: X Simpósio de Pesquisa dos Cafés do Brasil Vitória (ES)-Brazil. EMBRAPA, Vitória

Cenci A, Combes MC, Lashermes P (2012) Genome evolution in diploid and tetraploid *Coffea* species. Plant Mol Biol 78:135–145

Chaves SS, Fernandes-Brum CN, Silva GFF, Ferrara-Barbosa BC, Paiva LV, Nogueira FTS, Cardoso TCS, Amaral LR, de Souza Gomes M, Chalfun-Junior A (2015) New insights on *Coffea* miRNAs: features and evolutionary conservation. Appl Biochem Biotechnol 177:879–908

Cheng B, Furtado A, Henry RJ (2017) Long-read sequencing of the coffee bean transcriptome reveals the diversity of full-length transcripts. Gigascience 6:1–13

Cheng B, Furtado A, Henry RJ (2018) The coffee bean transcriptome explains the accumulation of the major bean components through ripening. Sci Rep 8:11414

Cheserek JJ, Gichimu BM (2012) Drought and heat tolerance in coffee: a review. Int Res J Agric Sci Soil Sci 2:498–501

Combes MC, Cenci A, Baraille H, Bertrand B, Lashermes P (2012) Homeologous gene expression in response to growing temperature in a recent allopolyploid (*Coffea arabica* L.). J Hered 103:36–46

Combes M-C, Dereeper A, Severac D, Bertrand B, Lashermes P (2013) Contribution of subgenomes to the transcriptome and their intertwined regulation in the allopolyploid *Coffea arabica* grown at contrasted temperatures. New Phytol 200:251–260

Combes M-C, Hueber Y, Dereeper A, Rialle S, Herrera J-C, Lashermes P (2015) Regulatory divergence between parental alleles determines gene expression patterns in hybrids. Genome Biol Evol 7:1110–1121

Costa TS (2014) Análise do perfil transcriptômico e proteômico de raízes de diferentes clones de *Coffea canephora* em condições de déficit hídrico. PhD dissertation, Federal University of Lavras, Brazil

Costa TS, Melo JAT, Carneiro FA, Vieira NG, Rêgo ECS, Block C Jr, Marraccini P, Andrade AC (2015) Drought effects on expression of genes involved in ABA signaling pathway in roots of susceptible and tolerant clones of *C. canephora*. In: 25th international conference on coffee science (ASIC), Armenia, Colombia (PB236), pp 42–45

Cotta MG (2017) Molecular mechanisms in the first step of ABA-mediated response in *Coffea* ssp. PhD dissertation, SupAgro, Montpellier, France, p 176

Cotta MG, Barros LMG, De Almeida JD, De Lamotte F, Barbosa EA, Vieira NG, Alves GSC, Vinecky F, Andrade AC, Marraccini P (2014) Lipid transfer proteins in coffee: isolation of *Coffea* orthologs, *Coffea arabica* homeologs, expression during coffee fruit development and promoter analysis in transgenic tobacco plants. Plant Mol Biol 85:11–31

Couttolenc-Brenis E, Carrión GL, Villain L, Ortega-Escalona F, Ramírez-Martínez D, Mata-Rosas-M, Méndez-Bravo A (2020) Prehaustorial local resistance to coffee leaf rust in a Mexican cultivar involves expression of salicylic acid-responsive genes. PeerJ 8:e8345

Cruz F, Kalaoun S, Nobile P, Colombo C, Almeida J, Barros LMG, Romano E, Grossi-De-Sá MF, Vaslin M, Alves-Ferreira M (2009) Evaluation of coffee reference genes for relative expression studies by quantitative real-time RT-PCR. Mol Breed 23:607–616

da Silva EAA, Acencio ML, Bovolenta LA, Lemke N, Varani AM, Bravo JP, Hoshino-Bezerra AA, Lemos EGM (2019) Gene expression during the germination of coffee seed. J Seed Sci 41:168–179

DaMatta FM, Ramalho JDC (2006) Impacts of drought and temperature stress on coffee physiology and production: a review. Braz J Plant Physiol 18:55–81

Davis AP, Gole TW, Baena S, Moat J (2012) The impact of climate change on indigenous arabica coffee (*Coffea arabica*): predicting future trends and identifying priorities. PLoS One 7:e47981

Davis AP, Chadburn H, Moat J, O'Sullivan R, Hargreaves S, Nic Lughadha E (2019) High extinction risk for wild coffee species and implications for coffee sector sustainability. Sci Adv 5:eaav3473

de Almeida JD, Barros LMG, Santos DBM, Cotta MG, Barbosa EA, Cação SB, Eira MTS, Alves GSC, Vinecky F, Pereira LFP, da Silva FR, Andrade AC, Marraccini P, Carneiro M (2008) Prospection of tissue specific promoters in coffee. In: 22th international conference on coffee science (ASIC), Campinas, Brazil, pp 954–957

de Aquino SO, Carneiro FA, Rêgo ECS, Alves GSC, Andrade AC, Marraccini P (2018) Functional analysis of different promoter haplotypes of the coffee (*Coffea canephora*) *CcDREB1D* gene through genetic transformation of *Nicotiana tabacum*. Plant Cell Tissue Organ Cult 132:279–294

de Aquino SO, Tournebize R, Marraccini P, Mariac C, Bethune K, Andrade AC, Kiwuka C, Crouzillat D, Anten NPR, de Kochko A, Poncet V (2019) Towards the identification of candidate gene nucleic polymorphisms to predict the adaptedness of Ugandense *C. canephora* populations to climate change. In: 27th international conference on coffee science (ASIC), Portland (OR), USA

de Carvalho K, Bespalhok Filho JC, dos Santos TB, de Souza SGH, Vieira LGE, Pereira LFP, Domingues DS (2013) Nitrogen starvation, salt and heat stress in coffee (*Coffea arabica* L.): identification and validation of new genes for qPCR normalization. Mol Biotechnol 53:315–325

de Carvalho K, Petkowicz CLO, Nagashima GT, Bespalhok Filho JC, Vieira LGE, Pereira LFP, Domingues DS (2014) Homeologous genes involved in mannitol synthesis reveal unequal contributions in response to abiotic stress in *Coffea arabica*. Mol Gen Genomics 289:951–963

De Castro R, Marraccini P (2006) Cytology, biochemistry and molecular changes during coffee fruit development. Braz J Plant Physiol 18:175–199

de Freitas Guedes FA, Nobres P, Ferreira DCR, Menezes-Silva PE, Ribeiro-Alves M, Correa RL, DaMatta FM, Alves-Ferreira M (2018) Transcriptional memory contributes to drought tolerance in coffee (*Coffea canephora*) plants. Environ Exp Bot 147:220–233

de Kochko A, Akaffou S, Andrade AC, Campa C, Crouzillat D, Guyot R, Hamon P, Ming R, Mueller LA, Poncet V, Tranchant Dubreuil C, Hamon S (2010) Advances in *Coffea* genomics. In: Kader JC, Delseny M (eds) Advances in botanical research, vol 53, pp 23–63

de Kochko A, Crouzillat D, Rigoreau M, Lepelley M, Bellanger L, Merot-L'Anthoene V, Vandecasteele C, Guyot R, Poncet V, Tranchant-Dubreuil C, Hamon P, Hamon S, Couturon E, Descombes P, Moine D, Mueller L, Strickler SR, Andrade AC, Protasio LFP, Marraccini P, Giuliano G, Fiore A, Pietrella M, Aprea G, Ming R, Wai J, Domingues DS,

Paschoal A, Kuhn G, Korlach J, Chin J, Sankoff D, Zheng C, Albert VA (2015) Dihaploid *Coffea arabica* genome sequencing and assembly. In: Plant and animal genome XXIII. San Diego (CA), USA. https://pag.confex.com/pag/xxiii/webprogram/Paper16983.html

de Kochko A, Hamon S, Guyot R, Couturon E, Poncet V, Dubreuil-Tranchant C, Crouzillat D, Rigoreau M, Hamon P (2017) Omics applications: coffee. In: Chowdappa P, Karun A, Rajesh MK, Ramesh SV (eds) Biotechnology of plantation crops. Daya Publishing House, New Delhi, pp 589–606

De Nardi B, Dreos R, Del Terra L, Martellossi C, Asquini E, Tornincasa P, Gasperini D, Pacchioni B, Rathinavelu R, Pallavicini A, Graziosi G (2006) Differential responses of *Coffea arabica* L. leaves and roots to chemically induced systemic acquired resistance. Genome 49:1594–1605

de Oliveira RR, Chalfun-Junior A, Paiva LV, Andrade AC (2010) In silico and quantitative analyses of MADS-box genes in *Coffea arabica*. Plant Mol Biol Rep 28:460–472

de Oliveira RR, Cesarino I, Mazzafera P, Dornelas MC (2014) Flower development in *Coffea arabica* L.: new insights into MADS-box genes. Plant Reprod 27:79–94

de Oliveira KC, de Souza GP, Bazioli JM, Martinati JC, dos Santos MM, Padilha L, Guerreiro-Filho O, Maluf MP (2019) Effects of somatic embryogenesis on gene expression of cloned coffee heterozygous hybrids. Acta Physiol Plant 41:118

Denoeud F, Carretero-Paulet L, Dereeper A, Droc G, Guyot R, Pietrella M, Zheng C, Alberti A, Anthony F, Aprea G, Aury J-M, Bento P, Bernard M, Bocs S, Campa C, Cenci A, Combes M-C, Crouzillat D, Da Silva C, Daddiego L, De Bellis F, Dussert S, Garsmeur O, Gayraud T, Guignon V, Jahn K, Jamilloux V, Joet T, Labadie K, Lan T, Leclercq J, Lepelley M, Leroy T, Li L-T, Librado P, Lopez L, Munoz A, Noel B, Pallavicini A, Perrotta G, Poncet V, Pot D, Priyono RM, Rouard M, Rozas J, Tranchant-Dubreuil C, VanBuren R, Zhang Q, Andrade AC, Argout X, Bertrand B, de Kochko A, Graziosi G, Henry RJ, Jayarama MR, Nagai C, Rounsley S, Sankoff D, Giuliano G, Albert VA, Wincker P, Lashermes P (2014) The coffee genome provides insight into the convergent evolution of caffeine biosynthesis. Science 345:1181–1184

Devi KJ, Chakraborty S, Deb B, Rajwanshi R (2016) Computational identification and functional annotation of microRNAs and their targets from expressed sequence tags (ESTs) and genome survey sequences (GSSs) of coffee (*Coffea arabica* L.). Plant Gene 6:30–42

Dinh SN, Sai TZT, Nawaz G, Lee K, Kang H (2016) Abiotic stresses affect differently the intron splicing and expression of chloroplast genes in coffee plants (*Coffea arabica*) and rice (*Oryza sativa*). J Plant Physiol 201:85–94

Diniz I, Figueiredo A, Loureiro A, Batista D, Azinheira H, Várzea V, Pereira AP, Gichuru E, Moncada P, Guerra-Guimarães L, Oliveira H, Silva MDC (2017) A first insight into the involvement of phytohormones pathways in coffee resistance and susceptibility to *Colletotrichum kahawae*. PLoS One 12:e0178159

Diola V, Brito GG, Caixeta ET, Pereira LFP, Loureiro ME (2013) A new set of differentially expressed signaling genes is early expressed in coffee leaf rust race II incompatible interaction. Funct Integr Genomics 13:379–389

Dong X, Jiang Y, Yang Y, Xiao Z, Bai X, Gao J, Tan S, Hur Y, Hao S, He F (2019a) Identification and expression analysis of the NAC gene family in *Coffea canephora*. Agronomy 9:670

Dong X, Yang Y, Zhang Z, Xiao Z, Bai X, Gao J, Hur Y, Hao S, He F (2019b) Genome-wide identification of WRKY genes and their response to cold stress in *Coffea canephora*. Forests 10:335

dos Santos AB, Mazzafera P (2013) Aquaporins and the control of the water status in coffee plants. Theor Exp Plant Physiol 25:79–93

dos Santos TB, Budzinski IGF, Marur CJ, Petkowicz CLO, Pereira LFP, Vieira LGE (2011) Expression of three *galactinol synthase* isoforms in *Coffea arabica* L. and accumulation of raffinose and stachyose in response to abiotic stresses. Plant Physiol Biochem 49:441–448 dos Santos TB, de Lima RB, Nagashima GT, Petkowicz CLO, Carpentieri-Pipolo V, Pereira LFP, Domingues DS, Vieira LGE (2015) *Galactinol synthase* transcriptional profile in two genotypes of *Coffea canephora* with contrasting tolerance to drought. Genet Mol Biol 38:182–190

dos Santos TB, Lima JE, Felicio MS, Soares JDM, Domingues DS (2017) Genome-wide identification, classification and transcriptional analysis of nitrate and ammonium transporters in *Coffea*. Genet Mol Biol 40:346–359

dos Santos TB, Soares JDM, Lima JE, Silva JC, Ivamoto ST, Baba VY, Souza SGH, Lorenzetti APR, Paschoal AR, Meda AR, Nishiyama Júnior MY, de Oliveira ÚC, Mokochinski JB, Guyot R, Junqueira-de-Azevedo ILM, Figueira AVO, Mazzafera P, Júnior OR, Vieira LGE, Pereira LFP, Domingues DS (2019) An integrated analysis of mRNA and sRNA transcriptional profiles in *Coffea arabica* L. roots: insights on nitrogen starvation responses. Funct Integr Genomics 19:151–169

Dussert S, Serret J, Bastos-Siqueira A, Morcillo F, Déchamp E, Rofidal V, Lashermes P, Etienne H, Joët T (2018) Integrative analysis of the late maturation programme and desiccation tolerance mechanisms in intermediate coffee seeds. J Exp Bot 69:1583–1597

Echeverria-Beirute F, Murray SC, Klein P, Kerth C, Miller R, Bertrand B (2018) Rust and thinning management effect on cup quality and plant performance for two cultivars of *Coffea arabica* L. J Agric Food Chem 66:5281–5292

Echeverría-Beirute F, Murray SC, Bertrand B, Klein PE (2019) Candidate genes in coffee (*Coffea arabica* L.) leaves associated with rust (*Hemileia vastatrix* Berk. & Br) stress. Peer J Preprints 7: e27923v1

Eira MTS, Amaral da Silva EA, de Castro RD, Dussert S, Walters C, Bewley JD, Hilhorst HWM (2006) Coffee seed physiology. Braz J Plant Physiol 18:149–163

Eskes A, Leroy T (2004) Coffee selection and breeding. In: Wintgens JN (ed) Coffee: growing, processing, sustainable production. A guidebook for growers, processors, traders, and researchers. Wiley, Weinheim, pp 57–86

Etienne H, Breton D, Breitler J-C, Bertrand B, Déchamp E, Awada R, Marraccini P, Léran S, Alpizar E, Campa C, Courtel P, Georget F, Ducos J-P (2018) Coffee somatic embryogenesis: how did research, experience gained and innovations promote the commercial propagation of elite clones from the two cultivated species? Front Plant Sci 9:1630

Eulgem T (2006) Dissecting the WRKY web of plant defense regulators. PLoS Pathog 2:1028–1030

Falco MC, Marbach PAS, Pompermayer P, Lopes FCC, Silva-Filho MC (2001) Mechanisms of sugarcane response to herbivory. Genet Mol Biol 24:113–122

Fernandes-Brum CN, Garcia BO, Moreira RO, Sagio SA, Barreto HG, Lima AA, Freitas NG, Lima RR, Carvalho CHS, Chalfun-Junior A (2017a) A panel of the most suitable reference genes for RT-qPCR expression studies of coffee: screening their stability under different conditions. Tree Genet Genomes 13:131

Fernandes-Brum CN, Rezende PM, Ribeiro THC, de Oliveira RR, Cardoso TCS, Amaral LR, Gomes MS, Chalfun-Junior A (2017b) A genome-wide analysis of the RNA-guided silencing pathway in coffee reveals insights into its regulatory mechanisms. PLoS One 12:e0176333

Fernandez D, Santos P, Agostini C, Bon MC, Petitot AS, Silva MC, Guerra-Guimarães L, Ribeiro A, Argout X, Nicole M (2004) Coffee (*Coffea arabica* L.) genes early expressed during infection by the rust fungus (*Hemileia vastatrix*). Mol Plant Pathol 5:527–536

Fernandez D, Tisserant E, Talhinhas P, Azinheira H, Vieira A, Petitot A-S, Loureiro A, Poulain J, Da Silva C, Silva MC, Duplessis S (2012) 454-pyrosequencing of *Coffea arabica* leaves infected by the rust fungus *Hemileia vastatrix* reveals in planta-expressed pathogen-secreted proteins and plant functions in a late compatible plant-rust interaction. Mol Plant Pathol 13:17–37

Figueiredo SA, Lashermes P, Aragão FJ (2011) Molecular characterization and functional analysis of the β-galactosidase gene during *Coffea arabica* (L.) fruit development. J Exp Bot 62:2691–2703

Figueiredo A, Loureiro A, Batista D, Monteiro F, Várzea V, Pais MS, Gichuru EK, Silva MC (2013) Validation of reference genes for normalization of qPCR gene expression data from *Coffea* spp. hypocotyls inoculated with *Colletotrichum kahawae*. BMC Res Notes 6:388

Florez JC, Mofatto LS, do Livramento Freitas-Lopes R, Ferreira SS, Zambolim EM, Carazzolle MF, Zambolim L, Caixeta ET (2017) High throughput transcriptome analysis of coffee reveals prehaustorial resistance in response to *Hemileia vastatrix* infection. Plant Mol Biol 95:607–623

Fortunato AS, Lidon FC, Batista-Santos P, Leitão AE, Pais IP, Ribeiro AI, Ramalho JC (2010) Biochemical and molecular characterization of the antioxidative system of *Coffea* sp. under cold conditions in genotypes with contrasting tolerance. J Plant Physiol 167:333–342

Freire LP, Marraccini P, Rodrigues GC, Andrade AC (2013) Analysis of the mannose 6 phosphate reductase gene expression in coffee trees submitted to water deficit. Coffee Sci 8:17–23

Freitas NC, Barreto HG, Fernandes-Brum CN, Moreira RO, Chalfun-Junior A, Paiva LV (2017) Validation of reference genes for qPCR analysis of *Coffea arabica* L. somatic embryogenesis-related tissues. Plant Cell Tissue Organ Cult 128:663–678

Frischknecht PM, Ulmer-Dufek J, Baumann TW (1986) Purine alkaloid formation in buds and developing leaflets of *Coffea arabica*: expression of an optimal defense strategy? Phytochemistry 25:613–616

Gaborit C, Caillet V, Deshayes A, Marraccini P (2003) Molecular cloning of a full-length cDNA and gene from *Coffea arabica* encoding a protein homologous to the yeast translation initiation factor SUI1: expression analysis in plant organs. Braz J Plant Physiol 15:55–58

Gaitan A, Cristancho MA, Gongora CE, Moncada P, Posada H, Gast F, Yepes M, Aldwinckle H (2015) Long-read deep sequencing and assembly of the allotetraploid *Coffea arabica* cv. Caturra and its maternal ancestral diploid species *Coffea eugenioides*. In: Plant and animal genome XXIII. San Diego, CA (USA). https://pag.confex.com/pag/xxiii/webprogram/Paper17662.html

Ganesh D, Petitot A-S, Silva MC, Alary R, Lecouls A-C, Fernandez D (2006) Monitoring of the early molecular resistance responses of coffee (*Coffea arabica* L.) to the rust fungus (*Hemileia vastatrix*) using real-time quantitative RT-PCR. Plant Sci 170:1045–1051

Gaspari-Pezzopane C, Bonturi N, Guerreiro Filho O, Favarin JL, Maluf MP (2012) Gene expression profile during coffee fruit development and identification of candidate markers for phenological stages. Pesqui Agropecu Bras 47:972–982

Georget F, Marie L, Alpizar E, Courtel P, Bordeaux M, Hidalgo JM, Marraccini P, Breitler J-C, Déchamp E, Poncon C, Etienne H, Bertrand B (2019) Starmaya: the first Arabica F1 coffee hybrid produced using genetic male sterility. Front Plant Sci 10:1344

Geromel C, Ferreira LP, Guerreiro SMC, Cavalari AA, Pot D, Pereira LFP, Leroy T, Vieira LGE, Mazzafera P, Marraccini P (2006) Biochemical and genomic analysis of sucrose metabolism during coffee (*Coffea arabica*) fruit development. J Exp Bot 57:3243–3258

Geromel C, Ferreira LP, Bottcher A, Pot D, Pereira LFP, Leroy T, Vieira LGE, Mazzafera P, Marraccini P (2008a) Sucrose metabolism during fruit development in *Coffea racemosa*. Ann Appl Biol 152:179–187

Geromel C, Ferreira LP, Davrieux F, Guyot B, Ribeyre F, dos Santos Scholz MB, Pereira LFP, Vaast P, Pot D, Leroy T, Androcioli Filho A, Vieira LGE, Mazzafera P, Marraccini P (2008b) Effects of shade on the development and sugar metabolism of coffee fruits. Plant Physiol Biochem 46:569–579

Ghini R, Hamada E, Pedro Júnior MJ, Marengo JA, Gonçalves RRV (2008) Risk analysis of climate change on coffee nematodes and leaf miner in Brazil. Pesqui Agropecu Bras 43:187–194

Ghini R, Hamada E, Pedro Júnior MJ, Gonçalves RRV (2011) Incubation period *of Hemileia vastatrix* in coffee plants in Brazil simulated under climate change. Summa Phytopathol 37:85–93

Ghini R, Torre-Neto A, Dentzien AFM, Guerreiro-Filho O, Iost R, FRA P, JSM P, Thomaziello RA, Bettiol W, FM DM (2015) Coffee growth, pest and yield responses to free-air CO_2 enrichment. Clim Chang 132:307–320

Gichuru EK (1997) Resistance mechanisms in Arabica coffee to coffee berry disease *Colletotrichum kahawae* sp. nov-a review. Kenya Coffee 727:2441–2445

Gichuru EK (2007) Histological comparison of susceptible and resistant interactions of coffee (*Coffea arabica* and *C. canephora* varieties) and *Colletotrichum kahawae*. Agron Afr 19:233–240

González-Mendoza VM, Sánchez-Sandoval ME, Munnik T, Hernández-Sotomayor SMT (2020) Biochemical characterization of phospholipases C from *Coffea arabica* in response to aluminium stress. J Inorg Biochem 204:110951

Goulao LF, Fortunato AS, Ramalho JC (2012) Selection of reference genes for normalizing quantitative real-time PCR gene expression data with multiple variables in *Coffea* spp. Plant Mol Biol Rep 30:741–759

Grover CE, Gallagher JP, Szadkowski EP, Yoo M-J, Flagel LE, Wendel JF (2012) Homoeolog expression bias and expression level dominance in allopolyploids. New Phytol 196:966–971

Guan Q, Lu X, Zeng H, Zhang Y, Zhu J (2013) Heat stress induction of miR398 triggers a regulatory loop that is critical for thermotolerance in Arabidopsis. Plant J 74:840–851

Guerreiro-Filho O (2006) Coffee leaf miner resistance. Braz J Plant Physiol 18:109–117

Guerreiro-Filho O, Mazzafera P (2000) Caffeine does not protect coffee against the leaf miner *Perileucoptera coffeella*. J Chem Ecol 26:1447–1464

Guerreiro-Filho O, Silvarolla MB, Eskes AB (1999) Expression and mode of inheritance of resistance in coffee to leaf miner *Perileucoptera coffeella*. Euphytica 105:7–15

Haile M, Kang WH (2018) Transcriptome profiling of the coffee (*C. arabica* L.) seedlings under salt stress condition. J Plant Biotechnol 45:45–54

Hajyzadeh M, Turktas M, Khawar KM, Unver T (2015) miR408 overexpression causes increased drought tolerance in chickpea. Gene 555:186–193

Hinniger C, Caillet V, Michoux F, Ben Amor M, Tanksley S, Lin C, McCarthy J (2006) Isolation and characterization of cDNA encoding three dehydrins expressed during *Coffea canephora* (Robusta) grain development. Ann Bot 97:755–765

ICO (2020) Coffee production by exporting countries (data at July 2020). www.ico.org/prices/po-production.pdf

Idárraga SM, Castro AM, Macea EP, Gaitán AL, Rivera LF, Cristancho MA, Góngora CE (2012) Sequences and transcriptional analysis of *Coffea arabica* var. Caturra and *Coffea liberica* plant responses to coffee berry borer *Hypothenemus hampei* (Coleoptera: Curculionidae: Scolytinae) attack. J Plant Interact 7:56–70

IPCC (2013) Climate change 2013: the physical science basis. In: Stocker TF, Qin D, Plattner G-K, Tignor M, Allen SK, Boschung J, Nauels A, Xia Y, Bex V, Midgley PM (eds) Contribution of working group i to the fifth assessment report of the intergovernmental panel on climate change. Cambridge University Press, Cambridge, New York, p 1535

Ivamoto ST, Reis O Jr, Domingues DS, dos Santos TB, de Oliveira FF, Pot D, Leroy T, Vieira LGE, Carazzolle MF, Pereira GAG, Pereira LFP (2017a) Transcriptome analysis of leaves, flowers and fruits perisperm of *Coffea arabica* L. reveals the differential expression of genes involved in raffinose biosynthesis. PLoS One 12:e0169595

Ivamoto ST, Sakuray LM, Ferreira LP, Kitzberger CSG, Scholz MBS, Pot D, Leroy T, Vieira LGE, Domingues DS, Pereira LFP (2017b) Diterpenes biochemical profile and transcriptional analysis of cytochrome P450s genes in leaves, roots, flowers, and during *Coffea arabica* L. fruit development. Plant Physiol Biochem 111:340–347

Jaramillo J, Muchugu E, Vega FE, Davis A, Borgemeister C, Chabi-Olaye A (2011) Some like it hot: the influence and implications of climate change on coffee berry borer (*Hypothenemus hampei*) and coffee production in East Africa. PLoS One 6:e24528

Jiménez-Guillen D, Pérez-Pascual D, Souza-Perera R, Godoy-Hernández G, Zúñiga-Aguilar JJ (2018) Cloning of the *Coffea canephora SERK1* promoter and its molecular analysis during the cell-to embryo transition. Electron J Biotechnol 36:34–46

Joët T, Laffargue A, Salmona J, Doulbeau S, Descroix F, Bertrand B, de Kochko A, Dussert S (2009) Metabolic pathways in tropical dicotyledonous albuminous seeds: *Coffea arabica* as a case study. New Phytol 182:146–162

Joët T, Salmona J, Laffargue A, Descroix F, Dussert S (2010) Use of the growing environment as a source of variation to identify the quantitative trait transcripts and modules of co-expressed genes that determine chlorogenic acid accumulation. Plant Cell Environ 33:1220–1233

Joët T, Pot D, Ferreira LP, Dussert S, Marraccini P (2012) Identification des déterminants moléculaires de la qualité du café par des approches de génomique fonctionnelle. Une revue. Cah Agric 21:125–133

Joët T, Laffargue A, Salmona J, Doulbeau S, Descroix F, Bertrand B, Lashermes P, Dussert S (2014) Regulation of galactomannan biosynthesis in coffee seeds. J Exp Bot 65:323–337

Kakegawa H, Shitan N, Kusano H, Ogita S, Yazaki K, Sugiyama A (2019) Uptake of adenine by purine permeases of *Coffea canephora*. Biosci Biotechnol Biochem 83:1300–1305

Kerepesi I, Galiba G (2000) Osmotic and salt stress induced alteration in soluble carbohydrate content in wheat seedlings. Crop Sci 40:482–487

Khraiwesh B, Zhu JK, Zhu J (2012) Role of miRNAs and siRNAs in biotic and abiotic stress responses of plants. Biochim Biophys Acta 1819:137–148

Kiba A, Saitoh H, Nishihara M, Omiya K, Yamamura S (2003) C-terminal domain of a hevein-like protein from *Wasabia japonica* has potent antimicrobial activity. Plant Cell Physiol 44:296–303

Klingler JP, Batelli G, Zhu JK (2010) ABA receptors: the START of a new paradigm in phytohormone signalling. J Exp Bot 61:3199–3210

Koshiro Y, Zheng XQ, Wang ML, Nagai C, Ashihara H (2006) Changes in content and biosynthetic activity of caffeine and trigonelline during growth and ripening of *Coffea arabica* and *Coffea canephora* fruits. Plant Sci 171:242–250

Kramer D, Breitenstein B, Kleinwächter M, Selmar D (2010) Stress metabolism in green coffee beans (*Coffea arabica* L.): expression of dehydrins and accumulation of GABA during drying. Plant Cell Physiol 51:546–553

Kumar A, Giridhar P (2015) Salicylic acid and methyl jasmonate restore the transcription of caffeine biosynthetic N-methyltransferases from a transcription inhibition noticed during late endosperm maturation in coffee. Plant Gene 4:38–44

Kumar A, Naik GK, Giridhar P (2017) Dataset on exogenous application of salicylic acid and methyl jasmonate and the accumulation of caffeine in young leaf tissues and catabolically inactive endosperms. Data Brief 13:22–27

Kutywayo D, Chemura A, Kusena W, Chidoko P, Mahoya C (2013) The impact of climate change on the potential distribution of agricultural pests: the case of the coffee white stem borer (*Monochamus leuconotus* P.) in Zimbabwe. PLoS One 8:e73432

Ky CL, Louarn J, Dussert S, Guyot B, Hamon S, Noirot M (2001) Caffeine, trigonelline, chlorogenic acids and sucrose diversity in wild *Coffea arabica* L. and *C. canephora* P. accessions. Food Chem 75:223–230

Lashermes P, Couturon E, Moreau N, Paillard M, Louarn J (1996) Inheritance and genetic mapping of self-incompatibility in *Coffea canephora* Pierre. Theor Appl Genet 93:458–462

Lashermes P, Combes MC, Robert J, Trouslot P, D'Hont A, Anthony F, Charrier A (1999) Molecular characterisation and origin of the *Coffea arabica* L. genome. Mol Gen Genet 261:259–266

Lashermes P, Andrade AC, Etienne H (2008) Genomics of coffee, one of the world's largest traded commodities. In: Moore H, Ming R (eds) Genomics of tropical crop plants. Springer, Berlin, pp 203–226

Lashermes P, Hueber Y, Combes M-C, Severac D, Dereeper A (2016) Inter-genomic DNA exchanges and homeologous gene silencing shaped the nascent allopolyploid coffee genome (*Coffea arabica* L.). G3 Genes Genom Genet 6:2937–2948

Lepelley M, Cheminade G, Tremillon N, Simkin A, Caillet V, McCarthy J (2007) Chlorogenic acid synthesis in coffee: an analysis of CGA content and real-time RT-PCR expression of *HCT*,

HQT, *C3H1*, and *CCoAOMT1* genes during grain development in *C. canephora*. Plant Sci 172:978–996

Lepelley M, Ben Amor M, Martineau N, Cheminade G, Caillet V, McCarthy J (2012a) Coffee cysteine proteinases and related inhibitors with high expression during grain maturation and germination. BMC Plant Biol 12:31

Lepelley M, Mahesh V, McCarthy J, Rigoreau M, Crouzillat D, Chabrillange N, de Kochko A, Campa C (2012b) Characterization, high-resolution mapping and differential expression of three homologous PAL genes in *Coffea canephora* Pierre (Rubiaceae). Planta 236:313–326

Leroy T, Ribeyre F, Bertrand B, Charmetant P, Dufour M, Montagnon C, Marraccini P, Pot D (2006) Genetics of coffee quality. Braz J Plant Physiol 18:229–242

Li B, Kadura I, Fu DJ, Watson DE (2004) Genotyping with TaqMAMA. Genomics 83:311–320

Lima EA (2015) Resistência múltipla do cafeeiro *Coffea canephora* população Conilon a *Meloidogyne* spp.: mecanismos e possíveis genes envolvidos. PhD dissertation, Federal University of Brasilia, Brazil, p 148

Lima EA, Carneiro FA, Costa TS, Rêgo ECS, Jorge A Jr, Furlanetto C, Marraccini P, Carneiro RMDG, Andrade AC (2014) Molecular characterization of resistance responses of *C. canephora* 'Clone 14' upon infection by *Meloidogyne paranaensis*. J Nematol 46:194

Lima EA, Furlanetto C, Nicole M, Gomes ACMM, Almeida MRA, Jorge-Júnior A, Correa VR, Salgado SM, Ferrão MAG, Carneiro RMDG (2015) The multi-resistant reaction of drought-tolerant coffee 'Conilon clone 14' to *Meloidogyne* spp. and late hypersensitive-like response in *Coffea canephora*. Phytopathology 105:805–814

Lin C, Mueller LA, Mc Carthy J, Crouzillat D, Pétiard V, Tanksley S (2005) Coffee and tomato share common gene repertoires as revealed by deep sequencing of seed and cherry transcripts. Theor Appl Genet 112:114–130

Loss-Morais G, Ferreira DCR, Margis R, Alves-Ferreira M, Corrêa RL (2014) Identification of novel and conserved microRNAs in *Coffea canephora* and *Coffea arabica*. Genet Mol Biol 37:671–682

Loureiro A, Nicole MR, Várzea V, Moncada P, Bertrand B, Silva MC (2012) Coffee resistance to *Colletotrichum kahawae* is associated with lignification, accumulation of phenols and cell death at infection sites. Physiol Mol Plant Pathol 77:23–32

Ma C, Burd S, Lers A (2015) miR408 is involved in abiotic stress responses in Arabidopsis. Plant J 84:169–187

Magalhães STV, Fernandes FL, Demuner AJ, Picanço MC, Guedes RNC (2010) Leaf alkaloids, phenolics, and coffee resistance to the leaf miner *Leucoptera coffeella* (Lepidoptera: Lyonetiidae). J Econ Entomol 103:1438–1443

Magrach A, Ghazoul J (2015) Climate and pest-driven geographic shifts in global coffee production: implications for forest cover, biodiversity and carbon storage. PLoS One 10:e0133071

Maluf MP, da Silva CC, de Oliveira MP, Tavares AG, Silvarolla MB, Filho OG (2009) Altered expression of the caffeine synthase gene in a naturally caffeine-free mutant of *Coffea arabica*. Genet Mol Biol 32:802–810

Marraccini P, Deshayes A, Pétiard V, Rogers WJ (1999) Molecular cloning of the complete 11S seed storage protein gene of *Coffea arabica* and promoter analysis in transgenic tobacco plants. Plant Physiol Biochem 37:273–282

Marraccini P, Rogers WJ, Allard C, André ML, Caillet V, Lacoste N, Lausanne F, Michaux S (2001) Molecular and biochemical characterization of endo-β-mannanases from germinating coffee (*Coffea arabica*) grains. Planta 213:296–308

Marraccini P, Courjault C, Caillet V, Lausanne F, Lepage B, Rogers WJ, Tessereau S, Deshayes A (2003) Rubisco small subunit of *Coffea arabica*: cDNA sequence, gene cloning and promoter analysis in transgenic tobacco plants. Plant Physiol Biochem 41:17–25

Marraccini P, Rogers WJ, Caillet V, Deshayes A, Granato D, Lausanne F, Lechat S, Pridmore D, Pétiard V (2005) Biochemical and molecular characterization of α-D-galactosidase from coffee beans. Plant Physiol Biochem 43:909–920

Marraccini P, Freire LP, Alves GSC, Vieira NG, Vinecky F, Elbelt S, Ramos HJO, Montagnon C, Vieira LGE, Leroy T, Pot D, Silva VA, Rodrigues GC, Andrade AC (2011) *RBCS1* expression in coffee: *Coffea* orthologs, *Coffea arabica* homeologs, and expression variability between genotypes and under drought stress. BMC Plant Biol 11:85

Marraccini P, Vinecky F, Alves GSC, Ramos HJO, Elbelt S, Vieira NG, Carneiro FA, Sujii PS, Alekcevetch JC, Silva VA, DaMatta FM, Ferrão MAG, Leroy T, Pot D, Vieira LGE, da Silva FR, Andrade AC (2012) Differentially expressed genes and proteins upon drought acclimation in tolerant and sensitive genotypes of *Coffea canephora*. J Exp Bot 63:4191–4212

Martinez M, Abraham Z, Gambardella M, Echaide M, Carbonero P, Diaz I (2005) The strawberry gene *Cyf1* encodes a phytocystatin with antifungal properties. J Exp Bot 56:1821–1829

Martins LD, Tomaz MA, Lidon FC, DaMatta FM, Ramalho JC (2014) Combined effects of elevated [CO_2] and high temperature on leaf mineral balance in *Coffea* spp. plants. Clim Chang 126:365–379

Martins MQ, Rodrigues WP, Fortunato AS, Leitão AE, Rodrigues AP, Pais IP, Martins LD, Silva MJ, Reboredo FH, Partelli FL, Campostrini E, Tomaz MA, Scotti-Campos P, Ribeiro-Barros AI, Lidon FJC, DaMatta FM, Ramalho JC (2016) Protective response mechanisms to heat stress in interaction with high [CO_2] conditions in *Coffea* spp. Front Plant Sci 7:947

Martins MQ, Fortunato AS, Rodrigues WP, Partelli FL, Campostrini E, Lidon FC, DaMatta FM, Ramalho JC, Ribeiro-Barros AI (2017) Selection and validation of reference genes for accurate RT-qPCR data normalization in *Coffea* spp. under a climate changes context of interacting elevated [CO_2] and temperature. Front Plant Sci 8:307

Matiello JB, de Almeida SR, da Silva MB, Ferreira IB, de Carvalho CHS (2015) Siriema AS1, coffee cultivar with resistance to leaf rust and leaf miner. In: IX Simpósio de Pesquisa dos Cafés do Brasil, Curitiba, Brazil, p 190

Mazzafera P, Eskes AB, Parvais JP, Carvalho A (1990) Male sterility detected in *Coffea arabica* and *C. canephora* in Brazil. In: 10st international conference on coffee science (ASIC), Paipa, Colombia, pp 466–473

Menezes-Silva PE, Sanglard LMVP, Ávila RT, Morais LE, Martins SCV, Nobres P, Patreze CM, Ferreira MA, Araújo WL, Fernie AR, DaMatta FM (2017) Photosynthetic and metabolic acclimation to repeated drought events play key roles in drought tolerance in coffee. J Exp Bot 68:4309–4322

Miniussi M, Del Terra L, Savi T, Pallavicini A, Nardini A (2015) Aquaporins in *Coffea arabica* L.: identification, expression, and impacts on plant water relations and hydraulics. Plant Physiol Biochem 95:92–102

Mizuno K, Kato M, Irino F, Yoneyama N, Fujimura T, Ashihara H (2003a) The first committed step reaction of caffeine biosynthesis: 7-methylxanthosine synthase is closely homologous to caffeine synthases in coffee (*Coffea arabica* L.). FEBS Lett 547:56–60

Mizuno K, Kuda A, Kato M, Yoneyama N, Tanaka H, Ashihara H, Fujimura T (2003b) Isolation of a new dual-functional caffeine synthase gene encoding an enzyme for the conversion of 7-methylxanthine to caffeine from coffee (*Coffea arabica* L.). FEBS Lett 534:75–81

Mizuno K, Matsuzaki M, Kanazawa S, Tokiwano T, Yoshizawa Y, Kato M (2014) Conversion of nicotinic acid to trigonelline is catalyzed by *N*-methyltransferase belonged to motif B′ methyltransferase family in *Coffea arabica*. Biochem Biophys Res Commun 452:1060–1066

Moat J, Williams J, Baena S, Wilkinson T, Gole TW, Challa ZK, Demissew S, Davis AP (2017) Resilience potential of the Ethiopian coffee sector under climate change. Nat Plants 19:17081

Moat J, Gole TW, Davis AP (2019) Least concern to endangered: applying climate change projections profoundly influences the extinction risk assessment for wild Arabica coffee. Glob Chang Biol 25:390–403

Mofatto LS, Carneiro FA, Vieira NG, Duarte KE, Vidal RO, Alekcevetch JC, Guitton MG, Verdeil J-L, Lapeyre-Montes F, Lartaud M, Leroy T, De Bellis F, Pot D, Rodrigues GC, Carazzolle MF, Pereira GAG, Andrade AC, Marraccini P (2016) Identification of candidate genes for drought tolerance in coffee by high-throughput sequencing in the shoot apex of different *Coffea arabica* cultivars. BMC Plant Biol 16:94

Moisyadi S, Stiles J (1995) A cDNA encoding a metallothionein I-like protein from coffee leaves (*Coffea arabica*). Plant Physiol 107:295–296

Mondego JMC, Guerreiro-Filho O, Bengtson MH, Drummond RD, Felix JM, Duarte MP, Ramiro D, Maluf MP, Sogayar MC, Menossi M (2005) Isolation and characterization of *Coffea* genes induced during coffee leaf miner (*Leucoptera coffeella*) infestation. Plant Sci 169:351–360

Mondego JMC, Vidal RO, Carazzolle MF, Tokuda EK, Parizzi LP, Costa GGL, Pereira LFP, Andrade AC, Colombo CA, Vieira LGE, Pereira GAG, Brazilian Coffee Genome Project Consortium (2011) An EST-based analysis identifies new genes and reveals distinctive gene expression features of *Coffea arabica* and *Coffea canephora*. BMC Plant Biol 11:30

Montoya G, Vuong H, Cristancho M, Moncada P, Yepes M (2007) Sequence analysis from leaves, flowers and fruits of *Coffea arabica* var. Caturra. In: 21st international conference on coffee science (ASIC), Montpellier, France (PB261), pp 717–721

Moraes MS, Teixeira AL, Ramalho AR, Espíndula MC, Ferrão MAG, Rocha RB (2018) Characterization of gametophytic self-incompatibility of superior clones of *Coffea canephora*. Genet Mol Res 17:gmr16039876

Morais H, Caramori PH, Koguishi MS, Ribeiro AMA (2008) Escala fenológica detalhada da fase reprodutiva de *Coffea arabica*. Bragantia 67:257–260

Morgante M, Scalabrin S, Scaglione D, Cattonaro F, Magni F, Jurman I, Cerutti M, Liverani FS, Navarini L, Del Terra L, Pellegrino G, Graziosi G, Vitulo N, Valle G (2015) Progress report on the sequencing and assembly of the allotetraploid *Coffea arabica* var. Bourbon genome. In: Plant and animal genome XXIII. San Diego, CA (USA). https://pag.confex.com/pag/xxiii/webprogram/Paper16802.html

Nellikunnumal SM, Chandrashekar A (2012) Computational identification of conserved microRNA and their targets in *Coffea canephora* by EST analysis. Dyn Biochem Process Biotech Mol Biol 6:70–76

Nguyen Dinh S, Sai TZT, Nawaz G, Lee K, Kang H (2016) Abiotic stresses affect differently the intron splicing and expression of chloroplast genes in coffee plants (*Coffea arabica*) and rice (*Oryza sativa*). J Plant Physiol 201:85–94

Nic-Can GI, López-Torres A, Barredo-Pool F, Wrobel K, Loyola-Vargas VM, Rojas-Herrera R, De-la-Peña C (2013) New insights into somatic embryogenesis: *leafy cotyledon1*, *baby boom1* and *WUSCHEL-related homeobox4* are epigenetically regulated in *Coffea canephora*. PLoS One 8:e72160

Nobile PM, Quecini V, Bazzo B, Quiterio G, Mazzafera P, Colombo CA (2010) Transcriptional profile of genes involved in the biosynthesis of phytate and ferritin in *Coffea*. J Agric Food Chem 58:3479–3487

Nobres P, Patreze CM, Waltenberg FP, Correa MF, Tavano ECR, Mendes BMJ, Alves-Ferreira M (2016) Characterization of the promoter of the homeobox gene *CaHB12* in *Coffea arabica*. Trop Plant Biol 9:50–62

Nowak MD, Davis AP, Anthony F, Yoder AD (2011) Expression and trans-specific polymorphism of self-incompatibility RNases in *Coffea* (Rubiaceae). PLoS One 6:e21019

Ogawa M, Herai Y, Koizumi N, Kusano T, Sano H (2001) 7-methylxanthine methyltransferase of coffee plants gene isolation and enzyme properties. J Biol Chem 276:8213–8218

Omondi CO, Agwanda CO, Gichuru EK (2004) Field expression of resistance to coffee berry disease (CBD) as affected by environmental and host-pathogen factors. In: 20th international conference on coffee science (ASIC), Bangalore, India

Ovalle-Rivera O, Läderach P, Bunn C, Obersteiner M, Schroth G (2015) Projected shifts in *Coffea arabica* suitability among major global producing regions due to climate change. PLoS One 10: e0124155

Park HC, Kim ML, Kang YH, Jeon JM, Yoo JH, Kim MC, Park CY, Jeong JC, Moon BC, Lee JH, Yoon HW, Lee S-H, Chung WS, Lim CO, Lee SY, Hong JC, Cho MJ (2004) Pathogen-and NaCl-induced expression of the *SCaM-4* promoter is mediated in part by a GT-1 box that interacts with a GT-1-like transcription factor. Plant Physiol 135:2150–2161

Pereira LFP, Galvão RM, Kobayashi AK, Cação SMB, Vieira LGE (2005) Ethylene production and acc oxidase gene expression during fruit ripening of *Coffea arabica* L. Braz J Plant Physiol 17:283–289

Pérez-Pascual D, Jiménez-Guillen D, Villanueva-Alonzo H, Souza-Perera R, Godoy-Hernández G, Zúñiga-Aguilar JJ (2018) Ectopic expression of the *Coffea canephora* SERK1 homolog-induced differential transcription of genes involved in auxin metabolism and in the developmental control of embryogenesis. Physiol Plant 163:530–551

Perrois C, Strickler SR, Mathieu G, Lepelley M, Bedon L, Michaux S, Husson J, Mueller L, Privat I (2015) Differential regulation of caffeine metabolism in *Coffea arabica* (Arabica) and *Coffea canephora* (Robusta). Planta 241:179–191

Petitot A-S, Lecouls AC, Fernandez D (2008) Sub-genomic origin and regulation patterns of a duplicated *WRKY* gene in the allotetraploid species *Coffea arabica*. Tree Genet Genomes 3:379–390

Petitot A-S, Barsalobres-Cavallari C, Ramiro D, Albuquerque Freire E, Etienne H, Fernandez D (2013) Promoter analysis of the WRKY transcription factors *CaWRKY1a* and *CaWRKY1b* homoeologous genes in coffee (*Coffea arabica*). Plant Cell Rep 32:1263–1276

Pezzopane JRM, Pedro Júnior MJ, Thomaziello RA, Camargo MBP (2003) Escala para avaliação de estádios fenológicos do cafeeiro Arábica. Bragantia 62:499–505

Pinto RT, Freitas NC, Máximo WPF, Cardoso TB, Prudente DO, Paiva LV (2019) Genome-wide analysis, transcription factor network approach and gene expression profile of *GH3* genes over early somatic embryogenesis in *Coffea* spp. BMC Genomics 20:812

Poncet V, Rondeau M, Tranchant C, Cayrel A, Hamon S, de Kochko A, Hamon P (2006) SSR mining in coffee tree EST databases: potential use of EST–SSRs as markers for the *Coffea* genus. Mol Gen Genomics 276:436–449

Pré M, Caillet V, Sobilo J, McCarthy J (2008) Characterization and expression analysis of genes directing galactomannan synthesis in coffee. Ann Bot 102:207–220

Privat I, Foucrier S, Prins A, Epalle T, Eychenne M, Kandalaft L, Caillet V, Lin C, Tanksley S, Foyer C, McCarthy J (2008) Differential regulation of grain sucrose accumulation and metabolism in *Coffea arabica* (Arabica) and *Coffea canephora* (Robusta) revealed through gene expression and enzyme activity analysis. New Phytol 178:781–797

Privat I, Bardil A, Gomez AB, Severac D, Dantec C, Fuentes I, Mueller L, Joët T, Pot D, Foucrier S, Dussert S, Leroy T, Journot L, De Kochko A, Campa C, Combes M-C, Lashermes P, Bertrand B (2011) The 'Puce Cafe' project: the first 15K coffee microarray, a new tool for discovering candidate genes correlated to agronomic and quality traits. BMC Genomics 12:5

Puranik S, Sahu PP, Srivastava PS, Prasad M (2012) NAC proteins: regulation and role in stress tolerance. Trends Plant Sci 17:369–381

Quintana-Escobar AO, Nic-Can GI, Galaz Avalos RM, Loyola-Vargas VM, Gongora-Castillo E (2019) Transcriptome analysis of the induction of somatic embryogenesis in *Coffea canephora* and the participation of ARF and Aux/IAA genes. PeerJ 7:e7752

Ramalho JC, Rodrigues AP, Semedo JN, Pais IP, Martins LD, Simões-Costa MC, Leitão AE, Fortunato AS, Batista-Santos P, Palos IM, Tomaz MA, Scotti-Campos P, Lidon FC, DaMatta FM (2013) Sustained photosynthetic performance of *Coffea* spp. under long-term enhanced [CO_2]. PLoS One 8:e82712

Ramalho JC, Rodrigues AP, Lidon FC, Marques LMC, Leitão AE, Fortunato AS, Pais IP, Silva MJ, Scotti-Campos P, Lopes A, Reboredo FH, Ribeiro-Barros AI (2018a) Stress cross-response of the antioxidative system promoted by superimposed drought and cold conditions in *Coffea* spp. PLoS One 13:e0198694

Ramalho JC, Pais IP, Leitão AE, Guerra M, Reboredo FH, Máguas CM, Carvalho ML, Scotti-Campos P, Ribeiro-Barros AI, Lidon FJC, DaMatta FM (2018b) Can elevated air [CO_2] conditions mitigate the predicted warming impact on the quality of coffee bean? Front Plant Sci 9:287

Ramiro D, Jalloul A, Petitot A-S, Grossi de Sá MF, Maluf MP, Fernandez D (2010) Identification of coffee WRKY transcription factor genes and expression profiling in resistance responses to pathogens. Tree Genet Genomes 6:767–781

Rebijith KB, Asokan R, Ranjitha HH, Krishna V, Nirmalbabu K (2013) *In silico* mining of novel microRNAs from coffee (*Coffea arabica*) using expressed sequence tags. J Hortic Sci Biotechnol 88:325–337

Reymond P, Weber H, Damond M, Farmer EE (2000) Differential gene expression in response to mechanical wounding and insect feeding in *Arabidopsis*. Plant Cell 12:707–720

Rodrigues WP, Martins MQ, Fortunato AS, Rodrigues AP, Semedo JN, Simões-Costa MC, Pais IP, Leitão AE, Colwell F, Goulao L, Máguas C, Maia R, Partelli FL, Campostrini E, Scotti-Campos P, Ribeiro-Barros AI, Lidon FC, DaMatta FM, Ramalho JC (2016) Long-term elevated air [CO_2] strengthens photosynthetic functioning and mitigates the impact of supra-optimal temperatures in tropical *Coffea arabica* and *C. canephora* species. Glob Chang Biol 22:415–431

Rogers WJ, Bézard G, Deshayes A, Meyer I, Pétiard V, Marraccini P (1999a) Biochemical and molecular characterization and expression of the 11S-type storage protein from *Coffea arabica* endosperm. Plant Physiol Biochem 37:261–272

Rogers WJ, Michaux S, Bastin M, Bucheli P (1999b) Changes to the content of sugars, sugar alcohols, myo-inositol, carboxylic acids, and inorganic anions in developing grains from different varieties of Robusta (*Coffea canephora*) and Arabica (*C. arabica*) coffees. Plant Sci 149:115–123

Romero JV, Cortina H (2004) Fecundidad y ciclo de vida de la coffee berry borer *Hypothenemus hampei* F. (Coleoptera: Curculionidae: Scolytinae) en introducciones silvestres de café. Rev Cenicafé 55:221–231

Romero JV, Cortina H (2007) Tablas de vida de *Hypothenemus hampei* (Coleoptera: Curculionidae: Scolytinae) sobre tres introducciones de café. Rev Colomb Entomol 33:10–16

Rushton PJ, Somssich IE, Ringler P, Shen QJ (2010) WRKY transcription factors. Trends Plant Sci 15:247–258

Ságio SA, Barreto HG, Lima AA, Moreira RO, Rezende PM, Paiva LV, Chalfun-Junior A (2014) Identification and expression analysis of ethylene biosynthesis and signaling genes provides insights into the early and late coffee cultivars ripening pathway. Planta 239:951–963

Salmona J, Dussert S, Descroix F, de Kochko A, Bertrand B, Joët T (2008) Deciphering transcriptional networks that govern *Coffea arabica* seed development using combined cDNA array and real-time RT-PCR approaches. Plant Mol Biol 66:105–124

Samson N, Bausher MG, Lee SB, Jansen RK, Daniell H (2007) The complete nucleotide sequence of the coffee (*Coffea arabica* L.) chloroplast genome: organization and implications for biotechnology and phylogenetic relationships amongst angiosperms. Plant Biotechnol J 5:339–353

Sant'Ana GC, Pereira LFP, Pot D, Ivamoto ST, Domingues DS, Ferreira RV, Pagiatto NF, da Silva BSR, Nogueira LM, Kitzberger CSG, Scholz MBS, de Oliveira FF, Sera GH, Padilha L, Labouisse JP, Guyot R, Charmetant P, Leroy T (2018) Genome-wide association study reveals candidate genes influencing lipids and diterpenes contents in *Coffea arabica* L. Sci Rep 8:465

Santos AB, Mazzafera P (2012) Dehydrins are highly expressed in water-stressed plants of two coffee species. Trop Plant Biol 5:218–232

Santos GC, Von Pinho EVR, Rosa SDVF (2013) Gene expression of coffee seed oxidation and germination processes during drying. Genet Mol Res 12:6968–6982

Santos JO, Santos MO, Torres LF, Matos CSM, Botelho A, Andrade AC, Carvalho GR, Silva VA (2019) Analysis of *APX* and *CaPYL8A* genes expression in *Coffea arabica* progenies under drought. In: X Simpósio de Pesquisa dos Cafés do Brasil, Vitoria, Brazil

Satyanarayana KV, Kumar V, Chandrashekar A, Ravishankar GA (2005) Isolation of promoter for N-methyltransferase gene associated with caffeine biosynthesis in *Coffea canephora*. J Biotechnol 119:20–25

Schluter U, Benchabane M, Munger A, Kiggundu A, Vorster J, Goulet MC, Cloutier C, Michaud D (2010) Recombinant protease inhibitors for herbivore pest control: a multitrophic perspective. J Exp Bot 61:4169–4183

Scotti-Campos P, Pais IP, Ribeiro-Barros AI, Martins LD, Tomaz MA, Rodrigues WP, Campostrini E, Semedo JN, Fortunato AS, Martins MQ, Partelli FL, Lidon FC, DaMatta FM, Ramalho JC (2019) Lipid profile adjustments may contribute to warming acclimation and to heat impact mitigation by elevated [CO_2] in *Coffea* spp. Environ Exp Bot 167:103856

Selmar D, Bytof G, Knopp SE, Breitenstein B (2006) Germination of coffee seeds and its significance for coffee quality. Plant Biol 8:260–264

Sera GH, Sera T, Ito DS, Filho CR, Villacorta A, Kanayama FS, Alegre CR, Del Grossi L (2010) Coffee berry borer resistance in coffee genotypes. Braz Arch Biol Technol 53:261–268

Severino FE, Brandalise M, Costa CS, Wilcken SRS, Maluf MP, Gonçalves W, Maia IG (2012) *CaPrx*, a *Coffea arabica* gene encoding a putative class III peroxidase induced by root-knot nematode infection. Plant Sci 191-192:35–42

Silva EA, Mazzafera P, Brunini O, Sakai E, Arruda FB, Mattoso LHC, Carvalho CRL, Pires RCM (2005) The influence of water management and environmental conditions on the chemical composition and beverage quality of coffee beans. Braz J Plant Physiol 17:229–238

Silva MC, Várzea V, Guerra-Guimarães L, Azinheira HG, Fernandez D, Petitot A-S, Bertrand B, Lashermes P, Nicole M (2006) Coffee resistance to the main diseases: leaf rust and coffee berry disease. Braz J Plant Physiol 18:119–147

Silva AT, Barduche D, do Livramento KG, Ligterink W, Paiva LV (2014) Characterization of a putative *Serk-Like* ortholog in embryogenic cell suspension cultures of *Coffea arabica* L. Plant Mol Biol Rep 32:176–184

Silva AT, Barduche D, do Livramento KG, Paiva LV (2015) A putative *BABY BOOM*-like gene (*CaBBM*) is expressed in embryogenic calli and embryogenic cell suspension culture of *Coffea arabica* L. In Vitro Cell Dev Biol Plant 51:93

Simkin AJ, Qian T, Caillet V, Michoux F, Ben Amor M, Lin C, Tanksley S, McCarthy J (2006) Oleosin gene family of *Coffea canephora*: quantitative expression analysis of five oleosin genes in developing and germinating coffee grain. J Plant Physiol 163:691–708

Simkin AJ, Moreau H, Kuntz M, Pagny G, Lin C, Tanksley S, McCarthy J (2008) An investigation of carotenoid biosynthesis in *Coffea canephora* and *Coffea arabica*. J Plant Physiol 165:1087–1106

Simkin AJ, Kuntz M, Moreau H, McCarthy J (2010) Carotenoid profiling and the expression of carotenoid biosynthetic genes in developing coffee grain. Plant Physiol Biochem 48:434–442

Singh R, Irikura B, Nagai C, Albert HH, Kumagai M, Paull RE, Moore PH, Wang M-L (2011) Characterization of prolyl oligopeptidase genes differentially expressed between two cultivars of *Coffea arabica* L. Trop Plant Biol 4:203–216

Sreedharan SP, Kumar A, Giridhar P (2018) Primer design and amplification efficiencies are crucial for reliability of quantitative PCR studies of caffeine biosynthetic N-methyltransferases in coffee. 3. Biotech 8:467

Thioune EH, McCarthy J, Gallagher T, Osborne B (2017) A humidity shock leads to rapid, temperature dependent changes in coffee leaf physiology and gene expression. Tree Physiol 37:367–379

Tiski I, Marraccini P, Pot D, Vieira LGE, Pereira LFP (2011) Characterization and expression of two cDNA encoding 3-hydroxy-3-methylglutaryl coenzyme A reductase isoforms in coffee (*Coffea arabica* L.). OMICS 15:719–727

Toniutti L, Breitler J-C, Etienne H, Campa C, Doulbeau S, Urban L, Lambot C, Pinilla JH, Bertrand B (2017) Influence of environmental conditions and genetic background of *Arabica* coffee (*C. arabica L*) on leaf rust (*Hemileia vastatrix*) pathogenesis. Front. Plant Sci 8:2025

Toniutti L, Bordeaux M, Klein PE, Bertrand B, Montagnon C (2019a) Association study of tree size and male sterility in a F2 *Coffea arabica* population. In: 27th international conference on coffee science (ASIC), Portland (OR), USA

Toniutti L, Breitler J-C, Guittin C, Doulbeau S, Etienne H, Campa C, Lambot C, Herrera Pinilla JC, Bertrand B (2019b) An altered circadian clock coupled with a higher photosynthesis efficiency could explain the better agronomic performance of a new coffee clone when compared with a standard variety. Int J Mol Sci 20:736

Torres LF, Diniz LEC, Do Livramento KG, Freire LL, Paiva LV (2015) Gene expression and morphological characterization of cell suspensions of *Coffea arabica* L. cv. Catiguá MG2 in different cultivation stages. Acta Physiol Plant 37:175

Torres LF, Reichel T, Déchamp E, de Aquino SO, Duarte KE, Alves GSC, Silva AT, Cotta MG, Costa TS, Diniz LEC, Breitler J-C, Collin M, Paiva LV, Andrade AC, Etienne H, Marraccini P (2019) Expression of *DREB*-like genes in *Coffea canephora* and *C. arabica* subjected to various types of abiotic stress. Trop Plant Biol 12:98–116

Tran HTM, Lee LS, Furtado A, Smyth H, Henry RJ (2016) Advances in genomics for the improvement of quality in coffee. J Sci Food Agric 96:3300–3312

Tran HTM, Ramaraj T, Furtado A, Lee LS, Henry RJ (2018) Use of a draft genome of coffee (*Coffea arabica*) to identify SNPs associated with caffeine content. Plant Biotechnol 16:1756–1766

Uefuji H, Ogita S, Yamaguchi Y, Koizumi N, Sano H (2003) Molecular cloning and functional characterization of three distinct N-methyltransferases involved in the caffeine biosynthetic pathway in coffee plants. Plant Physiol 132:372–380

Ülker B, Somssich IE (2004) WRKY transcription factors: from DNA binding towards biological functions. Curr Opin Plant Biol 7:491–498

Urwin P, Green J, Atkinson H (2003) Expression of a plant cystatin confers partial resistance to *Globodera*, full resistance is achieved by pyramiding a cystatin with natural resistance. Mol Breed 12:263–269

Valadez-González N, Colli-Mull JG, Brito-Argáez L, Muñoz-Sánchez JA, Zúñiga Aguilar JJ, Castaño E, Hernández-Sotomayor SMT (2007) Differential effect of aluminum on DNA synthesis and CDKA activity in two *Coffea arabica* cell lines. J Plant Growth Regul 26:69–77

Valeriano JC, Lima EA, de Aquino SO, Carneiro FA, Carneiro WWJ, Mattos NG, Carneiro RMDG, Andrade AC (2019) Allele specific expression in *Coffea canephora* associated with nematoides resistance. In: X Simpósio de Pesquisa dos Cafés do Brasil, Vitoria, Brazil

Van der Vossen HAM (1985) Coffee breeding and selection. In: Clifford MN, Wilson RC (eds) Coffee botany, biochemistry and production of beans and beverages. Croom Helm, London, pp 48–97

van der Vossen HAM, Walyaro DJ (2009) Additional evidence for oligogenic inheritance of durable host resistance to coffee berry disease (*Colletotrichum kahawae*) in arabica coffee (*Coffea arabica* L.). Euphytica 165:105–111

Van der Vossen H, Bertrand B, Charrier A (2015) Next generation variety development for sustainable production of arabica coffee (*Coffea arabica* L.): a review. Euphytica 204:243–256

Vasconcelos EAR, Santana CG, Godoy CV, Seixas CDS, Silva MS, Moreira LRS, Oliveira-Neto OB, Price D, Fitches E, Filho EXF, Mehta A, Gatehouse JA, Grossi-de-Sa MF (2011) A new chitinase-like xylanase inhibitor protein (XIP) from coffee (*Coffea arabica*) affects soybean Asian rust (*Phakopsora pachyrhizi*) spore germination. BMC Biotechnol 11:14

Vidal RO, Mondego JMC, Pot D, Ambrósio AB, Andrade AC, Pereira LFP, Colombo CA, Vieira LGE, Carazzolle MF, Pereira GAG (2010) A high-throughput data mining of single nucleotide polymorphisms in *Coffea* species expressed sequence tags suggests differential homeologous gene expression in the allotetraploid *Coffea arabica*. Plant Physiol 154:1053–1066

Vieira LGE, Andrade AC, Colombo CA, Moraes AAH, Metha A, Oliveira AC, Labate CA, Marino CL, Monteiro-Vitorello CB, Monte DC, Giglioti E, Kimura ET, Romano E, Kuramae EE, Lemos EGM, Almeida ERP, Jorge EC, Barros EVSA, da Silva FR, Vinecky F, Sawazaki HE, Dorry HFA, Carrer H, Abreu IN, Batista JAN, Teixeira JB, Kitajima JP, Xavier KG, Lima LM, Camargo LEA, Pereira LFP, Coutinho LL, Lemos MVF, Romano MR, Machado MA, Costa MMC, Grossi de Sá MF, Goldman MHS, Ferro MIT, Tinoco MLP, Oliveira MC, Sluys MAV, Shimizu MS, Maluf MP, Eira MTS, Guerreiro Filho O, Arruda P, Mazzafera P, Mariani PDSC,

Oliveira RL, Harakava R, Balbao SF, Tsai SM, Mauro SMZ, Santos SN, Siqueira WJ, Costa GGL, Formighieri EF, Carazzolle MF, Pereira GAG (2006) Brazilian coffee genome project: an EST-based genomic resource. Braz J Plant Physiol 18:95–108

Vieira NG, Carneiro FA, Sujii PS, Alekcevetch JC, Freire LP, Vinecky F, Elbelt S, Silva VA, DaMatta FM, Ferrão MAG, Marraccini P, Andrade AC (2013) Different molecular mechanisms account for drought tolerance in *Coffea canephora* var. Conilon. Trop Plant Biol 6:181–190

Vieira NG, Ferrari IF, Rezende JC, Mayer JLS, Mondego JMC (2019) Homeologous regulation of *Frigida-like* genes provides insights on reproductive development and somatic embryogenesis in the allotetraploid *Coffea arabica*. Sci Rep 9:8446

Vinecky F, da Silva FR, Andrade AC (2012) *In silico* analysis of cDNA libraries SH2 and SH3 for the identification of genes responsive to drought in coffee. Coffee Sci 7:1–19

Vinecky F, Davrieux F, Mera AC, Alves GSC, Lavagnini G, Leroy T, Bonnot F, Rocha OC, Bartholo GF, Guerra AF, Rodrigues GC, Marraccini P, Andrade AC (2017) Controlled irrigation and nitrogen, phosphorous and potassium fertilization affect the biochemical composition and quality of Arabica coffee beans. J Agric Sci 155:902–918

Walyaro DJ (1983) Considerations in breeding for improved yield and quality in arabica coffee (Coffea arabica L.). Thesis, University of Wageningen, Netherlands

Wang J, Tian L, Lee HS, Chen ZJ (2006) Nonadditive regulation of FRI and FLC loci mediates flowering-time variation in Arabidopsis allopolyploids. Genetics 173:965–974

Wang Z, Liu Q, Wang H, Zhang H, Xu X, Li C, Yang C (2016) Comprehensive analysis of trihelix genes and their expression under biotic and abiotic stresses in *Populus trichocarpa*. Sci Rep 6:36274

Xie F, Xiao P, Chen D, Xu L, Zhang B (2012) miRDeepFinder: a miRNA analysis tool for deep sequencing of plant small RNAs. Plant Mol Biol 80:75–84

Yepes M, Gaitan A, Cristancho MA, Rivera LF, Correa JC, Maldonado CE, Gongora CE, Villegas MA, Posada H, Zimin A, Yorke JA, Aldwinckle H (2016) Building high quality reference genome assemblies using PACBio long reads for the allotetraploid *Coffea arabica* and its diploid ancestral maternal species *Coffea eugenioides*. In: Plant and animal genome XXIII. San Diego, CA (USA). https://pag.confex.com/pag/xxiv/webprogram/Paper22250.html

Yuyama PM, Reis Júnior O, Ivamoto ST, Domingues DS, Carazzolle MF, Pereira GAG, Charmetant P, Leroy T, Pereira LFP (2016) Transcriptome analysis in *Coffea eugenioides*, an Arabica coffee ancestor, reveals differentially expressed genes in leaves and fruits. Mol Gen Genomics 291:323–336

Zheng XQ, Ashihara H (2004) Distribution, biosynthesis and function of purine and pyridine alkaloids in *Coffea arabica* seedlings. Plant Sci 166:807–813

Zheng XQ, Nagai C, Ashihara H (2004) Pyridine nucleotide cycle and trigonelline (N-methylnicotinic acid) synthesis in developing leaves and fruits of *Coffea arabica*. Physiol Plant 122:404–411

Zhu A, Goldstein J (1994) Cloning and functional expression of a cDNA encoding coffee bean α-galactosidase. Gene 140:227–231

Zhu C, Ding Y, Liu H (2011) MiR398 and plant stress responses. Physiol Plant 143:1–9

Crosstalk Between the Sporophyte and the Gametophyte During Anther and Ovule Development in Angiosperms

Jorge Lora and José I. Hormaza

Contents

Abstract A proper development of the male and female germlines is key to the reproductive success of plants. As a result of the development of the male and female germlines the male (pollen) and female (embryo sac) gametophytes will be produced. After pollination, pollen–pistil interaction, fertilization, embryogenesis, and finally, the formation of the persistent propagule – the seed will take place. During reproductive cell development in angiosperms, male and female germlines develop inside the sporophytic tissues. The male germline develops in the anther surrounded by the tapetum whereas the female germline initiates in the nucellus composed of a single or several layers of somatic cells of the ovule. Initially, the cells that will remain somatic and those that will develop in the germlines are morphologically identical. But, later on, cell differentiation starts with the transition from somatic to reproductive fate and remarkable differences arise during germline development and mainly after meiosis. Such differences are also observed in the somatic cells that surround the germline and are closely linked to the crosstalk between the sporophyte tissues and the germline, a key process for the formation of the male and female gametes.

J. Lora (✉) and J. I. Hormaza
Department of Subtropical Fruit Crops, Instituto de Hortofruticultura Subtropical y Mediterránea "La Mayora" (IHSM-UMA-CSIC), Málaga, Spain
e-mail: jlora@eelm.csic.es; ihormaza@eelm.csic.es

© Springer Nature Switzerland AG 2020

Progress in Botany (2021) 82: 113–130, https://doi.org/10.1007/124_2020_50,
Published online: 28 December 2020

1 Introduction

Alternation of generations between multicellular gametophytes and sporophytes is present in all land plants (Hofmesiter 1851). However, along the evolutionary line, the gametophytic phase gets reduced in terms of both size and lifespan compared to the sporophytic phase (Heslop-Harrison 1979). Thus, while in bryophytes the gametophytic generation is the most prominent phase and the sporophyte is nutritionally dependent on the gametophyte (Maciel-Silva and Porto 2014), this reduction reaches its extreme in seed plants and, especially, in angiosperms, in which the gametophytic generation develops, and spends most of its life, enclosed within the tissues of the sporophyte. In flowering plants, the mature male gametophyte (pollen) is reduced to three cells, the vegetative cell and the two sperm cells, whereas the female gametophyte (embryo sac) is generally formed by seven cells (two synergid cells, the egg cell, the central cell and three antipodal cells embedded in the sporophytic tissues of the ovule). The formation of the male and the female gametophytes is an essential process for the success of double fertilization and the resulting fruit set in plants. After pollen germination on the stigma, the pollen tube elongates and delivers a pair of sperm cells to the female gametes in the embryo sac in which one sperm cell fertilizes the egg cell, forming the zygote, whereas the other sperm cell fuses with the two polar nuclei, forming a triploid cell that gives raise to the endosperm. This pollen mitosis II can occur during pollen development, releasing tricellular pollen at anther dehiscence, or after anthesis, releasing bicellular pollen. In the latter case, the generative cell divides mitotically in the pollen tube after pollen germination on the stigma.

Gametophytic development is a highly conserved process in angiosperms. In the early developmental stages, during sporogenesis in the micro and macrosporangia, a somatic cell becomes a reproductive cell that develops to be either the microspore mother cell (MiMC) or the megaspore mother cell (MeMC) that will undergo meiosis. Later on, the microspore in the anther and the functional megaspore in the ovule develop through several rounds of mitosis during gametogenesis to form the male and female gametes, respectively. Concomitantly, somatic cells develop surrounding the germline, and a selective communication takes place between the sporophytic and gametophytic tissues. Although the cells that will produce the two lineages are initially similar, after the transition from somatic to germline identity in the anther and the ovule, remarkable differences arise during germline development and mainly after meiosis. They are reflected not only by cellular changes, but also by molecular and genetic regulation. In the anther, the communication between the male germline and the somatic cells is generally regulated by tapetum degeneration. Meanwhile, the crosstalk between the female germline and the neighbouring somatic cells shows a more direct cell–cell contact in the ovule.

2 Microsporogenesis and Microgametogenesis

The male reproductive organ of the flowers is the stamen, formed by the anther at the top and the filament that provides the vascular bundles. In the early flower developmental stages, the primordial stamen is formed by an epidermis and a mass of meristematic cells. Generally, four anther lobes which will become the microsporangia arise during anther development, in each of which a hypodermal archesporial cell divides periclinally to give rise to an outer primary parietal cell and an inner primary sporogenous cell. While the primary parietal cell divides mitotically to produce the somatic cell layers of the anther wall, such as the tapetum, endothecium, and epidermis, the mitosis of the primary sporogenous cell results in MiMCs (McCormick 1993). Following the formation of the MiMCs, the male germline develops within the anther in a highly conserved process in angiosperms (Gómez et al. 2015; McCormick 1993). MiMCs increase in size and undergo two meiotic divisions to form a tetrad of haploid microspores. The unicellular microspores experience an asymmetric mitotic division (pollen mitosis I) that results in a bicellular pollen with a large vegetative cell that hosts a smaller generative cell. The generative cell undergoes a second mitotic division (pollen mitosis II) producing two sperm cells that will be involved in the characteristic double fertilization process of angiosperms.

The somatic cells that develop surrounding the male germline are organized in different layers. The tapetum is the cell layer that is in direct contact with the MiMC and, consequently, shows an essential crosstalk with the male germline during gametophyte development. There is a high diversity of tapetum types in angiosperms that is even more diverse when tapetum degeneration starts. The different tapetum types can be generally grouped into two categories: secretory (also known as parietal, glandular or cellular) and amoeboid (also known as invasive or "genuine" periplasmodial) (Pacini et al. 1985). The main difference between both types of tapetum is related to the amount of locular fluid, which is abundant in the secretory tapetum but is very reduced or even absent in the ameboid tapetum. This is directly related to the cell–cell communication between the tapetum and the male germline, through the locular fluid in the secretory tapetum or showing a more direct cell to cell contact in the ameboid tapetum (Pacini 1990; Pacini et al. 1985). The secretory tapetum found in 175 families is the most common in flowering plants and it is believed to be the most primitive (Franchi and Pacini 1993; Pacini et al. 1985). It is present in the model plant *Arabidopsis thaliana* (Quilichini et al. 2014) and important crops such as *Oryza sativa*, rice (Raghavan 1988) or *Solanum lycopersicum*, tomato (Polowick and Sawhney 1993). The parietal tapetum has only been observed in 32 families of flowering plants (Pacini et al. 1985).

The differentiation of the tapetum is observed after the transition from the somatic cells to the male germinal cells and is concomitant with the increase in size of the MiMC. Thus, the role of the tapetum at these early developmental stages is mainly involved in male germline development, rather than in male germline identity. To this end, the tapetum provides nutrients, regulatory molecules, and components of the pollen cell wall (Franchi and Pacini 1993). A critical step for male germline development is the meiosis, during which a selective cell–cell communication between the MiMC and the tapetum takes place. Indeed, after the increase in size of the MiMC and just before meiosis, abundant amounts of callose, a β-1,4 glucan that acts as a molecular filter (Tucker et al. 2001), accumulate around the MiMC and this presence continues until the release of the four microspores (Blackmore et al. 2007). Other components such as cellulose, pectins, and arabinogalactan proteins have also been observed in this special cell wall (Lora and Hormaza 2018). This special cell wall limits the deposition of molecules on the MiMC/microspore wall and this is more evident at the future aperture sites of the microspores in which a higher amount of callose prevents the formation of the pollen cell wall showing a thinner wall (Albert et al. 2011).

As a result of all these processes a pollen grain is formed. The pollen grain wall is composed of an inner pectin-cellulosic layer named intine and an outer layer named exine. Sporopollenin, a chemically very stable mixture of biopolymers, is the main component of the exine and is essential for the species-specific ornamentation of the exine. During meiosis, the formation of the intine starts, together with the deposition of the sporopollenin precursor (Blackmore et al. 2007). The formation of the exine is completed during pollen development in which the role of the tapetum is more crucial than its role on the formation of the intine (Lora and Hormaza 2018).

The crosstalk between the tapetum and the MiMC/microspore is also involved in the release of the resulting four sibling microspores and the consequent dispersal of pollen after anther dehiscence. While most of the angiosperms release single pollen grains, some species release pollen grains in groups, mostly in groups of four pollen grains (Harder and Johnson 2008; Walker and Doyle 1975). Studies in several species of *Annona* show a delay in tapetum degeneration that results in holding the four sibling microspores very close to each other in the tapetal chamber and, additionally, the digestion of the special cell wall composed of callose and cellulose is delayed at the aperture site. Therefore, the combination of both features allows cell wall connections among the four sibling microspores that result in the formation of microspore tetrads (Lora et al. 2009, 2014). Thus, while the special cell wall around the MiMC/microspores holds a strict cell–cell communication between the sporophyte tissue (the tapetum) and the male germline, after meiosis the communication restriction is less strict. This results in a more important role of the tapetum during the rest of the male germline development process, the microgametogenesis.

Concomitant to tapetum degeneration, the unicellular microspores increase in size. The carbohydrates released from the tapetum degeneration are absorbed by the microspores (Clement and Audran 1995; Pacini et al. 2006) that use them as energy resources for metabolism or to develop the microspore/pollen cell wall (Pacini and Franchi 1988). Indeed, degeneration of tapetal cells induced by programmed cell death appears to be well coordinated with pollen cell wall development (Shi et al. 2015). It is during this phase when the major development of the microspore/pollen cell wall in species with secretory tapetum (Echlin 1971) such as in *Annona* (Lora et al. 2009, 2014) is also observed. During this increase of size, the microspore shows vacuolization and the nucleus migrates before the first haploid mitosis. Thus, the unicellular microspore becomes the bicellular pollen grain. After meiosis, the male germline can generally undergo one or two waves of amylogenesis-amylolysis. In some species, amylogenesis occurs during microspore development and its degradation after meiosis. However, in other species, amylogenesis-amylolysis occurs during microspore development and can also be observed after the first haploid mitosis, during pollen development (Pacini et al. 2006; Pacini and Franchi 1988). The role of carbohydrates is highly related to the internal turgor pressure that is even more crucial at anther dehiscence when the pollen grain is exposed to the environment. Thus, we can find partially dehydrated or hydrated pollen at anther dehiscence, a trait that is also closely related to pollen viability and longevity (Nepi et al. 2001).

3 Megasporogenesis and Megagametogenesis

The female reproductive organ is the pistil that can be formed by one or several carpels. The carpel can be divided in three different parts, the stigma, the style, and the ovary. The stigma is the landing place of the pollen grain. The style connects the stigma to the ovary and provides ample opportunity for pollen competition and selection. The ovary contains one or more ovules that will be the place where the female germline will develop (Lora et al. 2016; Maheshwari 1950).

The formation of the female germline derives directly from somatic cells and is surrounded by the diploid generation of the ovule, the female sporophyte. The sporophytic tissue of the ovule consists of the proximal funiculus, the central chalaza and the distal nucellus usually surrounded by two integuments. The transition of somatic to reproductive cells takes place within the nucellus. These somatic cells expand and are ambiguously termed as archesporial cells. The archesporial cell becomes the MeMC that develops surrounded by a variable number of nucellar cell layers producing two main types of ovules. In tenuinucellate ovules, such as the ovules of the model plant *Arabidopsis thaliana* (Bajon et al. 1999), a single, or sometimes even none, epidermal cell is present, whereas in crassinucellate ovules the nucellus can consist of two or more cell layers. While tenuinucellate ovules are mainly found in evolutionary divergent angiosperms, crassinucellate ovules are usually observed in early divergent angiosperms and can be considered as the ancestral ovule type in angiosperms (Lora et al. 2017; Sporne 1969).

The MeMC undergoes meiosis and, in contrast to the situation in the anther in which the four meiotic products eventually become pollen grains, in the ovule, generally, only the functional megaspore continues its development to become the embryo sac during megagametogenesis. Although usually only one MeMC is observed in each ovule of angiosperms, multiple MeMCs have been reported in some cases, such as in *Trimenia moorei*, a species that belongs to the early divergent angiosperm clade Austrobaileyales (Bachelier and Friedman 2011). This feature, although not generally reported in most gymnosperms, can also be observed in

Gnetum that shows up to 12 female gametophytes; some of them degenerate and only about five female gametophytes divide meiotically (Takaso and Bouman 1986).

During megasporogenesis and megagametogenesis, the female germline is surrounded by somatic nucellar cells. Generally, in contrast to the fate of the tapetum of the anther, the nucellus does not degenerate, but, in some species such as some monocots (Rudall 1997) or *Arabidopsis* (Schneitz et al. 1995), part of the nucellus undergoes degeneration at anthesis. Concomitantly, studies in the tenuinucellate ovule of *Arabidopsis* show that the innermost cell layer of the inner integument differentiates in the endothelium (Schneitz et al. 1995) that is also referred to as integumentary tapetum because of its tissue origin and presumed functional similarities with the tapetum of the anther (Kapil and Tiwari 1978). Integumentary tapetum has been generally observed in tenuinucellate ovules, but it has also been observed in some crassinucellate ovules (Kapil and Tiwari 1978). Although it has been considered as relatively evolutionary derived feature of seed development, recent fossil analyses indicate that the endothelium could be an ancient feature in angiosperm seed development (Friis et al. 2019).

Thus, during female germline development, a direct contact between the female germline and the nucellus occurs. The MeMC is surrounded by callose during meiosis, although its presence is not so conspicuous as in male meiosis. Callose is also present in the cell plates that divide the four megaspores (Rodkiewicz 1970) of which usually three will degenerate. Similarly, there is an intense crosstalk between the female germline and the neighbouring nucellar cells during meiosis. Evidence of this communication can be observed in aposporous species with anomalous crosstalk. In apospory, a somatic cell adjacent to the functional megaspore that is called the aposporous initial (AI) cell adopts a megaspore fate and undergoes megagametogenesis but without going through meiosis. Thus, while callose is observed around the MeMC/meiotic products, it is not observed around the AI in the aposporous *Hieracium* (Tucker et al. 2001), *Poa pratensis* and *Pennisetum* (Peel et al. 1997) and, interestingly, callose is not observed around the MeMC that fails to complete meiosis in diplosporous *Elymus rectisetus* (Carman et al. 1991).

4 Cell Wall Composition of the Germline Cells

In addition to the deposition of callose around the female and male germline cells during meiosis, cell wall components of the germline cells start to differentiate among the neighbouring cells. The main components of the cell wall such as polysaccharides, glycoproteins, and phenolic compounds are key components of the crosstalk between germline cells and the neighbouring somatic cells (Tucker and Koltunow 2014). Specific wall polymers such as pectins (methyl-esterified and unesterified, detected using the monoclonal antibody JIM5 and JIM7, respectively), arabinogalactan proteins (detected using the monoclonal antibody JIM8 and JIM13), and extensins (detected using the monoclonal antibody JIM11) have been observed in the cell wall of the female germline of *Annona cherimola* (cherimoya) (Fig. 1) and

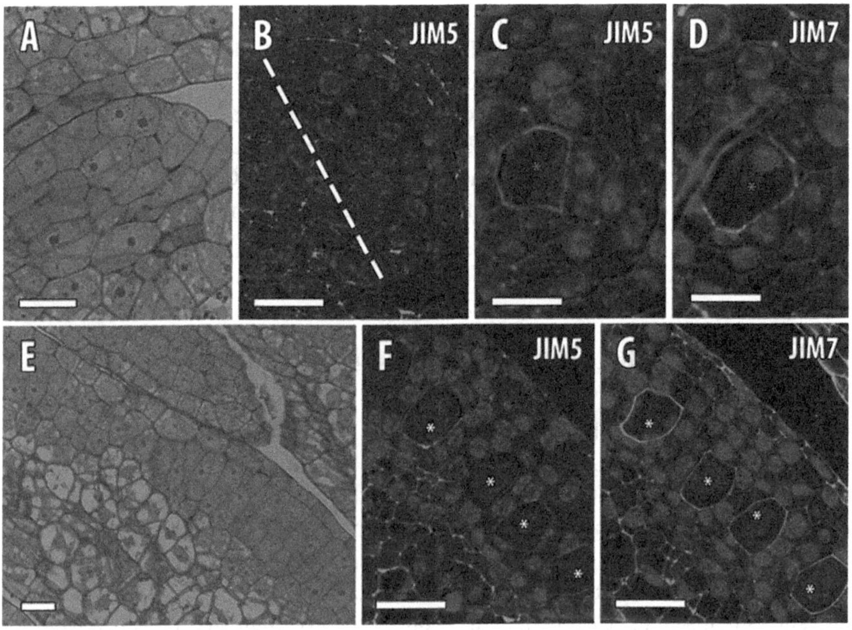

Fig. 1 The formation of the female and male germline in *Annona cherimola*. (**a**) Ovule primordium showing dense PAS reagent-positive cytoplasm and undifferentiated cells. (**b**) Ovule primordium showing a weak signal of unesterified signal detected by the antibody JIM5 that was observed mainly in the ovary walls. (**c–d**) The megaspore mother cell (red asterisk) was marked by antibodies against unesterified (**c**) and methyl-esterified (**d**) pectins. (**e**) Early stage of anther development showing a microsporogenous tissue with dense PAS reagent-positive cytoplasm with undifferentiated cells. (**f–g**) The microspore-mother-cell (white asterisks) was marked by antibodies against unesterified (**f**) and methyl-esterified (**g**) pectins. Bars (**a–b**, **e–g**) 20 μm; (**c–d**) 12.5 μm

Persea americana (avocado) (Lora et al. 2017). In those cases, the MeMC is surrounded by several somatic cell layers in crassinucellate ovules. The cell wall of the MeMC was not differently observed in the tenuinucellate ovules of *Arabidopsis* (Lora et al. 2017) but, recently, AGPs in the MeMC wall have been reported in the tenuinucellate ovules of *Hieracium* species (Juranic et al. 2018). Later on, during megagametogenesis, the cell wall of the functional megaspore has also been shown to be marked by pectins, AGPs, and extensins in *A. cherimola* and *P. americana* (Lora et al. 2017), and AGPs in *Arabidopsis* (Coimbra et al. 2007). Indeed, AGPs have also been observed in the cell wall of the functional AI (FAI) cell and the functional megaspore in *Hieracium*, constituting a similar molecular marker that reflects a functional megaspore-like fate (Juranic et al. 2018). Similarly, AGPs were observed in the apomictic female gametophyte of the facultative apomictic strawberry (Leszczuk and Szczuka 2018).

In the anther, the MiMC wall also develops differently and it is clearly marked by AGPs in *Arabidopsis thaliana* (Coimbra et al. 2007) and *Quercus suber* (Costa et al. 2015). Additionally, methyl-esterified and unesterified pectins were also observed in the MiMC wall of *Mangifera indica*, mango (Lora and Hormaza 2018) and in different *Annona* species (Fig. 1) (Lora et al. 2009, 2014). It seems, therefore, clear that the compositional changes of the cell wall of the male and female reproductive cells clearly differentiate the germline from the neighbouring somatic cells. This is even more evident after the meiosis of the MiMC. First, in the formation of the inner layer, the intine, showing pectins, cellulose, AGPs and extensins that continues during the development of the pollen. Second, in the formation of the exine.

5 Phytohormones During Pollen and Ovule Development

Cell wall biosynthesis has long been related to phytohormones (Lehman and Sanguinet 2019). Indeed, auxin can induce cell division or cell expansion (Perrot-Rechenmann 2010) and it has been observed in the tapetum and pollen grains (Feng et al. 2006). In the young anther, high auxin-dependent GUS expression was reported in the tapetum before it was observed in the developing pollen grains (Aloni et al. 2006) and it has been suggested to play a major role in pollen maturation and anther dehiscence (Cecchetti et al. 2008). Recently, high auxin levels have been observed in the middle layer of the tapetum, showing a role on pollen maturation, anther dehiscence, and filament elongation (Cecchetti et al. 2017).

In the ovule, during megasporogenesis, adjacent cells are initially identical in the nucellus, but are situated in different microenvironments and adopt specific fates in response to positional signals. The female germline shows a central position during this process (Lora et al. 2017). Auxin and cytokinin phytohormones may result in key positional signals that polarize megasporogenesis, inducing the changes of cell wall composition that are needed for cell expansion of the MeMC. Indeed, during gynoecium development, auxin and cytokinin distribution are complementary to each other; thus, while auxin is found distally, cytokinin distribution is proximal (close to the chalaza) (Marsch-Martínez et al. 2012; Schaller et al. 2015). Similarly, distal auxin and proximal cytokinin distribution polarize the ovule primordia development. There are several lines of evidence that reveal the role of cytokinin and auxin on the formation of the ovule primordium. The number of ovule primordia is reduced by inhibition of the regulator of polar auxin transport *Pin-formed 1* (*PIN1*) (Okada et al. 1991) or by mutation of a cytokinin receptor gene (Kinoshita-Tsujimura and Kakimoto 2011; Riefler et al. 2006). On the contrary, the number of ovule primordia increases in mutants in which the cytokinin oxidase/dehydrogenase enzyme is deactivated (Bartrina et al. 2011). Moreover, the relationship between cytokinin and auxin formation in ovule primordia was reported by Bencivenga et al. (2012) showing the control of *PIN1* expression by cytokinin. The polarized accumulation of both phytohormones continues later during megagametogenesis (Lora et al. 2019b; Zurcher et al. 2013).

Distal maximum auxin distribution during ovule development has also been reported in the eudicot *Hieracium* (Tucker et al. 2012b), in the monocot maize

(Forestan et al. 2012; Lituiev et al. 2013) and in the early divergent angiosperms *A. cherimola* and *P. americana* (Lora et al. 2017). These studies suggest a conserved microenvironment, in terms of phytohormones, around the female germline in flowering plants. Furthermore, it should be taken into account that maize, cherimoya, and avocado show crassinucellate ovules compared to the tenuinucellate ovules of *Arabidopsis* and *Hieracium*, and, despite the differences in the number of cell layers around the female germline, both types of ovules show a similar maximum distal auxin distribution. In addition to the polarized distribution of the phytohormones along the proximal-distal axis in the nucellus, MeMC also shows a polarized distribution of the organelles in *Arabidopsis* (Bajon et al. 1999). This polarization has also been observed in *Pisum sativum* and continues in the four meiotic products in which the functional megaspore incorporates most of the plastids and mitochondrias (Medina et al. 1981). Moreover, the functional megaspore is generally placed in the chalazal pole that continues its development during megagametogenesis (Maheshwari 1950).

6 Breaking the Crosstalk Between the Sporophyte and the Germline

Crosstalk between sporophyte and gametophyte has also clearly been demonstrated in mutants, mostly from the model plant *Arabidopsis* and the monocots rice and maize. Generally, most of the sporophytic mutants affecting reproduction are also defective in germline development by breaking the crosstalk between the sporophytic tissues and the germline.

During meiosis, mutations that alter the dissolution rate of the MiMC wall causes anomalies in the microspore division that result in the release of pollen in tetrads after anther dehiscence such as in *quartet* mutants (Francis et al. 2006; Preuss et al. 1994; Rhee et al. 2003). Similarly, microspores release and pollen development were affected in the mutants of callose synthase genes (Enns et al. 2005). Mutations in genes involved in tapetum development, such as MS1 (Wilson et al. 2001), AMS (Sorensen et al. 2003; Xu et al. 2010), *DYT1* (Zhang et al. 2006), *TDF1* (Zhu et al. 2008), and *MS188* (Zhang et al. 2007) which follow the same genetic pathway (Zhu et al. 2011), cause male sterility, reflecting the role of the tapetum on pollen development. *MS1* gene of *Arabidopsis* has a key role on primexine and exine pattern formation (Ariizumi et al. 2005), *AMS* is involved in sporopollenin biosynthesis and the secretion of materials for pollen wall patterning (Xu et al. 2014) and *MS188* is also required for callose dissolution and exine formation (Zhang et al. 2007). The mutation of *DYT1* causes anomalous tapetal cells and lower amount of callose deposition with incomplete meiotic cytokinesis (Zhang et al. 2006). *TDF1* is also involved in tapetum development and its mutation shows irregular division and dysfunction of the tapetum (Zhu et al. 2008).

In the ovule, most of the mutations that affect the sporophytic tissues show anomalies on integument development (Gasser et al. 1998; Kelley and Gasser 2009). *Arabidopsis* ovule mutants that lack integuments, such as *aintegument* (Elliott et al. 1996) and *wuchel* (Gross-Hardt et al. 2002), result in the lack of embryo sac development. Embryo sacs are also missing in the *Arabidopsis inner no outer* (*ino*) mutant (Baker et al. 1997; Gaiser et al. 1995; Lora et al. 2019a), but the disruption of an *Annona* ortholog of *INO* showed 87% of normal embryo sacs (Lora et al. 2011). The ovules of both *ino* mutants are orthotropous and unitegmic with a single inner integument, while the outer integument is missing (Baker et al. 1997; Gaiser et al. 1995; Lora et al. 2011). However, while, the female germline develops surrounded by 2–3 nucellar layers in *Annona*, it is only surrounded by one nucellar layer in *Arabidopsis*. The nucellus is not apparently affected in both *ino* mutants and, as consequence, the *Annona* female germline develops in a more protective environment (Lora et al. 2019a).

7 Similarities Between Male and Female Germline Development

As mentioned above, the development of the male and female germlines follows quite different processes, and this is also shown in their differential genetic and molecular regulation. Indeed, there are very few genes involved in both developmental process. However, the somatic cells are initially, at least morphologically, identical, before the transition from somatic to reproductive cell fate. Therefore, it is expected that similar genes could be found at least in the early stages during the sporogenesis in the anther and the ovule.

One of the earliest genes that showed a role in both male and female sporogenous cell differentiation was *Arabidopsis NOZZLE/SPOROCYTELESS* (*NZZ/SPL*) (Schiefthaler et al. 1999; Yang et al. 1999). The female and male sporogenesis of *nzz/spl* mutants are blocked and the selected somatic cells are unable to differentiate into microsporocytes or megasporocytes; in addition, the surrounding cells are also affected, the anther wall development is blocked and the nucellus forms a finger-like structure. The *NZZ/SPL* gene encodes a nuclear protein related to MADS box transcription factors that is expressed mainly in the germline during sporogenesis (Schiefthaler et al. 1999; Yang et al. 1999). Conversely, the mutation of the *EXTRA SPOROGENOUS CELL/ EXCESS MICROSPOROCYTE1* (*EXS/EMS1*) gene, which encodes a putative LRR-RPK protein and is expressed first in the anther primordia and later more strongly in the tapetum that in the microsporocyte, results in an excess of microsporocytes in *Arabidopsis* (Canales et al. 2002; Zhao et al. 2002). Similarly, the putative small secreted protein TAPETUM DETERMINANT1 (TPD1) of *Arabidopsis* is required for tapetal cell differentiation and development, and *tpd1* mutants generate an excessive number of microsporocytes instead of tapetum cells that are absent in the anther (Yang et al. 2003). Moreover, the ectopic expression of

TPD1 increases the number of cells in the carpel dependent on *EXS/EMS1* (Yang et al. 2005). TPD1 and EXS/EMS1 can physically interact in vivo and in vitro (Jia et al. 2008). The double mutant *somatic embryogenesis receptor-like kinase1* (*serk1*) *serk2* also produces an increase in the number of microsporocytes and *SERK1* and *SERK2* are also expressed in the locules of the anther (Albrecht et al. 2005; Colcombet et al. 2005).

Interestingly, while *exs/ems1* and *tpd1* mutants do not show effects on megasporogenesis, the mutation of *MSP1*, an ortholog to the *Arabidopsis EXS/EMS1* gene, generates extra male and female sporocytes in rice (Nonomura et al. 2003). The *TPD1* like gene of rice *TAPETUM DETERMINANT1/MULTIPLE ARQUESPORIAL CELL2* (*OsTLD1A/MIL2*) binds to *MSP1* in yeast and both are co-expressed in somatic cells surrounding the MiMC and MeMC but results in RNA interference against OsTLD1A phenocopies *msp1* only in the anther (Hong et al. 2012; Zhao et al. 2008). Another *TPD1* like gene, *OsTLD1B*, also co-expresses with *TPD1* but only in the anther (Zhao et al. 2008). Although the protein pairs *Arabidopsis* (*EXS/EMS1-TPD1*) and rice (*MSP1-OsTLD1A/MIL2*) show similarities, their discrepancies suggest differences in some aspects during microsporagium development (Hong et al. 2012) and reflects phylogenetic divergence. Similar protein pairs could also be present in maize, in which the *multiple archesporial cell 1* (*mac1*) also shows an increased number of archesporial cells (Sheridan et al. 1996, 1999). This gene was later reported as an ortholog of rice *TDL1A* (Wang et al. 2012) and transcriptome profiling of maize anthers revealed a Zm*MSP1* homologous to the rice *MSP1* gene (Ma et al. 2007).

In the ovule, the multiple-MMC trait is also present in cell cycle mutants of *Arabidopsis* (Cao et al. 2018; Yao et al. 2018; Zhao et al. 2017) and mutations that cause epigenetic changes (Olmedo-Monfil et al. 2010; Schmidt et al. 2011) (see recent reviews on molecular and genetic regulation of female germline fate in Lora et al. 2019b; Pinto et al. 2019). The data from several epigenetic mutants suggest an epigenetic control on the neighbouring somatic cells of the MeMC that limits their cell expansion (Armenta-Medina et al. 2011; Lora et al. 2019b). One example is the *argonaute9* (*ago9*) mutant of *Arabidopsis*, which appears to be defective in small-RNA related to crosstalk between the nucellar and germline cells (Olmedo-Monfil et al. 2010). A semi-dominant insertion in another argonaute, *AGO5-4*, that is also expressed around the MMC, produces a normal MeMC but, subsequently, fails to produce a gametophyte due to defects in the nucellus (Tucker et al. 2012a). Another study related to the disruption of crosstalk between the MMC and the neighbouring somatic cells was reported in *wuschel* (*wus*) ovule (Gross-Hardt et al. 2002), *windhose1-windhose2* (*wih1wih2*) and *tornado (trn)* mutants (Lieber et al. 2011). *NZZ/SPL* promotes the expression of the homeobox gene *WUS* in the nucellus, which in turn upregulates the expression of the small peptides WIH1 and WIH2 (Lieber et al. 2011). The homeobox gene *WUS* was wholly identified as a stem cell regulator in the shoot and floral meristems (Laux et al. 1996). *WIH1/2* genes act downstream of *WUSCHEL*, and double mutants of both genes show defective megasporogenesis in which around 30% ovules fail to produce an MeMC. A similar phenotype was also observed in *trn1* and *trn2* (Lieber et al. 2011).

8 Conclusion

The crosstalk between the germline and the neighbouring somatic cells is essential to the success of pollen and ovule development in angiosperms. In both processes, a key event is the meiosis, which shows restricted somatic-germline cell communication by deposition of callose. During gametogenesis, the cell walls of the male and female germline are clearly differentiated. In the pollen, the crosstalk between the pollen and the tapetum is key for the formation of the pollen cell wall. In this sense, the role of the male sporophyte is essential for pollination and subsequent pollen–pistil interaction that, in some species, is involved in the pollen-stigma-recognition in sporophytic self-incompatibility. In the ovule, the genetic and the epigenetic regulation of the female germline development suggests an essential cell–cell communication between the female germline and the neighbouring nucellar cells that results in the cell expansion only in the MeMC. Meanwhile, the cell expansion is apparently inhibited in the nucellar cells.

Acknowledgements This work was supported by Ministerio de Economía y Competitividad – European Regional Development Fund, European Union (AGL2015-74071-JIN, AGL2016-77267-R, PDI2019-109566RB-IOO) and Junta de Andalucía (P18-RT-3272).

References

Albert B, Ressayre A, Nadot S (2011) Correlation between pollen aperture pattern and callose deposition in late tetrad stage in three species producing atypical pollen grains. Am J Bot 98:189–196

Albrecht C, Russinova E, Hecht V, Baaijens E, de Vries S (2005) The *Arabidopsis thaliana* SOMATIC EMBRYOGENESIS RECEPTOR-LIKE KINASES1 and 2 control male sporogenesis. Plant Cell 17:3337–3349

Aloni R, Aloni E, Langhans M, Ullrich CI (2006) Role of auxin in regulating *Arabidopsis* flower development. Planta 223:315–328

Ariizumi T, Hatakeyama K, Hinata K, Sato S, Kato T, Tabata S, Toryyama K (2005) The *HKM* gene, which is identical to the *MS1* gene of *Arabidopsis thaliana*, is essential for primexine formation and exine pattern formation. Sex Plant Reprod 18:1–7

Armenta-Medina A, Demesa-Arévalo E, Vielle-Calzada JP (2011) Epigenetic control of cell specification during female gametogenesis. Sex Plant Reprod 24:137–147

Bachelier JB, Friedman WE (2011) Female gamete competition in an ancient angiosperm lineage. Proc Natl Acad Sci 108:12360–12365

Bajon C, Horlow C, Motamayor JC, Sauvanet A, Robert D (1999) Megasporogenesis in *Arabidopsis thaliana* L.: an ultrastructural study. Sex. Plant Reprod 12:99–109

Baker SC, Robinson-Beers K, Villanueva JM, Gaiser JC, Gasser CS (1997) Interactions among genes regulating ovule development in *Arabidopsis thaliana*. Genetics 145:1109–1124

Bartrina I, Otto E, Strnad M, Werner T, Schmülling T (2011) Cytokinin regulates the activity of reproductive meristems, flower organ size, ovule formation, and thus seed yield in *Arabidopsis thaliana*. Plant Cell 23:69–80

Bencivenga S, Simonini S, Benková E, Colombo L (2012) The transcription factors BEL1 and SPL are required for cytokinin and auxin signaling during ovule development in *Arabidopsis*. Plant Cell 24:2886–2897

Blackmore S, Wortley AH, Skvarla JJ, Rowley JR (2007) Pollen wall development in flowering plants. New Phytol 174:483–498
Canales C, Bhatt AM, Scott R, Dickinson H (2002) EXS, a putative LRR receptor kinase, regulates male germline cell number and tapetal identity and promotes seed development in *Arabidopsis*. Curr Biol 12:1718–1727
Cao L, Wang S, Venglat P, Zhao L, Cheng Y, Ye S, Ye S, Qin Y, Datla R, Zhou Y (2018) *Arabidopsis* ICK/KRP cyclin-dependent kinase inhibitors function to ensure the formation of one megaspore mother cell and one functional megaspore per ovule. PLoS Genet 14:e1007230
Carman JG, Crane CF, Riera-Lizarazu O (1991) Comparative histology of cell walls during meiotic and apomeiotic megasporogenesis in two hexaploid Australasian *Elymus* species. Crop Sci 31:1527
Cecchetti V, Altamura MM, Falasca G, Costantino P, Cardarelli M (2008) Auxin regulates *Arabidopsis* anther dehiscence, pollen maturation, and filament elongation. Plant Cell 20:1760–1774
Cecchetti V, Celebrin D, Napoli N, Ghelli R, Brunetti P, Costantino P, Cardarelli M (2017) An auxin maximum in the middle layer controls stamen development and pollen maturation in *Arabidopsis*. New Phytol 213:1194–1207
Clement C, Audran JC (1995) Anther wall layers control pollen sugar nutrition in *Lilium*. Protoplasma 187:172–181
Coimbra S, Almeida J, Junqueira V, Costa ML, Pereira LG (2007) Arabinogalactan proteins as molecular markers in *Arabidopsis thaliana* sexual reproduction. J Exp Bot 58:4027–4035
Colcombet J, Boisson-Dernier A, Ros-Palau R, Vera CE, Schroeder JI (2005) *Arabidopsis* somatic embryogenesis receptor kinases1 and 2 are essential for tapetum development and microspore maturation. Plant Cell 17:3350–3361
Costa ML, Sobral R, Ribeiro Costa MM, Amorim MI, Coimbra S (2015) Evaluation of the presence of arabinogalactan proteins and pectins during *Quercus suber* male gametogenesis. Ann Bot 115:81–92
Echlin P (1971) The role of the tapetum during microsporogenesis of angiosperms. In: Pollen. Butterworth-Heinemann, Oxford, pp 41–61
Elliott RC, Betzner AS, Huttner E, Oakes MP, Tucker WQJ, Gerentes D, Perez P, Smyth DR (1996) *AINTEGUMENTA*, an *APETALA2*-like gene of *Arabidopsis* with pleiotropic roles in ovule development and floral organ growth. Plant Cell 8:155–168
Enns LC, Kanaoka MM, Torii KU, Comai L, Okada K, Cleland RE (2005) Two callose synthases, GSL1 and GSL5, play an essential and redundant role in plant and pollen development and in fertility. Plant Mol Biol 58:333–349
Feng XL, Ni WM, Elge S, Mueller-Roeber B, Xu ZH, Xue HW (2006) Auxin flow in anther filaments is critical for pollen grain development through regulating pollen mitosis. Plant Mol Biol 61:215–226
Forestan C, Farinati S, Varotto S (2012) The maize *PIN* gene family of auxin transporters. Front Plant Sci 3:16
Franchi G, Pacini E (1993) Role of the tapetum in pollen and spore dispersal. Plant Syst Evol 7:1–11
Francis KE, Lam SY, Copenhaver GP (2006) Separation of *Arabidopsis* pollen tetrads is regulated by *QUARTET1*, a pectin methylesterase gene. Plant Physiol 142:1004–1013
Friis EM, Crane PR, Pedersen KR (2019) The endothelium in seeds of early angiosperms. New Phytol 224:1419–1424
Gaiser JC, Robinson-Beers K, Gasser CS (1995) The *Arabidopsis SUPERMAN* gene mediates asymmetric growth of the outer integument of ovules. Plant Cell 7:333–345
Gasser CS, Broadhvest J, Hauser BA (1998) Genetic analysis of ovule development. Annu Rev Plant Physiol Plant Mol Biol 49:1–24
Gómez JF, Talle B, Wilson ZA (2015) Anther and pollen development: a conserved developmental pathway. J Integr Plant Biol 57:876–891

Gross-Hardt R, Lenhard M, Laux T (2002) WUSCHEL signaling functions in interregional communication during *Arabidopsis* ovule development. Genes Dev 16:1129–1138

Harder L, Johnson S (2008) Function and evolution of aggregated pollen in angiosperms. Int J Plant Sci 169:59–78

Heslop-Harrison J (1979) The forgotten generation: some thoughts on the genetics and physiology of angiosperm gametophytes. In: The Bateson lecture: proceedings of the fourth John Innes symposium, pp 1–14

Hofmesiter W (1851) Vergleichende Untersuchungen der Keimung, Entfaltung und Fruchtbildung höherer Kryptogamen (Moose, Farrn, Equisetaceen, Rhizokarpeen and Lykopodiaceen) und der Samenbildung der Coniferen. Friedrich Hofmeister, Leipzig

Hong L, Tang D, Shen Y, Hu Q, Wang K, Li M, Lu T, Chang Z (2012) MIL2 (MICROSPORELESS2) regulates early cell differentiation in the rice anther. New Phytol 196:402–413

Jia G, Liu X, Owen HA, Zhao D (2008) Signaling of cell fate determination by the TPD1 small protein and EMS1 receptor kinase. Proc Natl Acad Sci U S A 105:2220–2225

Juranic M, Tucker MR, Schultz CJ, Shirley NJ, Taylor JM, Spriggs A, Johnson SD, Bulone V, Koltunow AM (2018) Asexual female gametogenesis involves contact with a sexually-fated megaspore in apomictic *Hieracium*. Plant Physiol 177:1027–1049.

Kapil RN, Tiwari SC (1978) The integumentary tapetum. Bot Rev 44:457–490

Kelley DR, Gasser CS (2009) Ovule development: genetic trends and evolutionary considerations. Sex Plant Reprod 22:229–234

Kinoshita-Tsujimura K, Kakimoto T (2011) Cytokinin receptors in sporophytes are essential for male and female functions in *Arabidopsis thaliana*. Plant Signal Behav 6:66–71

Laux T, Mayer KF, Berger J, Jurgens G (1996) The *WUSCHEL* gene is required for shoot and floral meristem integrity in *Arabidopsis*. Development 122:87–96

Lehman TA, Sanguinet KA (2019) Auxin and cell wall crosstalk as revealed by the *Arabidopsis thaliana* cellulose synthase mutant *radially swollen 1*. Plant Cell Physiol 60:1487–1503

Leszczuk A, Szczuka E (2018) Arabinogalactan proteins: immunolocalization in the developing ovary of a facultative apomict *Fragaria* x *ananassa* (Duch.). Plant Physiol Biochem 123:24–33

Lieber D, Lora J, Schrempp S, Lenhard M, Laux T (2011) *Arabidopsis WIH1* and *WIH2* genes act in the transition from somatic to reproductive cell fate. Curr Biol 21:1009–1017

Lituiev DS, Krohn NG, Müller B, Jackson D, Hellriegel B, Dresselhaus T, Grossniklaus U (2013) Theoretical and experimental evidence indicates that there is no detectable auxin gradient in the angiosperm female gametophyte. Development 140:4544–4553

Lora J, Hormaza JI (2018) Pollen wall development in mango (*Mangifera indica* L., Anacardiaceae). Plant Reprod 31(4):385–397, 1–13

Lora J, Testillano PS, Risueño MC, Hormaza JI, Herrero M (2009) Pollen development in *Annona cherimola* Mill. (Annonaceae). Implications for the evolution of aggregated pollen. BMC Plant Biol 9:129

Lora J, Hormaza JI, Herrero M, Gasser CS (2011) Seedless fruits and the disruption of a conserved genetic pathway in angiosperm ovule development. Proc Natl Acad Sci U S A 108:5461–5465

Lora J, Herrero M, Hormaza JI (2014) Microspore development in *Annona* (Annonaceae): differences between monad and tetrad pollen. Am J Bot 101:1508–1518

Lora J, Hormaza JI, Herrero M (2016) The diversity of the pollen tube pathway in plants: toward an increasing control by the sporophyte. Front Plant Sci 7:107

Lora J, Herrero M, Tucker MR, Hormaza JI (2017) The transition from somatic to germline identity shows conserved and specialized features during angiosperm evolution. New Phytol 216:495–509

Lora J, Laux T, Hormaza JI (2019a) The role of the integuments in pollen tube guidance in flowering plants. New Phytol 221:1074–1089. https://doi.org/10.1111/nph.15420

Lora J, Yang X, Tucker MR (2019b) Establishing a framework for female germline initiation in the plant ovule. J Exp Bot 70:2937–2949

Ma J, Duncan D, Morrow DJ, Fernandes J, Walbot V (2007) Transcriptome profiling of maize anthers using genetic ablation to analyze pre-meiotic and tapetal cell types. Plant J 50:637–648
Maciel-Silva AS, Porto KC (2014) Reproduction in bryophytes. In: Ramawat KG, Mérillon JM, Shivanna KR (eds) Reproductive biology of plants. Taylor & Francis, New York
Maheshwari P (1950) An introduction to the embryology of angiosperms. McGraw-Hill, New York
Marsch-Martínez N, Ramos-Cruz D, Irepan Reyes-Olalde J, Lozano-Sotomayor P, Zúñiga-Mayo VM, de Folter S (2012) The role of cytokinin during *Arabidopsis* gynoecia and fruit morphogenesis and patterning. Plant J 72:222–234
McCormick S (1993) Male gametophyte development. Plant Cell 5:1265–1275
Medina FJ, Risueño MC, Rodriguez-Garsia MI (1981) Evolution of the cytoplasmic organelles during female meiosis in *Pisum sativum* L. Planta 151:215–225
Nepi M, Franchi GG, Pacini E (2001) Pollen hydration status at dispersal: cytophysiological features and strategies. Protoplasma 216:171–180
Nonomura KI, Miyoshi K, Eiguchi M, Suzuki T, Miyao A, Hirochika H, Kurata N (2003) The *MSP1* gene is necessary to restrict the number of cells entering into male and female sporogenesis and to initiate anther wall formation in rice. Plant Cell 15:1728–1739
Okada K, Ueda J, Komaki MK, Bell CJ, Shimura Y (1991) Requirement of the auxin polar transport system in early stages of *Arabidopsis* floral bud formation. Plant Cell 3:677–684
Olmedo-Monfil V, Duran-Figueroa N, Arteaga-Vazquez M, Demesa-Arevalo E, Autran D, Grimanelli D, Slotkin RK, Martienssen RA, Vielle-Calzada JP (2010) Control of female gamete formation by a small RNA pathway in *Arabidopsis*. Nature 464:628–632
Pacini E (1990) Tapetum and microspore function. In: Microspores: evolution and ontogeny. Academic Press, London, pp 213–237
Pacini E, Franchi GG (1988) Amylogenesis and amylolysis during pollen grain development. In: Sexual reproduction in higher plants. Springer, Berlin, pp 181–186
Pacini E, Franchi GG, Hesse M (1985) The tapetum - its form, function, and possible phylogeny in embryophyta. Plant Syst Evol 149:155–185
Pacini E, Guarnieri M, Nepi M (2006) Pollen carbohydrates and water content during development, presentation, and dispersal: a short review. Protoplasma 228:73–77
Peel MD, Carman JG, Leblanc O (1997) Megasporocyte callose in apomictic buffelgrass, Kentucky bluegrass, *Pennisetum squamulatum* Fresen, *Tripsacum* L., and weeping lovegrass. Crop Sci 37:724
Perrot-Rechenmann C (2010) Cellular responses to auxin: division versus expansion. Cold Spring Harb Perspect Biol 2:a001446
Pinto SC, Mendes MA, Coimbra S, Tucker MR (2019) Revisiting the female germline and its expanding toolbox. Trends Plant Sci 24:455–467
Polowick PL, Sawhney VK (1993) Differentiation of the tapetum during microsporogenesis in tomato (*Lycopersicon esculentum* Mill.), with special reference to the tapetal cell wall. Ann Bot 72:595–605
Preuss D, Rhee SY, Davis RW (1994) Tetrad analysis possible in *Arabidopsis* with mutation of the *QUARTET* (*QRT*) genes. Science 264:1458–1460
Quilichini TD, Douglas CJ, Samuels AL (2014) New views of tapetum ultrastructure and pollen exine development in *Arabidopsis thaliana*. Ann Bot 114:1189–1201
Raghavan V (1988) Anther and pollen development in rice (*Oryza sativa*). Am J Bot 75:183–196
Rhee SY, Osborne E, Poindexter PD, Somerville CR (2003) Microspore separation in the *quartet 3* mutants of *Arabidopsis* is impaired by a defect in a developmentally regulated polygalacturonase required for pollen mother cell wall degradation. Plant Physiol 133:1170–1180
Riefler M, Novak O, Strnad M, Schmülling T (2006) *Arabidopsis* cytokinin receptor mutants reveal functions in shoot growth, leaf senescence, seed size, germination, root development, and cytokinin metabolism. Plant Cell 18:40–54
Rodkiewicz B (1970) Callose in cell walls during megasporogenesis in angiosperms. Planta 93:39–47

Rudall PJ (1997) The nucellus and chalaza in monocotyledons: structure and systematics. Bot Rev 63:140–181
Schaller GE, Bishopp A, Kieber JJ (2015) The yin-yang of hormones: cytokinin and auxin interactions in plant development. Plant Cell 27:44–63
Schiefthaler U, Balasubramanian S, Sieber P, Chevalier D, Wisman E, Schneitz K (1999) Molecular analysis of *NOZZLE*, a gene involved in pattern formation and early sporogenesis during sex organ development in Arabidopsis thaliana. Proc Natl Acad Sci U S A 96:11664–11669
Schmidt A, Wuest S, Vijverberg K, Baroux C, Grossniklaus U (2011) Transcriptome analysis of the *Arabidopsis* megaspore mother cell uncovers the importance of RNA helicases for plant germline development. PLoS Biol 9:e1001155
Schneitz K, Hulskamp M, Pruitt RE (1995) Wild-type ovule development in *Arabidopsis thaliana*: a light microscope study of cleared whole-mount tissue. Plant J 7:731–749
Sheridan WF, Avalkina NA, Shamrov II, Batygina TB, Golubovskaya IN (1996) The *mac1* gene: controlling the commitment to the meiotic pathway in maize. Genetics 142:1009–1020
Sheridan WF, Golubeva EA, Abrhamova LI, Golubovskaya IN (1999) The *mac1* mutation alters the developmental fate of the hypodermal cells and their cellular progeny in the maize anther. Genetics 153:933–941
Shi J, Cui M, Yang L, Kim YJ, Zhang D (2015) Genetic and biochemical mechanisms of pollen wall development. Trends Plant Sci 20:741–753
Sorensen AM, Krober S, Unte US, Huijser P, Dekker K, Saedler H (2003) The *Arabidopsis ABORTED MICROSPORES* (*AMS*) gene encodes a MYC class transcription factor. Plant J 33:413–423
Sporne KR (1969) The ovule as an indicator of evolutionary status in angiosperms. New Phytol 68:555–566
Takaso T, Bouman F (1986) Ovule and seed ontogeny in *Gnetum gnemon* L. Bot Mag (Tokyo) 99:241–266
Tucker MR, Koltunow AM (2014) Traffic monitors at the cell periphery: the role of cell walls during early female reproductive cell differentiation in plants. Curr Opin Plant Biol 17:137–145
Tucker MR, Paech NA, Willemse MT, Koltunow AM (2001) Dynamics of callose deposition and beta-1,3-glucanase expression during reproductive events in sexual and apomictic *Hieracium*. Planta 212:487–498
Tucker MR, Okada T, Hu Y, Scholefield A, Taylor JM, Koltunow AM (2012a) Somatic small RNA pathways promote the mitotic events of megagametogenesis during female reproductive development in *Arabidopsis*. Development 139:1399–1404
Tucker MR, Okada T, Johnson SD, Takaiwa F, Koltunow AM (2012b) Sporophytic ovule tissues modulate the initiation and progression of apomixis in *Hieracium*. J Exp Bot 63:3229–3241
Walker JW, Doyle JA (1975) The bases of angiosperm phylogeny: palynology. Ann Missouri Bot Gard 62:664
Wang CJ, Nan GL, Kelliher T, Timofejeva L, Vernoud V, Golubovskaya IN, Harper L, Egger R, Walbot V, Cande WZ (2012) *Maize multiple archesporial cells 1* (*mac1*), an ortholog of rice *TDL1A*, modulates cell proliferation and identity in early anther development. Development 139:2594–2603
Wilson ZA, Morroll SM, Dawson J, Swarup R, Tighe PJ (2001) The *Arabidopsis MALE STERILITY1* (*MS*1) gene is a transcriptional regulator of male gametogenesis, with homology to the PHD-finger family of transcription factors. Plant J 28:27–39
Xu J, Yang C, Yuan Z, Zhang D, Gondwe MY, Ding Z, Liang W, Zhang D, Wilson ZA (2010) The *ABORTED MICROSPORES* regulatory network is required for postmeiotic male reproductive development in *Arabidopsis thaliana*. Plant Cell 22:91–107
Xu J, Ding Z, Vizcay-Barrena G, Shi J, Liang W, Yuan Z, Werck-Reichhart D, Schreiber L, Wilson ZA, Zhang D (2014) *ABORTED MICROSPORES* acts as a master regulator of pollen wall formation in *Arabidopsis*. Plant Cell 26:1544–1556

Yang WC, Ye D, Xu J, Sundaresan V (1999) The *SPOROCYTELESS* gene of *Arabidopsis* is required for initiation of sporogenesis and encodes a novel nuclear protein. Genes Dev 13:2108–2117

Yang SL, Xie LF, Mao HZ, Puah CS, Yang WC, Jiang L, Sundaresan V, Ye D (2003) Tapetum determinant1 is required for cell specialization in the *Arabidopsis* anther. Plant Cell 15:2792–2804

Yang SL, Jiang L, Puah CS, Xie LF, Zhang XQ, Chen LQ, Yang WC, Ye D (2005) Overexpression of *TAPETUM DETERMINANT1* alters the cell fates in the *Arabidopsis* carpel and tapetum via genetic interaction with excess microsporocytes1/extra sporogenous cells. Plant Physiol 139:186–191

Yao X, Yang H, Zhu Y, Xue J, Wang T, Song T, Yang Z, Wang S (2018) The canonical E2Fs are required for germline development in *Arabidopsis*. Front Plant Sci 9:638

Zhang W, Sun Y, Timofejeva L, Chen C, Grossniklaus U, Ma H (2006) Regulation of *Arabidopsis* tapetum development and function by *DYSFUNCTIONAL TAPETUM1* (*DYT1*) encoding a putative bHLH transcription factor. Development 133:3085–3095

Zhang ZB, Zhu J, Gao JF, Wang C, Li H, Li H, Zhang HQ, Zhang S, Wang DM, Wang QX, Huang H, Xia HJ, Yanf ZN (2007) Transcription factor AtMYB103 is required for anther development by regulating tapetum development, callose dissolution and exine formation in *Arabidopsis*. Plant J 52:528–538

Zhao DZ, Wang GF, Speal B, Ma H (2002) The *EXCESS MICROSPOROCYTES1* gene encodes a putative leucine-rich repeat receptor protein kinase that controls somatic and reproductive cell fates in the *Arabidopsis* anther. Genes Dev 16:2021–2031

Zhao X, de Palma J, Oane R, Gamuyao R, Luo M, Chaudhury A, Hervé P, Xue Q, Bennett J (2008) OsTDL1A binds to the LRR domain of rice receptor kinase MSP1, and is required to limit sporocyte numbers. Plant J 54:375–387

Zhao X, Bramsiepe J, Van Durme M, Komaki S, Prusicki MA, Maruyama D, Forner J, Medzihradszky A, Wijnker E, Harashima H, Lu Y, Schmidt A, Guthörl D, Sahún-Logroño R, Guan Y, Pochon G, Grossniklaus U, Laux T, Higashiyama T, Lohmann JU, Nowack MK, Schnittger A (2017) Retinoblastoma related1 mediates germline entry in *Arabidopsis*. Science 356:eaaf6532

Zhu J, Chen H, Li H, Gao JF, Jiang H, Wang C, Guan YF, Yang ZN (2008) *Defective in Tapetal Development and Function 1* is essential for another development and tapetal function for microspore maturation in *Arabidopsis*. Plant J 55:266–277

Zhu J, Lou Y, Xu X, Yang ZN (2011) A genetic pathway for tapetum development and function in *Arabidopsis*. J Integr Plant Biol 53:892–900

Zurcher E, Tavor-Deslex D, Lituiev D, Enkerli K, Tarr PT, Muller B (2013) A robust and sensitive synthetic sensor to monitor the transcriptional output of the cytokinin signaling network in planta. Plant Physiol 161:1066–1075

The Photosynthetic System in Tropical Plants Under High Irradiance and Temperature Stress

G. Heinrich Krause and Klaus Winter

Contents

Abstract Plants in the tropics frequently experience stressful environmental conditions such as excessive sunlight including solar ultraviolet (UV) radiation or high leaf temperatures. In view of progressing climate change, the combination of various stress factors, particularly during extreme drought periods and heat waves, may cause damage to the photosynthetic system followed by cell death in leaves, resulting in reduction of total photosynthetic productivity. The present article reviews a series of investigations on tropical forest species in Panama. Recording of chlorophyll *a* (Chl *a*) fluorescence parameters served as a versatile method to

G. H. Krause (✉)
Institute of Plant Biochemistry, Heinrich Heine University Düsseldorf, Düsseldorf, Germany
e-mail: ghkrause@uni-duesseldorf.de

K. Winter
Smithsonian Tropical Research Institute, Balboa, Ancon, Republic of Panama
e-mail: winterk@si.edu

© Springer Nature Switzerland AG 2020
Progress in Botany (2021) 82: 131–170, https://doi.org/10.1007/124_2020_44,
Published online: 30 December 2020

assess the degree of damage and acclimation in chloroplasts. Analysis of chloroplast pigments and antioxidative cell constituents provided valuable information on the acclimation state of leaves. The studies indicate that tropical plants are capable of adjusting to potentially harmful conditions in their respective habitats. One important way of protecting photosystem II is the operation of the violaxanthin cycle and, in certain species, the lutein epoxide cycle. Responses to excess solar radiation in young and mature canopy sun leaves, in leaves of plants growing in treefall gaps and understory of the tropical forest are highlighted. The response of photosystem I to excessive visible light, as well as damaging and acclimatory processes induced by solar UV radiation in photosystems I and II, has been investigated. A reassessed method of Chl *a* fluorescence measurements was used to determine the limits of heat tolerance in sun and shade leaves of C_3 species and in sun leaves of species exhibiting crassulacean acid metabolism (CAM).

Keywords Climate change, Heat tolerance, Lutein epoxide cycle, Photoinhibition, Photoprotection, Tropical forest, UV radiation, Violaxanthin cycle

1 Introduction: Stress Imposed by Climate Conditions on Plants in the Humid Tropics

In the humid tropics, especially the humid, seasonally dry tropics, plants have to cope with various stressors such as intense solar radiation of both visible and ultraviolet (UV-A and UV-B) light and high leaf temperatures. Under clear sky in fully sun-exposed leaves, visible light can be highly in excess of saturating photosynthetic CO_2 assimilation and may cause overheating of leaves. When photoprotection is insufficient, visible and UV light might cause damage to the photosynthetic system. Shade leaves, representing the vast majority of leaves in the forest (Clark et al. 2008), need to optimize photosynthesis under conditions of limiting and fluctuating light.

Due to global climate change, besides rising mean land surface temperature, the number and intensity of extreme weather events such as severe heat waves are on the increase (IPCC 2014). More frequent and more intense drought periods combined with extreme heat have been termed "global change-type droughts" (see Choat et al. 2018) which markedly enhance evapotranspiration. Limited water supply and high leaf-air vapor pressure gradients can result in stomatal closure, minimizing transpirational cooling and restricting CO_2 flux into leaves (see Boyer 2017). Diminished CO_2 flux into leaves and increased leaf temperatures enhance photorespiration, leading to diminished carbon gain (Muraoka et al. 2000; Franco and Lüttge 2002; Demmig-Adams 1998; Takahashi and Badger 2011). Thus, under water stress, light energy absorbed by photosynthetic pigments becomes even more excessive causing photoinhibition of photosystem II (PSII). In situ, typical responses to such stress conditions include "midday depressions" of net CO_2 uptake (see Zotz et al. 1995 and Krause et al. 2006 for tropical plants). If cooling by ventilation is insufficient, leaves with closed stomata may experience tissue temperatures near the upper limit above which irreversible damage occurs. Drought combined with heat is regarded the major cause of forest decline and tree mortality (see Choat et al. 2018). For instance, in 2005 after a particularly intense drought period in the southern Amazon rain forest, increased tree mortality and a significant decrease in standing biomass was observed (Phillips et al. 2009). In 2015/2016 during a long-lasting drought combined with extreme heat in tropical northern Australia, an extensive dieback of mangroves occurred (Duke et al. 2017).

Plants have developed strategies to counteract adverse effects of stressful climatic conditions on the photosynthetic system (see Lüttge 2008; Yu et al. 2020). In the present review, a range of studies carried out at the Smithsonian Tropical Research Institute in Panama are discussed, dealing with processes of damage, protection, and acclimation in sun and shade leaves of tropical plants under potentially damaging solar radiation. Additionally, studies on leaf heat tolerance limits are reviewed.

2 Protection of Photosystem II Against Destructive Effects of Visible Light

2.1 *Chlorophyll* a *Fluorescence as Indicator of Protective and Damaging Processes in Photosystem II*

In the photosynthetic electron transport system, PSII responds most sensitively to the impact of environmental stress. In studies on leaves of tropical plants reviewed here, analysis of chlorophyll (Chl) *a*) fluorescence emission was used as a powerful noninvasive method to detect processes of protection and damage of PSII (see Krause and Jahns 2003, 2004; Wilson and Ruban 2019). Fluorescence is emitted predominantly by Chl *a* of the PSII light-harvesting pigment system, representing a very minor part of total absorbed photon energy. Development of the pulse-amplitude modulated technique to record Chl *a* fluorescence (Schreiber 1986; Schreiber et al. 1986) made it possible to routinely analyze fluorescence in vivo under sunlight or artificial white actinic light. In the absence of actinic light after dark adaptation of leaves, the low-intensity modulated "measuring light" induces a minimal basic fluorescence level, F_0, related to the state of PSII with fully oxidized Q_A, the primary quinone-type electron acceptor in the PSII reaction center. Part of F_0 is contributed by photosystem I (PSI). A superimposed saturating non-modulated light pulse causes full reduction of Q_A and a concomitant fluorescence rise from F_0 to maximum fluorescence emission, F_m. The fluorescence increment above F_0 is termed variable fluorescence, F_v. Reoxidation of Q_A via electron transport to PSI lowers F_v by means of "photochemical quenching," qP.

According to the model by Butler (1978), the ratio F_v/F_m, determined after dark adaptation of leaves, represents the potential quantum yield of the photochemical reaction in PSII. In non-stressed leaves of a large number of C_3 species, F_v/F_m was close to 0.83 (Björkman and Demmig 1987). Decline of F_v/F_m under stress conditions indicates thermal dissipation of excessively absorbed light energy in PSII, i.e., "nonphotochemical fluorescence quenching," qN, also termed NPQ (Horton et al. 1996; Demmig-Adams 1998; Krause and Jahns 2004).

Nonphotochemical quenching may consist of several components (see Krause and Jahns 2003, 2004; Jahns and Holzwarth 2012; Malnoë 2018). When under excess light, a high proton gradient (ΔpH) is built up across the thylakoid membrane of chloroplasts, and an intrathylakoid pH below 6 activates violaxanthin (Vx) de-epoxidase and the PSII protein PsbS (Li et al. 2004; Correa Galvis et al. 2016; Sacharz et al. 2017; Kaiser et al. 2019). In the violaxanthin cycle, located in the Chl *a* and *b* binding light-harvesting complexes (see Jahns et al. 2009), the diepoxide Vx is de-epoxidized in two steps via antheraxanthin (Ax) to zeaxanthin (Zx). A role of the violaxanthin cycle in thermal dissipation of excess energy was first postulated by Demmig et al. (1987). The ΔpH together with PsbS protein, Zx, and probably also Ax induce "energy-dependent quenching," qE. This qN component is built up rapidly under light stress and relaxes in dark or low light with a half-time of a few seconds. Thus qE appears to be important for protection of PSII under strong and rapidly fluctuating light.

The remaining "photoinhibitory fluorescence quenching," qI, consists of two components. The more rapidly relaxing component (1–2 h) is related to Zx (and Ax). Relaxation is correlated with epoxidation of Zx which has remained after decay of the ΔpH (Thiele et al. 1998). This type of qI has also been termed qZ (Nilkens et al. 2010). An antioxidative function of Zx in the thylakoid membrane, particularly in lipid protection, in addition to its role in qN has been proposed (Havaux et al. 2007; Johnson et al. 2007; Jahns and Holzwarth 2012; Kress and Jahns 2017). The second component of qI is related to inactivation of the D1 protein in the PSII reaction center. This qI component develops slowly within hours; time and extent depend on the intensity of high-light stress and acclimation state of leaves. In low light, slow recovery from D1-dependent qI is based on degradation of inactive D1 protein and its replacement by newly synthesized D1 protein (Aro et al. 1993a,b; Leitsch et al. 1994). In the literature, these processes have been originally regarded as "damage" and "repair" of PSII. However, it appears that in PSII units containing inactivated D1 protein, excitation energy is efficiently dissipated as heat, thus preventing photo-oxidative damage to PSII.

A further component of qN is caused by excitation energy transfer from PSII to PSI (qT) (Allen 2003). Such energy equilibration between the two photosystems is of importance under low and fluctuating light, i.e., in shade leaves, but does not play a major role in sun leaves stressed by excessive light.

Our studies have shown that F_v/F_m, i.e., potential efficiency of PSII, is a highly suitable fluorescence parameter to reveal stress impacts on the photosynthetic system in tropical plant species. F_v/F_m was determined after a 10 min dark period following exposure to high-light stress. This means the ΔpH together with qE was reversed during the dark period, but the two more slowly relaxing components of qN were still present.

2.2 *Function of the Violaxanthin Cycle in Acclimation and Photoprotection of Tropical Plants*

2.2.1 Young and Mature Sun Leaves of Trees

Photoinhibition was examined on sun leaves of potted saplings of *Calophyllum longifolium* Willd. (Calophyllaceae), *Ficus insipida* Willd. (Moraceae), and *Swietenia macrophylla* King (Meliaceae) grown outdoors and additionally on canopy sun leaves of mature trees of *Anacardium excelsum* (Bertero ex Kunth) Skeels (Anacardiaceae), *Antirhea trichantha* (Griseb.) Hemsl. (synonym of *Pittoniotis trichantha* Griseb.) (Rubiaceae), and *Castilla elastica* Cerv. (Moraceae) growing in a humid, semideciduous, seasonally dry forest near Panama City, where leaves were accessed by means of a construction crane. During measurements, maximum photosynthetically active radiation (PAR) was ~2,300 μmol photons $m^{-2}s^{-1}$. Additionally, detached canopy leaves of mature trees were studied under controlled light and temperature conditions (Krause et al. 1995; Thiele et al. 1996, 1997). Exposure to full sunlight in situ caused pronounced photoinhibition indicated by decline in F_v/F_m. Remarkably, young leaves exhibited a substantially faster and stronger decline in F_v/F_m than mature leaves. Reduction of incident light by passing clouds induced fast transient recovery of F_v/F_m. Gradual decline of light flux toward sunset led to restoration of high F_v/F_m ratios.

Based on leaf area, the young leaves studied contained about 50% less Chl *a+b* than mature ones, but Chl *a*/Chl *b* ratios were not different between the two leaf types. This means young leaves contained fewer photosynthetic units per unit leaf area but well-developed antenna systems. Light absorbance (~80%) was only slightly reduced compared to mature leaves, whereas rates of net CO_2 assimilation and PSII-driven electron transport, calculated from fluorescence data according to Krall and Edwards (1992), were much lower in young than mature leaves. Thus, under high light young leaves have to endure substantially higher excessive excitation of the light-harvesting system, which explains the increased photoinhibition.

Besides Chl, carotenoids were analyzed after controlled high-light pretreatment of detached canopy leaves. Based on Chl *a+b*, young leaves possessed a larger pool size of violaxanthin-cycle pigments than mature leaves; for instance, contents of Vx + Ax + Zx in young leaves of *A. excelsum* were ~160 and in mature leaves ~80 mmol mol^{-1} Chl *a+b* (Thiele et al. 1997). Under excessive light young leaves exhibited higher levels of Zx (and Zx + Ax). This indicates that the violaxanthin cycle provided enforced protection to the more severely stressed young leaves. The epoxidation state, calculated from pigment contents as (Vx + 0.5Ax)/(Vx + Ax + Zx), was similar in both leaf types, i.e., 0.9 in the dark and between ~0.2 and ~0.3 under high light, showing high activity of the violaxanthin cycle (Krause et al. 1995; cf Königer et al. 1995).

Following high-light treatment, detached leaves exhibited two distinct phases of F_v/F_m recovery in low light. In young and mature leaves of *A. excelsum* and *C. elastica*, two species characteristic of secondary forest, the fast phase, lasting ~1 h (half time ~30 min), was closely correlated with epoxidation of Zx to Ax and Vx. Correlation coefficients between rise of F_v/F_m and conversion of Zx were between −0.95 and −1.00 (Thiele et al. 1996). In leaves of *A. excelsum* treated with dithiothreitol (DDT) that blocks Zx formation, the fast recovery phase was absent. This shows the fast phase reflects relaxation of Zx-dependent qI (i.e., qZ).

Under low light, fast recovery of F_v/F_m was followed by a slow phase lasting several hours. Upon treatment with streptomycin (SM) to prevent chloroplast-encoded protein synthesis, the slow phase was absent and only the fast recovery phase remained. Quantification of the D1 protein (Thiele et al. 1996) indicated that the slow phase of F_v/F_m recovery is related to degradation of inactivated D1 protein and its de novo biosynthesis and replacement into PSII. In mature leaves incubated with SM after high-light treatment, a substantial proportion of D1 protein was degraded. In contrast, young leaves exhibited only marginal D1 degradation. Obviously, in the young, still developing leaves, enforced photoprotection by means of high Zx formation largely prevents costly D1 protein turnover (Thiele et al. 1997).

Parallel studies with spinach, *Spinacia oleracea* L. (Amaranthaceae), an annual species of temperate regions, supported the view that by means of high violaxanthin-cycle activity, PSII is stabilized and D1 inactivation minimized (Thiele et al. 1996). Our results are in agreement with those of Demmig-Adams et al. (1995) showing that high-light acclimation of leaves diminishes photoinactivation of the D1 protein. However, the observation of considerable D1 degradation in mature canopy leaves (Thiele et al. 1996) is not consistent with the suggestion by Demmig-Adams et al. (1995) that in fully acclimated leaves D1 inactivation is negligible.

Experiments to reveal a possible reduction of whole-plant biomass accumulation due to high-light stress were performed with tree seedlings of two species, *C. longifolium*, a common late-successional tree of Panamanian lowland and lower montane forests, and *Tectona grandis* L.f. (Verbenaceae), a pioneer species originating in Southeast Asia (Krause et al. 2006). In three parallel long-term experiments, the seedlings were grown under automatic neutral shading that reduced visible and UV light by ~50% whenever PAR surpassed 1,000, 1,200, or 1,600 μmol m^{-2} s^{-1}, respectively. Photoinhibition (i.e., decline in F_v/F_m) was studied in relatively young but fully expanded leaves. Plants shaded above 1,000 or 1,200 μmol photons m^{-2} s^{-1} exhibited considerably less photoinhibition than leaves of control plants grown permanently under full solar radiation. This effect was more pronounced in leaves of *C. longifolium* than in the pioneer *T. grandis* (teak), the latter showing generally less pronounced F_v/F_m decline. Characteristic midday depression of CO_2 assimilation (recorded with *C. longifolium*) was absent under partial shade. No significant loss of biomass accumulation was found in seedlings grown under full solar radiation. Leaves of these seedlings contained larger pool sizes of violaxanthin-cycle pigments; around midday substantially higher Zx levels were present. The content of α-tocopherol in *C. longifolium* was ~20% higher under full sun than shading above 1,200 μmol photons m^{-2} s^{-1}. Tocopherols are well-known antioxidants present in the thylakoid membrane and envelope of chloroplasts (Fryer 1992; García-Plazaola et al. 2004; Havaux et al. 2005).

These results were confirmed by a study (Krause et al. 2012) on shade-tolerant seedlings of an understory/forest gap species, *Piper reticulatum* L. (Piperaceae), and two shade-tolerant late-successional tree species, *Tetragastris panamensis* (Engl.) Kuntze (Burseraceae) and *Virola surinamensis* (Rol. ex Rottb.) Warb. (Myristicaceae). Growth under unshaded conditions did not affect biomass accumulation. In contrast to these three species, the late-successional tree *Ormosia macrocalyx* Ducke (Fabaceae) exhibited a significant reduction in relative growth rate under full sunlight as compared to growth under ~50% neutral shade (applied whenever ambient PAR exceeded 1,200 μmol m^{-2} s^{-1}); during a 4-month growth period under full sun, total biomass was reduced by ~27%. *O. macrocalyx* showed similar acclimatory responses of the violaxanthin cycle as the other tested species, but leaf extracts indicated substantially less UV protection, both in the UV-B and UV-A spectral region, than observed in leaves of *T. panamensis* and in mature canopy sun leaves of several other species (Krause et al. 2003a; see Sect. 4.2). Whereas leaves of *T. panamensis* showed highly increased α-tocopherol contents under full sunlight, tocopherol levels (β- and γ-tocopherol) were extremely low in *O. macrocalyx*.

In several but not all tested species, further acclimatory responses of the photosynthetic pigment system to full sunlight exposure included the reduction of α-carotene and increase in β-carotene and lutein contents of leaves. This was not observed in *O. macrocalyx*. It has been suggested that α-carotene which is present at high levels in shade leaves contributes to light harvesting (Krause et al. 2001; Matsubara et al. 2009), whereas β-carotene supports photoprotection by quenching excited triplet-state Chl *a* (^{3}Chl *a**) and singlet-state dioxygen ($^1O_2$*) in the core antennae (Edge et al. 1997; Scheer 2003). Lutein may act as an antioxidant by means of ^{3}Chl *a** quenching. Moreover, lutein possibly promotes qE as a quencher of singlet-state excited Chl *a* (^{1}Chl *a**) (Jahns and Holzwarth 2012).

As expected, seedling growth of *Psychotria marginata* Sw. (Rubiaceae), a highly shade-tolerant understory species, was strongly reduced under full sunlight. Total biomass after 6.8 months was ~58% lower, compared with plants cultivated at 40% of ambient PAR (Krause et al. 2012). Nevertheless, a number of individuals were capable of acclimatory responses, demonstrated by increases of violaxanthin-cycle pigments, as well as increases in β-carotene and decreases in α-carotene levels.

These results show that in most species tested, sun leaves are well protected from adverse effects of highly excessive sunlight. The violaxanthin cycle plays a major role in protection, aided by various antioxidative activities. No experimental evidence was gathered that midday depressions of CO_2 assimilation and the metabolic costs of acclimation would reduce biomass accumulation.

2.2.2 Plants Growing in Treefall Gaps

In the moist tropical lowland forest on Barro Colorado Island (Gatún Lake, Panama), photoinhibition was investigated in small natural treefall gaps of 60–90 m^2 (Krause and Winter 1996). Special light conditions prevail in such gaps. During most of the daytime, PAR is low, usually below 100 μmol m^{-2} s^{-1}, except for brief, 1–2 h periods of direct sun exposure of the gap floor, when under clear sky PAR suddenly rises to 1,700–1,800 μmol m^{-2} s^{-1}. Changes in F_v/F_m induced by such high-light exposure were monitored in situ in nine species. In all cases, an immediate decline in F_v/F_m ensued, indicating photoinhibition that was reversible upon return to low light. Recovery of F_v/F_m occurred in two phases. The fast phase (~1 h) started after the end of the high-light period and represented the major part of F_v/F_m increase. The following slow phase proceeded until sunset. These in situ observations of recovery confirmed the results obtained in the laboratory with detached canopy sun leaves (see Sect. 2.2.1). The similarity in response of nine species of different taxa and life forms to high-light periods indicates a general acclimation to gap conditions.

In mature leaves of eight species growing in three different gaps, photosynthetic pigment contents were determined (Thiele et al. 1998). Leaves were collected during the diurnal course and, after F_v/F_m recording, used for pigment analysis. Consistent with data by Königer et al. (1995), the pool size of violaxanthin-cycle pigments (Vx + Ax + Zx) was on average considerably lower in gap leaves than in canopy sun leaves (c.f. Thiele et al. 1997).

A striking result was that leaves of all gap plants tested exhibited a close correlation between changes in F_v/F_m and content of Zx, both during the high-light period and during the following fast recovery phase under low light. Highest correlation coefficients (between −0.90 and −0.99) were found between decline of F_v/F_m and increasing content of Zx under high light. Correlation coefficients between fast recovery of F_v/F_m and Zx epoxidation in the shade were only slightly lower (between −0.66 and −0.99). Overall, there was no clear correlation between slow F_v/F_m recovery and Zx. The data show that the fast recovery phase represents reversion of Zx-dependent photoinhibition (qZ), whereas the small slow phase probably indicates "repair" of PSII by means of D1 protein degradation and resynthesis. Apparently, under gap conditions, gross D1 protein inactivation does not occur, whereby demand for de novo biosynthesis is reduced.

It has been suggested that treefall gaps are important to maintain species diversity in the tropical forests (Yavitt et al. 1995). Typically, treefall gaps are first inhabited by pioneer species that germinate after gap formation, followed by late-successional species that may have been present as seedlings in the forest understory already before gap formation. In a model study using potted plants in artificial gaps (Krause et al. 2001), responses to light conditions of three pioneer and three late-successional tree species were determined during dry season. The simulated narrow, medium, and wide gaps allowed 0.5, 1.5, and 3 h, respectively, of daily direct sun exposure under clear sky. After cultivation of seedlings for 2 months, photosynthetic pigments were analyzed in leaf discs collected during the course of the day.

Interestingly, in leaves of pioneer species, high light exposure caused considerably less decline in F_v/F_m than in late-successional species. Recovery in pioneer species was completed in the late afternoon, reaching F_v/F_m morning vales of 0.8, independent of gap width. In late-successional species, recovery of F_v/F_m following the high-light period tended to be incomplete at sunset. The pool size of violaxanthin-cycle pigments (based on Chl *a+b*) and contents of UV-B absorbing substances (based on leaf area) strongly increased with gap size, but surprisingly did not differ significantly between the two life forms. Also the contents of Zx and Ax accumulating under high light depended on gap size and were similar in pioneer and late-successional species.

The reasons for reduced photoinhibition in pioneer species are not yet fully clarified, but might partly be explained by structural adjustment of the light-harvesting complex of PSII. In contrast to late-successional species, pioneer species reached Chl *a*/*b* ratios above 3.0 in wide gaps, and levels of β-carotene (based on Chl *a+b*) strongly increased with gap width, whereas levels of α-carotene decreased. Increased Chl *a*/*b* ratios indicate a reduced number of peripheral light-harvesting, Chl *a+b* binding complexes (Anderson and Osmond 1987), i.e., decreased ratios of these complexes to Chl *a* binding core complexes of PSII. The smaller size of the light-harvesting antenna system diminishes high-light stress. Higher maximum rates of CO_2 assimilation in pioneer species might contribute to reduced photoinhibition.

A special high-light response was observed in the late-successional species *Quararibea asterolepis* Pittier (Malvaceae – Bombacoideae). Leaves tilted toward the stem that reduced light exposure. In low light the original leaf position was restored. A similar effect had been noticed with the shrub *Psychotria limonensis* Krause (Rubiaceae) in a natural gap (Krause and Winter 1996). Leaves wilted upon exposure to high light and regained turgescence in the shade.

2.3 The Lutein Epoxide Cycle in Tropical Plants

Besides the ubiquitous violaxanthin cycle, the lutein epoxide cycle operates in certain taxa that are not phylogenetically closely related, as has been documented in a number of species of tropical and temperate climate. According to a study of *Inga sapindoides* Willd. (Fabaceae), the lutein epoxide cycle is located in the antennae of both PSII and PSI, but the major pool of lutein epoxide (Lx) can be found in the light-harvesting complexes of PSII (Matsubara et al. 2007). The lutein epoxide cycle is catalyzed by the enzymes of the violaxanthin cycle, namely, violaxanthin de-epoxidase and zeaxanthin epoxidase (Matsubara et al. 2001; García-Plazaola et al. 2007). In contrast to the β,β-carotenoids Zx, Ax, and Vx with two β-rings at the end of the phytoene chain, the β,ε-carotenoid lutein has only one β-ring that can be epoxidized to Lx. In leaves exhibiting high levels of Lx in the shade, Lx is de-epoxidized to lutein under high light in parallel with de-epoxidation of Vx to Ax and Zx. Reconversion of lutein to Lx occurs under low light or in the dark, but proceeds considerably slower than epoxidation of Zx via

Ax to Vx in the violaxanthin cycle (Matsubara et al. 2001, 2005; Snyder et al. 2005; Esteban et al. 2007).

In the neotropical genus *Inga*, shade leaves of nine tested species possessed large Lx pool sizes, between 18 and 38 mmol mol^{-1} Chl *a+b* (Matsubara et al. 2008). Sun leaves had very low Lx levels. When sun-acclimated seedlings of *I. marginata* Willd. were transferred into deep shade, leaf Lx content gradually increased within 4 days from <5 to >15 mmol mol^{-1} Chl *a+b*, while the lutein content decreased. Noticeably, the decrease of lutein was significantly stronger than the increase of Lx. Apparently, part of the lutein was degraded concomitant with lutein epoxidation.

By analyzing fast Chl *a* fluorescence induction (Strasser et al. 1995), the function of Lx in leaves of cultivated *I. marginata* seedlings could be partly clarified (Matsubara et al. 2008). Shade-acclimated leaves containing ~20 mmol Lx mol^{-1} Chl *a+b* were exposed briefly to high light that de-epoxidized Lx and Vx by ~50%. Upon returning to shaded conditions for 1 day, Lx levels remained relatively stable, while Vx levels were restored by epoxidation of Zx and Ax. Fluorescence induction was recorded in two epoxidation states: (1) when Lx pool size was high before high-light exposure and (2) when Lx pool size was reduced by ~50% 1 day later. Levels of Vx were identical under both conditions. Fluorescence rise in the first phase of induction (O-J), indicating the reduction kinetics of Q_A, was significantly faster in leaves with high compared to low Lx content. This result suggests a light-harvesting function of Lx. F_v/F_m ratios did not differ between both states, i.e., at decreased Lx content, there was no photoinhibition (qI) that would lower the potential efficiency of PSII. This conclusion was supported by two independent observations:

1. In the state of high Lx content, fast nonphotochemical (ΔpH-dependent) fluorescence quenching, qE, was diminished compared to the state of low Lx content.
2. The light compensation point of net photosynthetic CO_2 assimilation was lower in the state of high Lx level.

Apparently increased light-harvesting capacity mediated by Lx is diminishing thermal dissipation of excitation energy in PSII, thereby promoting CO_2 assimilation in limiting light. Thus the presence of Lx might be beneficial for light harvesting under deep shade. On the other hand, under excessive solar radiation, lutein may exert an antioxidative function and promote ΔpH-dependent fluorescence quenching, qE (see Sect. 2.2.1). Results of a diurnal time-course experiment in situ with a mature tree of *Inga spectabilis* (Vahl) Willd. (Fabaceae) were consistent with observations on cultivated *Inga* seedlings. Lx pool sizes were high (~20 mmol mol^{-1} Chl *a+b*) in shade leaves and low (<5 mmol mol^{-1} Chl *a+b*) in sun leaves. Changes in Lx content during the day were negligible in both leaf types, but in sun leaves violaxanthin-cycle pigments exhibited marked turnover correlated with changes in F_v/F_m.

Surprisingly, in dark-adapted leaves of *V. surinamensis*, extraordinarily high Lx pool sizes (~62 mmol mol^{-1} Chl *a+b*) were found both in sun and shade leaves. Such high levels were never reported before. Similarly, two other *Virola* species, *V. elongata* (Benth.) Warb. and *V. sebifera* Aubl., exhibited high Lx pool sizes in both leaf types, as revealed by a broad survey of photosynthetic pigments in leaves of 86 neotropical species of various life forms and habitats from 64 families (Matsubara et al. 2009). In the species *Otoba novogranatensis* Moldenke (Myristicaceae), an intermediate Lx content was detected (~10 mmol mol^{-1} Chl *a* +*b* in sun and ~25 mmol mol^{-1} Chl *a*+*b* in shade leaves). In potted seedlings of *V. surinamensis* grown at four different intensities (100%, 70%, 40%, and 12%) of natural PAR (Krause et al. 2012), no distinct differences in Lx contents were observed in samples collected at pre-dawn, but levels of violaxanthin-cycle pigments, lutein, and β-carotene had decreased in response to lowered PAR, whereas α-carotene levels had increased. Generally, low α/β-carotene ratios and high levels of β-carotene were found in sun leaves of tropical plants (Matsubara et al. 2009). These results are in agreement with the supposed function of Lx in light harvesting under limiting light and of Zx, Ax, lutein, and β-carotene in photoprotection under excessive solar radiation.

In contrast to *Inga*, a diel course of the Lx cycle was discovered in situ in sun leaves of a mature tree of *V. elongata* growing in the primary, moist lowland forest of San Lorenzo National Park close to Colón/Panama (Matsubara et al. 2009). Lx and Vx were de-epoxidized in parallel, as PAR rose from morning to midday up to ~2,000 μmol m^{-2} s^{-1}. No clear turnover of Lx and Vx was seen in shade leaves. Upon dark adaptation of sun leaves, restoration of Lx was slower than that of Vx, but in the following morning, both pigments were found to be fully restored (Fig. 1). Diel operation of the Lx cycle was also observed in sun-acclimated cultivated seedlings of *V. surinamensis* and *Virola multiflora* (Standl.) A.C.Sm. (Matsubara et al. unpublished).

In the survey of 86 species by Matsubara et al. (2009), pool sizes of Lx >10 mmol mol^{-1} Chl *a*+*b* were found in shade leaves or both shade and sun leaves of only 7 species; 12 further species contained 5–10 mmol Lx mol^{-1} Chl *a*+*b* in shade leaves. Similar to *Inga*, two further Fabaceae species tested, *Albizia guachapele* (Kunth) Dugand and *Dalbergia monetaria* L.f., contained much lower Lx in sun than in shade leaves. Moreover, the lutein epoxide cycle is operating in Lauraceae, a family with tropical and subtropical species (Esteban et al. 2007, 2008; Förster et al. 2009), e.g., in the tropical tree *Persea americana* Mill. (avocado).

So far, only in *Virola* species the Lx pool size has been found to be extremely high in both sun and shade leaves, and only in sun leaves of this genus, operation of the Lx cycle in a diel fashion has been observed. Overall, Lx accumulation in shade leaves appears to be widespread among tropical plants and is particularly frequent in certain families such as Fabaceae, Myristicaceae, and Lauraceae. Obviously, Lx exerts a light-harvesting function under low light, whereas lutein formed by de-epoxiation of Lx under high light is supposed to participate in photoprotection.

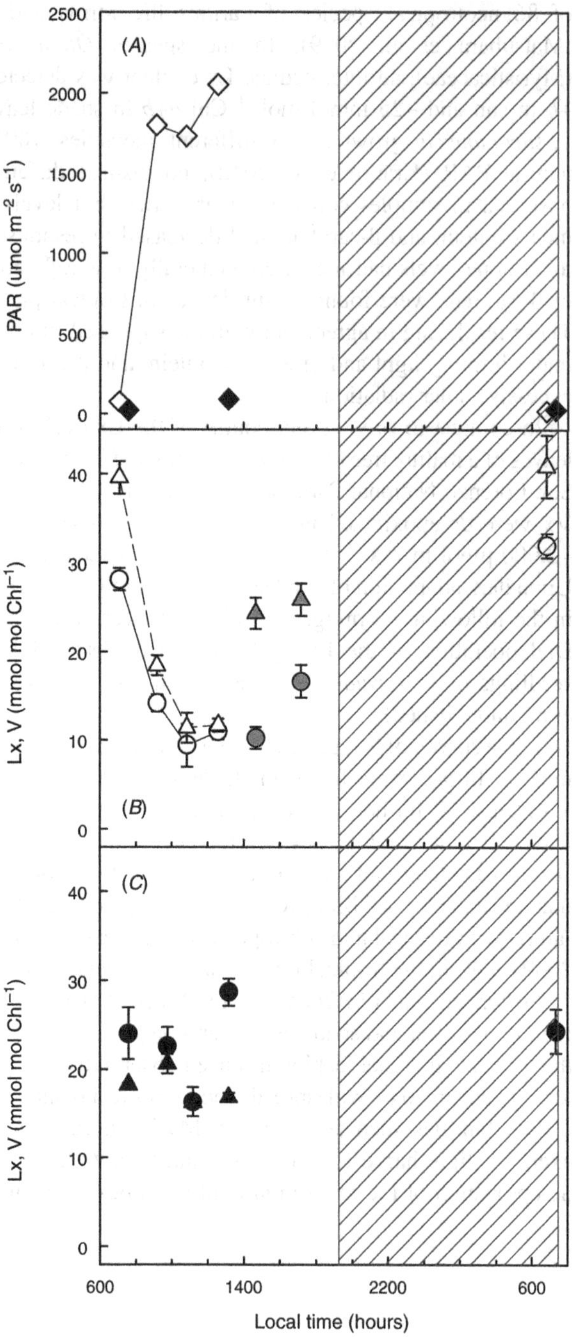

Fig. 1 Diel changes in lutein epoxide and violaxanthin in sun leaves of *V. elongata*. Upper panel: PAR incident on sun (open symbols) and shade leaves (closed symbols). Middle panel: changes in levels of lutein epoxide (○) and violaxanthin (Δ) in sun leaves in situ from morning until 12:40 h and in detached dark-adapted sun leaves in the afternoon and in situ in the following morning. Lower panel: levels of lutein epoxide (●) and violaxanthin (▲) in shade leaves in situ. Different sets of five leaves each were used. Error bars (shown when larger than symbols) indicate s.e. ($n = 5$). Reproduced from Matsubara et al. (2009) with permission from CSIRO Publishing

3 Photoinhibition of Photosystem I

Electrons are transferred from PSII via the electron transport chain to plastocyanin that donates an electron to cation radical $P700^+$, a Chl *a* dimer, in the reaction center complex of photosystem I (PSI). Light is absorbed by PSI preferentially in the long-wavelength range around 700 nm. From excited P700 an electron is transmitted via the PSI electron transport system, comprising two Chl *a* binding proteins, phylloquinone, and three iron sulfur centers, to the PSI electron acceptor ferredoxin (Amunts et al. 2007; Kozuleva and Ivanov 2016).

It has been assumed that under high-light stress, PSI is scarcely affected compared to PSII. However, substantial PSI photoinhibition has been observed under specific stress conditions, e.g., under high light at low (above freezing) temperatures in leaves of chilling-sensitive and chilling-tolerant plants (Havaux and Davaud 1994; Terashima et al. 1994; Tjus et al. 1998; Barth and Krause 1999; Teicher et al. 2000). Presumably, when the electron transport system beyond PSI is saturated, electrons are transferred to molecular oxygen forming superoxide radicals (O_2^-) that can be transformed to other reactive oxygen species (ROS) damaging iron sulfur centers and proteins A and B in PSI (Teicher et al. 2000; Sonoike 2011). Insufficient turnover rates of ROS (Asada 1999) or generally low antioxidative activities may lead to PSI damage.

In sun and shade leaves of several tropical species, photoinhibition of PSI was investigated by Barth et al. (2001). The method of Klughammer and Schreiber (1994) was applied to record the maximum absorbance change at 810 nm (ΔA_{810max}) representing the capacity to oxidize P700 to $P700^+$. This method is supposed to minimize the contribution of plastocyanin to the absorbance change. In parallel, F_v/F_m was recorded to monitor PSII efficiency. Detached shade leaves (*A. excelsum* and *V. surinamensis*) were exposed to 1,800 μmol photons m^{-2} s^{-1}, and sun leaves of *A. excelsum* and *C. elastica* were exposed to 2,000 μmol photons m^{-2} s^{-1} white light. In agreement with data by Königer et al. (1995), the shade leaves had substantially lower levels of violaxanthin-cycle pigments than sun leaves. Moreover, a smaller proportion of Vx became de-epoxidized to Ax and Zx under high light (Barth et al. 2001). After 75 min high-light exposure, ΔA_{810max}, i.e., potential PSI activity, was not significantly or only slightly affected in both leaf types. In contrast, PSII efficiency was strongly reduced compared to controls. F_v/F_m declined by 30–35% in sun leaves and by 65–75% in shade leaves. Similar results were obtained with shade leaves of *Dieffenbachia longispatha* Engl. & K.Krause (Araceae) and *Piper carrilloanum* C.DC. (Piperaceae), two forest understory species.

Several mechanisms probably contribute to PSI protection. Photoinhibition of PSII reduces electron flow to PSI, thereby diminishing electron pressure on PSI and reducing formation of ROS. Moreover, electron transport from PSII to PSI is known to be controlled at the cytochrome b_6f complex (see Tikhonov 2014), regulated by the high ΔpH built up under excess light, the latter being mediated by the "proton gradient regulating 5" protein (pgr5). In the *pgr5* mutant of *Arabidopsis thaliana* (L.)

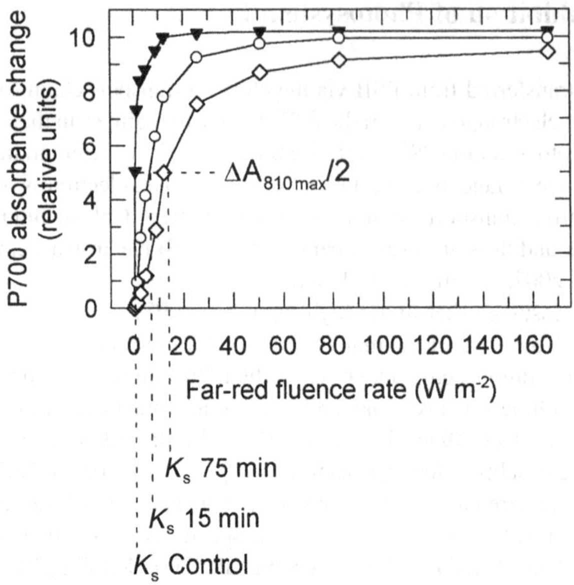

Fig. 2 Effects of pre-illumination of leaves with high white light on light saturation of P700 photooxidation. Samples of shade leaves of *V. surinamensis* were pre-illuminated at 1,800 μmol photons m^{-2} s^{-1}. After dark adaption, absorbance changes at ~810 nm were measured as a function of increasing far-red fluence rate. The "saturation constant," K_s, represents the far-red intensity at which half of the maximum absorbance change, ΔA_{820max}, is reached in the steady state. The graph demonstrates light saturation curves of P700 oxidation and indicates K_s of untreated controls (▼) and of samples pre-illuminated for 15 (o) and 75 min (◊). Reproduced from Barth et al. (2001) with permission from John Wiley and Sons

Heynh. (Brassicaceae), PSI was described to be highly susceptible to excess light, proving that the pgr5 protein is essential for photoprotection of PSI (Munekage et al. 2002; Tiwari et al. 2016; Tikkanen and Grebe 2018; Lima-Melo et al. 2019). Wild-type *A. thaliana* grown at 125 μmol photons m^{-2} s^{-1} and exposed to 1,000 μmol photons m^{-2} s^{-1} (Lima-Melo et al. 2019) did not exhibit significant reduction in the capacity of P700 oxidation (ΔA_{810max}), i.e., PSI was highly stable, confirming the results of Barth et al. (2001) on sun and shade leaves of tropical plants.

Although ΔA_{810max} was not or only slightly affected in leaf samples pretreated in high light, saturation curves of P700 oxidation under far-red light ($\lambda > 720$ nm) exhibited a conspicuously diminished efficiency of P700 oxidation (Barth et al. 2001). The far-red intensity required to induce 50% of P700 oxidation in the steady state (i.e., 50% of ΔA_{810max}), designated as "saturation constant" (K_s), was markedly increased. The extent of increase in K_s depended on the duration of high-light pretreatment (Fig. 2) and was much more pronounced in shade than in sun leaves (Fig. 3). During 75 min under excess light, K_s increased by more than 40 times in shade but only about 3 times in sun leaves. A similar but kinetically different effect has been reported in a study of the alpine species *Geum montanum* L. (Rosaceae)

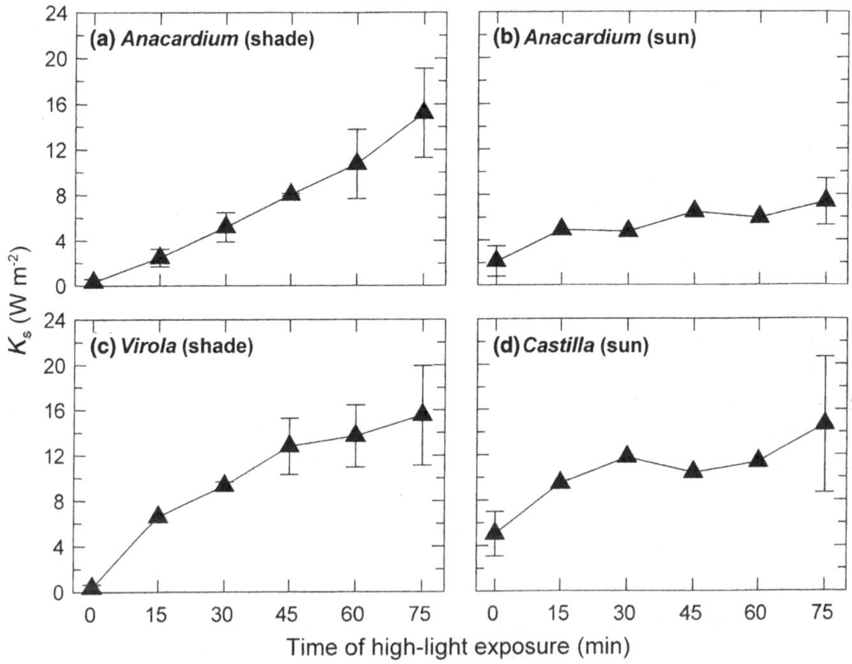

Fig. 3 Effects of pre-illumination of leaves with high white light on the saturation constant, K_s, of P700 oxidation under far-red light (c.f. Fig. 2). K_s is presented as function of pre-illumination time. Shade leaves were pre-illuminated with 1,800 and sun leaves with 2,000 μmol photons m^{-2} s^{-1}; (**a**, **b**) shade and sun leaves of *A. excelsum*; (**c**) shade leaves of *V. surinamensis*; (**d**) sun leaves of *C. elastica*. Means ± s.d. from three independent experiments [except for only two experiments at 15–65 min in (**b**) and at 15 min in (**c**)]; s.d. not shown when smaller than symbols. Reproduced from Barth et al. (2001) with permission from John Wiley and Sons

(Manuel et al. 1999). Upon return to shade conditions, the increase in K_s was reversible within hours. Recovery of K_s was much slower in shade than in sun leaves similar to the slower recovery from PSII photoinhibition (diminished F_v/F_m) in shade leaves. However, there was no close correlation between the two parameters (Barth et al. 2001).

It has been suggested that the increase in K_s, indicating decreased efficiency of P700 oxidation, is caused by enhanced charge recombination between $P700^+$ and reduced electron acceptors in the PSI reaction center (Barth et al. 2001). According to Brettel (1997) and Polm and Brettel (1998), such charge recombination yields reduced P700 in the ground state, partly via triplet P700 (^{3}P700). Thereby, excess energy in PSI is harmlessly dissipated as heat and potentially harmful reduction of O_2 to superoxide radicals (O_2^-) diminished.

After high-light pretreatment of shade and sun leaves with 1,800 and 2,000 μmol photons m^{-2} s^{-1}, respectively, the proportion of $P700^+$ in the steady state during subsequent illumination with white light (120 and 1,750 μmol photons m^{-2} s^{-1}) was monitored (Barth et al. 2001). The proportion of $P700^+$ was larger under high than under low light conditions. Particularly when monitored under low light (120 μmol photons m^{-2} s^{-1}), the percentage of $P700^+$ increased with duration of preceding high-light exposure. This accumulation of $P700^+$ can be interpreted to result from control of electron transport from PSII to PSI, as discussed above (see also Harbinson et al. 1989). $P700^+$ may convert excitation energy to heat and thus contribute to avoid PSI damage.

In PSI, violaxanthin-cycle pigments are present in similar amounts as in PSII and can be de-epoxidized to Ax and Zx under high light (Thayer and Björkman 1992: Färber et al. 1997). However, thus far a protective effect of the violaxanthin cycle in PSI has not been confirmed. Cyclic electron transport from PSI to the plastoquinone pool and back to PSI has been suggested to contribute to PSI protection (Heber and Walker 1992; Fork and Herbert 1993; Endo et al. 1999; Teicher et al. 2000). Two pathways of cyclic electron transport have been described: (1) via NADPH using chloroplastic NAD(P)H dehydrogenase (Teicher and Scheller 1998; Joet et al. 2001) and (2) via ferredoxin by means of ferredoxin plastoquinone oxidoreductase (Bendall and Manasse 1995). Electron cycling is supposed to increase the trans-thylakoid ΔpH, thereby diminishing electron pressure on PSI (see above). However, a comparison of PSI photoinhibition (decrease in ΔA_{810max}) between wild-type plants of the chilling-sensitive species *Nicotiana tabacum* L. (Solanaceae) and transformants (ΔndhCKJ) deficient in functional NAD(P)H dehydrogenase (Barth and Krause 2002) did not show a difference in PSI sensitivity to high light. Apparently, electron transport around PSI catalyzed by NAD(P)H dehydrogenase does not significantly contribute to PSI stability.

In summary, in the sun and shade leaves of tropical plants that have been examined, the high stability of PSI under excess irradiation may be based on several factors: (1) thermal dissipation of excitation energy in PSII as indicated by nonphotochemical fluorescence quenching (qE and qI), possibly promoted by cyclic electron flow around PSI via ferredoxin and plastoquinone, which diminishes electron flow to PSI and leads to accumulation of $P700^+$; (2) control of linear electron transport by a high trans-thylakoid proton gradient (ΔpH) and the pgr5 protein; and (3) enhanced charge recombination within PSI as indicated by decrease in the efficiency of P700 oxidation under far-red light.

It should be noted that solar UV-B radiation may affect PSI (Krause et al. 2003a), as reviewed below.

4 Responses of Leaves to High Solar Ultraviolet Radiation

4.1 Damage by Ultraviolet Light

Although ambient UV-B radiation exerts a number of beneficial functions in plants, e.g., defense against herbivores by accumulation of UV-absorbing substances, it may cause damage when protection by UV-screening and antioxidative systems is insufficient (see Bornman et al. 2019). Because of the small solar zenith angle in the tropics, plants encounter increased UV radiation even in the lowlands that may adversely affect photosynthesis (Ziska 1996). From an assessment of global stratospheric O_3 levels between 1964 and 2000, there is no evidence of O_3 decline over the tropical belt between 25°N and 25°S that would increase impact of UV radiation on plants (Fioletov et al. 2002). Currently it is not expected that solar UV-B (280–315 nm) and UV-A (315–400 nm) at the earth's surface will increase as a consequence of climate change. However, effects of UV radiation on plants may be altered in combination with various climate change factors. For instance, more frequently occurring extreme weather events (see Introduction) such as prolonged hot drought periods with reduced cloud cover may increase the impact of both UV and visible radiation on tropical plants. Most studies on effects of UV radiation on plants have been carried out using artificial supplementary UV light, often much above ambient levels, whereas investigations on plants exposed to ambient UV light are relatively scarce (see Bornman et al. 2019).

Seedlings developing in the forest understory may become exposed to high solar UV and visible radiation in sun-flecks (Watling et al. 1997) or upon formation of gaps. To simulate such situations, shade-grown tree seedlings and, for comparison, detached sun leaves of mature trees were exposed for brief periods (10–75 min) to direct sunlight (Krause et al. 1999). During light exposure, whole plants or detached leaves were shielded by plastic filters transmitting most of the visible light but either excluding ("Mylar") or transmitting ("Aclar") UV-B radiation. In addition, cutoff glass filters transmitting ~90% of visible light and absorbing either most UV-B or both UV-B and UV-A radiation were placed above central areas (25 cm^2) of leaf blades. Shade-grown seedlings of *A. excelsum* and *V. surinamensis* were tested. *A. excelsum* is a pioneer tree characteristic to secondary forests, but seedlings require shade conditions for early development. Seedlings of *V. surinamensis*, a widespread late-successional tree species, are also shade tolerant. In both cases, exposure to full sunlight including natural UV-B caused a very strong decline in F_v/F_m (from ~0.8 to ~0.2) that was markedly less rapid when UV-B was absent, demonstrating that UV-B substantially contributed to PSII photoinhibition. Additional experiments with *A. excelsum* by using cutoff filters proved that UV-A, besides visible light and UV-B, caused part of the F_v/F_m decline observed under full sunlight. Upon returning the seedling to shade conditions, the recovery from PSII photoinhibition, examined in shade-grown leaves of *V. surinamensis*, was extremely slow, requiring about 2 weeks to reach F_v/F_m ratios close to control values (Fig. 4). Obviously, severe protein damage in PSII occurs when shade-acclimated leaves are suddenly exposed to direct sunlight. According to Aro et al. (1993a,b), shade leaves are incapable of fast D1 protein degradation and de novo biosynthesis. Besides D1 protein inactivation, the D2 protein in PSII might be damaged, as suggested by studies that applied artificial UV-B (Friso et al. 1994; Jansen et al. 1998).

In detached sun leaves, enhancement of PSII photoinhibition by solar UV-B and UV-A was observed in a number of tests, but the response to UV radiation varied or was even absent (Krause et al. 1999). Possibly, variable sensitivity to UV stress depended on differences in the acclimation state of individual leaves. Full recovery in the shade occurred within about 1 h, indicating that the F_v/F_m decline in sun leaves

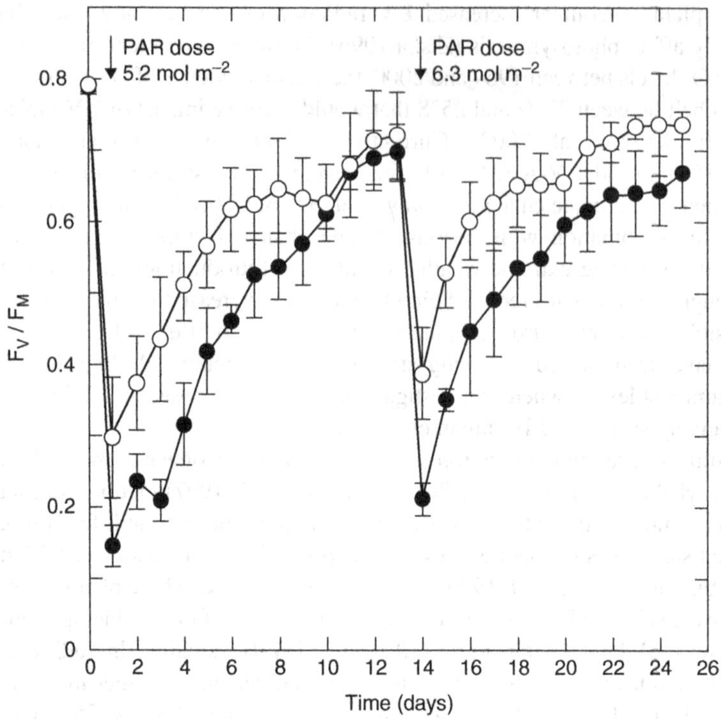

Fig. 4 Photoinhibition and following recovery in mature leaves of shade-grown *V. surinamensis* seedlings (age 7–8 months) recorded upon exposure to full sunlight in the presence (●) and absence (o) of ambient UV-B. Plants were sun-exposed for 1 h on days 1 and 14, respectively (doses of PAR noted in the graph). Subsequently, plants were transferred back to shade conditions. F_v/F_m, indicating potential efficiency of PSII, was determined before sunrise or, following sun exposure, after 10 min dark adaptation. Means ± s.e. of three leaves from three different seedlings. Reproduced from Krause et al. (1999) with permission (www.plantphysiol.org). Copyright American Society of Plant Biologists

caused both by visible and UV radiation was largely related to formation of Zx in the violaxanthin cycle.

Further studies on shade-grown tree seedlings revealed effects of solar UV-B radiation on PSI and photosynthetic CO_2 assimilation (Krause et al. 2003a). Exposure of seedlings to full sunlight did not affect the capacity of P700 photooxidation ($\Delta_{A810max}$), as had been observed by exposure to high artificial white light (see Sect. 3, Barth et al. 2001). This high stability of PSI is remarkable in view of the strong effect of full sunlight on PSII in shade leaves (see Fig. 4). However, the decrease in efficiency of P700 oxidation (determined by means of Δ_{A810}) was much more pronounced upon sun exposure in the presence than in the absence of ambient UV-B radiation, indicating a strong effect of UV-B on PSI. It has been hypothesized that the decreased efficiency of P700 oxidation might be caused by charge recombination between $P700^+$and reduced electron acceptors of P700 (see Sect. 3).

Full solar radiation including UV-B inhibited net CO_2 assimilation in shade leaves of *A. excelsum* more strongly than full sun exposure in the absence of UV-B. Similar results, although less pronounced, were obtained with shade leaves of *V. surinamensis* and *C. longifolium* (Krause et al. 2003a). Gas exchange analysis after sunlight exposure did not show increased stomatal control of CO_2 uptake. There was no direct relationship between decline in F_v/F_m and inhibition of CO_2 assimilation both under full sunlight and in absence of most UV-B. Probably, damage of enzymes and inhibition of light activation of the carbon reduction cycle are involved in the inhibition of CO_2 assimilation. Inactivation of ribulose-1,5-bisphospate carboxylase/oxygenase (Rubisco) in *Brassica napus* L. (Brassicaceae) by high supplementary UV-B has been reported by Allen et al. (1997). Noticeably, de-epoxidation of Vx to Zx and Ax in the violaxanthin cycle was not affected by ambient UV-B radiation (Krause et al. 1999).

4.2 Acclimation

Effective protection of leaves from solar UV radiation is achieved by vacuolar, mostly epidermal phenolic compounds such as flavonoids and derivatives of hydroxycinnamic acid (Reuber et al. 1993, 1996; Burchard et al. 2000; Kolb et al. 2001; Markstädter et al. 2001). In studies on acclimation of tropical plants to ambient light conditions reviewed below, absorbance of ethanolic-aqueous leaf extracts in the UV spectral region served as a relative measure of UV screening.

In numerous species from two lowland forests and one montane cloud forest (~1,000 m above sea level), the degree of acclimation to UV and visible solar radiation was investigated (Krause et al. 2003b). Canopy sun leaves of large trees were compared with shade leaves of tree crowns and leaves of plants in the forest understory. Sun leaves of one mangrove species, *Laguncularia racemosa* (L.) C.F. Gaertn. (Combretaceae), were included in the study. PAR around midday under clear sky was about 2,000–2,300, 200–400, and 20–30 μmol photons m^{-2} s^{-1} for canopy sun leaves, for shade leaves of tree crowns, and for leaves in the understory, respectively. UV-A shielding at 375 nm of the adaxial and abaxial leaf epidermis was recorded in situ with a UV-A PAM Chlorophyll Fluorometer (Kolb et al. 2001; see Barnes et al. 2015). In part of the species studied, composition of chloroplast pigments and UV-A absorbance at 375 nm and UV-B absorbance at 305 nm in leaf extracts were analyzed.

Very high UV-A shielding (estimated at >90%) by the adaxial epidermis was observed in sun leaves of most species tested. UV-A shielding of the abaxial epidermis was lower but varied widely. Generally, in shade leaves of tree crowns and more so in leaves from the understory, epidermal UV-A screening was markedly lower than in sun leaves. Similar differences between the three leaf types were seen in UV-B and UV-A absorbance of leaf extracts at 305 and 375 nm, respectively. In simulated treefall gaps (see Sect. 2.2.2, Krause et al. 2001), UV-B absorbance of leaf extracts indicated significantly strengthened UV-B protection of plants growing in wide compared to small gaps, i.e., leaves responded to longer daily high-light periods with increased production of UV-shielding substances.

Composition of chloroplast pigments (Krause et al. 2003b) was similar to known differences between sun and shade leaves (see Demmig-Adams 1998). Mean Chl *a*/*b* ratios were highest (>3.0) in sun leaves, lower in shade leaves of tree crowns, and even lower in the understory, indicating a reduced number of light-harvesting Chl *a* +*b* binding antenna complexes in sun leaves (see Sect. 2.2.2). Contents of violaxanthin-cycle pigments and β-carotene were highest and contents of α-carotene lowest in sun leaves. Extremely high β-carotene and low α-carotene levels were found in sun leaves of the mangrove *L. racemosa*. Further studies are needed to evaluate whether this is related to salt stress.

In conclusion, the data indicate with regard to protection against UV and visible solar radiation that leaves are generally well adapted to cope with their respective environment. In plants acclimated to a shaded environment, protection is however weak, and sudden strong increase of solar radiation, especially of UV light, may cause severe damage (see Sect. 4.1).

Long-term acclimation of shade-grown tree seedlings (two pioneer and two late-successional species) to full solar radiation was investigated by daily transfer to high sunlight for defined, stepwise prolonged periods (Krause et al. 2004). After either 7 weeks or 3 months of daily sun exposure, absorbance of leaf extracts indicated a substantial increase in UV-B shielding, although the in situ levels of canopy sun leaves were not reached. In parallel, general acclimatory responses regarding chloroplast pigments of shade-grown seedlings exposed to full sunlight were observed. Furthermore, the content of the antioxidant ascorbate (and the sum of ascorbate and dehydroascorbate) was increased. Formation of Zx and Ax under high light was enhanced, photoinhibition of PSII strongly reduced, and the recovery period under low light shortened. Maximum rates of net CO_2 assimilation were significantly increased upon exposure to high sunlight, but reached only about half the rates of canopy sun leaves. Certain acclimatory changes in shade-grown plants of several tropical species by high-light exposure have also been reported by Demmig-Adams et al. (1989), Mulkey and Pearcy (1992), and Lovelock et al. (1994). Taken together, the experimental data show that shade-grown pioneer and late-successional species can acclimate to some extent to high sunlight including UV radiation when light flux strongly increases, as happens, e.g., upon treefall gap formation. This allows stabilization and adjustment of photosynthetic activities to changed light conditions until fully sun-acclimated leaves develop.

To single out the role of UV-absorbing compounds in acclimation and protection of leaves under ambient solar radiation, seedlings of the late-successional species *T. panamensis* and *C. longifolium* were grown for 3–4 months at an open site under plastic films that either absorb or transmit most UV-B radiation (Krause et al. 2007). Under strongly reduced UV-B, absorbance of leaf extracts at 305 nm, tested in *T. panamensis*, was close to that of species developed in deep shade of the forest understory (see above). Similar results were obtained by Searles et al. (1995) in a study on several tropical species. Interestingly, exclusion of most UV-B strongly reduced also absorbance of UV-A at 360 and 375 nm. Otherwise, leaves grown under reduced UV-B were indistinguishable from leaves grown under near-ambient UV-B regarding carotenoid and Chl *a+b* contents and Chl *a/b* ratios. Moreover, there were no significant differences in maximum net CO_2 assimilation, total dry mass, and other growth parameters between the two sets of seedlings (Krause et al. 2007). This indicates that under full solar radiation seedlings were perfectly protected against adverse effects of UV-B and under reduced UV-B were fully acclimated to excessive visible light. It should be noted that these results may not apply to plant species in general. Inhibition of dry mass accumulation by solar UV-B has been observed in several studies. For instance, Liu et al. (2005) tested four tropical *Acacia* (Fabaceae) and two *Eucalyptus* (Myrtaceae) species and reported reduction in biomass formation under solar UV-B in several but not all of them.

When growth conditions of seedlings were altered from strongly reduced to near-ambient UV-B, enhanced photoinhibition of PSII (decline in F_v/F_m) around midday and, independently, strong inhibition of maximum net CO_2 assimilation occurred on the first day of exposure. The inhibitory effect on PSII vanished within 1–3 days, but recovery of CO_2 assimilation required about 1 week in *C. longifolium* and was still incomplete after 1 week in *T. panamensis*. During this period, UV-B absorbance at 305 nm had increased in *T. panamensis* to 73.5% of the level in plants grown under near-ambient UV-B. In addition, there was a substantial increase in UV-A absorbance at 360 and 375 nm (Krause et al. 2007).

The study shows that leaves grown in the absence of most solar UV-B but acclimated to high visible light are strongly UV-B sensitive, as is obvious from inhibition of potential PSII efficiency and CO_2 assimilation occurring upon transfer to solar radiation that includes natural UV-B. However, these leaves are capable of repairing damage and acclimating to high UV exposure by synthesis of UV-absorbing compounds.

5 Heat Tolerance

5.1 Sun Leaves

In early studies, the temperature limit of heat tolerance of plants was determined by means of visible tissue damage observed after heat treatments. Sachs (1864) already documented thermal limits of 50–51°C, similar to many later investigations. Schreiber and Berry (1977) introduced the first method to determine heat tolerance of plant leaves by using Chl *a* fluorescence recording. The critical temperature causing onset of F_0 increase, $T_c(F_0)$, during continuous heating of leaves (1°C min^{-1} temperature increase) served as indicator of heat damage. More recently, pulse-amplitude modulated recording of Chl *a* fluorescence (Schreiber et al. 1986; see Sect. 2.1) provided a convenient tool to determine heat tolerance. Electron transport driven by PSII is known to be a highly heat-sensitive process (Krause and Santarius 1975; Smillie and Nott 1979; Berry and Björkman 1980; Havaux et al. 1991). The heat-induced F_0 increase results from damage to PSII and appears to be a complex process leading to accumulation of Q_A^- (see Kouřil et al. 2004; Ducruet et al. 2007).

By using the method of Schreiber and Berry (1977), heat tolerance might be underestimated, particularly since recovery from heat stress is not considered. Nonetheless, this method has been shown to be suitable for comparative screening of heat tolerance of a large number of species from different biomes. O'Sullivan et al. (2017) documented $T_c(F_0)$ values between 41.5°C and 50.8°C in 218 species from the Arctic in Alaska to the lowland rain forest in the Amazon basin. Alternatively, heat treatments at a series of fixed temperatures followed by a recovery period may provide more reliable data of heat tolerance. Correlation between $T_c(F_0)$ and 50% necrotic leaf damage after 30 min heat treatment followed by long-term storage of leaves was reported by Bilger et al. (1984). Many later studies used the decline in F_v/F_m upon heat treatment to document heat damage, e.g., in heat tolerance tests on Australian rain forest trees by Cunningham and Read (2006).

In an investigation on sun leaves of the pioneer tree *F. insipida* and the late-successional tree *V. sebifera*, the methods using F_0 increase and decline of F_v/F_m as indicators of heat damage were reassessed (Krause et al. 2010). The results of heat treatment (15 min) of leaf sections in a water bath at controlled temperatures showed that the decline of F_v/F_m by 50%, $T_{50}(F_v/F_m)$, recorded after 24 h storage of leaf sections under low light, is a more suitable parameter than $T_c(F_0)$. Heat tests have been performed during the rainy season (December), early dry season (January), and late dry season (March). Results did not significantly differ between seasons. Means of both $T_c(F_0)$ and $T_{50}(F_v/F_m)$ were ~2°C higher after 24 h compared to 15 min of storage, indicating significant recovery. In canopy sun leaves of *F. insipida*, the mean value (± s.d.) after 24 h recovery of $T_{50}(F_v/F_m)$ reached 52.1 ± 0.6°C and of $T_c(F_0)$ 49.5 ± 0.5°C. In accordance with $T_{50}(F_v/F_m)$ recorded after 24 h, mean visible necrotic tissue damage after 11 days of storage of leaf sections under low light was ~52.8% of leaf area. No tissue damage was observed after heating at 51.3°C, but total damage occurred at 53.3°C.

The above results obtained with leaf sections were confirmed by experiments with whole plants. Potted seedlings of *F. insipida*, roots and stem bases insulated, were heat-treated at controlled leaf temperatures for 15 min in an incubator (Krause et al. 2010). After a recovery period of 24 h, $T_{50}(F_v/F_m)$ values were around 52.0°C, in accordance with strong leaf damage seen after 8 days of further culture. Almost total leaf damage occurred upon heating to 52.8°C, but the plants survived and grew new leaves during the following 2 weeks.

Recording of leaf temperatures on fully sun-exposed leaves of *F. insipida* at a wind-shielded side of the canopy (ambient maximum temperature 33.7°C) showed mostly values of 40–42°C but occasionally top temperatures of 46–48°C (see also Slot et al. 2016), indicating that under extreme light exposure, canopy leaves operate close to their limit of heat tolerance.

To test whether sun leaves are capable of acclimating to increased air temperatures, potted seedlings of *F. insipida* and the late-successional species *V. surinamensis* were grown outdoors either under ambient conditions or under Aclar cover at day temperatures elevated by 10–15°C (Krause et al. 2010). Heat tests on both species did reveal only marginal, but not significant, increases in $T_{50}(F_v/F_m)$ upon growth at elevated compared to ambient day temperatures.

In further experiments it was investigated whether increased day and night temperatures would promote heat resistance. *F. insipida* seedlings were cultivated in controlled-environment chambers with 12 h light/12 h dark periods at light/dark temperatures either at 39/32°C or 39/22°C (Krause et al. 2013). The latter night temperature (22°C) was chosen to be close to natural conditions. Heat tests showed after 24 h recovery periods $T_{50}(F_v/F_m)$ values of ~51.5°C and ~51.0°C, respectively, close to values obtained with canopy sun leaves. The difference in $T_{50}(F_v/F_m)$ between seedlings grown under high and low night temperature was not significant. Also the high day temperature (39°C) did not raise the heat tolerance limit.

Obviously, the two neotropical species tested possess a very low capacity for raising heat tolerance of leaves in response to increased day or night temperatures. This is in clear contrast to species adapted to climates with strong seasonal and diel temperature changes. Significant heat acclimation potentials of temperate rain forest trees but limited heat acclimation of several tree species of the tropical rain forest in Australia have been described by Cunningham and Read (2006). Leaves of tropical trees in a seasonally dry forest in India exhibited markedly increased limits of heat tolerance in the hot-dry season compared to the cool-wet monsoon season (Sastry and Barua 2017). Experiments by Drake et al. (2018) in New South Wales/Australia showed heat tolerance was increased during a simulated heatwave in leaves of *Eucalyptus parramattensis* E.C. Hall (Myrtaceae). In desert (Downton et al. 1984) and alpine (Braun et al. 2002) species, considerable capacities of heat acclimation have been documented. Apparently lowland species in the humid tropics lack heat acclimation because of the relatively constant seasonal temperature conditions. This hypothesis requires further investigation.

A noticeable result of the climate-chamber experiment discussed above was that the seedlings grown at the high night temperature (32°C) exhibited a drastically increased growth rate compared to growth at a night temperature of 22°C. After 39 days of culture, total biomass of seedlings was more than three times higher upon growth at high than at low night temperature. In addition, light-saturated net CO_2 assimilation was enhanced and the increase in dark respiration with rising temperature was reduced in plants grown at the night temperature of 32°C (Krause et al. 2013). This means the high nighttime temperature caused thermal acclimation of important physiological processes, while the limit of heat tolerance remained essentially unchanged. Increased biomass accumulation and acclimation of CO_2 fixation and respiration at moderately elevated day and night temperatures was also observed in seedlings of a variety of neotropical trees, although acclimation potential varied between species (Cheesman and Winter 2013a, b). These results are in contrast to reports that in the tropics thermal acclimation of photosynthesis is limited due to low seasonal and day-to-day temperature variation (Cunningham and Read 2003) and that increased night temperature inhibits tree growth in a neotropical forest (Clark et al. 2010). Nonetheless, nighttime warming has been shown to enhance growth of *N. tabacum* (Camus and Went 1952) and *Gossypium hirsutum* L. (Malvaceae) (Königer and Winter 1993). Slot et al. (2014) reported that dark respiration is downregulated by increased night temperatures in canopy sun leaves of tropical tree and liana species in situ. However, moderate nighttime warming did not increase biomass accumulation of saplings of the tropical pioneer tree *Ochroma pyramidale* (Cav. ex Lam.) Urb. (Malvaceae) (Slot and Winter 2018). The reasons for these contrasting observations still need clarification.

Under ambient conditions, leaves are usually subjected to heat stress under excessive solar radiation combined with high air temperature. However, heat tolerance tests have been mainly performed on leaf sections, intact leaves, or whole seedlings kept in the dark or low light. In a study on canopy sun leaves of two tropical tree species, *F. insipida* and *C. longifolium*, leaf disks were kept during heat treatment under white light (500 μmol photons m^{-2} s^{-1}) and for comparison in the dark (Krause et al. 2015). Chl *a* fluorescence was recorded 24 and 48 h after heat treatment to allow for long-term recovery of possible heat-enhanced photoinhibition of PSII. It has been shown that under excessive light, high leaf temperature promotes photoinhibition of photosynthesis (Gamon and Pearcy 1990; Dongsansuk et al. 2013). On the other hand, high light activates antioxidative processes, e.g., by formation of Zx in the violaxanthin cycle (see Sect. 2.2). Therefore, it was not possible to predict how light would interact with heat tolerance of tropical plants.

The results by Krause et al. (2015) revealed a significantly higher temperature limit of heat tolerance when heat treatment was performed under light as compared to heating in the dark. The F_v/F_m decline in leaf disks of *F. insipida* (Fig. 5) shows the protective effect of light more clearly after 48 h than 24 h recovery periods. Values of $T_{50}(F_v/F_m)$ were at ~52.5°C upon heating in the dark and at ~53.5°C under high light. Such increase in the heat tolerance limit by 1°C appears to be small at first sight but in view of climate change might be crucial for survival of leaves during extreme heat periods. In addition to F_v/F_m, the rise in F_0 and decline in F_m also indicated significant heat protection by light. Moreover, visible tissue damage in the critical temperature range was markedly reduced when leaf sections were illuminated during temperature treatments. Protection by light against heat damage was confirmed with leaf disks of *C. longifolium* and with intact attached seedling leaves of both species. It should be noticed that photosynthetic carbon metabolism responded more sensitively to heat stress than PSII (Krause et al. 2015). At stepwise increasing temperatures of intact attached seedling leaves, net CO_2 assimilation started to decline at ~38°C in *F. insipida* and at ~35°C in *C. longifolium* (cf Slot et al. 2016). The upper temperature compensation points of net CO_2 exchange were significantly below $T_{50}(F_v/F_m)$.

Light-stimulated heat tolerance has been also described for leaves of species adapted to hot, dry climate (Schreiber and Berry 1977) and alpine plants (Buchner et al. 2015). The exact mechanism of the protective action of light is still unknown. Sun leaves of tropical plants possess high pool sizes of violaxanthin-cycle pigments. Under excessive light, there is a fast conversion of most Vx to Zx (see Sect. 2.2). As a hypothesis, the high level of Zx present in the photosynthetic pigment systems may, in addition to dissipation of excessively absorbed light energy, exert protection against oxidative heat-induced PSII inactivation and indirectly against general tissue damage.

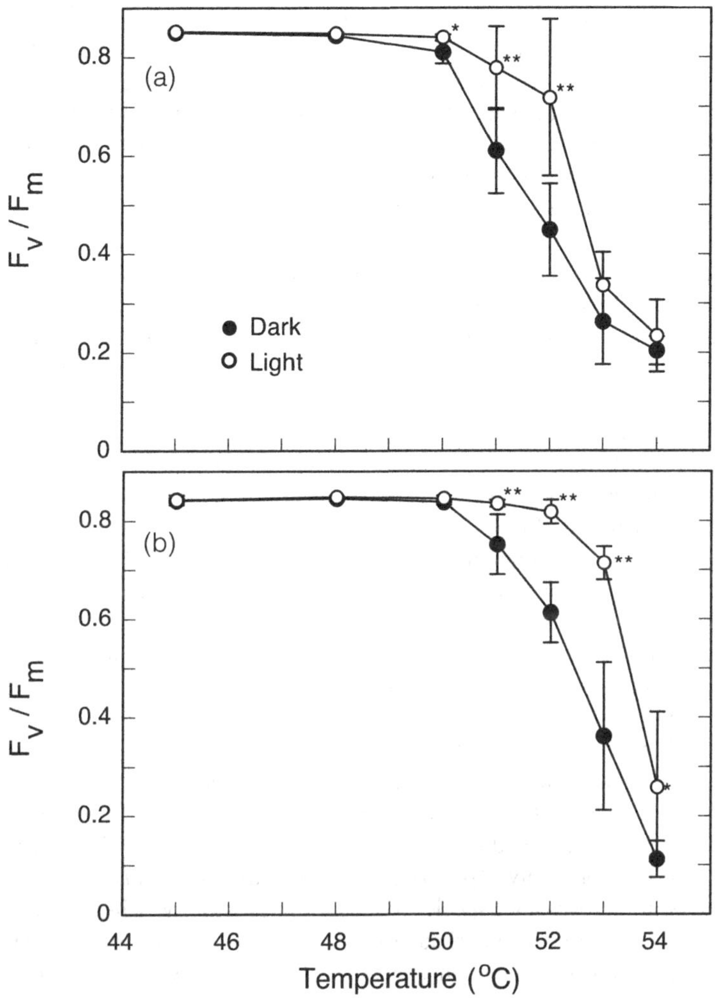

Fig. 5 Response of F_v/F_m to 15 min heat exposure of leaf disks of *F. insipida* canopy sun leaves at leaf temperatures between 45 and 54°C. Heat treatment was done in the dark (●) and under 500 μmol m^{-2} s^{-1} white light (o). F_v/F_m recording was done (**a**) 24 h and (**b**) 48 h after heat treatment. Means ± s.d. from six different leaves (error bars shown when larger than symbols); significant differences between data from treatments in dark and light are indicated: **, $P < 0.01$; *, $P < 0.05$. Reproduced from Krause et al. (2015) with permission from CSIRO Publishing

5.2 *Shade Leaves*

Heat tolerance has been studied predominantly in sun leaves. However, in forests, the great majority of leaves is more or less shaded, as is obvious from a global leaf area index (LAI), i.e., the projected leaf area per unit ground area, of 5 m^2 m^{-2}

(Asner et al. 2003). In tropical forests, the LAI tends to be larger (6.0 m^2 m^{-2}) than in the global average (Clark et al. 2008). This means there are about four or five times more partly or fully shaded leaves than outer-canopy sun leaves. Thus, despite lower rates of CO_2 assimilation, shade leaves contribute significantly to total carbon gain. Shade leaves usually experience lower maximum temperatures than sun leaves (Rey-Sánchez et al. 2016). Therefore, it could be expected that plants downregulate heat tolerance of shade leaves, thereby saving metabolic costs of maintaining high heat resistance (Wahid et al. 2007).

To test this hypothesis, the temperature limit of heat tolerance, $T_{50}(F_v/F_m)$, in shade and sun leaves was compared in a study by Slot et al. (2019). Tests were performed on three tropical tree species, *Calophyllum inophyllum* L. (Calophyllaceae), a high-light-demanding, slow-growing species native to Asian and Pacific coastal and lowland forests; *I. spectabilis*, a medium-sized Central American species (belonging to the group of species possessing lutein epoxide cycle activity; see Sect. 2.3); and *O. macrocalyx*, a late-successional neotropical species. Shade and sun leaves were characterized by recording CO_2 exchange in situ and by their specific leaf area (SLA). Rates of net photosynthetic CO_2 uptake at optimum temperature (A_{Opt}) were more than 50% lower in shade compared to sun leaves. The high-temperature CO_2 compensation points were markedly lower than $T_{50}(F_v/F_m)$ in both shade and sun leaves (see also Hernández et al. 2020).

Heat treatment was performed with leaf disks in the dark, but not in the light, to avoid photoinhibition which could be particularly severe in shade leaves. Chl *a* fluorescence was recorded after 48 h recovery periods. The results revealed significant but relatively small differences in heat tolerance between shade and sun leaves in *C. inophyllum* and *I. spectabilis* (Fig. 6). The value of $T_{50}(F_v/F_m)$ was ~1.4°C lower in shade than sun leaves of *C. inophyllum*, but only ~0.7°C lower in *I. spectabilis*, whereas no difference at all was detected in the F_v/F_m decline of *O. macrocalyx*. Noticeably, $T_{50}(F_v/F_m)$ in shade leaves of *O. macrocalyx* was not higher than in the two other species, but $T_{50}(F_v/F_m)$ was lower in sun leaves. This might be related to the distinct shade tolerance of this species. Among seedlings of four late-successional species, only in *O. macrocalyx* growth was inhibited under full sunlight compared to partial shading (Krause et al. 2012; see Sect. 2.2). Moreover, leaves of *O. macrocalyx* had the lowest content of the antioxidant tocopherol.

Overall, in the three species studied, differences in the temperature limit of heat tolerance between shade and sun leaves were small or absent. These results are in accordance with the general notion of a low capacity of heat acclimation in species adapted to the low seasonal temperature changes in the humid tropics. The relatively high heat tolerance of shade leaves could be advantageous when leaf temperatures suddenly rise during sun-flecks or upon opening of canopy gaps.

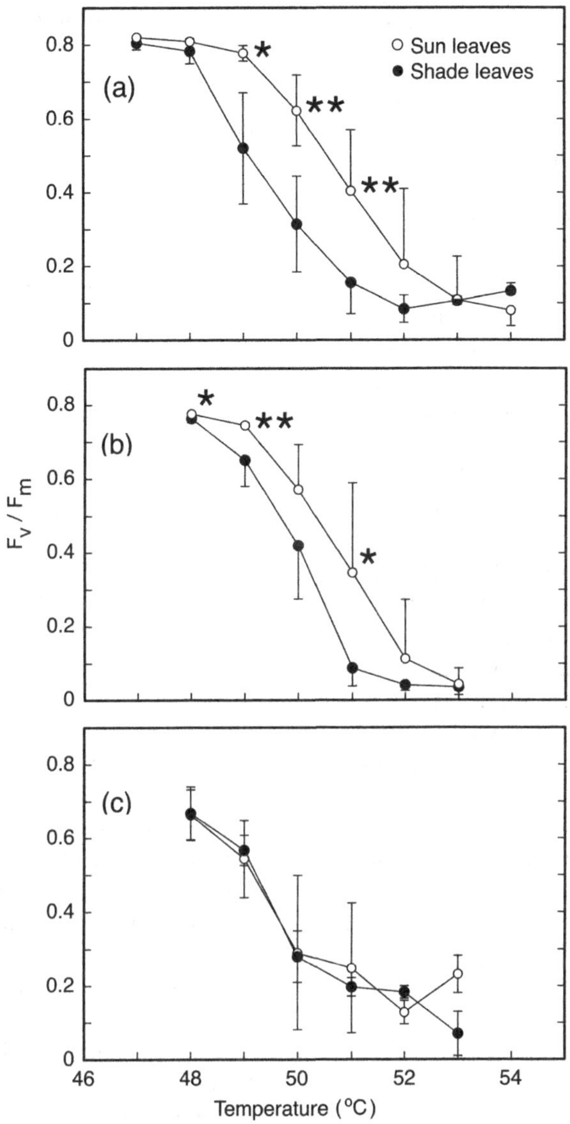

Fig. 6 Heat tolerance of shade (●) and sun leaves (o) as shown by the response of F_v/F_m to 15 min heat treatment of leaf disks at leaf temperatures given in the graph. Tested tree species were (**a**) *C. inophyllum*, (**b**) *I. spectabilis*, (**c**) *O. macrocalyx*. Means ± s.d. from six different leaves (error bars shown when larger than symbols); significant differences of F_v/F_m between shade and sun leaves are indicated: **, $P < 0.01$; *, $P < 0.05$. Redrawn from Krause et al. (2015). Copyright Springer International Publishing AG

5.3 Sun Leaves of Plants with Crassulacean Acid Metabolism

In early studies, heat tolerance of plants exhibiting crassulacean acid metabolism (CAM) has been described to be related to vacuolar acid contents, predominantly of malic acid (Schwemmle and Lange 1959; Lösch and Kappen 1983; Lehrum et al. 1987). For instance, in greenhouse-grown *Aeonium* (Crassulaceae) species (Lösch and Kappen 1983), extremely low heat tolerance limits were observed at the end of the dark period, when acid accumulation was highest. Tissue damage was initiated already at 39°C, but during the course of acid turnover in the light period, the limit of heat tolerance increased substantially. Similar heat tolerance dependence on acid contents has been reported by Chaves et al. (2015) in *Aechmea blanchetiana* (Baker) L.B.Sm. (Bromeliaceae), a tropical CAM species. These findings have been generally interpreted as a result of heat-induced acid release from the vacuoles to the cytoplasm, causing cell damage.

In the above studies, heat treatment was performed with darkened leaf materials collected at various times of the day-night CAM cycle. Since in the natural environment, extreme leaf temperatures result from excessively absorbed light energy, heat tolerance of CAM plants was investigated by heat treatment in light compared to darkness (Krause et al. 2016). Three species were studied: *Clusia rosea* Jacq., *Clusia pratensis* Seem. (Clusiaceae), and *Agave angustifolia* Haw. (Asparagaceae). *C. rosea* and *A. angustifolia* are constitutive CAM species, whereas *C. pratensis* is a facultative CAM species (Popp et al. 1987; Winter et al. 2008, 2014; Winter and Holtum 2014). The two *Clusia* species are native to humid, seasonably dry neotropical forests. *A. angustifolia* grows in relatively dry areas of Central America and Mexico.

Heat tolerance was determined in the rainy and dry seasons. For heat treatments, leaves were harvested in the morning (07:00 h), when acids had accumulated overnight, and in the afternoon (16:00 h) when acid content was low. After heat exposure, extended periods of storage under dim light were required for complete recovery. For morning leaves, heat tolerance limits, $T_{50}(F_v/F_m)$, showed extreme differences between dark and light heat treatment. For instance, in *C. rosea* during the rainy season, $T_{50}(F_v/F_m)$ was ~40°C in the dark and ~51°C in the light. A similar difference was found for morning samples during the following dry season. In contrast, upon harvest in the afternoon, heat tolerance was not different between dark and light treatment; $T_{50}(F_v/F_m)$ was ~54°C in both seasons.

In *C. pratensis*, operating in C_3 mode during the late rainy season (November), leaves harvested in the morning did not exhibit a significantly different heat tolerance upon heating in dark or light; $T_{50}(F_v/F_m)$ was ~52°C, close to values typically observed in sun leaves of C_3 plants (see Sect. 5.1). However, in the dry season (February/March), when CAM was active in *C. pratensis*, similar temperature limits of heat tolerance were found as for *C. rosea*, i.e., in the morning there was a high

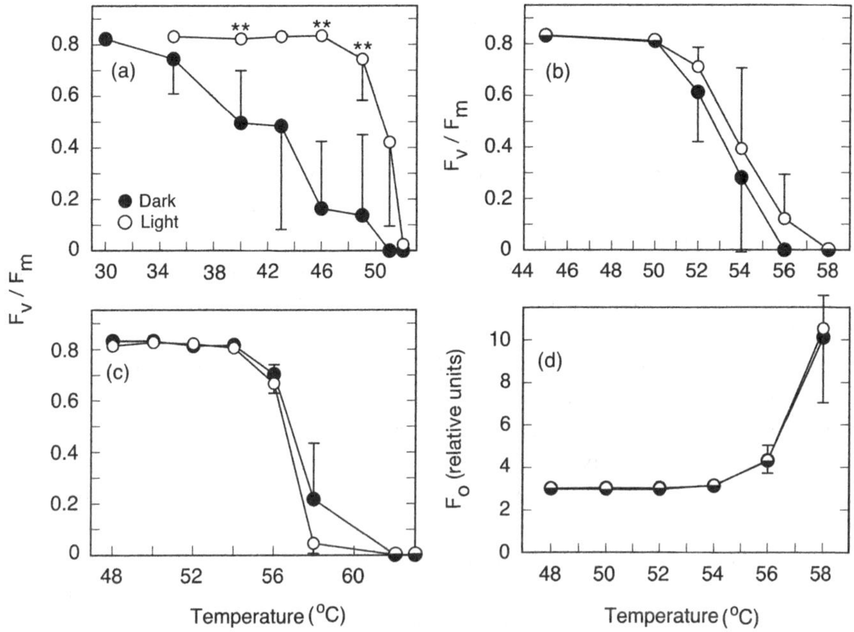

Fig. 7 Heat tolerance of the CAM plant *A. angustifolia* tested during dry season. Plants were grown for 2–3 years under ambient conditions. Mean maximum day temperature during ~6 weeks before heat tests was ~33°C. Sections of fully developed leaves were tested; (**a**, **b**) response of F_v/F_m to 20 min heat treatments in the dark (●) and under 500 μmol m^{-2} s^{-1} white light (o). Leaves harvested either in the highly acidified state at 7:00 h (**a**) or in the de-acidified state at 16:00 h (**b**). Tests performed in the acidified state show highly significant differences between treatments in dark and light (**, $P < 0.01$), i.e., very low heat tolerance when tested in the dark (**a**) in contrast to tests performed in the de-acidified state (**b**). In this state, $T_{50}(F_v/F_m)$ was ~54°C, without significant difference between tests in dark and light. Upon acclimation for 6 weeks to elevated temperatures (mean maximum day temperature ~44°C), tests performed in dark and light in the de-acidified state (leaf harvest at 16:00 h) showed an increase of $T_{50}(F_v/F_m)$ to ~57°C (**c**). A similarly high heat tolerance was apparent from the rise of initial fluorescence, F_o (**d**). Note different scales of heating temperature. Means ± s.d. of six leaf sections from different leaves (error bars shown when larger than symbols). Reproduced from Krause et al. (2016) with permission from CSIRO Publishing

difference between dark and light treatments, but no significant difference in the afternoon with $T_{50}(F_v/F_m)$ at 50–51°C.

Experiments with *A. angustifolia* confirmed the results obtained with the two *Clusia* species (Fig. 7). As shown by the decline of F_v/F_m, leaves harvested in the morning during the dry season were extremely heat sensitive when heated in the dark. The high heat sensitivity was largely abolished when heating occurred under light (Fig. 7a). The leaves were particularly heat stable, without differences between heating in dark and light in the afternoon, when acid content was very low (Fig. 7b). In contrast to tropical C_3 species (see Sect. 5.1), *A. angustifolia* exhibited a marked potential of heat acclimation. Upon culture of plants for 6 weeks under elevated day temperatures (maxima ~11°C above ambient), $T_{50}(F_v/F_m)$ rose from ~54°C to

~57°C, tested in the afternoon in light and dark during the dry season (Fig. 7c). The high heat tolerance was also evident from the F_0 increase (Fig. 7d). However, the tolerance limits at 62 ± 2.1°C previously reported for 15 heat-acclimated species of *Agave* adapted to desert climate (Nobel 1988) were not reached.

Tissue damage visible by brownish coloration corresponded to the above fluorescence results in the three species tested. From light-brown-colored leaf tissue of *C. rosea*, ethanolic pigment extracts exhibited the *Q*-band of the pheophytin *a* absorbance spectrum. The presence of pheophytin in the absence of Chl in the extracts was proven by reaction with Cu^{2+} solution, forming the acid-stable Cu-pheophytin *a* complex with the *Q*-band shifted to shorter wavelength (cf White et al. 1977; Gerola et al. 2011).

In summary, the data of the study by Krause et al. (2016) clearly show that acids accumulated overnight are released from the vacuoles under heat stress. In the dark, the acids penetrate the chloroplast envelope and transform the acid-labile Chl to the brownish-colored pheophytin by exchange of Mg^{2+} against $2H^+$ in the porphyrin ring, depending on the degree of acidification of the chloroplast stroma. A "demetalation pH" of 3.5 for Chl *a* in ethanolic solution was determined by Gerola et al. (2011). Evidently during heat stress under light, released malic acid is rapidly turned over in the CAM cycle, thereby protecting the leaves against heat damage of the photosynthetic system. It should be noted that protection by light might not be perfect, since values of $T_{50}(F_v/F_m)$ determined under high light were slightly lower in the morning, at high acid levels, than at largely depleted acid pools in the late afternoon.

The investigation confirms suggestions in earlier studies that heat stress induces release of vacuolar acids. Clearly, in CAM plants containing high acid levels, heat tests performed in the dark strongly underestimate intrinsic heat tolerance. On the other hand, such experiments may provide useful information on mechanism of malic acid storage in vacuoles, efflux to the cytosol, and related physiological processes (cf Lüttge and Smith 1984; Smith et al. 1996; Cernusak et al. 2008).

6 Conclusion

The reviewed studies reveal the complexities of PSII protection in leaves of tropical plants against potential damage by high solar visible and UV radiation. A major factor of leaf acclimation to full sunlight is the violaxanthin cycle. Compared to shade leaves, sun leaves exhibit higher contents of violaxanthin-cycle pigments and under excess light increased de-epoxidation of Vx to Zx. Additionally, in a number of species, the lutein epoxide cycle takes part in photoprotection. In most cases, as in the investigated species of *Inga* (Fabaceae), Lx was found predominantly in shade leaves, where it possibly promotes light harvesting. Upon high-light exposure, Lx was rapidly de-epoxidized to lutein, probably exerting a protective function; restoration of Lx required several days in the shade. Noteworthy is the discovery of very high levels of Lx in shade and sun leaves and the operation of the lutein epoxide cycle on a diel basis in sun leaves of *Virola* species (Myristicaceae). Various antioxidative systems contribute to acclimation under excess solar radiation, e.g., accumulation of α-tocopherol and ascorbate, as well as alterations of carotenoid composition in the light-harvesting systems, such as increases in the β/α-carotene ratio. Protection against UV radiation is predominantly achieved by the accumulation of UV-absorbing compounds in the vacuoles of leaf epidermal cells.

Remarkable stability of PSI under full solar radiation was observed in sun leaves and particularly in shade leaves upon full light exposure, while PSII efficiency was strongly reduced. The capacity of P700 oxidation (recorded by absorbance changes at ~820 nm) was not or only marginally affected by exposure to high sunlight both in the presence and absence of solar UV-B. However, a specific photoinhibitory response of PSI was discovered, i.e., a decrease in efficiency of P700 oxidation, while the capacity of P700 oxidation remained unchanged. This process was caused by both visible and UV light; it was substantially more pronounced in shade than in sun leaves and might contribute to PSI protection. The causes of high PSI stability and the role of reduction in P700 oxidation efficiency need further clarification.

Temperature limits of heat tolerance in sun leaves of tropical C_3 plants, indicated by inactivation of PSII and followed by visible tissue damage, were slightly above 50°C in the species tested. There was only a marginal capacity for increased heat tolerance in response to growth at elevated temperatures. Possibly this is related to the relatively high seasonal temperature stability to which plants are adapted in the humid tropics. Of interest is the observation that light (500 μmol photons m^{-2} s^{-1}) does not aggravate heat damage but rather has a significant protective role. Since in situ leaf temperatures up to 48°C, i.e., close to the thermal tolerance limits, have already been measured in tropical forest canopies, future leaf-temperature monitoring in the face of global warming would be useful. Although shade leaves experience on average lower leaf temperatures than sun leaves, they did not exhibit substantially reduced heat tolerance. In the special case of CAM plants, heat tolerance values were extremely low when temperature treatments were performed in the dark on highly acidified leaf tissue collected at the end of the night. Heat treatments of illuminated end-of-night samples showed substantially increased heat tolerance values. Evidently, high treatment temperatures applied in the dark cause acids to be rapidly released from the vacuoles leading to tissue damage: the acids penetrate the chloroplast envelope and transform chlorophyll to pheophytin as shown by absorbance spectroscopy of ethanolic leaf extracts. By contrast, under illumination, released acids are metabolized without significantly damaging cell constituents. Therefore, heat tolerance limits of CAM plants are severely underestimated when heat tests are performed in the dark on acidified tissues.

Acknowledgments We thank Barbara Krause for competent assistance in preparation of the manuscript, Shizue Matsubara for critical reading of the text, and Aurelio Virgo for preparing the illustrations.

The research was supported by the Deutsche Forschungsgemeinschaft (GHK), the Andrew W. Mellon Foundation (KW), and the Smithsonian Tropical Research Institute (KW).

References

Allen JF (2003) State transitions – a question of balance. Science 299:1530–1532

Allen DJ, McKee IF, Farage PK, Baker NR (1997) Analysis of limitations to CO_2 assimilation on exposure of leaves of two *Brassica napus* cultivars to UV-B. Plant Cell Environ 20:633–640

Amunts A, Drory O, Nelson N (2007) The structure of a plant photosystem I supercomplex at 3.4 Å resolution. Nature 447:58–63

Anderson JM, Osmond CB (1987) Shade-sun responses: compromises between acclimation and photoinhibition. In: Kyle DJ, Osmond CB, Arntzen CJ (eds) Photoinhibition, Topics in photosynthesis, vol 9. Elsevier, Amsterdam, pp 1–38

Aro EM, McCaffery S, Anderson JM (1993a) Photoinhibition and D1 protein degradation in peas acclimated to different growth irradiances. Plant Physiol 103:835–843

Aro EM, Virgin I, Andersson B (1993b) Photoinhibition of photosystem II. Inactivation, protein damage and turnover. Biochim Biophys Acta 1143:113–134

Asada K (1999) The water-water cycle in chloroplast: scavenging of active oxygens and dissipation of excess photons. Annu Rev Plant Physiol Plant Mol Biol 50:601–639

Asner GB, Scurlock JM, Hicke JA (2003) Global synthesis of leaf area index observations: implications for ecological and remote sensing studies. Glob Ecol Biogeogr 12:191–205

Barnes PW, Flint SD, Ryel RJ, Tobler MA, Barkley AE, Wargent JJ (2015) Rediscovering leaf optical properties: new insights into plant acclimation to solar UV radiation. Plant Physiol Biochem 93:94–100

Barth C, Krause GH (1999) Inhibition of photosystems I and II in chilling-sensitive and chilling-tolerant plants under light and low-temperature stress. Z Naturforsch 54c:645–657

Barth C, Krause GH (2002) Study of tobacco transformants to assess the role of chloroplastic NAD (P)H dehydrogenase in photoprotection of photosystems I and II. Planta 216:273–279

Barth C, Krause GH, Winter K (2001) Responses of photosystem I compared with photosystem II to high-light stress in tropical shade and sun leaves. Plant Cell Envir 24:163–176

Bendall DS, Manasse RS (1995) Cyclic photophosphorylation and electron-transport. Biochim Biophys Acta 1229:23–38

Berry J, Björkman O (1980) Photosynthetic response and adaptation to temperature in higher plants. Annu Rev Plant Physiol 31:491–543

Bilger H-W, Schreiber U, Lange OL (1984) Determination of leaf heat resistance: comparative investigation of chlorophyll fluorescence changes and tissue necrosis methods. Oecologia 63:256–262

Björkman O, Demmig B (1987) Photon yield of O_2 evolution and chlorophyll fluorescence characteristics at 77K among vascular plants of diverse origins. Planta 179:489–504

Bornman JF, Barnes PW, Robson TM, Robinson SA, Jansen MAK, Ballaré CL, Flint SD (2019) Linkages between stratospheric ozone, UV radiation and climate change and their implications for terrestrial ecosystems. Photochem Photobiol Sci 18:681–716

Boyer JS (2017) Plant water relations: a whirlwind of change. Progr Bot 79:1–31

Braun V, Buchner O, Neuner G (2002) Thermotolerance of photosystem II of three alpine plant species under field conditions. Photosynthetica 40:587–595

Brettel K (1997) Electron transfer and arrangement of the redox cofactors in photosystem I. Biochim Biophys Acta 1318:322–373

Buchner O, Stoll M, Karadar M, Kranner I, Neuner G (2015) Application of heat stress in situ demonstrates a protective role of irradiation on photosynthetic performance in alpine plants. Plant Cell Envir 38:812–826

Burchard P, Bilger B, Weissenböck G (2000) Contribution of hydroxycinnamates and flavonoids to epidermal shielding of UV-A and UV-B radiation in developing rye primary leaves as assessed by ultraviolet-induced chlorophyll fluorescence measurements. Plant Cell Environ 23:1373–1380

Butler WJ (1978) Energy distribution in the photosynthetic apparatus of photosynthesis. Annu Rev Plant Physiol 29:345–378

Camus GC, Went FW (1952) Thermoperiodicity of three varieties of *Nicotiana tabacum*. Amer J Bot 38:521–528

Cernusak LA, Mejia-Chang M, Winter K, Griffith H (2008) Oxygen isotope composition of CAM and C_3 *Clusia* species: non-steady-state dynamics control leaf water ^{18}O enrichment in succulent leaves. Plant Cell Envir 31:1644–1662

Chaves CJN, Santos Leal BS, de Lemos-Filho JP (2015) Temperature modulation of thermal tolerance of a CAM-tank bromeliad and the relationship with acid accumulation in different leaf regions. Physiol Plant 154:500–510

Cheesman AW, Winter K (2013a) Elevated night-time temperatures increase growth in seedlings of two tropical pioneer species. New Phytol 197:1185–1192

Cheesman AW, Winter K (2013b) Growth response and acclimation of CO_2 exchange characteristics to elevated temperatures in tropical tree seedlings. J Exp Bot 64:3817–3828

Choat B, Brodribb TJ, Brodersen CR, Duursma RA, López R, Medlyn BL (2018) Triggers of tree mortality under drought. Nature 558:531–539

Clark DB, Olivas PC, Oberbauer SF, Clark DA, Ryan MG (2008) First direct landscape-scale measurement of tropical rain forest leaf area index, a key driver of global primary productivity. Ecol Lett 11:163–172

Clark DB, Clark DA, Oberbauer SF (2010) Annual wood production in a tropical rain forest in NE Costa Rica linked to climatic variation but not to increasing CO_2. Glob Change Biol 16:747–759

Correa Galvis V, Poschmann G, Melzer M, Stühler K, Jahns P (2016) PsbS interactions involved in the activation of energy dissipation in *Arabidopsis*. Nature Plants 2:15225

Cunningham SC, Read J (2003) Do temperate rainforest trees have a greater ability to acclimate to changing temperatures than tropical rainforest trees? New Phytol 157:55–64

Cunningham SC, Read J (2006) Foliar temperature tolerance of temperate and tropical evergreen rain forests in Australia. Tree Physiol 26:1435–1443

Demmig B, Winter K, Krüger A, Czygan F-C (1987) Photoinhibition and zeaxanthin formation in intact leaves. A possible role of the xanthophyll cycle in the dissipation of excess light energy. Plant Physiol 84:599–626

Demmig-Adams B (1998) Survey of thermal energy dissipation and pigment composition in sun and shade leaves. Plant Cell Physiol 39:474–482

Demmig-Adams B, Winter K, Winkelmann E, Krüger A, Czygan F-C (1989) Photosynthetic characteristics and the ratios of chlorophyll, β-carotene, and the components of the xanthophyll cycle upon a sudden increase in growth light regime in several plant species. Plant Biol (formerly Bot Acta) 102:319–325

Demmig-Adams B, Adams WW III, Logan BA, Verhoeven AS (1995) Xanthophyll cycle-dependent energy dissipation and flexible photosystem II efficiency in plants acclimated to light stress. Aust J Plant Physiol 22:249–260

Dongsansuk A, Lütz C, Neuner G (2013) Effects of temperature and irradiance on quantum yield of PSII photochemistry and xanthophyll cycle in a tropical and a temperate species. Photosynthetica 51:13–21

Downton WJS, Berry JA, Seemann JR (1984) Tolerance of photosynthesis to high temperature in desert plants. Plant Physiol 74:786–790

Drake JE, Tjoelker MG, Vårhammar A, Medlyn BE, Reich PB, Leigh A, Pfautsch S, Blackman CJ, López R, Aspinwall MJ, Crous KY, Duursma RA, Kumarathunge D, DeKauwe MG, Jiang M, Nicotra AB, Tissue DT, Choat B, Atkin OK, Barton CVM (2018) Trees tolerate an extreme heatwave via sustained transpirational cooling and increased leaf thermal tolerance. Glob Change Biol 24:2390–2402

Ducruet J-M, Peeva V, Havaux M (2007) Chlorophyll thermofluorescence and thermoluminescence as complementary tools for the study of temperature stress in plants. Photosynth Res 93:159–171

Duke NC, Kovacs JM, Griffith AD, Preece L, Hill DJE, van Oosterzee P, Mackenzie J, Morning HS, Burrows D (2017) Large-scale dieback of mangroves in Australia's Gulf of Carpentaria: a

severe ecosystem response, coincidental with an unusually extreme weather event. Mar Freshw Res 68:1816–1829
Edge R, McGarvey DJ, Truscott TG (1997) The carotenoids as anti-oxidants – a review. J Photochem Photobiol B Biol 41:189–200
Endo T, Shikanai T, Takabayashi A, Asada K, Sato F (1999) The role of chloroplastic NAD(P)H dehydrogenase in photoprotection. FEBS Lett 457:5–8
Esteban R, Jiménez ET, Jiménez MS, Morales D, Hormaetxe K, Becerril JM, García-Plazaola JI (2007) Dynamics of violaxanthin and lutein epoxide xanthophyll cycles in Lauraceae tree species under field conditions. Tree Physiol 27:1407–1414
Esteban R, Jiménez MS, Morales D, Jiménez ET, Hormaetxe K, Becerril JM, Osmond B, García-Plazaola JI (2008) Short- and long-term modulation of the lutein epoxide and violaxanthin cycle in two species of the Lauraceae: sweet bay laurel (*Laurus nobilis* L.) and avocado (*Persea americana* Mill.). Plant Biol 10:288–297
Färber A, Young AJ, Ruban AV, Horton P, Jahns P (1997) Dynamics of xanthophyll-cycle activity in different antenna subcomplexes in the photosynthetic membranes of higher plants. Plant Physiol 115:1609–1618
Fioletov VE, Bodeker GE, Miller AJ, McPeters RD, Stolarski R (2002) Global and zonal total ozone variations estimated from ground-based and satellite measurements: 1964-2000. J Geophys Res-Atmos 107(D22):4647
Fork DC, Herbert SK (1993) Electron-transport and photophosphorylation by photosystem-I in vivo in plants and cyanobacteria. Photosynth Res 36:149–168
Förster B, Osmond CB, Pogson BJ (2009) De novo synthesis and degradation of Lx and V cycle pigments during shade and sun acclimation in avocado leaves. Plant Physiol 149:1179–1195
Franco AC, Lüttge U (2002) Midday depression in savanna trees: coordinated adjustments in photosynthetic efficiency, photorespiration, CO_2 assimilation and water use efficiency. Oecologia 131:356–365
Friso G, Barbato R, Giacometti GM, Barber J (1994) Degradation of the D1 protein due to UV-B irradiation of the reaction centre of photosystem II. FEBS Lett 339:217–221
Fryer MJ (1992) The antioxidative effect of thylakoid Vitamin E (α-tocopherol). Plant Cell Environ 15:381–392
Gamon JA, Pearcy RW (1990) Photoinhibition in *Vitis californica*. The role of temperature during high-light treatment. Plant Physiol 92:487–494
García-Plazaola JI, Becerril JM, Hernández A, Niinemets U, Kollist H (2004) Acclimation of antioxidant pools to the light environment in a natural forest canopy. New Phytol 163:87–97
García-Plazaola JI, Matsubara S, Osmond CB (2007) The lutein epoxide cycle in higher plants: its relationship to other xanthophyll cycles and possible functions. Funct Plant Biol 34:759–773
Gerola AP, Tsubone TM, Santana A, De Oliveira HPM, Hioka N, Caetano W (2011) Properties of chlorophyll and derivatives in homogeneous and microheterogeneous systems. J Phys Chem B 115:7364–7373
Harbinson J, Genty B, Baker NR (1989) Relationship between quantum efficiencies of photosystems I and II in pea leaves. Plant Physiol 90:1029–1034
Havaux M, Davaud A (1994) Photoinhibition of photosynthesis in chilled potato leaves is not correlated with a loss of photosystem-II activity. Photosynth Res 40:75–92
Havaux M, Greppin H, Strasser RJ (1991) Functioning of photosystems I and II in pea leaves exposed to heat stress in the presence or absence of light: analysis using *in vivo* fluorescence, absorbance, oxygen and photoacoustic measurements. Planta 186:88–98
Havaux M, Eymery F, Porfirova S, Rey P, Dörmann P (2005) Vitamin E protects against photoinhibition and photooxidative stress in *Arabidopsis thaliana*. Plant Cell 17:3451–3469
Havaux M, Dall'Ostro L, Bassi R (2007) Zeaxanthin has enhanced antioxidative capacity with respect to all other xanthophylls in *Arabidopsis* leaves and functions independent of binding to PSII antennae. Plant Physiol 145:1506–1520
Heber U, Walker D (1992) Concerning a dual function of coupled cyclic electron-transport in leaves. Plant Physiol 100:1621–1626

Hernández GG, Winter K, Slot M (2020) Similar temperature dependence of photosynthetic parameters in sun and shade leaves of three tropical tree species. Tree Physiol 40:637–651

Horton P, Ruban AV, Walters RG (1996) Regulation of light harvesting in green plants. Annu Rev Plant Physiol Plant Mol Biol 47:655–684

IPCC (2014) Climate Change 2014: Synthesis Report. Contribution of Working Groups I, II and III to the Fifth Assessment Report of the Intergovernmental Panel on Climate Change [Core Writing Team, R.K. Pachauri and L.A. Meyer (eds.)]. IPCC, Geneva, Switzerland, p 151

Jahns P, Holzwarth AR (2012) The role of the xanthophyll cycle and of lutein in photoprotection of photosystem II. Biochim Biophys Acta 1817:182–193

Jahns P, Latowski D, Strzalka K (2009) Mechanism and regulation of the violaxanthin cycle: the role of antenna proteins and membrane lipids. Biochim Biophys Acta 1787:3–14

Jansen MAK, Gaba V, Greenberg BM (1998) Higher plants and UV-B radiation: balancing damage, repair and acclimation. Trends Plant Sci 3:131–135

Joet T, Cournac L, Horvath EM, Medgyesy P, Peltier G (2001) Increased sensitivity of photosynthesis to antimycin A induced by inactivation of the chloroplast ndhB gene. Evidence for a participation of the NADH-dehydrogenase complex to cyclic electron flow around photosystem I. Plant Physiol 125:1919–1929

Johnson MP, Havaux M, Triantaphylidès C, Ksas B, Pascal AA, Robert B, Davison PA, Ruban AV, Horton P (2007) Elevated zeaxanthin bound to oligomeric LHCII enhances the resistance of *Arabidopsis* to photooxidative stress by a lipid-protective, antioxidant mechanism. J Biol Chem 282:22605–22618

Kaiser E, Correa Galvis V, Armbruster U (2019) Efficient photosynthesis in dynamic light environments: a chloroplast's perspective. Biochem J 476:2725–2741

Klughammer C, Schreiber U (1994) An improved method, using saturating light pulses, for the determination of photosystem I quantum yield via $P700^+$-absorbance changes at 830 nm. Planta 192:261–268

Kolb CA, Käser MA, Kopecky J, Zotz G, Riederer M, Pfündel EE (2001) Effects of natural intensities of visible and ultraviolet radiation on epidermal ultraviolet screening and photosynthesis in grape leaves (*Vitis vinifera* cv. Silvaner). Plant Physiol 127:863–875

Königer M, Winter K (1993) Growth and photosynthesis of *Gossypium hirsutum* L. at high photon flux densities: effects of soil temperatures and nocturnal air temperatures. Agronomie 13:423–431

Königer M, Harris GC, Virgo A, Winter K (1995) Xanthophyll-cycle pigments and photosynthetic capacity in tropical forest species: a comparative field study on canopy, gap and understory plants. Oecologia 104:280–290

Kouřil R, Lazár D, Ilík P, Skotnica J, Krchňák P, Nauš J (2004) High-temperature induced chlorophyll fluorescence rise in plants at 40-50°C: experimental and theoretical approach. Photosynth Res 81:49–66

Kozuleva MA, Ivanov BN (2016) The mechanisms of oxygen reduction in the terminal reducing segment of the chloroplast photosynthetic electron transport chain. Plant Cell Physiol 67:1397–1404

Krall JP, Edwards GE (1992) Relationship between photosystem II activity and CO_2 fixation in leaves. Physiol Plant 86:180–187

Krause GH, Jahns P (2003) Pulse amplitude modulated chlorophyll fluorometry and its application in plant science. In: Green BR, Parson WW (eds) Light harvesting antennas in photosynthesis, Advances in photosynthesis and respiration, vol 13. Kluwer Academic Publishers, Dordrecht, pp 373–399

Krause GH, Jahns P (2004) Non-photochemical energy dissipation determined by chlorophyll fluorescence quenching. Characterization and function. In: Papageorgiou GC, Govindjee (eds) Chlorophyll fluorescence: a signature of photosynthesis. Advances in photosynthesis and respiration, vol 14. Springer, Heidelberg, pp 463–495

Krause GH, Santarius KA (1975) Relative thermostability of the chloroplast envelope. Planta 127:285–299

Krause GH, Winter K (1996) Photoinhibition of photosynthesis in plants growing in natural tropical forest gaps. A chlorophyll fluorescence study. Bot Acta 109:456–462

Krause GH, Virgo A, Winter K (1995) High susceptibility to photoinhibition of young leaves of tropical forest trees. Planta 197:583–591

Krause GH, Schmude C, Garden H, Koroleva OY, Winter K (1999) Effects of solar ultraviolet radiation on the potential efficiency of photosystem II in leaves of tropical plants. Plant Physiol 121:1349–1358

Krause GH, Koroleva OY, Dalling JW, Winter K (2001) Acclimation of tropical tree seedlings to excessive light in simulated tree-fall gaps. Plant Cell Environ 24:1345–1352

Krause GH, Grube E, Virgo A, Winter K (2003a) Sudden exposure to solar UV-B radiation reduces net CO_2 uptake and photosystem I efficiency in shade-acclimated tropical tree seedlings. Plant Physiol 131:745–752

Krause GH, Gallé A, Gademann R, Winter K (2003b) Capacity of protection against ultraviolet radiation in sun and shade leaves of tropical forest plants. Funct Plant Biol 30:533–542

Krause GH, Grube E, Koroleva OY, Barth C, Winter K (2004) Do mature shade leaves of tropical tree seedlings acclimate to high sunlight and UV radiation? Funct Plant Biol 31:743–756

Krause GH, Gallé A, Virgo A, García M, Bucic P, Jahns P, Winter K (2006) High-light stress does not impair biomass accumulation of sun-acclimated tropical tree seedlings (*Calophyllum longifolium* Willd. and *Tectona grandis* L.f.). Plant Biol 8:31–41

Krause GH, Jahns P, Virgo A, García M, Aranda J, Wellmann E, Winter K (2007) Photoprotection, photosynthesis and growth of tropical tree seedlings under near-ambient and strongly reduced solar ultraviolet-B radiation. J Plant Physiol 164:1311–1322

Krause GH, Winter K, Krause B, Jahns P, García M, Aranda J, Virgo A (2010) High-temperature tolerance of a tropical tree, *Ficus insipida*: methodological reassessment and climate change considerations. Funct Plant Biol 37:890–900

Krause GH, Winter K, Matsubara S, Krause B, Jahns P, Virgo A, Aranda J, García M (2012) Photosynthesis, photoprotection, and growth of shade-tolerant tropical tree seedlings under full sunlight. Photosynth Res 113:273–285

Krause GH, Cheesman AW, Winter K, Krause B, Virgo A (2013) Thermal tolerance, net CO_2 exchange and growth of a tropical tree species, Ficus *insipida*, cultivated at elevated daytime and nighttime temperatures. J Plant Physiol 170:822–827

Krause GH, Winter K, Krause B, Virgo A (2015) Light-stimulated heat tolerance in leaves of two neotropical tree species, *Ficus insipida* and *Calophyllum longifolium*. Funct Plant Biol 42:42–51. https://doi.org/10.1071/FP14095

Krause GH, Winter K, Krause B, Virgo A (2016) Protection by light against heat stress in leaves of tropical crassulacean acid metabolism plants containing high acid levels. Funct Plant Biol 43:1061–1069. https://doi.org/10.1071/FP16093

Kress E, Jahns P (2017) The dynamics of energy dissipation and xanthophyll conversion in *Arabidopsis* indicate an indirect photoprotective role of zeaxanthin in slowly inducible and relaxing components of non-photochemical quenching of excitation energy. Front Plant Sci 8:2094

Lehrum W, Kappen L, Lösch R (1987) Zusammenhang zwischen Hitzeresistenz und Säuregehalt in sukkulenten Pflanzen. Verhandlungen Gesellschaft Ökologie 16:207–212

Leitsch J, Schnettger B, Critchley C, Krause GH (1994) Two mechanisms of recovery from photoinhibition in vivo: reactivation of photosystem II related and unrelated to D1-prtein turnover. Planta 194:15–21

Li X-P, Gilmore AM, Caffari S, Bassi R, Golan T, Kramer D, Niyogi KK (2004) Regulation of photosynthetic light harvesting involves intrathylakoid lumen pH sensing of the PsbS protein. J Biol Chem 279:22866–22874

Lima-Melo Y, Gollan PJ, Tikkanen M, Silveira JAG, Aro EM (2019) Consequences of photosystem-I damage and repair on photosynthesis and carbon use in *Arabidopsis thaliana*. Plant J 97:1061–1072

Liu L-X, Xu S-M, Woo KC (2005) Solar UV-B radiation on growth, photosynthesis and the xanthophyll cycle in tropical acacias and eucalyptus. Environ Exp Bot 54:121–130

Lösch R, Kappen L (1983) Die Temperaturresistenz makaronesischer *Sempervivoideae*. Verhandlungen Gesellschaft Ökologie 10:521–528

Lovelock CE, Jebb M, Osmond CB (1994) Photoinhibition and recovery in tropical plant species: response to disturbance. Oecol 97:297–307

Lüttge U (2008) Physiological ecology of tropical plants. Springer, Berlin

Lüttge U, Smith JAC (1984) Mechanisms of passive malic-acid efflux from vacuoles of the CAM plant *Kalanchoë daigremontiana*. J Membr Biol 81:149–158

Malnoë A (2018) Photoinhibition or photoprotection of photosynthesis? Update of the (newly termed) sustained quenching component qH. Envir Exp Bot 154:123–133

Manuel N, Cornic G, Aubert S, Choler P, Bligny R, Heber U (1999) Protection against photoinhibition in the alpine plant *Geum montanum*. Oecologia 119:149–158

Markstädter C, Queck I, Baumeister J, Riederer M, Schreiber U, Bilger W (2001) Epidermal transmittance of leaves of *Vicia faba* for UV radiation as determined by two different methods. Photosynth Res 67:17–25

Matsubara S, Gilmore AM, Osmond CB (2001) Diurnal and acclimatory responses of violaxanthin and lutein epoxide in the Australian mistletoe *Amyema miquelii*. Aust J Plant Physiol 28:793–800

Matsubara S, Naumann M, Martin R, Nichol C, Rascher U, Morosinotto T, Bassi R, Osmond B (2005) Slowly reversible de-epoxidation of lutein epoxide in deep shade leaves of a tropical tree legume may 'lock in' lutein-based photoprotection during acclimation to strong light. J Exper Bot 56:461–468

Matsubara S, Morosinotto T, Osmond B, Bassi R (2007) Short- and long-term operation of the lutein epoxide cycle in light-harvesting antenna complexes. Plant Phsiol 144:926–941

Matsubara S, Krause GH, Seltmann M, Virgo A, Kursar TA, Jahns P, Winter K (2008) Lutein epoxide cycle, light harvesting and photoprotection in species of the tropical tree genus *Inga*. Plant Cell Environ 31:548-561

Matsubara S, Krause GH, Aranda J, Virgo A, Beisel KG, Jahns P, Winter K (2009) Sun-shade patterns of leaf carotenoid composition in 86 species of neotropical forest plants. Funct Plant Biol 36:20–36. https://doi.org/10.1071/FP08214

Mulkey SS, Pearcy RW (1992) Interactions between acclimation and photoinhibition of photosynthesis of a tropical understory herb, *Alocasia macrorrhiza*, during simulated canopy gap formation. Funct Ecol 6:719–729

Munekage Y, Hojo M, Meurer J, Endo T, Tasaka M, Shikanai T (2002) PGR5 is involved in cyclic electron flow around photosystem I and is essential for photoprotection in *Arabidopsis*. Cell 110:361–371

Muraoka H, Tang Y, Terashima I, Koizumi H, Washitani I (2000) Contributions of diffusional limitation, photoinhibition and photorespiration to midday depression in *Arisaema heterophyllum* in natural light. Plant Cell Environ 23:235–250

Nilkens M, Kress E, Lambrev P, Miloslavina Y, Müller M, Holzwath AR, Jahns P (2010) Identification of a slowly inducible zeaxanthin-dependent component on non-photochemical quenching of chlorophyll fluorescence generated under steady-state conditions in *Arabidopsis*. Biochim Biophy Acta 1797:466–475

Nobel PS (1988) Environmental biology of agaves and cacti. Cambridge University Press, Cambridge

O'Sullivan OS, Heskel MA, Reich PB, Tjoelker MG, Weerasinghe LK, Penillard A, Zhu L, Egerton JJ, Bloomfield KJ, Creek D, Bahar NH, Griffin KL, Hurry V, Meir P, Turnbull MH, Atkin OK (2017) Thermal limits of leaf metabolism across biomes. Glob Change Biol 23:209–223

Phillips OL, Aragao LEOC, Luiz SL et al (2009) Drought sensitivity of the Amazon rainforest. Science 323:1344–1347

Polm M, Brettel K (1998) Secondary pair charge recombination in photosystem I under strongly reducing conditions: temperature dependence and suggested mechanism. Biophys J 74:3173–3181

Popp M, Kramer C, Lee H, Diaz M, Ziegler H, Lüttge U (1987) Crassulacean acid metabolism in tropical dicotyledonous trees of the genus *Clusia*. Trees – Struct Funct 1:238–247

Reuber S, Leitsch J, Krause GH, Weissenböck G (1993) Metabolic reduction of phenylpropanoid compounds in primary leaves of rye (*Secale cereale* L.) leads to increased UV-B sensitivity of photosynthesis. Z Naturforsch 48c:749–756

Reuber S, Bornman JF, Weissenböck G (1996) Phenylpropanoid compounds in primary leaf tissue of rye (*Secale cereale*): light response of their metabolism and the possible role in UV-B protection. Physiol Plant 97:160–168

Rey-Sánchez AC, Slot M, Posada JM, Kitajima K (2016) Spatial and seasonal variation of leaf temperature within the canopy of a tropical forest. Climate Res 71:75–89

Sacharz J, Giovagnetti V, Ungerer P, Mastroianni G, Ruban AV (2017) The xanthophyll cycle affects reversible interactions between PsbS and light-harvesting complex II to control non-photochemical quenching. Nature Plants 3:16225

Sachs J (1864) Ueber die obere Temperatur-Gränze der Vegetation. Flora 47:5–12, 24–29, 33–39, 65–75

Sastry A, Barua D (2017) Leaf thermotolerance in tropical trees from a seasonally dry climate varies along the slow-fast resource acquisition spectrum. Sci Rep 7:1–11

Scheer H (2003) The pigments. In: Green R, Parson WW (eds) Light-harvesting antennas in photosynthesis. Kluwer Academic Publishers, Dordrecht, pp 29–81

Schreiber U (1986) Detection of rapid induction kinetics with a new type of high-frequency modulated chlorophyll fluorometer. Photosynth Res 9:261–272

Schreiber U, Berry JA (1977) Heat-induced changes in chlorophyll fluorescence in intact leaves correlated with damage of the photosynthetic apparatus. Planta 136:233–238

Schreiber U, Schliwa U, Bilger W (1986) Continuous recording of photochemical and non-photochemical chlorophyll fluorescence quenching with a new type of modulation fluorometer. Photosynth Res 10:51–62

Schwemmle B, Lange OL (1959) Endogen-tagesperiodische Schwankungen der Hitzeresistenz bei *Kalanchoë blossfeldiana*. Planta 53:134–144

Searles PS, Caldwell MM, Winter K (1995) The response of five tropical dicotyledon species to solar ultraviolet-B radiation. Amer J Bot 82:445–453

Slot M, Winter K (2018) High tolerance of tropical sapling growth and gas exchange to moderate warming. Funct Ecol 32:599–611

Slot M, Rey-Sánchez C, Gerber S, Lichstein JW, Winter K, Kitajima K (2014) Thermal acclimation of leaf respiration of tropical trees and lianas: response to experimental canopy warming, and consequences for tropical carbon balance. Global Change Biol 20:2915–2926

Slot M, García MN, Winter K (2016) Temperature responses of CO_2 exchange in three tropical tree species. Funct Plant Biol 43:468–478

Slot M, Krause GH, Krause B, Hernández GG, Winter K (2019) Photosynthetic heat tolerance of shade and sun leaves of three tropical tree species. Photosynth Res 141:119–130

Smillie RM, Nott R (1979) Heat injury in leaves of alpine, temperate and tropical plants. J Plant Physiol 6:135–141

Smith JAC, Ingram J, Tsiantis MS, Barkla BJ, Bartholomew DM, Bettey M, Pantoja O, Pennington AJ (1996) Transport across the vacuolar membrane in CAM plants. In: Winter K, Smith JAC (eds) Ecological studies. Crassulacean acid metabolism, Biochemistry, ecophysiology and evolution, vol 114. Springer, Berlin, pp 53–71

Snyder AM, Clark BM, Bungard RA (2005) Light-dependent conversion of carotenoids in the parasitic angiosperm *Cuscuta reflexa* L. Plant Cell Environ 28:1326–1333

Sonoike K (2011) Photoinhibition of photosystem I. Physiol Plant 142:56–64

Strasser RJ, Srivastava A, Govindjee (1995) Polyphasic chlorophyll *a* fluorescence transients in plants and cyanobacteria. Photochem Photobiol 61:32–42

Takahashi S, Badger MR (2011) Photoprotection in plants: a new light on photosystem II damage. Trends Plant Sci 16:53–60
Teicher HB, Scheller HV (1998) The NAD(P)H dehydrogenase in barley thylakoids is photoactivatable and uses NADPH as well as NADH. Plant Physiol 117:525–532
Teicher HB, Møller BL, Scheller HV (2000) Photoinhibition of photosystem I in field-grown barley (*Hordeum vulgare* L.): induction, recovery and acclimation. Photosynth Res 64:53–61
Terashima I, Funayama S, Sonoike K (1994) The site of photoinhibition in leaves of *Cucumis sativus* L. at low temperatures is photosystem I, not photosystem II. Planta 193:300–306
Thayer SS, Björkman O (1992) Carotenoid distribution and de-epoxidation in thylakoid pigment-protein complexes from cotton leaves and bundle-sheath cells of maize. Photosynth Res 33:213–225
Thiele A, Schirwitz K, Winter K, Krause GH (1996) Increased xanthophyll cycle activity and reduced D1 protein inactivation related to photoinhibition in two plant systems acclimated to excess light. Plant Sci 115:237–250
Thiele A, Winter K, Krause GH (1997) Low inactivation of D1 protein of photosystem II in young canopy leaves of *Anacardium excelsum* under high-light stress. J Plant Physiol 151:286–292
Thiele A, Krause GH, Winter K (1998) In situ study of photoinhibition of photosynthesis and xanthophyll cycle activity in plants growing in natural gaps of the tropical forest. Aust J Plant Physiol 25:189–195
Tikhonov AN (2014) The cytochrome b_6f complex at the crossroad of photosynthetic electron transport pathways. Plant Physiol Biochem 81:163–183
Tikkanen M, Grebe S (2018) Switching off photoprotection of photosystem I – a novel tool for gradual PSI photoinhibition. Physiol Plant 162:156–161
Tiwari A, Mamedov F, Grieco M, Suorsa M, Jajoo A, Styring S, Tikkanen M, Aro EM (2016) Photodamage of iron-sulphur clusters in photosystem I induces non-photochemical energy dissipation. Nat Plants 2:16035
Tjus SE, Møller BL, Scheller HV (1998) Photosystem I is an early target of photoinhibition in barley illuminated at chilling temperatures. Plant Physiol 116:755–764
Wahid A, Gelani S, Ashraf M, Foolad MR (2007) Heat tolerance in plants: an overview. Environ Exp Bot 61:199–223
Watling JR, Robinson SA, Woodrow IE, Osmond CB (1997) Responses of rainforest understorey plants to excess light during sunflecks. Aust J Plant Physiol 24:17–25
White RC, Gibbs E, Butler LS (1977) Estimation of copper pheophytins, chlorophylls and pheophytins in mixtures in diethyl ether. J Agric Food Chem 25:143–145
Wilson S, Ruban AV (2019) Quantitative assessment of the high-light tolerance in plants with impaired photosystem II donor side. Biochem J 476:1377–1386
Winter K, Holtum JAM (2014) Facultative crassulacean acid metabolism (CAM) plants: powerful tools for unravelling the functional elements of CAM photosynthesis. J. Exp Bot 65:3425–3441
Winter K, García M, Holtum JAM (2008) On the nature of facultative and constitutive CAM: environmental and developmental control of CAM expression during early growth of *Clusia*, *Kalanchoë* and *Opuntia*. J Exp Bot 59:1829–1840
Winter K, García M, Holtum JAM (2014) Nocturnal versus diurnal CO_2 uptake: how flexible is *Agave angustifolia*? J Exp Bot 65:3695–3703
Yavitt JB, Battles JJ, Lang GE, Knight DH (1995) The canopy gap regime in a secondary neotropical forest in Panama. J Trop Ecol 11:391–402
Yu Z-C, Zheng X-T, Lin W, Cai M-L, Zhang Q-L, Peng C-L (2020) Different photoprotection strategies for mid- and late-successional dominant tree species in a high-light environment in summer. Environ Exp Bot 171:103927
Ziska LH (1996) The potential sensitivity of tropical plants to increased ultraviolet-B radiation. J Plant Physiol 148:35–41
Zotz G, Harris G, Königer M, Winter K (1995) High rates of photosynthesis in the tropical pioneer tree, *Ficus insipida* Willd. Flora 190:265–272

Plant Peroxisomes and Their Metabolism of ROS, RNS, and RSS

Luis A. del Río

Contents

Abstract Peroxisomes are subcellular organelles with a single membrane and devoid of DNA, with an essentially oxidative type of metabolism, and are probably the major loci of intracellular H_2O_2 production. A general property of peroxisomes is that they contain as basic enzymatic constituents catalase and hydrogen peroxide-producing flavin oxidases and that carry out the fatty acid ß-oxidation, but in recent years, it has been established that peroxisomes are involved in a range of important cellular functions in nearly all eukaryotic cells. Research developed in the last 30 years has indicated the existence of cellular functions of peroxisomes related to reactive oxygen species (ROS), like H_2O_2, superoxide radicals ($O_2^{\cdot-}$), singlet

L. A. del Río (✉)
Group of Antioxidants, Free Radicals and Nitric Oxide in Biotechnology, Food and Agriculture, Department of Biochemistry and Cell and Molecular Biology of Plants, Estación Experimental del Zaidín, Consejo Superior de Investigaciones Científicas (CSIC), Granada, Spain
e-mail: luisalfonso.delrio@eez.csic.es

© Springer Nature Switzerland AG 2020
Progress in Botany (2021) 82: 171–210, https://doi.org/10.1007/124_2020_37,
Published online: 30 December 2020

oxygen, etc., and reactive nitrogen species (RNS), like nitric oxide (NO) and other thereof derived compounds, and a function for peroxisomes as important centers of the cellular signaling apparatus has been postulated. More recently, evidence has been obtained suggesting new roles of peroxisomes involving reactive sulfur species (RSS), sulfur compounds with different higher oxidation states.

In this review, the generation and/or metabolism of ROS, RNS, and RSS in peroxisomes and its regulation, as well as the different antioxidant systems present in these organelles, will be analyzed in the context of distinct ROS-, RNS-, and RSS-mediated functions of plant peroxisomes that can be involved in the metabolism of plant cells under both physiological and stress conditions.

Keywords Abiotic stress, Antioxidants, Peroxisomes, Plant metabolism, Reactive nitrogen species (RNS), Reactive oxygen species (ROS), Reactive sulfur species (RSS), RNS function, ROS function, RSS function

Abbreviations

$^{\cdot}OH$	Hydroxyl radical
$^{1}O_2$	Singlet oxygen
2,4-D	2,4-dichlorophenoxyacetic acid
APF	Aminophenyl fluorescein
APX	Ascorbate peroxidase
cGMP	Cyclic guanosine monophosphate
CLSM	Confocal laser scanning microscopy
CuAO	Cu-containing amine oxidase
EM	Electron microscopy
GOX	Glycolate oxidase
GR	Glutathione reductase
H_2O_2	Hydrogen peroxide
HPLC	High-pressure liquid chromatography
ICDH	Isocitrate dehydrogenase
MDAR	Monodehydroascorbate reductase
NO	Nitric oxide
$O_2^{\cdot-}$	Superoxide radical
O_3	Ozone
PAO	Polyamine oxidase
PMP	Peroxisomal membrane polypeptide
Prx	Peroxiredoxin
PTM	Posttranslational modification
PTS	Peroxisomal targeting signal
RNS	Reactive nitrogen species
ROS	Reactive oxygen species
RSS	Reactive sulfur species
SOD	Superoxide dismutase

XDH Xanthine dehydrogenase
XOD Xanthine oxidase

1 Introduction

The existence of peroxisomes was first reported in 1954, in electron microscopy studies carried out in mouse kidney tubules, and these organelles were initially designated with the morphological name "microbodies" (Rhodin 1954). In 1960 Prof. Christian De Duve accomplished the biochemical characterization of peroxisomes from mammalian tissues and their recognition as distinct and new cell organelles (De Duve et al. 1960). By using biochemical and morphological methods, the identity of microbodies and peroxisomes was unequivocally demonstrated. The occurrence of different hydrogen peroxide-producing oxidases, and the hydrogen peroxide-scavenging enzyme catalase in microbodies, suggested De Duve to designate these new cell organelles with the name "peroxisomes" (De Duve and Baudhuin 1966). Years later these organelles were found to be present in nearly all eukaryotic cells.

Plant peroxisomes usually have a granular matrix but can have amorphous or crystalline inclusions composed of catalase (del Río et al. 2002). Figure 1 shows electron micrographs of plant peroxisomes from tissues of three different plant species, olive, pea, and pepper. The subcellular organelles peroxisomes are bounded by a single membrane and are devoid of DNA, they have an essentially oxidative type of metabolism, and are probably the major loci of intracellular H_2O_2 generation. A general property of peroxisomes is that they contain as basic enzymatic constituents catalase and hydrogen peroxide-producing flavin oxidases and are present in almost all eukaryotic cells (Fahimi and Sies 1987; Baker and Graham 2002; del Río 2013; del Río and Schrader 2018). For this reason initially it was thought that the main role of peroxisomes was the removal by catalase of toxic H_2O_2 generated inside peroxisomes by different oxidases.

A general feature of virtually all types of peroxisomes is fatty acid β-oxidation, but in later years, it was found that peroxisomes were involved in a range of important cellular functions in nearly all eukaryotic cells (Erdmann 2016; del Río et al. 2006; del Río 2013; del Río and López-Huertas 2016; del Río and Schrader 2018; Wanders and Waterham 2006; Wanders 2013; Baker and Graham 2002; Hu et al. 2012; Kao et al. 2018). The main roles of peroxisomes actually known were elucidated on the basis of peroxisome purification and analysis by cell biology and biochemical methods (Baker and Graham 2002; Wanders and Waterham 2006; Palma et al. 2009; del Río 2013; del Río and López-Huertas 2016). However, in recent years the application of proteomic analysis to peroxisomes has confirmed the presence of many proteins previously described in peroxisomes by classical methods but has also disclosed many new peroxisomal proteins. The development of sensitive proteomics and mass spectrometry (MS) technologies allows the identification of

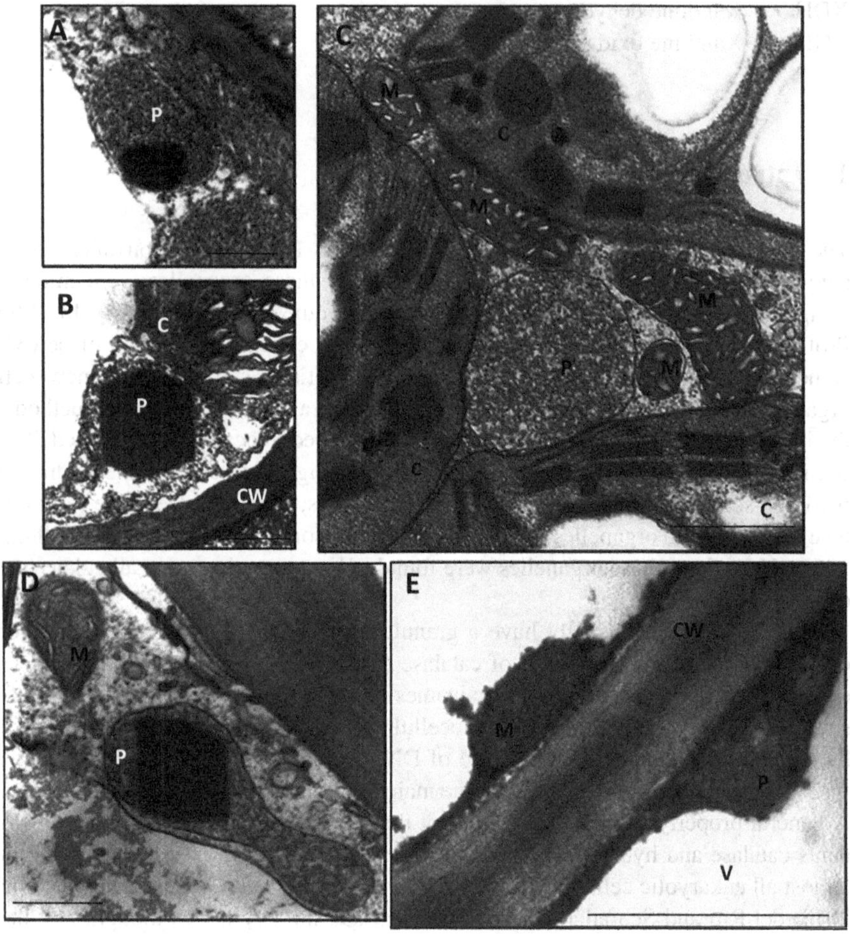

Fig. 1 Electron micrographs of peroxisomes from different plant tissues. (**a**) Peroxisome from olive tree leaf (*Olea europaea* L.) showing an amorphous core inside the peroxisome. (**b**) Peroxisome from pepper leaf (*Capsicum annuum* L.) with a crystalline inclusion. (**c**) Peroxisome from pea leaf (*Pisum sativum* L.) showing a granular matrix and in close contact with chloroplasts and mitochondria. (**d**) Peroxisome from ripe pepper fruit (*Capsicum annuum* L.). (**e**) Peroxisome from mature olive fruit (*Olea europaea* L.). C, chloroplast. CW, cell wall. M, mitochondrion. P, Peroxisome. V, vacuole. Scale bar = 1 μm. Reproduced from del Río and López-Huertas (2016) Plant Cell Physiol. 57: 1364–76

low-abundance and transient peroxisomal proteins with the concomitant increase of our knowledge of peroxisome roles and their metabolic and regulatory networks (Reumann 2011; Erdmann 2016; Gronemeyer et al. 2013; Kaur and Hu 2011; del Río and Schrader 2018). Some of the major functions that have been postulated so far for peroxisomes in plant cells are shown in Table 1.

Peroxisomes of plant cells are highly dynamic compartments that for its subcellular distribution and movements are dependent upon the actin cytoskeleton, not

Table 1 Main functions of peroxisomes in plant cells

Function	References
ß-oxidation of fatty acids	Baker and Graham (2002), Nyathi and Baker (2006)
Photorespiration	Tolbert (1997)
Glyoxylate cycle	Pracharoenwattana and Smith (2008)
Metabolism of ureides	Schubert (1986), Baker and Graham (2002)
Purine catabolism	Corpas et al. (1997)
Sulfur metabolism	Corpas et al. (2009a), Corpas and Barroso (2015)
Polyamine catabolism	Moschou et al. (2008), Reumann et al. (2009), Wimalasekera et al. (2011a), Corpas et al. (2019a)
Metabolism of ROS and RNS	del Río et al. (2002, 2006), del Río (2011), del Río and López-Huertas (2016), Corpas et al. (2017a)
Metabolism of RSS	Corpas et al. (2009a), Corpas and Barroso (2015)
Biosynthesis of phytohormones (auxin, jasmonic acid, salicylic acid)	Hu et al. (2012), del Río (2013)
Photomorphogenesis	Kaur et al. (2013)
Leaf senescence	del Río et al. (1998)
Plant defense against fungal infection	Koh et al. (2005)
Protection against herbivores	Shabab (2013)

Modified from del Río and López-Huertas (2016) Plant Cell Physiol 57: 1364–76

microtubules (Mathur et al. 2002; Mano et al. 2002; Rodríguez-Serrano et al. 2009; Oikawa et al. 2015; Sparkes and Gao 2014). There are three key characteristic properties of peroxisomes: (1) their oxidative type of metabolism; (2) their power for sharing metabolic pathways with other cell compartments; and (3) their metabolic plasticity, since their enzymatic complement can change depending on the organism, cell/tissue type, and environmental conditions (Fahimi and Sies 1987; del Río et al. 2002; Baker and Graham 2002; del Río 2013). The occurrence in peroxisomes of different regulatory proteins, such as kinases, phosphatases, and heat-shock proteins, has been described (Hayashi and Nishimura 2003; Reumann et al. 2009; Hu et al. 2012; del Río and Schrader 2018).

The group of JK Reddy in the USA was the first to report the induction of the peroxisomal population as well as the activity of the H_2O_2-producing acyl-CoA oxidase in animals by different xenobiotics (Reddy et al. 1987). In plants, the proliferation of the cellular population of peroxisomes has been reported both under natural and different abiotic stress conditions. The cases and the agents responsible for the proliferation of peroxisomes in different plant species are presented in Table 2. The induction of peroxisome biogenesis genes (*PEX*) by H_2O_2 has been demonstrated in plant and animal cells, suggesting that the signal molecule H_2O_2 is responsible for the proliferation of peroxisomes (López-Huertas et al. 2000). The peroxisome proliferator-activated receptor (PPAR), the transcription factor implied in peroxisomal proliferation and induction of peroxisomal fatty acid β-oxidation in animal tissues, was demonstrated to be functional in transgenic tobacco plants, and its expression induced the proliferation of peroxisomes, as described in animal tissues (Nila et al. 2006).

Table 2 Peroxisome proliferation in plants

Proliferating agent	Plant tissue	Reference
Senescence	Carnation petals	Droillard and Paulin (1990)
	Pea leaves	del Río et al. (1998)
Light	*Arabidopsis* seedlings	Desai and Hu (2008)
	Lemna minor fronds	Ferreira et al. (1989)
H_2O_2	*Arabidopsis* leaves	López-Huertas et al. (2000)
Isoproturon (herbicide)	Ryegrass leaves	de Felipe et al. (1988)
Clofibrate (hypolipidemic drug)	Pea leaves	Palma et al. (1991)
	Tobacco leaves	Nila et al. (2006)
	Arabidopsis leaves	Castillo et al. (2008)
Ozone	Norway spruce needles	Morré et al. (1990)
	Aspen and birch leaves	Oksanen et al. (2003)
Cadmium	Pea leaves	Romero-Puertas et al. (1999)
		del Río et al. (2003b)
Salt stress	*Arabidopsis* leaves	Mitsuya et al. (2010)

Research carried out in the last 30 years has indicated the existence of cellular functions for peroxisomes related to reactive oxygen species (ROS) and reactive nitrogen species (RNS) (del Río et al. 2002, 2006; Schrader and Fahimi 2004, 2006; Corpas et al. 2009a, 2019a; del Río 2011; Fransen et al. 2012; Fransen and Lismont 2018; Sandalio and Romero-Puertas 2015), and a function for peroxisomes as important centers of the cellular signaling apparatus has been postulated (del Río 2013). More recently, new roles of peroxisomes in the oxidative metabolism involving reactive sulfur species (RSS) have been proposed (Corpas et al. 2009a, 2019a; Corpas and Barroso 2015). As shown in Table 1, the diverse physiological functions that have been demonstrated for peroxisomes from different sources strongly indicate the interest of these organelles as a cellular source of different signaling molecules.

In this review, the generation and metabolism of ROS, RNS, and RSS in peroxisomes and its regulation, as well as the different antioxidant systems present in these organelles, will be analyzed in the context of different ROS-, RNS-, and RSS-mediated functions of plant peroxisomes that could be involved in plant cell metabolism under both physiological and stress conditions.

2 Generation of Reactive Oxygen Species (ROS), Reactive Nitrogen Species (RNS), and Reactive Sulfur Species (RSS)

2.1 ROS

In peroxisomes the generation of distinct reactive oxygen species (ROS), including hydrogen peroxide (H_2O_2), superoxide radicals ($O_2^{\cdot-}$), and singlet oxygen ($^1O_2$), has been reported. For the generation of H_2O_2, the main metabolic processes responsible in different types of peroxisomes are, in decreasing order, (1) the photorespiratory glycolate oxidase (GOX) reaction (in green tissues); (2) the main enzyme of fatty acid β-oxidation, acyl-CoA oxidase; (3) the enzymatic reaction of different flavin oxidases, like polyamine oxidases (PAOs), among others; (4) the Cu-containing amine oxidases (CuAOs); and (5) the spontaneous or enzymatic dismutation of $O_2^{\cdot-}$ radicals (del Río et al. 2002, 2006, 2018; Baker and Graham 2002; del Río 2013; Foyer 2018).

Peroxisomes contain distinct H_2O_2-producing flavin oxidases depending on the organism and tissue origin, like GOX, sulfite oxidase, xanthine oxidase (XOD), sarcosine oxidase, as well as enzymes of polyamine catabolism (diamine oxidase and polyamine oxidase), among others (Corpas et al. 2009a, 2019a; del Río 2013; Sandalio and Romero-Puertas 2015; Hänsch and Mendel 2005). The peroxisomal GOX has a rate of H_2O_2 production about 2- and 50-fold higher than that reported for chloroplasts and mitochondria, respectively (Foyer et al. 2009). The accumulation of H_2O_2 in peroxisomes can be imaged in vivo by confocal microscopy, using fluorescent probes such as 2′,7′-dichlorofluorescein diacetate (Sandalio et al. 2008), or by expressing specific H_2O_2 biosensors, such as HyPer, targeted to peroxisomes (Costa et al. 2010). H_2O_2 also can be detected by cytochemistry using $CdCl_3$ and visualization by electron microscopy (Romero-Puertas et al. 2004a).

Peroxisomes, like mitochondria and chloroplasts, also generate superoxide radicals ($O_2^{\cdot-}$) as a result of their normal metabolism. Using biochemical and electron spin resonance spectroscopy (ESR) methods in peroxisomes from pea leaves and watermelon cotyledons, the existence of, at least, two sites of $O_2^{\cdot-}$ production was found: one in the organelle matrix (soluble fraction), in which the generating system was identified as xanthine oxidase (XOD), and another locus in the peroxisomal membranes dependent on NAD(P)H (Sandalio et al. 1988; del Río et al. 1989, 2002; del Río and Donaldson 1995; Corpas et al. 2008). XOD catalyzes the oxidation of xanthine and hypoxanthine to uric acid and is a characteristic producer of superoxide radicals (Halliwell and Gutteridge 2015). Xanthine oxidoreductases (XORs) occur in two forms, depending on their electron acceptors. In experiments incubating peroxisomal matrices from pea leaves with microbial XOR, it was shown that peroxisomal endoproteases (EPs) could produce the irreversible conversion of xanthine dehydrogenase (XDH) into the $O_2^{\cdot-}$-generating XOD (Distefano et al. 1999; Palma et al. 2002). In the matrix of peroxisomes, urate oxidase is also present, and this enzyme was also found to be a $O_2^{\cdot-}$ producer although weaker than XOD, judging by the fainter $O_2^{\cdot-}$ ESR signals obtained (Sandalio et al. 1988). The peroxisomal XOD from pea leaves has been characterized, and the occurrence of this enzyme in plant peroxisomes has been confirmed by immunogold electron microscopy (Corpas et al. 2008a).

The peroxisomal membrane is the other site of $O_2^{\cdot-}$ generation, where a small electron transport chain appears to be involved. This consists of a flavoprotein NADH-ferricyanide reductase of about 32 kDa and a Cyt *b* (Bowditch and Donaldson 1990). The integral peroxisomal membrane polypeptides (PMPs) of pea leaf peroxisomes were identified, and three of these membrane polypeptides, with molecular masses of 18, 29, and 32 kDa, have been characterized and found to be responsible for $O_2^{\cdot -}$ generation (López-Huertas et al. 1999). The main producer of superoxide radicals in the peroxisomal membrane was the 18 kDa PMP which was identified as a Cyt *b* (López-Huertas et al. 1999). The 18 and 32 kDa PMPs use NADH as the electron donor for $O_2^{\cdot -}$ production, but the 29 kDa PMP was dependent on NADPH and was able to reduce Cyt *c* with NADPH as electron donor (López-Huertas et al. 1999; del Río et al. 2002). The PMP32 very probably corresponds to the enzyme of the ascorbate-glutathione cycle, monodehydroascorbate reductase (MDAR) (López-Huertas et al. 1999), whose activity was previously detected in pea leaf peroxisomal membranes by Jiménez et al. (1997). The third superoxide-generating polypeptide, PMP29, could be related to the peroxisomal NADPH-cytochrome P450 reductase (López-Huertas et al. 1999).

The production of superoxide radicals by peroxisomal membranes could be an obligatory consequence of NADH reoxidation by the peroxisomal electron transport chain, directed to regenerate NAD^+ to be reused in peroxisomal metabolic processes (del Río et al. 1990, 1992; del Río and Donaldson 1995). The $O_2^{\cdot -}$ accumulation in peroxisomes can be imaged in vivo by confocal laser microscopy, using fluorescent probes such as dihydroethidium (Sandalio et al. 2008).

Singlet oxygen ($^1O_2$) is a non-radical ROS normally produced in chloroplasts through different photodynamic reactions (Asada 2006; Triantaphylidés and Havaux 2009), but the light-independent generation of singlet oxygen has been reported in mitochondria, peroxisomes, and the nucleus of *Arabidopsis thaliana* non-photosynthetic tissue. In vivo imaging with a singlet oxygen-specific probe and confocal microscopy demonstrated the production of singlet oxygen in peroxisomes, mitochondria, and the nucleus, and $^1O_2$ accumulation was enhanced in roots of plants, in the dark, under various biotic and abiotic stresses (Mor et al. 2014). The authors indicated that the origin of singlet oxygen could be the reaction of $O_2^{\cdot -}$ with H_2O_2 via a Haber-Weiss mechanism.

In conclusion, peroxisomes are probably the major sites of intracellular H_2O_2 production and are an important source of $O_2^{\cdot -}$ radicals. A model of the function of peroxisomes in the generation of the ROS H_2O_2, $O_2^{\cdot -}$, and $^1O_2$ is presented in Fig. 2.

2.2 RNS

Reactive nitrogen species (RNS) is another collective name, like ROS, but including radicals like nitric oxide ($NO^{\cdot}$) and nitric dioxide ($NO_2^{\cdot}$), as well as non-radicals such as nitrous acid (HNO_2) and dinitrogen tetroxide (N_2O_4) among others (Halliwell and Gutteridge 2007; del Río 2015). Both RNS and ROS have been demonstrated to have an important role in biology and medicine (Schrader and Fahimi 2004; Fransen et al. 2012; Halliwell and Gutteridge 2015; del Río 2015).

The gaseous free radical nitric oxide ($NO^{\cdot}$) is a widespread intracellular and intercellular messenger with a broad spectrum of regulatory functions in many physiological processes of animal and plant systems (del Río et al. 2004; Neill

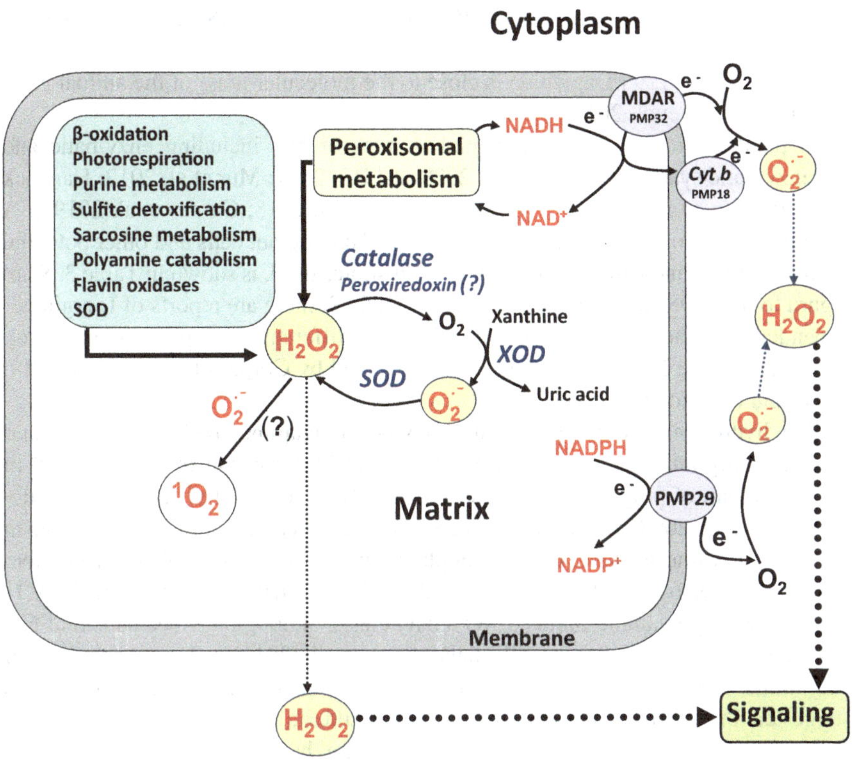

Fig. 2 Scheme of the different sources of the ROS superoxide ($O_2^{\cdot-}$), hydrogen peroxide (H_2O_2), and singlet oxygen ($^1O_2$) in plant peroxisomes. PMP, peroxisomal membrane polypeptides. Cyt *b*, a *b* type cytochrome (PMP18). MDAR, monodehydroascorbate reductase (PMP32). PMP29, peroxisomal membrane polypeptide of 29 kDa probably related to the NADPH-cytochrome P450 reductase. SOD, superoxide dismutase. XOD, xanthine oxidase. Broken arrows indicate signaling. Reproduced from del Río and López-Huertas (2016) Plant Cell Physiol. 57: 1364–76

et al. 2008; Martínez-Ruiz and Lamas 2009; Astier et al. 2018). In animal systems, most of the NO produced is due to the enzyme nitric oxide synthase (NOS; EC 1.14.13.39) (Alderton et al. 2001). NOS catalyzes the oxygen- and NADPH-dependent oxidation of L-arginine to NO and citrulline, in a complex reaction requiring FAD, FMN, tetrahydrobiopterin (BH_4), Ca^{2+}, and calmodulin (Knowles and Moncada 1994; Alderton et al. 2001). However, in plants a gene or a protein with homology to mammalian NOS enzymes has not been found in *Arabidopsis thaliana* (the *Arabidopsis* genome initiative, 2000). The different molecular approaches developed so far to clone a higher plant NOS based on the sequence of animal NOS have always given negative results (del Río 2011; del Río et al. 2014; Astier et al. 2018). The only case reported to date in the plant kingdom of a NOS that has been characterized is that of a unicellular species of marine green alga, *Ostreococcus tauri* (Foresi et al. 2010). The length sequence of *O. tauri* NOS showed a similarity of 42,43, and 34% with respect to eNOS, iNOS, and nNOS, respectively. The authors suggested that the active form of *O. tauri* NOS is a dimer with a subunit of 119 kDa, which is close to the molecular mass of the animal NOS subunits (Foresi et al. 2010).

In plants there are several potential sources of NO including enzymatic and nonenzymatic systems (Wilson et al. 2008; del Río 2011; Mur et al. 2012; Hancock 2012; del Río et al. 2014; del Río 2015; Astier et al. 2018; Corpas et al. 2019a). A summarized list of some established sources of NO in plant cells and other potential sources, with an indication of the different substrates used, is shown in Table 3. With regard to the NOS-like activity indicated in Table 3, there are reports of L-arginine-dependent NOS activity in extracts of, at least, 11 different plant species (Cueto et al. 1996; Jasid et al. 2006; Zhao et al. 2007; reviewed by Corpas et al. 2009b, del Río 2011, and del Río et al. 2014).

The hydrogen peroxide-producing enzymes polyamine oxidases (PAOs) and Cu-containing amine oxidases (CuAOs), present in peroxisomes, are involved in the catabolism of polyamines (Tiburcio et al. 2014; Corpas et al. 2019a; Kamada-Nobusada et al. 2008), and in recent years, it has reported that polyamine catabolism induces the synthesis of NO in different plant organs (Tun et al. 2006; Wimalasekera et al. 2011a; Diao et al. 2016; Agurla et al. 2018). In *Arabidopsis*, isoform CuAO1 participates in polyamine-induced NO biosynthesis, although the mechanism of NO production remains unknown (Wimalasekera et al. 2011b). The occurrence of polyamine oxidases and Cu-containing amine oxidases in plant peroxisomes has suggested that perhaps these two enzymes could be additional sources of NO generation in peroxisomes, although this has not been demonstrated yet (Kaur et al. 2013; Corpas et al. 2019a).

In plant systems, there is little information on the subcellular sites where NO is produced. Besides peroxisomes the only cell compartments where the generation of NO has been clearly demonstrated are mitochondria and chloroplasts. The first biochemical characterization of a NOS-like activity in higher plants was accomplished in isolated peroxisomes (Barroso et al. 1999). In peroxisomes purified from pea leaves, the NOS activity was determined using L-arginine as substrate plus all the NOS cofactors. Four different assays were used: (1) monitoring the conversion of L-[^{3}H]arginine into L-[^{3}H]citrulline; (2) fluorometric detection with 4,5-diaminofluorescein diacetate (DAF-2 DA) of NO produced in the enzymatic reaction; (3) ozone chemiluminescence detection of NO produced with a nitric oxide analyzer (NOA); and (4) spin trapping electron paramagnetic resonance (EPR) spectroscopy of NO generated during the enzymatic reaction, using the spin trap $Fe(MGD)_2$ (Barroso et al. 1999; Corpas et al. 2004a, 2009b; del Río 2011).

Using the arginine-citrulline method, it was found that the NOS activity was strictly dependent on L-arginine and NADPH and required Ca^{2+}, calmodulin, FAD, FMN, and BH_4, the same cofactors necessary for the animal NOS (Alderton et al. 2001). Additionally, the peroxisomal NOS activity was sensitive to archetype inhibitors of the three NOS isoforms (Barroso et al. 1999; del Río 2011). Since the validity of the arginine-citrulline method has been questioned due to interferences by arginase and arginine decarboxylase – two enzymes which also use L-arginine as

Table 3 Main sources of NO in plant cells and other potential enzymatic sources

Source	Substrates	References
Nonenzymatic		
NO_2^-	Acid pH (ASC)	Reviewed by del Río et al. (2004)
Enzymatic		
Nitrate reductase	NO_2^- + NADH	Dean and Harper (1988), Yamasaki et al. (1999), Gupta and Kaiser (2010)
NOS-like activity	L-Arg + NOS cofactors	Reviewed by Corpas et al. (2009b, 2019c)
Plasma membrane	NO_2^- + reduced Cyt *c*	Stöhr et al. (2001), Stöhr and Stremlau (2006)
Catabolism of polyamines		
PAOs and CuAOs	Unknown mechanism	Reviewed by Corpas et al. (2019c)
Xanthine oxidoreductase	NO_2^- + NADH	[a]Reviewed by Harrison (2002)
Catalase	NaN_3	[a]Nicholls (1964)
Peroxidase	NOHA + H_2O_2	[a]Boucher et al. (1992a)
	Hydroxyurea + H_2O_2	[a]Huang et al. (2002)
Hemeproteins	NOHA + H_2O_2/ROOH	[a]Boucher et al. (1992a)
Cytochrome P450	NOHA + NADPH + O_2	[a]Boucher et al. (1992b)
Cell organelles		
Peroxisomes	L-Arg + NOS cofactors	Barroso et al. (1999), Corpas et al. (2004a)
Mitochondria	NO_2^- + NADH	Gupta and Kaiser (2010)
Chloroplasts	NO_2^-; L-Arg	Jasid et al. (2006), Jasid et al. (2006)

NOHA *N*-hydroxyarginine; ROOH alkyl hydroperoxides, *PAO* polyamine oxidase, *CuAO* Cu-containing amine oxidase
[a]Not reported in plants so far

substrate and mimic NOS activity (Tischner et al. 2007) – another two alternative methods of NOS activity determination were used, spectrofluorimetry with the fluorescence probe DAF-2 and an ozone chemiluminescence assay (Corpas et al. 2004a, b, 2008a, b). The biochemical characterization of NOS activity in peroxisomes purified from pea leaves using the ozone chemiluminescence method is shown in Fig. 3. Results obtained showed that the peroxisomal NOS activity depended on the same cofactors as those found by the arginine-citrulline assay, where the conversion of L-[^{3}H]arginine into L-[^{3}H]citrulline was monitored.

The localization of the NOS-like activity in peroxisomes was also studied by immunological methods. By Western blotting, using a polyclonal antibody to murine iNOS, the presence in peroxisomes from pea leaves of an immunoreactive polypeptide of about 130 KDa was detected (Barroso et al. 1999). The electron microscopy (EM) immunolocalization of NOS activity showed the presence of the

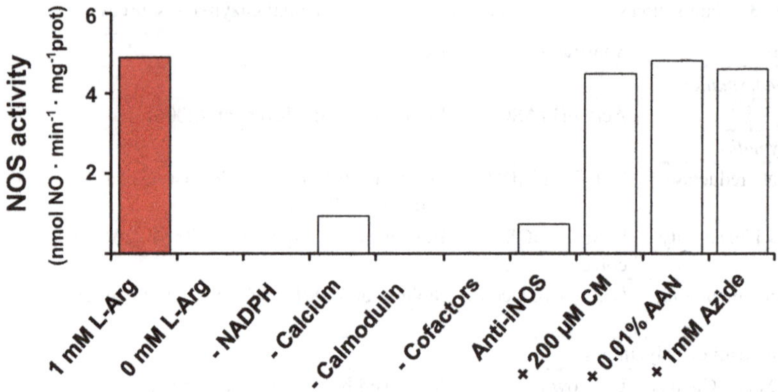

Fig. 3 Biochemical characterization of L-arginine-dependent NOS activity in peroxisomes purified from pea leaves using the ozone chemiluminescence assay which measures the NO generated in the enzymatic reaction. Reaction mixtures containing peroxisomal fractions were incubated in the absence and presence of L-arginine (1 mM), NADPH (1 mM), EGTA (0.5 mM), calmodulin (10 μg/mL), cofactors (10 μM FAD, 10 μM FMN, and 10 μM BH_4), antibody to iNOS, 200 μM CM, 0.01% AAN, and 1 mM azide. Then the NO production was assayed using a 1 mM L-arginine concentration and an incubation time of 30 min. The NO produced was quantified by ozone chemiluminescence using a nitric oxide analyzer (Corpas et al. 2008a). Reproduced from Corpas et al. (2009b) New Phytol 184: 9–14

enzyme in the matrix of peroxisomes and also in chloroplasts, and no immunogold labeling was detected in mitochondria (Barroso et al. 1999). The peroxisomal localization of NOS was confirmed by confocal laser scanning microscopy (CLSM) using antibodies against catalase, a characteristic marker enzyme of peroxisomes, and murine iNOS. The punctuate patterns of both immunofluorescent markers colocalized indicating that NOS was present in peroxisomes (del Río et al. 2003b; Corpas et al. 2004a). In conclusion, the results obtained using different experimental approaches indicate that there are clear and unequivocal evidence of the presence of L-arginine-dependent NOS activity in plant peroxisomes (Corpas et al. 2001, 2009b; del Río 2011). The results obtained on the presence of NOS in plant peroxisomes were some years later extended to animal peroxisomes. In rat hepatocytes the occurrence of inducible nitric oxide synthase (iNOS) in peroxisomes was reported (Stolz et al. 2002), while in the cytosol, both the iNOS active dimer and monomer exist (Loughran et al. 2005). The import of the NOS-like protein into plant peroxisomes has been studied, and it was found to be dependent on peroxins PEX12 and PEX13 (Corpas et al. 2009c), and it seems to have a peroxisomal targeting signal type 2 (PTS2) (Corpas and Barroso 2014), similarly to peroxisomal NOS from animal origin (Loughran et al. 2013).

Regarding the possible identity of the L-arginine-dependent NOS activity detected in plant peroxisomes, this activity is not a canonical NOS enzyme, and this suggests that in higher plant cells, perhaps NOS activity is carried out by several proteins functioning together from L-arginine and using the same substrate and cofactors as the animal NOS. Nevertheless, the possibility of a peroxisomal enzyme that can generate NO from L-arginine as a by-product of an unknown secondary reaction cannot be ruled out. For example, it has been suggested that catalase in the presence of H_2O_2 perhaps could transform peroxidatically L-arginine into NO, using NADPH and Ca^{2+} as cofactors (del Río 2011), and the enzymes of the catabolism of polyamines, PAOs and CuAOs, could be another possibility (Corpas et al. 2019a).

In an attempt to identify the protein responsible for the L-arginine-dependent NOS activity found in peroxisomes, the activity of three enzymes known to be present in plant peroxisomes was assayed: (1) animal xanthine oxidoreductase (XOR); (2) catalase purified from pea leaf peroxisomes; and (3) recombinant monodehydroascorbate reductase from cucumber (MDAR), an enzyme of the ascorbate-glutathione cycle which is known to produce $O_2^{\cdot-}$ (del Río et al. 2014). The NOS activity was determined using either the ozone chemiluminescence method (Corpas et al. 2008b) or by spin trapping EPR (Corpas et al. 2004a). Under the experimental conditions used, catalase, xanthine oxidase, and MDAR did not produce any NO in the L-arginine-dependent NOS reaction, and, accordingly, they do not seem to be responsible for the NO generation in peroxisomes, at least under the experimental conditions used (del Río et al. 2014).

The import of the NOS-like protein into plant peroxisomes has been studied, and it was found to be dependent on peroxins PEX12 and PEX13 (Corpas et al. 2009c), and it seems to have a peroxisomal targeting signal type 2 (PTS2) (Corpas and Barroso 2014), similarly to peroxisomal NOS from animal origin (Loughran et al. 2013).

Although multiple attempts have been made to find a characteristic NOS enzyme in plants, the genetic data available at present clearly indicate that a NOS enzyme similar to animal NOS does not exist in higher plants (for a review see Hancock and Neill 2019). For some reason plants do not have the necessity of a classical NOS enzyme (Frölich and Durner, 2011). However, this does not exclude the existence of an L-arginine-dependent enzymatic system in higher plants, and there are multiple references supporting this finding, as it was mentioned before. Higher plants do not appear to have an essential NOS-dependent system of NO production which likely occur in mammals but multiple systems distributed in different cell compartments (Corpas et al. 2004b) whose relevance might vary depending on the specialized plant tissue and its physiological stage as well as the abiotic and biotic stress situations that the plant might have to face. It could be a similar situation to that of the $O_2^{\cdot-}$ radicals production which is distributed in multiple cell compartments including plasma membrane, chloroplasts, peroxisomes, mitochondria, nuclei, etc., as a result of different oxidative and electron transport reactions (del Río 2015; del Río et al. 2018). There is not a centralized generating system of superoxide but multiple systems, strategically distributed in distinct cell sites and whose importance might depend on the plant development and its physiological stage and/or different stress conditions. The characterized systems of NO production, indicated in Table 3, oscillate from the simple NO_2^- anion to different enzymatic sources like nitrate reductase, the L-arginine-dependent enzyme system so-called NOS-like activity for its substrate and cofactors similarity to animal NOS, and a plasma-membrane enzyme, as well as enzymes involved in the catabolism of polyamines, apart from other potential candidates, like XOD, catalase, peroxidases, hemeproteins, and cytochrome P450. These systems may be more or less important under certain conditions, but they could acquire a protagonist role when the physiological stage of plants requires source(s) of NO in particular tissues. Which is the necessity for plants of a centralized enzymatic generator of NO, instead of different producing systems distributed in different cellular compartments like in the case of the free radical species $O_2^{\cdot-}$? Perhaps these multiple independent systems could be more easily activated and the NO released more accessible within the plant under certain physiological conditions and different biotic and abiotic stresses. The challenge now is, as indicated by Fröhlich and Durner (2011), to decipher how these multiple systems could act, together or independently, to constitute the elaborated NO signaling network observed in plants.

The presence of glutathione and ascorbate with the whole enzymes of the antioxidative ascorbate-glutathione cycle have been demonstrated in leaf peroxisomes (Jiménez et al. 1997). NO in the presence of O_2 can react with reduced glutathione (GSH) to form the reactive nitrogen species *S*-nitrosoglutathione (GSNO) (Wink et al. 1996). GSNO could also be formed by reaction of reduced glutathione with peroxynitrite (Moro et al. 1994). The presence of GSNO in leaf peroxisomes has been demonstrated in olive and pea plants by EM immunocytochemistry. Using a commercial antibody to GSNO, immunogold labeling was found in peroxisomes and chloroplasts of pea leaves (Rodríguez-Serrano 2007; Barroso et al. 2013).

Peroxynitrite ($ONOO^-$) is a class of reactive nitrogen species which is produced by a very quick chemical reaction between superoxide radicals ($O_2^{\cdot-}$) and NO ($k = 1.9 \times 10^{10}\ M^{-1}\ s^{-1}$) (Kissner et al. 1997). Peroxynitrite is a powerful oxidant and nitrating species that produces the oxidation and nitration of proteins and other biomolecules (Radi 2013). Peroxynitrite has a very short half-life, and its action must take place at the site of generation of both superoxide and NO (Szabó et al. 2007). The presence of peroxynitrite in peroxisomes has been studied by CLSM in *Arabidopsis thaliana* transgenic plant expressing CFP-PTS1 which allows the in vivo visualization of peroxisomes (Corpas and Barroso 2014). Using the specific fluorescent probe APF, it was found that peroxynitrite was generated endogenously in peroxisomes of root and guard cells.

The generation and function of the main components whose presence has been demonstrated in plants peroxisomes (NO, GSNO, and $ONOO^-$) are shown in Fig. 4. This implies that single membrane-bound peroxisomes can function in plant cells as a source of RNS signal molecules, besides $O_2^{\cdot-}$ and H_2O_2. The production of NO and GSNO in peroxisomes implies that important posttranslational changes can take place in these organelles, such as *S*-nitrosylation and nitration of proteins.

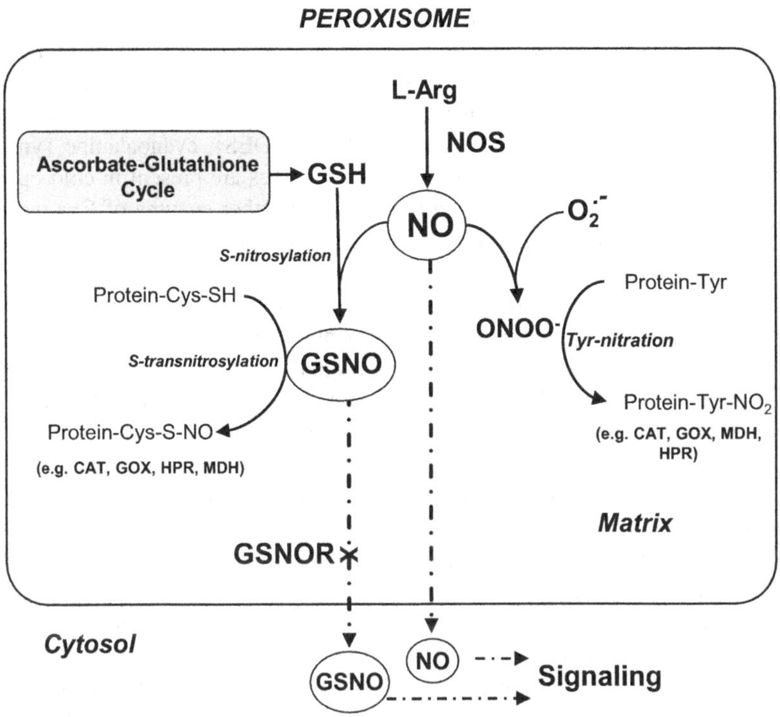

Fig. 4 Scheme proposed for the different sources of RNS in peroxisomes and the metabolism and signaling function of NO and *S*-nitrosoglutathione (GSNO). L-arginine-dependent nitric oxide synthase (NOS-like activity) produces NO which can react with reduced glutathione (GSH) to form *S*-nitrosoglutathione (GSNO), a process designated as *S*-nitrosylation. This metabolite can interact with SH-containing proteins by a reaction of *S*-transnitrosylation, affecting their function, or can be metabolized by the enzyme GSNO reductase (GSNOR). Nitric oxide can also react with superoxide radicals ($O_2^{\cdot-}$) to generate the powerful oxidant and nitrating species peroxynitrite ($ONOO^-$) which can produce the tyrosine nitration of proteins. Interestingly, NO and GSNO can be released to the cytosol to participate in signaling cascades. CAT, catalase. GOX, glycolate oxidase. MDH, malate dehydrogenase. HPR, hydroxypyruvate reductase. Reproduced from del Río et al. (2014)

2.3 RSS

It is known that in animal systems under conditions of oxidative stress, sulfur may exist in different higher oxidation states, and for these redox active molecules, the term reactive sulfur species (RSS) was coined in analogy with ROS and RNS (Giles et al. 2001). These molecules include thiyl radicals ($RS^{\cdot}$), sulfenic acids (RSOH), and disulfide S-oxides [$RS(O)_2SR$] and are strong oxidizing agents that by preference attack the thiol functionality (Giles et al. 2001, 2002). In recent years the occurrence of RSS in plant peroxisomes has been postulated as well as their interaction with the physiologically important signaling molecules ROS and RNS (Corpas and Barroso 2015).

The generation of the gas hydrogen sulfide (H_2S), the simplest thiol compound, was first demonstrated in animal cells, and a role for endogenous H_2S as neuromodulator has been proposed (Abe and Kimura 1996). In plants, several enzymes of the cysteine metabolism can generate H_2S, including sulfite reductase (SiR), L- and D-cysteine desulfhydrase (L-DES/D-DES), cyanoalanine synthase (CAS), and cysteine synthase (CS), and these enzymes are present in chloroplasts, mitochondria, and cytosol (Corpas et al. 2019b). Another enzyme of Cys metabolism that has been shown to produce H_2S is the cytosolic isoform of *O*-acetylserine (thiol)lyase (OASTL), which catalyzes the desulfuration of L-Cys to sulfide plus ammonia and pyruvate (Alvarez et al. 2010). The production of hydrogen sulfide has been found in bacteria, fungi, worms, and humans, besides plants (Kabil et al. 2014), and it has been proposed that H_2S can act as a signaling molecule together with other signaling molecules like RNS, ROS, and carbon monoxide (CO) (Lisjak et al. 2013; Hancock and Whiteman 2014; Kimura 2015). The role of hydrogen sulfide in plant biology has been reviewed by different authors (Zhang 2016; Hancock and Whiteman 2016; Filipovic and Jovanovic 2017; and Corpas et al. 2019b, among others).

Recent data obtained in *Arabidopsis* plants have provided evidence of the presence of H_2S in peroxisomes and the inhibition of catalase activity by H_2S possibly by persulfidation. As catalase activity is one of the main antioxidant enzymes in peroxisomes, this finding suggests that H_2S could regulate the peroxisomal H_2O_2 metabolism (Corpas et al. 2019c). In plant peroxisomes the occurrence of key sulfur compounds has been demonstrated, such as the antioxidant glutathione (GSH) (Jiménez et al. 1997) and the RNS *S*-nitrosoglutathione (GSNO), an intercellular and intracellular NO carrier (Barroso et al. 2013). In addition, the presence of enzymes involved in sulfur metabolism, such as glutathione reductase (GR) (Jiménez et al. 1997; Romero-Puertas et al. 2006), *S*-nitrosoglutathione reductase (GSNOR) (Barroso et al. 2013), and sulfite oxidase (SO) (Nowak et al. 2004), has also been described. All these findings led Corpas and Barroso (2015) to postulate that peroxisomes might play a function in the RSS metabolism, similarly to ROS and RNS.

3 Antioxidant Systems in Peroxisomes

Catalase is a characteristic enzyme of peroxisomes whose presence in these organelles is known since their initial characterization in mammalian tissues (De Duve and Baudhuin 1966). Today it is known that catalase, besides its function in the control of H_2O_2 produced in peroxisomes, can also have a role in the maintenance of cellular redox homeostasis (Mhamdi et al. 2012; Inzé et al. 2012; Sandalio and Romero-Puertas 2015; Smirnoff and Arnaud 2019). In *Arabidopsis thaliana* and other plants, three genes encoding catalase have been identified: *CAT1*, whose expression is related to fatty acid β-oxidation; *CAT2*, related to the photorespiration pathway; and *CAT3* associated to senescence processes (Mhamdi et al. 2010, 2012). But, in addition to catalase, a complex variety of antioxidant systems has been demonstrated in plant peroxisomes.

3.1 Superoxide Dismutases

The family of metalloenzymes superoxide dismutases (SODs) catalyze the dismutation or disproportionation of superoxide radicals ($O_2^{\cdot -}$) into molecular oxygen (O_2) and hydrogen peroxide (H_2O_2) (Fridovich 1995; Imlay 2011; del Río et al. 2018). In plants essentially there are three groups of SODs depending on the prosthetic metals in their active sites: either copper and zinc (Cu,Zn-SODs), manganese (Mn-SODs), or iron (Fe-SODs). Different plant SODs have been isolated and characterized, and many cDNAs and genes for SODs have been identified and characterized (see del Río et al. 2018).

SODs occur mainly in chloroplasts, cytoplasm, mitochondria, apoplast, and nuclei (Asada 2006; Rodríguez-Serrano et al. 2007), but the presence of SOD in peroxisomes was demonstrated for the first time in plant tissues. Using protoplasts from pea leaves, by immunofluorescence and immunoelectron microscopy, with a polyclonal antibody to Mn-SOD purified from pea leaves, the metalloenzyme Mn-SOD was found localized in peroxisomes and was absent in chloroplasts (del Río et al. 1983). The presence of Mn-SOD in peroxisomes was confirmed in organelles isolated from pea leaves by classical cell biology methods (Sandalio et al. 1987). After that, the occurrence of different classes of SODs in plant peroxisomes has been reported in at least 11 distinct plant species, and in 5 of these plants, the localization of SOD in peroxisomes has been confirmed by immunogold electron microscopy (Sandalio et al. 1997; del Río et al. 2002; Mateos et al. 2003; Corpas et al. 2006; Rodríguez-Serrano et al. 2007; Corpas et al. 2017a). The results obtained on the presence of SOD in plant peroxisomes stimulated studies of other authors in human, animal, and yeast peroxisomes where SODs were also found to be present (Keller et al. 1991; Petrova et al. 2009; Corpas et al. 2017a), and now SOD is widely accepted as a characteristic antioxidant enzyme in all types of peroxisomes.

Three SODs of peroxisomal origin have been purified and characterized, a Cu,Zn-SOD and a Mn-SOD from watermelon cotyledons and a Mn-SOD from pea leaves (Bueno et al. 1995; Palma et al. 1998; Rodríguez-Serrano et al. 2007). Nevertheless, the genes coding for the peroxisomal SODs have not been characterized yet. As Mn-SOD is distributed in peroxisomes and mitochondria, it has been suggested that peroxisomal Mn-SOD could be produced from a single gene by alternative splicing or alternative targeting (del Río et al. 2003a).

3.2 Ascorbate-Glutathione Cycle

The ascorbate-glutathione cycle, also known as the Foyer-Halliwell-Asada pathway after the names of the three major contributors, is a reaction pathway in animals and plants considered to be an important mechanism for H_2O_2 metabolism (Asada 2006; Foyer and Noctor 2011). This cycle that is present in chloroplasts, cytoplasm, and mitochondria (Foyer and Noctor 2011) has also been demonstrated in peroxisomes (Jiménez et al. 1997). The cycle has four enzymes, ascorbate peroxidase (APX), monodehydroascorbate reductase (MDAR), dehydroascorbate reductase (DAR), and glutathione reductase (GR), and all these enzymes were found to be present in peroxisomes purified from pea leaves (Jiménez et al. 1997) and tomato leaves and roots (Mittova et al. 2004; Kuzniak and Sklodowska 2005). APX is perhaps the most thoroughly studied enzyme of this cycle (Ishikawa et al. 1998). The peroxisomal GR of pea leaves has been purified and characterized (Romero-Puertas et al. 2006). In purified peroxisomes the presence of reduced ascorbate and glutathione and their oxidized forms was demonstrated by HPLC analysis (Corpas et al. 1997; Jiménez et al. 1997), and in *Arabidopsis thaliana*, the relative concentration of glutathione and ascorbate in peroxisomes was also quantified by immunocytochemical analysis with antibodies to glutathione and ascorbate and electron microscopy (Zechmann et al. 2008; Fernández-García et al. 2009).

The intraperoxisomal distribution of the ascorbate-glutathione cycle enzymes in membranes and matrices was studied in organelles isolated from pea leaves, and results are shown in Table 4 (Jiménez et al. 1997). In pea and *Arabidopsis thaliana*, the peroxisomal MDAR was found to contain a peroxisomal targeting signal (PTS) type 1 (Leterrier et al. 2005; Lisenbee et al. 2005). The function of APX and MDAR in the leaf peroxisomal membrane could be to re-oxidize endogenous NADH to maintain a steady supply of NAD^+ for peroxisomal metabolism (del Río and Donaldson 1995). However, MDAR and APX could also have a regulating role of H_2O_2 leaking from peroxisomes, as well as the H_2O_2 that is formed by dismutation of the $O_2^{\cdot-}$ that is generated by the NAD(P)H-dependent electron transport system of the peroxisomal membrane (López-Huertas et al. 1999; del Río et al. 2002, 2006). This membrane scavenging of H_2O_2 could also avoid a cytoplasmic accumulation of this metabolite, especially under plant stress situation, when the level of H_2O_2 generated in peroxisomes can be largely increased (del Río et al. 1996).

3.3 NADPH-Generating Dehydrogenases

NADPH is a key element in cell redox homeostasis, and its regeneration is essential for reductive biosynthesis and detoxification pathways. NADPH has an important role in the protection against oxidative stress mainly due to its participation in the ascorbate-glutathione cycle and the water-water cycle (Asada 2006; Foyer and Noctor 2011). This evidence has supported the consideration that NADP-dependent

Table 4 Main antioxidants present in plant peroxisomes

Antioxidant	Intraperoxisomal locus	Reference
Enzymatic		
Catalase	Matrix and core	Huang et al. (1983), Kleff et al. (1997)
Superoxide dismutase (SOD)		
Mn-SOD	Matrix	del Río et al. (1983, 2003a, 2018)
	Membrane	Rodríguez-Serrano et al. (2007)
Cu,Zn-SOD	Matrix	Sandalio et al. (1997), Corpas et al. (1998a)
Fe-SOD	?	Droillard and Paulin (1990); Palma et al. (2015)
Ascorbate-glutathione cycle		
Ascorbate peroxidase	Membrane	Jiménez et al. (1997)
Monodehydroascorbate reductase	Matrix	Jiménez et al. (1997)
	Membrane and matrix	Leterrier et al. (2005)
Dehydroascorbate reductase	Matrix	Jiménez et al. (1997)
Glutathione reductase	Matrix	Jiménez et al. (1997), Romero-Puertas et al. (2006)
Peroxiredoxin-like protein	Matrix	Corpas et al. (2017b)
NADP-dehydrogenases		
Glucose-6-phospate dehydrogenase	Matrix	Corpas et al. (1998b)
6-phosphogluconate dehydrogenase	Matrix	Corpas et al. (1998b)
Isocitrate dehydrogenase	Matrix	Leterrier et al. (2016)
Nonenzymatic or low molecular weight		
Glutathione (GSH)	Matrix	Fernández-García et al. (2009)
Ascorbic acid (vitamin C)	Matrix	Fernández-García et al. (2009)

dehydrogenases are also antioxidant enzymes which can be included in the group of SOD, catalase, APX, and GR/peroxidase (Corpas et al. 1998b, 1999). In isolated plant peroxisomes, the occurrence of three NADP-dehydrogenases was demonstrated: glucose-6-phosphate dehydrogenase (G6PDH), 6-phosphogluconate dehydrogenase (G6PDH), and isocitrate dehydrogenase (ICDH) (Corpas et al. 1998b, 1999; Valderrama et al. 2006). In *A. thaliana* NADP-ICDH was demonstrated to be localized in peroxisomes by immunocytochemistry, and this dehydrogenase has been proposed to be involved in stomatal movement (Leterrier et al. 2016).

The presence of several NADP dehydrogenases in peroxisomes indicates that these organelles have the capacity to reduce NADP to NADPH for its reuse in their metabolism. NADPH is necessary in peroxisomal membranes for the function of the NADPH-cytochrome P450 reductase (Baker and Graham 2002) and the $O_2^{\cdot-}$-producing polypeptide, PMP29 (López-Huertas et al. 1999). The peroxisomal NO-generating activity, nitric oxide synthase-like, requires NADPH for its activity

(Corpas et al. 2004). On the other hand, catalase has been reported to be protected by NADPH from oxidative damage (Kirkman et al. 1999). NADPH is also required for the GR activity of the ascorbate-glutathione cycle which recycles reduced glutathione (GSH) from its oxidized form (GSSG) (Foyer and Noctor 2011) and for the activity of 2,4-dienoyl-CoA reductase which reduces double bonds of unsaturated fatty acids (Reumann et al. 2007).

3.4 Glutathione Peroxidase and Glutathione S-Transferase

The presence of glutathione peroxidase has been described in leaf peroxisomes of tomato plants (Kuzniak and Sklodowska 2005). In *Arabidopsis*, three families of glutathione *S*-transferases have been localized in peroxisomes, and it has been suggested that they could be involved in the removal of toxic hydroperoxides due to their glutathione peroxidase activity (Dixon et al. 2009).

3.5 Peroxiredoxins

Peroxiredoxins are thioredoxin-dependent peroxidases, a family of thiol-specific antioxidant enzymes, which are present in bacteria, yeasts, plants, and mammals (Dietz 2003). In peroxisomes from yeasts and animals, the presence of peroxiredoxins has been demonstrated (Walbrecq et al. 2015) in contrast to plants. In plants, recently the presence of a protein of about 50 kDa immunorelated to peroxiredoxins was localized in the matrix of pea leaf peroxisomes (Corpas et al. 2017b). The presence in the matrix of peroxisomes of a protein immunorelated to Prx raises new questions on the molecular properties of Prxs, as well as on their role in the metabolism of ROS and RNS. The expression of this peroxiredoxin-like protein was differentially modulated under oxidative stress conditions, and this suggests that it could be involved in the regulation of hydrogen peroxide and/or peroxynitrite (Corpas et al. 2017b).

The antioxidant systems of peroxisomes from two different fruits, pepper and olive, have been characterized (Mateos et al. 2003; López-Huertas and del Río 2014). In peroxisomes from pepper fruits, the metabolism of antioxidants was investigated during the ripening process, and proteomic analysis showed no changes between the antioxidant metabolism of immature (green) and ripe (red) fruits (Palma et al. 2015).

4 Role of ROS, RNS, and RSS Produced in Peroxisomes

ROS are oxidizing species, particularly the hydroxyl radicals (˙OH) and singlet oxygen ($^1O_2$) (del Río 2015). These species are very powerful oxidants that can react with most components of living cells producing severe damage to lipids, nucleic acids, and proteins (oxidative stress situations) (Bailey-Serres and Mittler 2006; Halliwell and Gutteridge 2007; del Río and Puppo 2009; Sies 2014). Nonetheless, to avoid oxidative stress situations, plant cells have a battery of enzymatic and nonenzymatic antioxidants, which were indicated in Sect. 3, that under normal conditions can scavenge the excess oxidants produced and so prevent their deleterious effects on plant cells.

However, the concept of "oxidative stress," which strictly implies a state to be avoided, was re-evaluated, and the term "oxidative signaling" or "redox signaling" was coined (Foyer and Noctor 2005; Foyer 2018). This means that ROS production, which was initially considered as an exclusively harmful and dangerous process, is also a relevant component of the signaling network that plants use for their development and for responding to environmental challenges (del Río and Puppo 2009; Mittler et al. 2011; Foyer 2018). Therefore, ROS have a double role, and unfavorable environmental conditions result in excessive ROS production that at high concentrations lead to oxidative cell injuries. But, as it was mentioned above, to prevent ROS-induced cellular damage, plants are equipped with a broad variety of antioxidative systems in order to use ROS simultaneously as a signal within distinct biological processes (Vanderauwera et al. 2009).

Now it is widely admitted that ROS play an important signaling role in plants, as key regulators of processes like growth, development, response to environmental and biotic stimuli, plant metabolism, and programmed cell death (del Río and Puppo 2009; Mittler et al. 2011; Inzé et al. 2012; Sandalio et al. 2012; Baxter et al. 2014; Mittler 2017). In plants, it has been identified the homologue of the respiratory-burst NADPH oxidase of leukocytes, and this has led to the finding that plant cells, like mammalian cells, can initiate and very probably amplify ROS production for signaling purposes (Suzuki et al. 2011; Marino et al. 2012). The fact that ROS are produced in many compartments of plant cells, including chloroplasts, mitochondria, peroxisomes, apoplasts, and nuclei, is very important from the point of view of initiating signaling cascades. However, it must also be considered that besides single increases in ROS production, the controlled downregulation of antioxidant enzymes can also be involved in the signaling mechanisms during plant stress.

Plant peroxisomes are one of the main cellular sources of ROS, and it has been demonstrated their involvement in the oxidative stress induced by xenobiotics like clofibrate or 2,4-dichlorophenoxyacetic acid (Palma et al. 1991; Nila et al. 2006; Romero-Puertas et al. 2004b; McCarthy et al. 2011), heavy metals (Romero-Puertas et al. 1999, 2004a), salinity (del Río et al. 2002; Mittova et al. 2004), ozone (Pellinen et al. 1999), or senescence (del Río et al. 1998; Rosenwasser et al. 2011). Moreover, in the last decade, the key role played by ROS in the complex signaling network which regulates essential cell processes including stress response has been demonstrated (del Río and Puppo 2009; Mittler et al. 2011; del Río 2013; Considine et al. 2015). The accumulation of ROS in specific cell compartments, like peroxisomes, can be perceived by the cell as an alarm and so trigger a cascade of events to promote specific defense responses (Mittler et al. 2011).

The idea that peroxisomes could be a cellular source of the signal molecules ROS and NO was proposed in the last two decades (del Río and Donaldson 1995; del Río et al. 1996; Corpas et al. 2001; Turkan 2018). Years later, the idea of a signaling function for plant peroxisomes was extended to other molecules also produced in peroxisomes and derived from β-oxidation, like jasmonic acid and its derivatives, salicylic acid and IAA (Nyathi and Baker 2006; see del Río 2013).

Hydrogen peroxide (H_2O_2) is an important transduction signal in plant-pathogen interactions, stomatal closure, response to wounding, excess light stress, and osmotic stress, processes where H_2O_2 leads to the induction of defense genes encoding different cellular protectants

(Inzé et al. 2012). In catalase loss-of-function *A. thaliana* mutants, microarrays studies have evidenced that a total of 783 transcripts changed their expression in response to high levels of photorespiratory H_2O_2, and most of transcripts were related to abiotic stress responses (Inzé et al. 2012). Different studies have also demonstrated a close association between peroxisomal H_2O_2, oxidative stress, and phytohormone-dependent signaling involving ethylene, jasmonic acid, salicylic acid, auxins, and ABA, indicating that redox homeostasis (the GSH/GSSG ratio), together with NAD and NADP systems, might control this relationship (Jones and Sies 2015; Wang et al. 2015; Sandalio and Romero-Puertas 2015). On the other hand, studies on the important peroxisomal H_2O_2-producer glycolate oxidase (GOX) using *A. thaliana* loss-of-function mutants have revealed the contribution of H_2O_2 from each GOX isoform to the regulation of cell response to the plant infection by the bacterium *Pseudomonas* (Rojas et al. 2012; Sandalio et al. 2013).

An important advantage of peroxisomes as a source of signaling molecules is their metabolic plasticity that allows metabolic adjustments depending on developmental and environmental situations (del Río et al. 2006), as well as the ability to rapidly change their motility and population in response to plant environmental conditions (Palma et al. 1991; Romero-Puertas et al. 1999; López-Huertas et al. 2000; Castillo et al. 2008; Rodríguez-Serrano et al. 2009, 2016; Hu et al. 2012). A scheme of the different sources of ROS in plant peroxisomes and their role in the cellular metabolism is shown in Fig. 2.

Posttranslational modifications (PTMs) have an important function in the regulation of protein functionality, their stability, interactions with other proteins, and subcellular localization (Van Bentem et al. 2006). Perhaps the most studied ROS-dependent PTM is protein carbonylation which mainly targets lysine, proline, threonine, and arginine (Rao and Møller 2011), although the H_2O_2-dependent protein sulfenylation has also been proposed as a mechanism to regulate protein function in living cells in response to changing redox states (Young et al. 2019; Sandalio et al. 2019). Carbonylation is produced by nucleophilic attacks on the carbonyl groups [C=O] by ·OH radicals formed in Fenton-type reactions involving H_2O_2, $O_2^{\cdot-}$, and metals (mainly Fe and Cu) (Halliwell and Gutteridge 2007; Sies

2014). In plants under certain stress conditions, there is a high production of ROS in peroxisomes, and peroxisomal proteins can be oxidized by carbonylation. In pea plants it has been demonstrated that Cd induces oxidative stress and produces the carbonylation of a number of peroxisomal proteins, including catalase, glutathione reductase, and Mn-SOD (Romero-Puertas et al. 2002), and these proteins were more efficiently degraded by the peroxisomal proteases induced by the Cd treatment (Distefano et al. 1997; McCarthy et al. 2001; Romero-Puertas et al. 2002). In peroxisomes isolated from castorbean endosperm, malate synthase, isocitrate lyase, malate dehydrogenase, and catalase have been found to be carbonylated in peroxisomes (Nguyen and Donaldson 2005). It is important to take into account that the activity of these carbonylated peroxisomal proteins was considerably reduced by this type of protein modification (Romero-Puertas et al. 2002; Nguyen and Donaldson 2005). Nonetheless, the nature of the proteases involved in degrading oxidatively modified proteins in peroxisomes and the genes encoding these proteins are not well-known.

On the other hand, nitric oxide (NO) and other reactive nitrogen species (RNS) can also produce posttranslational modifications, including oxidation, nitration, and *S*-nitrosylation, that can affect peroxisomal protein function. *S*-Nitrosylation, also called *S*-nitrosation, consists in the reaction of NO with an –SH group of a Cys residue to form a nitrosothiol (Martínez-Ruiz and Lamas 2004). Nitration is another PTM where nitro groups (-NO_2) are added to C atoms of organic substrates, and the most common form of nitration is the addition of a nitro group to an aromatic ring system like that of a tyrosine residue (Radi 2004; Corpas et al. 2015). The number of identified plant peroxisomal proteins susceptible to undergo a specific RNS-derived PTM has increased with the use of specific proteomic methodologies combined with biochemical analyses, such as the biotin switch method and labeling with isotope-coded affinity tags (ICAT) (Corpas et al. 2019c). Several peroxisomal proteins have been identified as putative targets of *S*-nitrosylation, and they are involved in photorespiration, β-oxidation, and ROS detoxification (Ortega-Galisteo et al. 2012; Corpas et al. 2017a, 2019c). The activities of catalase, glycolate oxidase, and malate dehydrogenase were inhibited by NO donors in peroxisomes from pea leaves, and the *S*-nitrosylation level of catalase and glycolate oxidase changed in plants treated with Cd and the herbicide 2,4-D, indicating that under abiotic stress situations, this posttranslational modification could be involved in regulating the H_2O_2 accumulation in peroxisomes, as well as the flux of metabolites between different metabolic pathways (Ortega-Galisteo et al. 2012). In peroxisomes from pepper fruits, tyrosine nitration and *S*-nitrosylation inhibited catalase activity (Chaki et al. 2015), and in *Arabidopsis* catalase, it has been proposed that the potential target of *S*-nitrosylation is Cys86 (Puyaubert et al. 2014), although this still has to be corroborated by specific mass spectrometry analyses.

The activity of the peroxisomal enzyme of the ascorbate-glutathione cycle, monodehydroascorbate reductase (MDAR) (Leterrier et al. 2005), was inhibited in the presence of *S*-nitrosylating or nitrating agents (GSNO or $ONOO^-$, respectively) (Begara-Morales et al. 2015), and this suggests a possible modulation in peroxisomes by RNS of the ascorbate regeneration and H_2O_2 scavenging. As to the SOD,

Table 5 Some peroxisomal proteins from plant origin that have been identified to be susceptible of posttranslational modifications mediated by RNS

Peroxisomal enzyme	RNS-derived posttranslational modification	Reference
3-ketoacyl-CoA thiolase	*S*-nitrosylation	Fares et al. (2011)
Hydroxypyruvate reductase	*S*-nitrosylation/ nitration	Ortega-Galisteo et al. (2012), Corpas et al. (2013), Puyaubert et al. (2014)
Glycolate oxidase	*S*-nitrosylation/ nitration	Ortega-Galisteo et al. (2012), Abat et al. (2008), Tanou et al. (2009)
Malate dehydrogenase	*S*-nitrosylation/ nitration	Ortega-Galisteo et al. (2012), Lozano-Juste et al. (2011)
Catalase	*S*-nitrosylation/ nitration	Ortega-Galisteo et al. (2012), Puyaubert et al. (2014), Begara-Morales et al. (2013), Chaki et al. (2015)
Cu,Zn-superoxide dismutase	Nitration	Holzmeister et al. (2015)
Monodehydroascorbate reductase	*S*-nitrosylation/ nitration	Begara-Morales et al. (2015)

Modified from Corpas et al. (2019c)

in vitro assays carried out with the *Arabidopsis* peroxisomal Cu,Zn-SOD showed that the nitrating agent $ONOO^-$ partially inhibited the Cu,Zn-SOD activity, whereas the *S*-nitrosylating agent GSNO did not cause any effect (Holzmeister et al. 2015). This indicates that from the viewpoint of NO-mediated PTMs of SODs, it is interesting to know that Cu,Zn-SOD could discriminate between nitration and *S*-nitrosylation processes (Corpas et al. 2019c).

A summary of identified proteins from higher plant peroxisomes that undergo RNS-derived PTMs, either by *S*-nitrosylation or tyrosine nitration, is presented in Table 5.

Reactive sulfur species (RSS), including hydrogen sulfide (H_2S), have been proposed to function as signaling molecules in diverse organisms in conjunction with other reactive molecules, mainly carbon monoxide (CO), nitric oxide and other RNS, and ROS (Zhang 2016). On the other hand, the postulation of RSS as possible new actors in the oxidative metabolism of plant peroxisomes has opened new avenues in this research area (Corpas and Barroso 2015). This hypothesis has been supported by the recent detection of the presence of H_2S in plant peroxisomes (Corpas et al. 2019d) although it is not known if the H_2S accumulated in *Arabidopsis* peroxisomes was generated inside peroxisomes or imported from the cytosol or chloroplasts where H_2S is also present. Further research is necessary to know whether plant peroxisomes can generate H_2S endogenously and which would be the enzymes involved in this production.

It is interesting to consider that gaseous molecules, like NO, carbon monoxide (CO), and H_2S, at high levels are well-known to be toxic for animals and plants, and at low levels, these molecules can also have beneficial functions by acting as signaling molecules regulating different processes essential for plant performance, similarly to the multifunctional signaling molecule H_2O_2 (Zhang 2016; Hancock and Whiteman 2016; Hancock 2019; Filipovic and Jovanovic 2017; Sandalio et al. 2019; Smirnoff and Arnaud 2019). The main processes in which H_2S plays a signaling role are photosynthesis, programmed cell death, and autophagy, but H_2S also has a function in plant stress responses to adverse conditions, affecting seed germination, root elongation, and overall plant survival (Sandalio et al. 2019).

H_2S can mediate a PTM called persulfidation that modulates the function of target proteins (Ju et al. 2017). Persulfidation (formerly known as *S*-sulfhydration) consists in the oxidation of the cysteine thiol groups (R-SH) to persulfide groups (R-S-SH). Proteomic studies carried out in *Arabidopsis* have shown that persulfidation is involved in diverse biological functions and in the regulation of important plant processes (Aroca et al. 2017, 2018). Recent data obtained in *Arabidopsis* have shown the presence of H_2S in peroxisomes and the inhibition of catalase activity by this signaling gasotransmitter (Corpas et al. 2019d). Considering that catalase is perhaps the main antioxidant enzyme of peroxisomes, this means that the H_2S present in these oxidative organelles could participate in the regulation of catalase activity and, therefore, the peroxisomal H_2O_2 metabolism (Corpas et al. 2019d).

The co-existence in peroxisomes of physiologically important signaling molecules like ROS, RNS, and RSS suggests the possibility of diverse interactions between them that could have important metabolic consequences. A scheme of some of the potential interactions between different RNS, ROS, and RSS that could occur in plant peroxisomes are presented in Fig. 5. NO is produced by the L-arginine-dependent nitric oxide synthase activity (NOS-like activity) (Corpas et al. 2009b) and can react with superoxide radicals generated by xanthine oxidase to form peroxynitrite ($ONOO^-$). This RNS is a very strong oxidant and can catalyze the conversion of xanthine dehydrogenase to xanthine oxidase (Corpas et al. 2008a) apart from bringing about protein nitration (Radi 2004). On the other hand, NO in the presence of O_2 can react with reduced glutathione (GSH), by an *S*-nitrosylation reaction, to yield *S*-nitrosoglutathione (GSNO), an important mobile reservoir of NO bioactivity. GSNO can be decomposed by the enzyme *S*-nitrosoglutathione reductase (GSNOR) with generation of oxidized glutathione (GSSG) and ammonia (NH_3). GSSG can be reduced to GSH by the enzyme glutathione reductase (GR), a component of the ascorbate-glutathione cycle. H_2O_2, generated in peroxisomes mainly by different flavin-containing oxidases, is removed either by catalase or ascorbate peroxidase (APX). Sulfite oxidase (SO) catalyzes the oxidation of sulfite to sulfate, and sulfite has been reported to inhibit catalase activity as well as H_2S. This suggests that sulfite and H_2S could cooperate in regulating the production of H_2O_2 in peroxisomes.

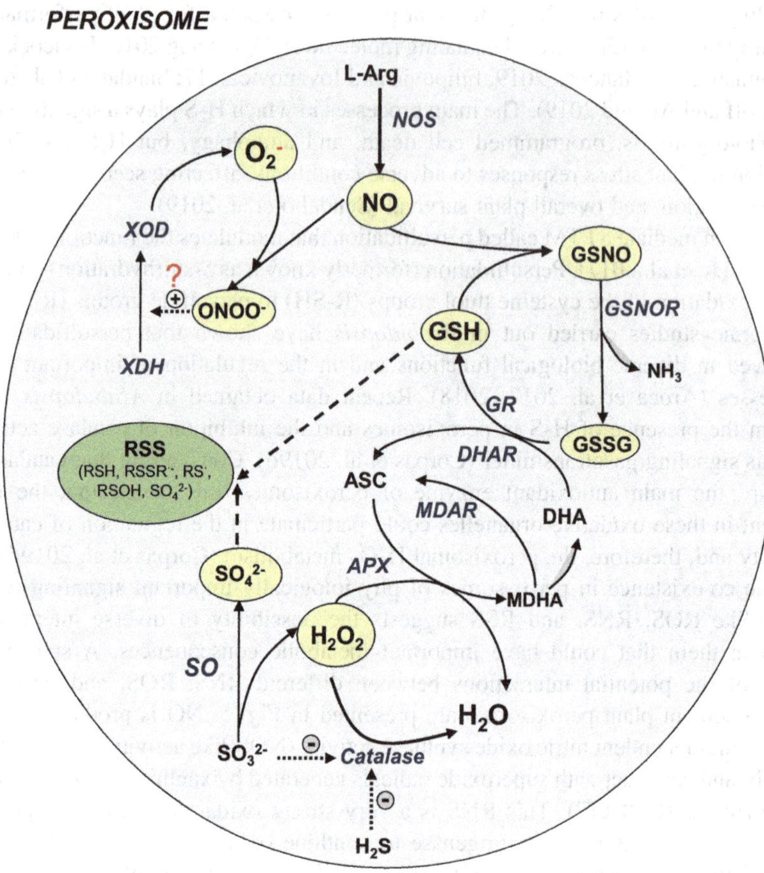

Fig. 5 Hypothetical model of the interaction or cross talk between ROS, RNS, and RSS in plant peroxisomes. ASC, ascorbate reduced form; DHA (dehydroascorbate), ascorbate oxidized form; GSH, glutathione reduced form; GSNOR, *S*-nitrosoglutathione reductase; GSSG, glutathione oxidized form; NO, nitric oxide; NOS, L-arginine-dependent nitric oxide synthase-like; MDAR, monodehydroascorbate reductase; $ONNO^-$, peroxynitrite; SO, sulfite oxidase; XOD, xanthine oxidase; XDH, xanthine dehydrogenase; RSS, reactive sulfur species; R-SH, thiol group; $RSSR^{\cdot-}$, disulfide radical; $RS^{\cdot}$, thiyl radical; ROSH, sulfenic acid; SO_4^{2-}, sulfate. Reproduced and modified from Corpas and Barroso (2015)

5 Effect of Abiotic Stress on ROS, RNS, and RSS Metabolism of Peroxisomes

In general, when plants are subjected to biotic and environmental stresses, a rapid overproduction of ROS and RNS takes place (del Río and Puppo 2009; Mittler et al. 2011; Airaki et al. 2012; Sandalio et al. 2012; del Río 2013; Baxter et al. 2014; Nasir Khan et al. 2014; Yu et al. 2014). The induction of ROS and RNS overproduction has been reported mainly in the following stress situations: infection by pathogens;

high light intensities; UV radiation; high and low temperatures; drought and salt stress; heavy metals; atmospheric pollutants; and physical and mechanical wounding (del Río and Puppo 2009; Baxter et al. 2014; Nasir Khan et al. 2014; Yu et al. 2014; Corpas et al. 2017a; del Río et al. 2018). Usually, the responses of the rapid production of ROS and RNS trigger a designated programmed cell death (PCD) process, where ROS and NO play important functions. PCD is an important mechanism to control different sides of growth and development and also to eliminate damaged or infected cells during responses to environmental stresses and pathogen attacks (Wang et al. 2013).

As a consequence of the presence in plant peroxisomes of H_2O_2, NO, GSNO, and the generation of $ONOO^-$, important covalent posttranslational modifications (PTMs) can occur in plants under natural and stress conditions, such as protein carbonylation and sulfenylation, *S*-nitrosylation, and protein nitration (Corpas et al. 2013, 2017a; Romero-Puertas et al. 2013; Ortega-Galisteo et al. 2012; Young et al. 2019; Rao and Møller 2011; Chaki et al. 2015; Sandalio et al. 2019). This can produce important changes in the activity of different peroxisomal enzymes, as it was mentioned in Sect. 4 and, as a result of it, also in the level of ROS and RNS signal molecules which can diffuse to the cytoplasm and initiate signaling cascades or interactions with other biomolecules. This can finally regulate gene expression of defense molecules in response to various types of stresses and facilitate cellular adaptation to environmental changes (Corpas et al. 2017a).

6 Concluding Remarks

The existence of a ROS and RNS metabolism in plant peroxisomes and the presence in these organelles of a complex variety of antioxidative enzymes emphasizes the importance of these organelles in cellular oxidative metabolism.

Plant peroxisomes have a ROS- and RNS-mediated metabolic function in different physiological processes and certain types of abiotic stress. More recently, reactive sulfur species (RSS) have also been proposed to have a role in the oxidative metabolism of peroxisomes. The demonstration of the presence of the gasotransmitter hydrogen sulfide (H_2S) in these organelles and its inhibition of catalase lends support to this proposal, although the physiological function of H_2S in the metabolism of plant peroxisomes still has to be investigated. The main ROS and RNS whose generation have been demonstrated in plant peroxisomes include H_2O_2, $O_2^{\cdot-}$ radicals, NO, *S*-nitrosoglutathione (GSNO), and peroxynitrite ($ONOO^-$), and these single membrane-bound organelles can function in plant cells as a source of ROS and RNS signaling molecules as well as some phytohormone-related molecules, and possibly also of RSS.

Peroxisomes could act as subcellular sensors of plant stress by releasing the signaling molecules ROS, RNS, and RSS, as well as other second messengers, into the cytoplasm and triggering specific changes in the expression of defense genes (stress signaling), and they could also have a role in cross talk between peroxisomes, chloroplasts, and mitochondria. This signal molecule-producing function of plant peroxisomes is also favored by the fact that the cellular population of these organelles can proliferate in plants under different physiological and stress conditions, and their dynamics/motility can be modified by some of those situations.

The presence of H_2O_2, NO, GSNO, and the generation of $ONOO^-$ indicates that important posttranslational modifications can take place in peroxisomes, such as carbonylation, *S*-nitrosylation, and nitration of proteins. It is known that some of these PTMs affect the functionality of key peroxisomal enzymes like catalase, glycolate oxidase, hydroxypyruvate reductase, Cu,Zn-superoxide dismutase, and monodehydroascorbate reductase, among others. Consequently, those PTMs might have an important function in the regulation of peroxisomal metabolism, particularly in plants under environmental or abiotic stress conditions. Further research is necessary to study the mechanisms of all these processes and their interconnection with cellular metabolism where results obtained in recent years suggest that peroxisomes might play an important role in the maintenance of redox homeostasis.

Acknowledgments The valuable help of Dr. Javier Corpas in the modification of Fig. 5 is appreciated. The author apologizes to the many colleagues whose work could not be cited because of space limitations. This work was supported by ERDF-cofinanced grants from the Ministry of Economy and Competitiveness and *Junta de Andalucía* (Group BIO-192), Spain.

References

Abat JK, Mattoo AK, Deswal R (2008) *S*-nitrosylated proteins of a medicinal CAM plant *Kalanchoe pinnata* ribulose-1,5-bisphosphate carboxylase/oxygenase activity targeted for inhibition. FEBS J 275:2862–2872

Abe K, Kimura H (1996) The possible role of hydrogen sulfide as an endogenous neuromodulator. J Neurosci 16:1066–1071

Agurla S, Gayatri G, Raghavendra AS (2018) Polyamines increase nitric oxide and reactive oxygen species in guard cells of *Arabidopsis thaliana* during stomatal closure. Protoplasma 255:153–162

Airaki M, Leterrier M, Mateos RM, Valderrama R, Chaki M, Barroso JB, del Río LA, Palma JM, Corpas FJ (2012) Metabolism of reactive oxygen species and reactive nitrogen species in pepper (*Capsicum annuum*L.) plants under low temperature stress. Plant Cell Physiol 35:281–295

Alderton WK, Cooper CE, Knowles RG (2001) Nitric oxide synthases: structure, function and inhibition. Biochem J 357:593–615

Alvarez C, Calo L, Romero LC, García I, Gotor C (2010) An *O*-acetylserine(thiol)lyase homolog with L-cysteine desulfhydrase activity regulates cysteine homeostasis in *Arabidopsis*. Plant Physiol 152:656–669

Aroca A, Benito JM, Gotor C, Romero LC (2017) Persulfidation proteome reveals the regulation of protein function by hydrogen sulfide in diverse biological processes in *Arabidopsis*. J Exp Bot 68:4915–4927

Aroca A, Gotor C, Romero LC (2018) Hydrogen sulfide signaling in plants: emerging roles of protein persulfidation. Front Plant Sci 9:1369

Asada K (2006) Production and scavenging of reactive oxygen species in chloroplasts and their functions. Plant Physiol 141:391–396

Astier J, Gross I, Durner J (2018) Nitric oxide production in plants: an update. J Exp Bot 69:3401–3411

Bailey-Serres J, Mittler R (2006) Special issue on reactive oxygen species. Plant Physiol 141:311–508

Baker A, Graham I (2002) Plant peroxisomes. Biochemistry, cell biology and biotechnological applications. Kluwer, Dordrecht

Barroso JB, Corpas FJ, Carreras A, Sandalio LM, Valderrama R, Palma JM, Lupiáñez JA, del Río LA (1999) Localization of nitric oxide synthase in plant peroxisomes. J Biol Chem 274:36729–36733

Barroso JB, Valderrama R, Corpas FJ (2013) Immunolocalization of *S*-nitrosoglutathione, *S*-nitrosoglutathione reductase and tyrosine nitration in pea leaf organelles. Acta Physiol Plant 35:2635–2640

Baxter A, Mittler R, Suzuki N (2014) ROS as key players in plant stress signaling. J Exp Bot 65:1229–1240

Begara-Morales JC, López-Jaramillo FJ, Sánchez-Calvo B, Carreras A, Ortega-Muñoz M, Santoyo-González F, Corpas FJ, Barroso JB (2013) Vinyl sulfone silica: application of an open preactivated support to the study of transnitrosylation of plant proteins by *S*-nitrosoglutathione. BMC Plant Biol 13:61

Begara-Morales JC, Sánchez-Calvo B, Chaki M, Mata-Pérez C, Valderrama R, Padilla MN, López-Jaramillo J, Luque F, Corpas FJ, Barroso JB (2015) Differential molecular response of monodehydroascorbate reductase and glutathione reductase by nitration and *S*-nitrosylation. J Exp Bot 66:5983–5996

Boucher JL, Genet A, Vadon S et al (1992a) Formation of nitrogen oxides and citrulline upon oxidation of N-omega-hydroxy-L-arginine by hemeproteins. Biochem Biophys Res Commun 184:1158–1164

Boucher JL, Genet A, Vadon S et al (1992b) Cytochrome P450 catalyzes the oxidation of N-omega-hydroxy-L-arginine by NADPH and O_2 to nitric oxide and citrulline. Biochem Biophys Res Commun 187:880–886

Bowditch MY, Donaldson RP (1990) Ascorbate free radical reduction by glyoxysomal membranes. Plant Physiol 94:531–537

Bueno P, Varela J, Giménez-Gallego G, del Río LA (1995) Peroxisomal copper,zinc superoxide dismutase: characterization of the isoenzyme from watermelon cotyledons. Plant Physiol 108:1151–1160

Castillo MC, Sandalio LM, del Río LA, León J (2008) Peroxisome proliferation, wound-activated responses and expression of peroxisome-associated genes are cross-regulated but uncoupled in *Arabidopsis thaliana*. Plant Cell Environ 31:492–505

Chaki M, Álvarez de Morales P, Ruiz C, Begara-Morales JC, Barroso JB, Corpas FJ et al (2015) Ripening of pepper (*Capsicum annuum*) fruit is characterized by an enhancement of protein tyrosine nitration. Ann Bot 116:637–647

Considine MJ, Sandalio LM, Foyer CH (2015) Unravelling how plants benefit from ROS and NO reactions, while resisting oxidative stress. Ann Bot 116:469–473

Corpas FJ, Barroso JB (2014) Peroxisomal plant nitric oxide synthase (NOS) protein is imported by peroxisomal targeting signal type 2 (PTS2) in a process that depends on the cytosolic receptor PEX7 and calmodulin. FEBS Lett 588:2049–2054

Corpas FJ, Barroso JB (2015) Reactive sulfur species (RSS): possible new players in the oxidative metabolism of plant peroxisomes. Front Plant Sci 6:116

Corpas FJ, de la Colina C, Sánchez-Rasero F, del Río LA (1997) A role for leaf peroxisomes in the catabolism of purines. J Plant Physiol 151:246–250

Corpas FJ, Sandalio LM, del Río LA, Trelease RN (1998a) Copper-zinc superoxide dismutase is a constituent enzyme of the matrix of peroxisomes in the cotyledons of oilseed plants. New Phytol 138:307–314

Corpas FJ, Barroso JB, Sandalio LM, Distefano S, Palma JM, Lupiáñez JA, del Río LA (1998b) A dehydrogenase-mediated recycling system of NADPH in plant peroxisomes. Biochem J 330:777–784

Corpas FJ, Barroso JB, Sandalio LM, Palma JM, Lupiáñez JA, del Río LA (1999) Peroxisomal NADP-dependent isocitrate dehydrogenase. Characterization and activity regulation during natural senescence. Plant Physiol 121:921–928

Corpas FJ, Barroso JB, del Río LA (2001) Peroxisomes as a source of reactive oxygen species and nitric oxide signal molecules in plant cells. Trends Plant Sci 6:145–150

Corpas FJ, Barroso JB, Carreras A, Quirós M, León AM, Romero-Puertas MC, Esteban FJ, Valderrama R, Palma JM, Sandalio LM, Gómez M, del Río LA (2004a) Cellular and subcellular localization of endogenous nitric oxide in young and senescent pea plants. Plant Physiol 136:2722–2733

Corpas FJ, Barroso JB, del Río LA (2004b) Enzymatic sources of nitric oxide in plant cells – beyond one protein-one function. New Phytol 162:243–251

Corpas FJ, Fernández-Ocaña A, Carreras A, Valderrama R, Luque F, Esteban FJ et al (2006) The expression of different superoxide dismutase forms is cell-type dependent in olive (*Olea europaea* L.) leaves. Plant Cell Physiol 47:984–994

Corpas FJ, Carreras A, Esteban FJ, Chaki M, Valderrama R, del Río LA, Barroso JB (2008a) Localization of *S*-Nitrosothiols and assay of nitric oxide synthase and *S*-Nitrosoglutathione reductase activity in plants. Methods Enzymol 437:561–574

Corpas FJ, Palma JM, Sandalio LM, Valderrama R, Barroso JB, del Río LA (2008b) Peroxisomal xanthine oxidoreductase: characterization of the enzyme from pea (*Pisum sativum* L.) leaves. J Plant Physiol 165:1319–1330

Corpas FJ, Barroso JB, Palma JM, del Río LA (2009a) Peroxisomes as key organelles in the metabolism of reactive oxygen species, reactive nitrogen species, and reactive sulfur species in plants. In: Terlecky SR, Titorenko VI (eds) Emergent functions of the peroxisome. Research Signpost, Kerala, pp 97–124

Corpas FJ, Palma JM, del Río LA, Barroso JB (2009b) Evidence supporting the existence of L-arginine-dependent nitric oxide synthase activity in plants. New Phytol 184:9–14

Corpas FJ, Hayashi M, Mano S, Nishimura M, Barroso JB (2009c) Peroxisomes are required for *in vivo* nitric oxide accumulation in the cytosol following salinity stress of *Arabidopsis* plants. Plant Physiol 151:2083–2094

Corpas FJ, Palma JM, del Río LA, Barroso JB (2013) Protein tyrosine nitration in higher plants grown under natural and stress conditions. Front Plant Sci 4:29

Corpas FJ, Begara-Morales JC, Sánchez-Calvo B, Chaki M, Barroso JB (2015) Nitration and *S*-nitrosylation: two post-translational modifications (PTMs) mediated by reactive nitrogen species (RNS) which participate in signaling processes of plant cells. In: Kagapunti J, Gupta KJ, Igamberdiev AU (eds) Reactive oxygen and nitrogen species signaling and communication in plants. Springer, Berlin, pp 267–281

Corpas FJ, Barroso JB, Palma JM, Rodríguez-Ruiz M (2017a) Plant peroxisomes: a nitro-oxidative cocktail. Redox Biol 11:535–542

Corpas FJ, Pedrajas JR, Palma JM, Valderrama R, Rodríguez-Ruiz M, Chaki M, del Río LA, Barroso JB (2017b) Immunological evidence for the presence of peroxiredoxin in pea leaf peroxisomes and response to oxidative stress conditions. Acta Physiol Plant 39:57–69

Corpas FJ, del Río LA, Palma JM (2019a) Plant peroxisomes at the crossroad of NO and H_2O_2 metabolism. J Integr Plant Biol 61:803–816

Corpas FJ, González-Gordo S, Cañas A, Palma JM (2019b) Nitric oxide and hydrogen sulfide in plants: which comes first? J Exp Bot 70:4391–4404

Corpas FJ, del Río LA, Palma JM (2019c) Impact of nitric oxide (NO) on the ROS metabolism of peroxisomes. Plants 8:37. https://doi.org/10.3390/plants8020037

Corpas FJ, Barroso JB, González-Gordo S, Muñoz-Vargas MA, Palma JM (2019d) Hydrogen sulfide: a novel component in *Arabidopsis* peroxisomes which triggers catalase inhibition. J Integr Plant Biol 61:871–883

Costa A, Drago IO, Behera S, Zottini M, Pizzo P, Schroeder JI et al (2010) H_2O_2 in plant peroxisomes: an *in vivo* analysis uncovers a Ca^{2+}-dependent scavenging system. Plant J 62:760–772

Cueto M, Hernández-Perea O, Martín R, Ventura ML, Rodrigo J, Lamas S, Golvano PM (1996) Presence of nitric oxide synthase activity in roots and nodules of *Lupinus albus*. FEBS Lett 398:159–164

de Duve C, Baudhuin P (1966) Peroxisomes (microbodies and related particles). Physiol Rev 46:323–357

de Duve C, Beaufay H, Jacques P, Rahman-Li Y, Selinger OZ, Wattiaux R, de Connick S (1960) Intracellular localization of catalase and of some oxidases in rat liver. Biochim Biophys Acta 40:186–187

de Felipe MR, Lucas MM, Pozuelo JM (1988) Cytochemical study of catalase and peroxidase in the mesophyll of *Lolium rigidum* plants treated with isoproturon. J Plant Physiol 132:67–73

Dean JV, Harper JE (1988) The conversion of nitrite to nitrogen oxide(s) by the constitutive NAD (P)H-nitrate reductase enzyme from soybean. Plant Physiol 88:389–395

del Río LA (2011) Peroxisomes as a source of reactive nitrogen species signal molecules. Arch Biochem Biophys 506:1–11

del Río LA (2013) Peroxisomes and their key role in cellular signaling and metabolism. Subcellular biochemistry, vol 69. Springer, Dordrecht

del Río LA (2015) ROS and RNS in plant physiology: an overview. J Exp Bot 66:2827–2837

del Río LA, Donaldson RP (1995) Production of superoxide radicals in glyoxysomal membranes from castor bean endosperm. J Plant Physiol 146:283–287

del Río LA, López-Huertas E (2016) ROS generation in peroxisomes and its role in cell signaling. Plant Cell Physiol 57:1364–1376

del Río LA, Puppo A (2009) Reactive oxygen species in plant signaling. Springer, Dordrecht

del Río LA, Schrader M (2018) Proteomics of peroxisomes. Identifying novel functions and regulatory networks. Subcellular biochemistry, vol 89. Springer, Dordrecht

del Río LA, Lyon DS, Olah I, Glick B, Salin ML (1983) Immunocytochemical evidence for a peroxisomal localization of manganese superoxide dismutase in leaf protoplasts from a higher plant. Planta 158:216–224

del Río LA, Fernández VM, Rupérez FL, Sandalio LM, Palma JM (1989) NADH induces the generation of superoxide radicals in leaf peroxisomes. Plant Physiol 89:728–731

del Río LA, Sandalio LM, Palma JM (1990) A new cellular function for peroxisomes related to oxygen free radicals? Experientia 46:989–992

del Río LA, Sandalio LM, Palma JM, Bueno P, Corpas FJ (1992) Metabolism of oxygen radicals in peroxisomes and cellular implications. Free Radic Biol Med 13:557–580

del Río LA, Palma JM, Sandalio LM, Corpas FJ, Pastori GM, Bueno P, López-Huertas E (1996) Peroxisomes as a source of superoxide and hydrogen peroxide in stressed plants. Biochem Soc Trans 24:434–438

del Río LA, Pastori GM, Palma JM, Sandalio LM, Sevilla F, Corpas FJ, Jiménez A, López-Huertas E, Hernández JA (1998) The activated oxygen role of peroxisomes in senescence. Plant Physiol 116:1195–1200

del Río LA, Corpas FJ, Sandalio LM, Palma JM, Gómez M, Barroso JB (2002) Reactive oxygen species, antioxidant systems and nitric oxide in peroxisomes. J Exp Bot 53:1255–1272

del Río LA, Sandalio LM, Altomare DA, Zilinskas BA (2003a) Mitochondrial and peroxisomal manganese superoxide dismutase: differential expression during leaf senescence. J Exp Bot 54:923–933

del Río LA, Corpas FJ, Sandalio LM, Palma JM, Barroso JB (2003b) Plant peroxisomes, reactive oxygen metabolism and nitric oxide. IUBMB Life 55:71–81

del Río LA, Corpas FJ, Barroso JB (2004) Nitric oxide and nitric oxide synthase activity in plants. Phytochemistry 65:783–792 del Río LA, Sandalio LM, Corpas FJ, Palma JM, Barroso JB (2006) Reactive oxygen species and reactive nitrogen species in peroxisomes. Production, scavenging and role in cell signaling. Plant Physiol 141:330–335

del Río LA, Corpas FJ, Barroso JB, López-Huertas E, Palma JM (2014) Function of peroxisomes as a cellular source of nitric oxide and other reactive nitrogen species. In: Nasir Khan M, Mobin M, Mohammad F, Corpas FJ (eds) Nitric oxide in plants: metabolism and role in stress physiology. Springer, Berlin, pp 33–55

del Río LA, Corpas FJ, López-Huertas PJM (2018) Plant superoxide dismutases: function under abiotic stress conditions. In: Gupta DK, Palma JM, Corpas FJ (eds) Antioxidants and antioxidant enzymes in higher plants. Springer, Cham, pp 1–26

Desai M, Hu J (2008) Light induces peroxisome proliferation in *Arabidopsis* seedlings through the photoreceptor phytochrome A, the transcription factor HYS HOMOLOG, and the peroxisomal protein PEROXIN11B. Plant Physiol 146:1117–1127

Diao QN, Song YJ, Shi DM, Qi HY (2016) Nitric oxide induced by polyamines involves antioxidant systems against chilling stress in tomato (*Lycopersicon esculentum* Mill.) seedling. J Zhejiang Univ Sci B 17:916–930

Dietz KJ (2003) Plant peroxiredoxins. Annu Rev Plant Biol 54:93–107

Distefano S, Palma JM, Gómez M, del Río LA (1997) Characterization of endoproteases from plant peroxisomes. Biochem J 327:399–405

Distefano S, Palma JM, McCarthy I, del Río LA (1999) Proteolytic cleavage of plant proteins by peroxisomal endoproteases from senescent pea leaves. Planta 209:308–313

Dixon DP, Hawkins T, Hussey PJ, Edwards R (2009) Enzyme activities and subcellular localization of members of the *Arabidopsis* glutathione transferase superfamily. J Exp Bot 60:1207–1218

Droillard MJ, Paulin A (1990) Isoenzymes of superoxide dismutase in mitochondria and peroxisomes isolated from petals of carnation (*Dianthus caryophyllus*) during senescence. Plant Physiol 94:1187–1192

Erdmann R (2016) Assembly, maintenance and dynamics of peroxisomes. Special issue. Biochim Biophys Acta 1863:787–789

Fahimi HD, Sies H (1987) Peroxisomes in biology and medicine. Springer, Berlin

Fares A, Rossignol M, Peltier JB (2011) Proteomics investigation of endogenous *S*-nitrosylation in *Arabidopsis*. Biochem Biophys Res Commun 416:331–336

Fernández-García N, Martí MC, Jiménez A, Sevilla F, Olmos E (2009) Sub-cellular distribution of glutathione in an *Arabidopsis* mutant (*vtc1*) deficient in ascorbate. J Plant Physiol 166:2004–2012

Ferreira RMB, Bird B, Davies DD (1989) The effect of light on the structure and organization of *Lemna* peroxisomes. J Exp Bot 40:1029–1035

Filipovic MR, Jovanovic VM (2017) More than just an intermediate: hydrogen sulfide signalling in plants. J Exp Bot 68:4733–4736

Foresi N, Correa-Aragunde N, Parisi G et al (2010) Characterization of a nitric oxide synthase from the plant kingdom: NO generation from the green alga *Ostreococcus tauri* is light irradiance and growth phase dependent. Plant Cell 22:3816–3830

Foyer CH (2018) Reactive oxygen species, oxidative signaling and the regulation of photosynthesis. Environ Exp Bot 154:134–142

Foyer CH, Noctor G (2005) Oxidant and antioxidant signalling in plants: a re-evaluation of the concept of oxidative stress in a physiological context. Plant Cell Environ 28:1056–1071

Foyer CH, Noctor G (2011) Ascorbate and glutathione: the heart of the redox hub. Plant Physiol 155:2–18

Foyer CH, Bloom AJ, Queval G, Noctor G (2009) Photorespiratory metabolism: genes, mutants, energetic, and redox signaling. Annu Rev Plant Biol 60:455–484

Fransen M, Lismont C (2018) Peroxisomes and cellular oxidant/antioxidant balance: protein redox modifications and impact on inter-organelle communication. Subcell Biochem 89:435–461

Fransen M, Nordgren M, Wang B, Apanasets O (2012) Role of peroxisomes in ROS/RNS-metabolism: implications for human disease. Biochim Biophys Acta 1822:1363–1373

Fridovich I (1995) Superoxide radical and superoxide dismutases. Annu Rev Biochem 64:97–112
Frölich A, Durner J (2011) The hunt for a plant nitric oxide synthase (NOS): is one really needed? Plant Sci 181:401–404
Giles GI, Tasker KM, Jacob C (2001) Hypothesis: the role of reactive sulfur species in oxidative stress. Free Radic Biol Med 31:1279–1283
Giles GI, Tasker KM, Collins C, Giles NM, O'Rourke E, Jacob C (2002) Reactive Sulphur species: an *in vitro* investigation of the oxidation properties of disulphide S-oxides. Biochem J 364:579–585
Gronemeyer T, Wiese S, Ofman R et al (2013) The proteome of human liver peroxisomes: identification of five new peroxisomal constituents by a label-free quantitative proteomics survey. PLoS One 8(2):e57395
Gupta KJ, Kaiser WM (2010) Production and scavenging of nitric oxide by barley root mitochondria. Plant Cell Physiol 51:576–584
Halliwell B, Gutteridge JMC (2007) Free radicals in biology and medicine. Oxford University Press, Oxford
Halliwell B, Gutteridge JMC (2015) Free radicals in biology and medicine, 5th edn. Oxford University Press, Oxford
Hancock JT (2012) NO synthase? Generation of nitric oxide in plants. Period Biol 114:19–24
Hancock JT (2019) Hydrogen sulfide and environmental stresses. Environ Exp Bot 161:50–56
Hancock JT, Neill SJ (2019) Nitric oxide: its generation and interactions with other reactive signaling compounds. Plan Theory 8:41. https://doi.org/10.3390/plants8020041
Hancock JT, Whiteman M (2014) Hydrogen sulfide and cell signaling: team player or referee. Plant Physiol Biochem 78:37–42
Hancock JT, Whiteman M (2016) Hydrogen sulfide signaling: interactions with nitric oxide and reactive oxygen species. Ann N Y Acad Sci 1365:5–14
Hänsch R, Mendel RR (2005) Sulfite oxidase in plant peroxisomes. Photosynth Res 86:337–343
Harrison R (2002) Structure and function of xanthine oxidoreductase: where are we now? Free Radic Biol Med 33:774–797
Hayashi M, Nishimura M (2003) Entering a new era of research on plant peroxisomes. Curr Opin Plant Biol 6:577–582
Holzmeister C, Gaupels F, Geerlof A, Sarioglu H, Sattler M, Durner J, Lindermayr C (2015) Differential inhibition of *Arabidopsis* superoxide dismutases by peroxynitrite-mediated tyrosine nitration. J Exp Bot 66:989–999
Hu J, Baker A, Bartel B, Linka N, Mullen RT, Reumann S, Zolman BK (2012) Plant peroxisomes: biogenesis and function. Plant Cell 24:2279–2303
Huang AHC, Trelease RN, Moore TS Jr (1983) Plant peroxisomes. Academic Press, New York
Huang J, Sommers EM, Kim-Shapiro DB, King SB (2002) Horseradish peroxidase catalyzed nitric oxide formation from hydroxyurea. J Am Chem Soc 124:3473–3480
Imlay JA (2011) Redox pioneer: professor Irwin Fridovich. Antioxid Redox Signal 14:335–340
Inzé A, Vanderauwera S, Hoeberichts FA, Vandorpe M, van Gaever T, van Breusegem F (2012) A subcellular localization compendium of hydrogen peroxide-induced proteins. Plant Cell Environ 35:308–320
Ishikawa T, Yoshimura K, Sakai K, Tamoi M, Takeda T, Shigeoka S (1998) Molecular characterization and physiological role of a glyoxysome-bound ascorbate peroxidase from spinach. Plant Cell Physiol 39:23–34
Jasid S, Simontacchi M, Bartoli CG, Puntarulo S (2006) Chloroplasts as a nitric oxide cellular source. Effect of reactive nitrogen species on chloroplastic lipids and proteins. Plant Physiol 142:1246–1255
Jiménez A, Hernández JA, del Río LA, Sevilla F (1997) Evidence for the presence of the ascorbate-glutathione cycle in mitochondria and peroxisomes of pea leaves. Plant Physiol 114:275–284
Jones DP, Sies H (2015) The redox code. Antioxid Redox Signal 23:734–746
Ju Y, Fu M, Stokes E, Wu L, Yang G (2017) H_2S-mediated protein *S*-sulfhydration: a prediction for its formation and regulation. Molecules 22:E1334

Kabil O, Vitvitsky V, Banerjee R (2014) Sulfur as a signaling nutrient through hydrogen sulfide. Annu Rev Nutr 34:171–205

Kamada-Nobusada T, Hayashi M, Fukazawa M, Sakakibara H, Nishimura M (2008) A putative peroxisomal polyamine oxidase, AtPAO4, is involved in polyamine catabolism in *Arabidopsis thaliana*. Plant Cell Physiol 49:1272–1282

Kao Y-T, González KL, Bartel B (2018) Peroxisome function, biogenesis, and dynamics in plants. Plant Physiol 176:162–177

Kaur N, Hu J (2011) Defining the plant peroxisomal proteome from *Arabidopsis* to rice. Front Plant Sci 2:103

Kaur N, Li J, Hu J (2013) Peroxisomes and photomorphogenesis. Subcell Biochem 69:195–211

Keller GA, Warner TG, Steimer KS, Hallewell RA (1991) Cu, Zn superoxide dismutase is a peroxisomal enzyme in human fibroblasts and hepatoma cells. Proc Natl Acad Sci U S A 88:7381–7385

Kimura H (2015) Signaling molecules: hydrogen sulfide and polysulfide. Antioxid Redox Signal 22:362–376

Kirkman HN, Rolfo M, Ferraris AM, Gaetani GF (1999) Mechanisms of protection of catalase by NADPH. Kinetics and stoichiometry. J Biol Chem 274:13908–13914

Kissner R, Nauser T, Bugnon P et al (1997) Formation and properties of peroxynitrite as studied by laser flash photolysis, high pressure stopped flow technique, and pulse radiolysis. Chem Res Toxicol 10:1285–1292

Kleff S, Sander S, Mielke G, Eising R (1997) The predominant protein in peroxisomal cores of sunflower cotyledons is a catalase that differs in primary structure from the catalase in the peroxisomal matrix. Eur J Biochem 245:402–410

Knowles RG, Moncada S (1994) Nitric oxide synthases in mammals. Biochem J 298:249–258

Koh S, André A, Edwards H, Ehrhardt D, Somerville S (2005) *Arabidopsis thaliana* subcellular responses to compatible *Erysiphe cichoracearum* infections. Plant J 44:516–529

Kuzniak E, Sklodowska M (2005) Fungal pathogen-induced changes in the antioxidant systems of leaf peroxisomes from infected tomato plants. Planta 222:192–200

Leterrier M, Corpas FJ, Barroso JB, Sandalio LM, del Río LA (2005) Peroxisomal monodehydroascorbate reductase genomic clone characterization and functional analysis under environmental stress conditions. Plant Physiol 138:2111–2123

Leterrier M, Barroso JB, Valderrama R, Begara-Morales JC, Sánchez-Calvo B, Chaki M et al (2016) Peroxisomal NADP-isocitrate dehydrogenase is required for *Arabidopsis* stomatal movement. Protoplasma 253:403–415

Lisenbee CS, Lingard MJ, Trelease RN (2005) Arabidopsis peroxisomes possess functionally redundant membrane and matrix isoforms of monodehydroascorbate reductase. Plant J 43:900–914

Lisjak M, Teklic T, Wilson ID, Whiteman M, Hancock JT (2013) Hydrogen sulfide: environmental factor or signalling molecule? Plant Cell Environ 36:1607–1616

López-Huertas E, del Río LA (2014) Characterization of antioxidant enzymes and peroxisomes of olive (*Olea europaea* L.) fruits. J Plant Physiol 171:1463–1471

López-Huertas E, Corpas FJ, Sandalio LM, del Río LA (1999) Characterization of membrane polypeptides from pea leaf peroxisomes involved in superoxide radical generation. Biochem J 337:531–536

López-Huertas E, Charlton WL, Johnson B, Graham I, Baker A (2000) Stress induces peroxisome biogenesis genes. EMBO J 19:6770–6777

Loughran PA, Stolz DB, Vodovotz Y et al (2005) Monomeric inducible nitric oxide synthase localizes to peroxisomes in hepatocytes. Proc Natl Acad Sci U S A 102:13837–13842

Loughran PA, Stolz DB, Barrick SR, Wheeler DS, Friedman PA, Rachubinski RA, Watkins SC, Billiar TR (2013) PEX7 and EBP50 target iNOS to the peroxisome in hepatocytes. Nitric Oxide 31:9–19

Lozano-Juste J, Colom-Moreno R, León J (2011) In vivo protein tyrosine nitration in *Arabidopsis thaliana*. J Exp Bot 62:3501–3517

Mano S, Nakamori C, Hayashi M et al (2002) Distribution and characterization of peroxisomes in *Arabidopsis* by visualization with GFP: dynamic morphology and actin-dependent movement. Plant Cell Physiol 43:331–341
Marino D, Dunand C, Puppo A, Pauly N (2012) A burst of plant NADPH oxidases. Trends Plant Sci 17:9–15
Martínez-Ruiz A, Lamas S (2004) *S*-Nitrosylation: a potential new paradigm in signal transduction. Cardiovasc Res 62:43–52
Martínez-Ruiz A, Lamas S (2009) Two decades of new concepts in nitric oxide signaling: from the discovery of a gas messenger to the mediation of nonenzymatic posttranslational modifications. IUBMB Life 61:91–98
Mateos RM, León AM, Sandalio LM, Gómez M, del Río LA, Palma JM (2003) Peroxisomes from pepper fruits (*Capsicum annuum* L.): purification, characterization and antioxidant activity. J Plant Physiol 160:1507–1516
Mathur J, Mathur N, Hulskamp M (2002) Simultaneous visualization of peroxisomes and cyto-skeletal elements reveals actin and not microtubule-based peroxisome motility in plants. Plant Physiol 128:1031–1045
McCarthy I, Romero-Puertas MC, Palma JM, Sandalio LM, Corpas FJ, Gómez M, del Río LA (2001) Cadmium induces senescence symptoms in leaf peroxisomes of pea plants. Plant Cell Environ 24:1065–1073
McCarthy I, Gómez M, del Río LA, Palma JM (2011) Role of peroxisomes in the oxidative injury induced by 2,4-dichlorophenoxyacetic acid in leaves of pea plants. Biol Plant 55:485–492
Mhamdi A, Queval G, Chaouch S, Vanderauwera S, van Breusegem F, Noctor G (2010) Catalase function in plants: a focus on *Arabidopsis* mutants as stress-mimic models. J Exp Bot 61:4197–4220
Mhamdi A, Noctor G, Baker A (2012) Plant catalases: peroxisomal redox guardians. Arch Biochem Biophys 525:181–194
Mitsuya S, El Shami M, Sparkes IA, Charlton WL, de Marcos LC, Johnson B et al (2010) Salt stress causes peroxisome proliferation, but inducing peroxisome proliferation does not improve NaCl tolerance in *Arabidopsis thaliana*. PLoS One 5:e9408
Mittler R (2017) ROS are good. Trends Plant Sci 22:11–19
Mittler R, Vanderauwera S, Suzuki N, Miller G, Tognetti VB, Vandepoele K, Gollery M, Shulaev V, van Breusegem F (2011) ROS signaling: the new wave? Trends Plant Sci 16:300–309
Mittova V, Guy M, Tal M, Volokita M (2004) Salinity up-regulates the antioxidative system in root mitochondria and peroxisomes of the wild salt-tolerant tomato species *Lycopersicon pennellii*. J Exp Bot 55:1105–1113
Mor A, Koh E, Weiner L, Rosenwasser S, Sibony-Benyamini H, Fluhr R (2014) Singlet oxygen signatures are detected independent of light or chloroplasts in response to multiple stresses. Plant Physiol 165:249–261
Moro MA, Darley-Usmar VM, Goodwin DA, Read NG, Zamora-Pino R, Feelisch M, Radomski MW, Moncada S (1994) Paradoxical fate and biological action of peroxynitrite on human platelets. Proc Natl Acad Sci U S A 91:6702–6706
Morré DJ, Sellden G, Ojanperae K, Sandelius AS, Egger A, Morré DM, Chalko CM, Chalko RA (1990) Peroxisome proliferation in Norway spruce induced by ozone. Protoplasma 155:58–65
Moschou PN, Sanmartin M, Andriopoulou AH, Rojo E, Sánchez-Serrano JJ, Kalliopi K et al (2008) Bridging the gap between plant and mammalian polyamine catabolism: a novel peroxisomal polyamine oxidase responsible for a full back-conversion pathway in *Arabidopsis*. Plant Physiol 147:1845–1857
Mur LAJ, Mandon J, Persijn S, Crisstescu SM, Moshkov IE, Novikova GV, Hall MA, Harren FJM, Hebelstrup KM, Gupta KJ (2012) Nitric oxide in plants: an assessment of the current state of knowledge. AoB Plants 5:pls052
Nasir Khan M, Mobin M, Mohammad F, Corpas FJ (eds) (2014) Nitric oxide in plants: metabolism and role in stress physiology. Springer, Berlin

Neill S, Bright J, Desikan R, Hancock JT, Harrison J, Wilson I (2008) Nitric oxide evolution and perception. J Exp Bot 59:25–35

Nguyen AT, Donaldson RP (2005) Metal-catalyzed oxidation induces carbonylation of peroxisomal proteins and loss of enzymatic activities. Arch Biochem Biophys 439:25–31

Nicholls P (1964) The reactions of azide with catalase and their significance. Biochem J 40:331–343

Nila AG, Sandalio LM, López MG, Gómez M, del Río LA, Gómez-Lim MA (2006) Expression of a peroxisomes proliferator-activated receptor gene (*xPPARα*) from *Xenopus laevis* in tobacco (*Nicotiana tabacum*) plants. Planta 224:569–558

Nowak K, Luniak N, Witt C, Wüstefeld Y, Wachter A, Mendel RR, Hänsch R (2004) Peroxisomal localization of sulfite oxidase separates it from chloroplast-based sulfur assimilation. Plant Cell Physiol 45:1889–1894

Nyathi Y, Baker A (2006) Plant peroxisomes as a source of signaling molecules. Biochim Biophys Acta 1763:1478–1495

Oikawa KS, Matsunaga S, Mano S, Kondo M, Yamada K, Hayashi M et al (2015) Physical interaction between peroxisomes and chloroplasts elucidated by *in situ* laser analysis. Nat Plants 1:15035

Oksanen E, Haikio E, Sober J, Karnosky DF (2003) Ozone-induced H_2O_2 accumulation in field-grown aspen and birch is linked to foliar ultrastructure and peroxisomal activity. New Phytol 161:791–799

Ortega-Galisteo AP, Rodríguez-Serrano M, Pazmiño DM, Gupta DK, Sandalio LM, Romero-Puertas MC (2012) *S*-nitrosylated proteins in pea (*Pisum sativum* L.) leaf peroxisomes: changes under abiotic stress. J Exp Bot 63:2089–2103

Palma JM, Garrido M, Rodríguez-García MI, del Río LA (1991) Peroxisome proliferation and oxidative stress mediated by activated oxygen species in plant peroxisomes. Arch Biochem Biophys 287:68–74

Palma JM, López-Huertas E, Corpas FJ, Sandalio LM, Gómez M, del Río LA (1998) Peroxisomal manganese superoxide dismutase: purification and properties of the isozyme from pea leaves. Physiol Plant 104:720–726

Palma JM, Sandalio LM, Corpas FJ, Romero-Puertas MC, McCarthy I, del Río LA (2002) Plant proteases, protein degradation and oxidate stress: role of peroxisomes. Plant Physiol Biochem 40:521–530

Palma JM, Corpas FJ, del Río LA (2009) Proteome of plant peroxisomes: new perspectives on the role of these organelles in cell biology. Proteomics 9:2301–2312

Palma JM, Sevilla F, Jiménez A, del Río LA, Corpas FJ, Álvarez de Morales P, Camejo DM (2015) Physiology of pepper fruit and the metabolism of antioxidants: chloroplasts, mitocondria and peroxisomes. Ann Bot 116:627–636

Pellinen R, Palva T, Kangasjärvi J (1999) Subcellular localization of ozone-induced hydrogen peroxide production in birch (*Betula pendula*) leaf cells. Plant J 20:349–356

Petrova V, Uzunov Z, Kujumdzieva A (2009) Peroxisomal localization of Mn-SOD enzyme in *Saccharomyces cerevisiae* yeasts: *in silico* analysis. Biotechnol Biotechnol Equip 23:1531–1536

Pracharoenwattana I, Smith SM (2008) When is a peroxisome not a peroxisome? Trends Plant Sci 13:522–525

Puyaubert J, Fares A, Rézé N, Peltier JB, Baudouin E (2014) Identification of endogenously *S*-nitrosylated proteins in *Arabidopsis* plantlets: effect of cold stress on cysteine nitrosylation level. Plant Sci 215–216:150–156

Radi R (2004) Nitric oxide, oxidants, and protein tyrosine nitration. Proc Natl Acad Sci U S A 101:4003–4008

Radi R (2013) Protein tyrosine nitration: biochemical mechanisms and structural basis of functional effects. Acc Chem Res 46:550–559

Rao RS, Møller IM (2011) Pattern of occurrence and occupancy of carbonylation sites in proteins. Proteomics 11:4166–4173

Reddy JK, Rao MS, Lalwani ND, Reddy MK, Nemali MR, Alvares K (1987) Induction of hepatic peroxisome proliferation by xenobiotics. In: Fahimi HD, Sies H (eds) Peroxisomes in biology and medicine. Springer, Berlin, pp 254–262

Reumann S (2011) Toward a definition of the complete proteome of plant peroxisomes: where experimental proteomics must be complemented by bioinformatics. Proteomics 11:1764–1779

Reumann S, Babujee L, Ma C, Wienkoop S, Siemsen T, Antonicelli GE et al (2007) Proteome analysis of *Arabidopsis* leaf peroxisomes reveals novel targeting peptides, metabolic pathways and defense mechanisms. Plant Cell 19:3170–3193

Reumann S, Quan S, Aung K, Yang P, Manandhar-Shrestha K, Holbrook D et al (2009) In-depth proteome analysis of *Arabidopsis* leaf peroxisomes combined with *in vivo* subcellular targeting verification indicates novel metabolic and regulatory functions of peroxisomes. Plant Physiol 150:125–143

Rhodin J (1954) Correlation of ultrastructural organization and function in normal and experimentally changed proximal convoluted tubule cells of the mouse kidney. Doctoral thesis, Karolinska Institut, Aktiebolaget Godvil, Stockholm

Rodríguez-Serrano M (2007) Molecular mechanisms of response to cadmium in *Pisum sativum* L. plants: function of reactive oxygen and nitrogen species. PhD thesis, University of Granada, Spain

Rodríguez-Serrano M, Romero-Puertas MC, Pastori GM, Corpas FJ, Sandalio LM, del Río LA, Palma JM (2007) Peroxisomal membrane manganese superoxide dismutase: characterization of the isozyme from watermelon (*Citrullus lanatus* Schrad.) cotyledons. J Exp Bot 58:2417–2427

Rodríguez-Serrano M, Romero-Puertas MC, Sparkes I, Hawes C, del Río LA, Serrano LM (2009) Peroxisome dynamics in *Arabidopsis* plants under oxidative stress induced by cadmium. Free Radic Biol Med 47:1632–1639

Rodríguez-Serrano M, Romero-Puertas MC, Sanz-Fernández M, Hu J, Sandalio LM (2016) Peroxisomes extend peroxules in a fast response to stress via a reactive oxygen species-mediated induction of the peroxin PEX11a. Plant Physiol 171:1665–1674

Rojas CM, Senthil-Kumar M, Wang K, Ryu C-M, Kaundal A, Mysorek K (2012) Glycolate oxidase modulates reactive oxygen species-mediated signal transduction during nonhost resistance in *Nicotiana benthamiana* and *Arabidopsis*. Plant Cell 24:336–352

Romero-Puertas MC, McCarthy I, Sandalio LM, Palma JM, Corpas FJ, Gómez M, del Río LA (1999) Cadmium toxicity and oxidative metabolism of pea leaf peroxisomes. Free Radic Res 31: S25–S31

Romero-Puertas MC, Palma JM, Gómez M, del Río LA, Sandalio LM (2002) Cadmium causes the oxidative modification of proteins in pea plants. Plant Cell Environ 25:677–686

Romero-Puertas MC, Rodríguez-Serrano M, Corpas FJ, Gómez M, del Río LA, Sandalio LM (2004a) Cadmium-induced subcellular accumulation of $O_2^{\cdot-}$ and H_2O_2 in pea leaves. Plant Cell Environ 27:1122–1134

Romero-Puertas MC, McCarthy I, Gómez M, Sandalio LM, Corpas FJ, del Río LA, Palma JM (2004b) Reactive oxygen species-mediated enzymatic systems involved in the oxidative action of 2,4-dichlorophenoxyacetic acid. Plant Cell Environ 27:1135–1148

Romero-Puertas MC, Corpas FJ, Sandalio LM, Leterrier M, Rodríguez-Serrano M, del Río LA et al (2006) Glutathione reductase from pea leaves: response to abiotic stress and characterization of the peroxisomal isozyme. New Phytol 170:43–52

Romero-Puertas MC, Rodríguez-Serrano M, Sandalio LM (2013) Protein *S*-nitrosylation in plants under abiotic stress: an overview. Front Plant Sci 4:1–6

Rosenwasser S, Rot I, Sollner E, Meyer AJ, Smith Y, Leviatan N, Fluhr R, Friedman H (2011) Organelles contribute differentially to reactive oxygen species-related events during extended darkness. Plant Physiol 156:185–201

Sandalio LM, Romero-Puertas MC (2015) Peroxisomes sense and respond to environmental cues by regulating ROS and RNS signaling networks. Annu Bot 116:475–485

Sandalio LM, Palma JM, del Río LM (1987) Localization of manganese superoxide dismutase in peroxisomes isolated from *Pisum sativum*. Plant Sci 51:1–8

Sandalio LM, Fernández VM, Rupérez FL, del Río LA (1988) Superoxide free radicals are produced in glyoxysomes. Plant Physiol 87:1–4
Sandalio LM, López-Huertas E, Bueno P, del Río LA (1997) Immunocytochemical localization of copper,zinc superoxide dismutase in peroxisomes from watermelon (*Citrullus vulgaris* Schrad.) cotyledons. Free Radic Res 26:187–194
Sandalio LM et al (2008) Imaging of reactive oxygen species and nitric oxide *in vivo* in plant tissues. Methods Enzymol 440:397–409
Sandalio LM, Rodríguez-Serrano M, Gupta DK, Archilla A, Romero-Puertas MC, del Río LA (2012) Reactive oxygen species and nitric oxide in plants under cadmium stress: from toxicity to signaling. In: Ahmad P, Prasad MNV (eds) Environmental adaptations and stress tolerance of plants in the era of climate change. Springer, Berlin, pp 199–215
Sandalio LM, Gotor C, Romero LC, Romero-Puertas MC (2019) Multilevel regulation of peroxisomal proteome by post-translational modifications. Int J Mol Sci 20:4881. https://doi.org/10.3390/ijms20194881
Schrader M, Fahimi HD (2004) Mammalian peroxisomes and reactive oxygen species. Histochem Cell Biol 122:383–393
Schrader M, Fahimi HD (2006) Peroxisomes and oxidative stress. Biochim Biophys Acta 1763:1755–1766
Schubert KR (1986) Products of biological nitrogen fixation in higher plants: synthesis, transport, and metabolism. Annu Rev Plant Physiol 37:539–574
Shabab M (2013) Role of plant peroxisomes in protection against herbivores. Subcell Biochem 69:315–328
Sies H (2014) Role of metabolic H_2O_2 generation: redox signaling and oxidative stress. J Biol Chem 289:8735–8741
Smirnoff N, Arnaud D (2019) Hydrogen peroxide metabolism and functions in plants. New Phytol 221:1197–1214
Sparkes I, Gao H (2014) Plant peroxisome dynamics: movement, positioning and connections. In: Brocard C, Hartig A (eds) Molecular machines involved in peroxisome biogenesis and maintenance. Springer, Wien, pp 461–477
Stöhr C, Stremlau S (2006) Formation and possible roles of nitric oxide in plant roots. J Exp Bot 57:463–470
Stöhr C, Strube F, Marx G, Ullrich WR, Rockel P (2001) A plasma membrane-bound enzyme of tobacco roots catalyses the formation of nitric oxide from nitrite. Planta 212:835–841
Stolz DB, Zamora R, Vodovotz Y, Loughran PA, Billiar TR, Kim Y-M, Simmons RL, Watkins SM (2002) Peroxisomal localization of inducible nitric oxide synthase in hepatocytes. Hepatology 36:81–93
Suzuki N, Miller G, Morales J, Shulaev V, Torres MA, Mittler R (2011) Respiratory burst oxidases: the engines of ROS signaling. Curr Opin Plant Biol 14:691–699
Szabó C, Ischiropoulos H, Radi R (2007) Peroxynitrite: biochemistry, pathophysiology and development of therapeutics. Nat Rev Drug Discov 6:662–680
Tanou G, Job C, Rajjou L, Arc E, Belghazi M, Diamantidis G, Molassiotis A, Job D (2009) Proteomics reveals the overlapping roles of hydrogen peroxide and nitric oxide in the acclimation of citrus plants to salinity. Plant J 60:795–804
Tiburcio AF, Altabella T, Bitrián M, Alcázar R (2014) The roles of polyamines during the lifespan of plants: from development to stress. Planta 240:1–18
Tischner R, Galli M, Heimer YM et al (2007) Interference with the citrulline-based nitric oxide synthase assay by argininosuccinate lyase activity in Arabidopsis extracts. FEBS J 274:4238–4245
Tolbert NE (1997) The C_2 oxidative photosynthetic carbon cycle. Annu Rev Plant Physiol Plant Mol Biol 48:1–25
Triantaphylidés C, Havaux M (2009) Singlet oxygen in plants: production, detoxification and signaling. Trends Plant Sci 14:219–228

Tun NN, Santa-Catarina C, Begum T, Silveira V, Handro W, Floh EL, Scherer GF (2006) Polyamines induce rapid biosynthesis of nitric oxide (NO) in *Arabidopsis thaliana* seedlings. Plant Cell Physiol 47:346–354

Turkan I (2018) ROS and RNS: key signalling molecules in plants. J Exp Bot 69(14):3313

Valderrama R, Corpas FJ, Carreras A, Gómez-Rodríguez MV, Chaki M, Pedrajas JR, Fernández-Ocaña A, del Río LA, Barroso JB (2006) The dehydrogenase-mediated recycling of NADPH is a key antioxidant system against salt-induced oxidative stress in olive plants. Plant Cell Environ 29:1449–1459

van Bentem S, Anrather D, Roitinger E et al (2006) Phosphoproteomics reveals extensive *in vivo* phosphorylation of *Arabidopsis* proteins involved in RNS metabolism. Nucleic Acids Res 34:3267–3278

Vanderauwera S, Hoeberichts FA, Van Breusegem F (2009) Hydrogen peroxide-responsive genes in stress acclimation and cell death. In: del Río LA, Puppo A (eds) Reactive oxygen species in plant signaling. Springer, Berlin, pp 149–164

Walbrecq G, Wang B, Becker S, Hannotiau A, Fransen M, Knoops B (2015) Antioxidant cytoprotection by peroxisomal peroxiredoxin-5. Free Radic Biol Med 84:215–226

Wanders RJA (2013) Peroxisomes in human health and disease: metabolic pathways, metabolite transport, interplay with other organelles and signal transduction. Subcell Biochem 69:23–44

Wanders RJA, Waterham HR (2006) Biochemistry of mammalian peroxisomes revised. Annu Rev Biochem 75:295–332

Wang Y, Loake GJ, Chu C (2013) Cross-talk of nitric oxide and reactive oxygen species in plant programmed cell death. Front Plant Sci 4:314

Wang X, Li S, Liu Y, Ma C et al (2015) Redox regulated peroxisome homeostasis. Redox Biol 4:104–108

Wilson ID, Neill SJ, Hancock JT (2008) Nitric oxide synthesis and signaling in plants. Plant Cell Environ 31:622–631

Wimalasekera R, Tebartz F, Scherer GF (2011a) Polyamines, polyamine oxidases and nitric oxide in development, abiotic and biotic stresses. Plant Sci 181:593–603

Wimalasekera R, Villar C, Begum T, Scherer GF (2011b) *COPPER AMINE OXIDASE1* (*CuAO1*) of *Arabidopsis thaliana* contributes to abscisic acid- and polyamine-induced nitric oxide biosynthesis and abscisic acid signal transduction. Mol Plant 4:663–678

Wink DA, Hanbauer I, Grisham MB et al (1996) Chemical biology of nitric oxide: regulation and protective and toxic mechanisms. Curr Top Cell Regul 34:159–187

Yamasaki H, Sakihama Y, Takahashi S (1999) An alternative pathway for nitric oxide production in plants: new features of an old enzyme. Trends Plant Sci 4:128–129

Young D, Pedre B, Ezerina D, de Smet B, Lewandowska A, Tossounian M-A, Bodra N, Huang J, Astolfi-Rosado L, Van Breusegem F, Messens J (2019) Protein promiscuity in H_2O_2 signaling. Antioxid Redox Signal 30:1285–1324

Yu M, Lamattina L, Spoel SH, Loake GJ (2014) Nitric oxide function in plant biology: a redox cue in deconvolution. New Phytol 202:1142–1156

Zechmann B, Mauch F, Sticher L, Müller M (2008) Subcellular immunocytochemical analysis detects the highest concentrations of glutathione in mitochondria and not in plastids. J Exp Bot 59:4017–4027

Zhang H (2016) Hydrogen sulfide in plant biology. In: Lamattina L, García-Mata C (eds) Gasotransmitters in plants. The rise of a new paradigm in cell signaling. Springer, Berlin, pp 23–51

Zhao MG, Tian QY, Zhang WH (2007) Nitric oxide synthase-dependent nitric oxide production is associated with salt tolerance in *Arabidopsis*. Plant Physiol 144:206–217

Ammonium Assimilation and Metabolism in Rice

Soichi Kojima, Keiki Ishiyama, Marcel Pascal Beier, and Toshihiko Hayakawa

Contents

Abstract Rice sustains the world. It is a staple food in most Asian countries. Both a high yield and healthy growth of rice are highly dependent on nitrogen fertilizer. Ammonium is the most important nitrogen source for rice growth. Rice is grown in paddy fields in which nitrification is inhibited due to anaerobic conditions. Ammonium transporters (AMTs) transport ammonium from the soil into cells in plant roots. Glutamine synthetase (GS) ligates ammonium with glutamate in an ATP-dependent manner and is a key enzyme of ammonium assimilation. Glutamate synthase (GOGAT) transfers the amide residue of glutamine to 2-oxoglutarate to yield two molecules of glutamate. As GS and GOGAT provide substrates for each other, their reaction is coupled. The GS/GOGAT cycle is the major pathway in

S. Kojima (✉), K. Ishiyama, and T. Hayakawa
Graduate School of Agricultural Science, Tohoku University, Sendai, Japan
e-mail: soichi.kojima.a2@tohoku.ac.jp

M. P. Beier
Graduate School of Agricultural Science, Tohoku University, Sendai, Japan

Graduate School of Agricultural and Life Sciences, The University of Tokyo, Tokyo, Japan

© Springer Nature Switzerland AG 2020

Progress in Botany (2021) 82: 211–232, https://doi.org/10.1007/124_2020_40,
Published online: 29 December 2020

ammonium assimilation in plants, and the rice genome encodes several isozymes for GS/GOGAT. In this chapter, we describe the physiological functions and "job-sharing" of three cytosolic GS isoenzymes (GS1;1, GS1;2, and GS1;3) and two NADH-GOGAT enzymes (NADH-GOGAT1 and NADH-GOGAT2) using both a reverse genetic and a biochemical approach.

Keywords AMT, Asparagine synthetase, Glutamate, Glutamine, GOGAT, Grain protein content, GS, Metabolism, Nitrogen translocation, Rice, Seedlings, Tillering, Yield

1 Introduction: Ammonium Is a Primary Nitrogen Source for Paddy Rice

Rice is a staple food for more than half of the world's population. According to an investigation by the Food and Agriculture Organization of the United Nations, rice is cultivated in 118 countries, and the production amount is approximately 770 million tons, which is near equal to the wheat production in 2017. Although Asia contributes to more than 90% of the world's rice production, Africa increased its production approximately to 2.9-fold in 2017, compared in 1990 (from 12.7 million tons in 1990 to 36.6 million tons in 2017). The world's population is increasing exponentially and is expected to reach 10 billion by the middle of this century (United Nations, World population prospects 2019; https://population.un.org/wpp/Graphs/Probabilistic/POP/TOT/900). For this reason, further increases in rice production are expected within a relatively short period. With little scope for expanding the land area, an increase in rice production must be achieved by increasing the yield from the land currently used for rice cultivation. In addition, interest in the damage caused by nitrogen management practices to the environment is increasing (Canfield et al. 2010; Good and Beatty 2011). Therefore, it is important to increase the grain yield while limiting damage to the environment by nitrogen management practices in rice cultivation. One solution to this problem is to develop a high-yielding rice cultivar that produces a higher grain yield per unit amount of plant nitrogen absorption.

Nitrogen is a macronutrient that directly affects the biomass production of plants. In most soils, ammonium and nitrate are the predominant sources of inorganic nitrogenous mineral nutrients for plants. However, irrigated rice paddy fields are usually subject to anaerobic conditions in which the majority of inorganic nitrogen is ammonium, which is easily protonated under acidic environments (Kronzucker et al. 1997). Rice plants preferentially utilize ammonium rather than nitrate (Yamaya and Oaks 2004), and the development and productivity of rice cultivation are determined by the efficiency of inorganic ammonium uptake and assimilation in the plants' roots, as well as nitrogen recycling during the natural senescence of aerial parts (Yamaya and Kusano 2014).

The uptake of ammonium from soils is predominantly mediated by ammonium transporters (AMTs). The expression of AMTs is affected by nitrogen conditions and strictly regulates the influx of ammonium in plants including rice (Loque and von Wiren 2004). The glutamine synthetase (GS)/glutamate synthase (GOGAT) cycle, as defined by Lea and Miflin (1974), represents the major pathway involved in the assimilation of ammonium under normal metabolic conditions in plants. GS (EC 6.3.1.2) is a key enzyme involved in ammonium assimilation in plants and catalyzes the ATP-dependent conversion of glutamate into glutamine using ammonium (Ireland and Lea 1999; Yamaya and Kusano 2014). Most plants contain two isoenzymes of GS: cytosolic GS1 and chloroplastic/plastidic GS2 (Ireland and Lea 1999). GS2, the major isoenzyme in leaves, is responsible for the re-assimilation of ammonium that is released from photorespiration in leaves (Wallsgrove et al. 1987), while GS1 is necessary for normal growth and development (Yamaya and Oaks 2004; Tabuchi et al. 2007). In general, plants possess a small gene family encoding two to six GS1 isoenzymes (Ireland and Lea 1999; Yamaya and Kusano 2014). GOGAT catalyzes the transfer of the amide group in glutamine formed by GS to 2-oxoglutarate to yield two molecules of glutamate. One of the glutamate molecules is cycled back as a substrate for the GS reaction, while the other is used for many synthetic reactions (Yamaya and Kusano 2014). Two molecular species of GOGAT are found in both green and nongreen tissues. One species requires NADH as a reductant (NADH-GOGAT; EC 1.4.1.14), and the other requires ferredoxin (Fd-GOGAT; EC 1.4.7.1) (Yamaya and Kusano 2014). The major role of Fd-GOGAT, which is located in the chloroplast stroma, is in the re-assimilation of NH_4^+ released from photorespiration in a reaction coupled with GS2 (Yamaya and Kusano 2014).

In this study, we present a summary of the research results on AMTs, three isoenzymes of cytosolic GS (GS1;1, GS1;2, and GS1;3), and NADH-GOGAT1 and NADH-GOGAT2, characterized using a reverse genetic approach in rice (*Oryza sativa* L.).

2 Ammonium Assimilation

2.1 *Primary Ammonium Assimilation in Roots*

Paddy field-grown rice plants take up ammonium via their roots as a major nitrogen source for growth. Short-term influxes of ammonium have been studied with radioactive tracers (Wang et al. 1993a, b). Ammonium uptake is biphasic in rice roots. Below an external ammonium concentration of 1 mM, a high-affinity transport system (HATS) is responsible for ammonium uptake in rice roots. Between external ammonium concentrations of 1 and 40 mM, a low-affinity transport system contributes to ammonium uptake. AMTs are membrane proteins that transport ammonium from the environment into plant cells over the plasma membrane. AMT family proteins belong to the HATS. There are 12 genes similar to AMT in the rice genome

(Li et al. 2009). AMT1 is a plant-specific AMT, while other AMT genes of plants are homologous to those of other organisms. Three AMT1 genes have been isolated from rice, and their expression has been studied (Sonoda et al. 2003a, b). In rice seedlings, AMT1;1 is ubiquitously expressed in both roots and shoots, while AMT1;2 and AMT1;3 are only expressed in roots. AMT1 expression is responsive to nitrogen in roots. AMT1;1 and AMT1;2 are highly expressed under ammonium supply, while AMT1;3 is induced by nitrogen deficiency and suppressed by ammonium supply. The supply of glutamine mimics the ammonium supply. The activity of AMT is regulated at posttranscriptional and posttranslational levels in tobacco and *Arabidopsis*, respectively (Yuan et al. 2007a; Loque et al. 2007). In rice roots, the STY protein kinase, ACTPK1, is induced by ammonium supply and phosphorylates the C-terminal conserved threonine residue of AMT1;2 to decrease its activity (Beier et al. 2018).

After the transport of ammonium into cells, GS conjugates ammonium with glutamate to synthesize glutamine. The rice genome encodes four GS isozymes, which are categorized into two groups. The first is GS1 that localizes to the cytoplasm, and the other is GS2 that localizes to the plastid/chloroplast. GS2 assimilates ammonium derived from the conversion of two molecules of glycine into one molecule of serine during photorespiration in leaves. The GS1 protein is more abundant than GS2 protein in rice roots, while the shoots contain both GS1 and GS2 (Ishiyama et al. 2004a). The rice genome encodes three GS1 isozymes, GS1;1, GS1;2, and GS1;3. Among the GS1 isozymes, the expression of GS1;2 is increased by ammonium supply in roots of rice seedlings (Tabuchi et al. 2007). In situ hybridization analysis revealed that GS1;2 is expressed in the dermatogen, epidermis, and exodermis (Ishiyama et al. 2004a). The enzymatic kinetics of recombinant GS1 isozymes have been compared between rice and *Arabidopsis* (Ishiyama et al. 2004a, b). It has been shown that the K_m values for ammonium in rice and *Arabidopsis* range from 0.01 to 2.5 mM, while the V_{max} values range from 20 to 200 nanokatal/mg protein. Rice GS1;2 protein has a high affinity to ammonium, with higher V_{max} values among GS1 isozymes. The characteristics of GS1;2 point to the importance of GS1;2 in the primary ammonium assimilation in rice roots. A reverse genetic approach has confirmed that GS1;2 plays a key role in ammonium assimilation (Funayama et al. 2013). The total GS activity is reduced by half in the knockout mutant *gs1;2*, and the number of active tillers is decreased in *gs*1;2, which is a typical symptom of nitrogen deficiency. The concentration of free glutamine is reduced by 70% in roots and by 75% in xylem sap in *gs*1;2 mutants when ammonium is supplied as the sole nitrogen source. Complementation of *GS1;2* under the control of its own promoter into *gs*1;2 mutants was shown to restore glutamine concentrations in the xylem sap to wild-type levels. Although *GS1;1* is expressed in roots, it does not compensate for GS1;2 functions. The GS reaction requires glutamate as a substrate, and in order to assimilate ammonium, GS must be provided with sufficient glutamate by GOGAT. The rice genome encodes three GOGAT genes: two NADH-dependent GOGAT and one ferredoxin (Fd)-dependent GOGAT. Ammonium supply triggers the accumulation of NADH-GOGAT1 in rice seedling roots (Yamaya et al. 1995). The mRNA and protein of NADH-GOGAT1 are increased markedly within 12 h after supply of ammonium, and a supply of 50 μM ammonium has been shown to be sufficient to induce NADH-GOGAT1 protein accumulation in roots.

Immunohistochemistry has revealed that the NADH-GOGAT1 protein is accumulated in the epidermis and exodermis, the cell layers of the root surface (Ishiyama et al. 1998). *NADH-GOGAT1* expression is induced by ammonium as well as glutamine. Methionine sulfoximine (MSX), an inhibitor of GS, inhibits ammonium-induced *NADH-GOGAT1* expression (Hirose et al. 1997). Therefore, glutamine and/or its metabolites act as signal molecules for the *NADH-GOGAT1* expression. Interestingly, okadaic acid, an inhibitor of the protein phosphatases 1 and 2a, mimics the effect of ammonium on *NADH-GOGAT1* expression (Hirose and Yamaya 1999). NADH-GOGAT1 in rice roots plays an important role in the generation of glutamate for the assimilation of ammonium by supplying the substrate for the GS reaction, which suggests it could be controlled by a signal transduction network for ammonium use. Furthermore, completion of the rice genome sequence has revealed an additional *NADH-GOGAT2* gene besides the ammonium-induced *NADH-GOGAT1* gene (Tabuchi et al. 2007). The *Tos17* insertion rice mutant for NADH-GOGAT1, *nadh-gogat1*, has been isolated, and its characteristics have confirmed the contribution of NADH-GOGAT1 to the primary ammonium assimilation in roots (Tamura et al. 2010). The *nadh-gogat1* mutant shows inhibition of root elongation in an ammonium concentration-dependent manner in the medium. Six-day-old seedlings of *nadh-gogat1* mutants show root inhibition when a hydroponic solution contains more than 50 μM ammonium, and the inhibition is enhanced as ammonium concentrations increased in the medium (Tamura et al. 2010). In contrast, a nitrate supply did not inhibit the elongation of the mutant roots. Compared with roots, the shoot length is influenced less by a loss of NADH-GOGAT1. The glutamate concentration was shown to be reduced by half in *nadh-gogat1* mutants, regardless of whether or not the medium contained nitrogen. Ammonium is highly accumulated in *nadh-gogat1* mutants, in which ammonium concentration has been found to be ten times higher than in wild types in 1 mM ammonium supply. This indicates the essential role of NADH-GOGAT1 in primary ammonium assimilation. As glutamate serves as the substrate for the syntheses of other amino acids, aspartate, asparagine, and alanine are reduced in mutant roots under ammonium conditions. The expression of *NADH-GOGAT2* and *Fd-GOGAT* genes in *nadh-gogat1* mutants is no different to wild-type plants, which suggests that the function of NADH-GOGAT1 in primary ammonium assimilation cannot be substituted with other GOGATs.

The cell-specific localization of GS1;2 and NADH-GOGAT1 in rice roots indicates that the primary ammonium assimilation mainly occurs in the epidermis and exodermis, the two cell layers at the roots' surface. The Casparian strip, deposited between the root exodermis cells, acts as a physical barrier to prevent apoplastic transfer of water and solutes into the cortex. Ammonium is transported into the cells to be moved in the symplast at either the epidermis or the exodermis in rice roots. The localization of the ammonium-induced GS1;2/NADH-GOGAT1 cycle is responsible for effective ammonium assimilation.

2.2 Secondary Ammonium Assimilation in Shoots

The main nitrogen sources delivered from roots to shoots are amino acids, particularly glutamine, as the bulk of the primary assimilation of environmental ammonium occurs in rice roots (Kiyomiya et al. 2001). In photosynthetic active tissue, the plastid-/chloroplast-localized GS2/Fd-GOGAT cycle carries the workload for the photorespirational flux (Tobin and Yamaya 2001). Other than ammonium derived from primary uptake in roots and from nitrate assimilation and the photorespiration in photosynthetic active tissue, ammonium is set free during protein degradation and subsequent amino acid catabolism. Protein degradation and amino acid oxidation are important during the plant life cycle steps, as shown by Hildebrandt et al. (2015). Of particular interest is the autophagy observed in *Arabidopsis*, which provides energy during the night (Izumi et al. 2013), and developmental leaf senescence, which plays an important role in nitrogen redistribution, as discussed in Chap. 4 in this paper. Other than the role of GS2/Fd-GOGAT in photorespiration, the isoforms GS1;1 and NADH-GOGAT2 are of particular importance in shoot tissue. A *Tos17* insertion knockout line for NADH-GOGAT2 has been shown to have a reduced yield and whole-plant biomass (Tamura et al. 2011). A more extreme reduction of shoot biomass has been observed in *Tos17* insertion knockout lines for GS1;1 (Tabuchi et al. 2005). The localization of both isoforms in the same tissues, mainly in the vascular bundles in rice leaves (Sakurai et al. 1996; Tamura et al. 2011), suggests that both isoforms form a GS1;1/NADH-GOGAT2 cycle. The importance of this cycle, which is mainly responsible for ammonium assimilation in photosynthetic tissues besides photorespiration, is supported by metabolomic analysis that suggests GS1;1 coordinates in the metabolic balance in rice (Kusano et al. 2011).

3 Tillering

The rice tiller number is a critical agronomic trait because grain yield is defined as the active tiller (panicle) number per unit land area, spikelet number per panicle, the ratio of filled spikelets, and single grain weight (Mae 1997; Sakamoto and Matsuoka 2008). Tiller axillary buds emerge from the basal portions of rice shoots, which include internodes, axillary buds, and the shoot apical meristem (Hoshikawa 1989). The vascular tissues in the basal portions of rice shoots are important for the translocation of assimilation products from roots to either leaves or axillary buds for tillering (Hoshikawa 1989).

A physiological function of GS1;2 located in the basal portion of rice shoots is its involvement in the re-assimilation of vast amounts of ammonium released from the phenylalanine ammonia-lyase reaction, which is the first step in lignin biosynthesis (Sakurai et al. 2001; Ohashi et al. 2015a). An in situ hybridization analysis using the basal portions of wild-type rice plants showed that *OsGS1;2* is expressed in phloem companion cells of the nodal vascular anastomoses and large vascular bundles of axillary buds (Ohashi et al. 2015a). A rice mutant lacking GS1;2 (*gs1;2*) showed severe outgrowth suppression of tiller axillary buds and hence a substantial decrease in active tiller numbers and grain yield (Funayama et al. 2013; Ohashi et al. 2015a). The *gs1;2* mutant exhibits a large decrease in both glutamine and asparagine in the basal portions of shoots and roots (Ohashi et al. 2017). Both amide-formed amino acids play crucial roles in plant growth and development as the primary compounds involved in the long-distance transportation of nitrogen in various plants (Ireland and Lea 1999; Lea and Azevedo 2007; Gaufichon et al. 2010). In addition, glutamine is known to act as a signaling molecule.

At least three classes of phytohormones, cytokinins (CKs), strigolactones (SLs), and auxins, are involved in the regulation of plant axillary bud outgrowth (Domagalska and Leyser 2011; Evers et al. 2011; Waldie et al. 2014; Rameau et al. 2015). In ammonium-fed rice seedlings, major CKs in the basal portions of the shoots, namely, N^6-(Δ^2isopentenyl) adenosine (iP) and *trans*-zeatin types (tZ), have much higher contents compared to the roots and shoots (Ohashi et al. 2017). In the basal portions of rice shoots, the adenosine phosphate-isopentenyltransferase 4 (*OsIPT4*) gene, which is one of the major iso-genes for IPT, is upregulated by glutamine or related metabolites (Ohashi et al. 2017). In the *gs1;2* mutant, expression of the glutamine-responsive *OsIPT4*, which predominantly synthesizes CKs, and the content of the active-form of CKs are significantly decreased compared to wild-type rice plants (Ohashi et al. 2017). As previously reported, phosphate deficiency increases one kind of SLs, *ent-2′-epi*-5-deoxystrigol (*epi*-5DS), in the roots, which suppresses outgrowth of the axillary buds and reduces the tiller number of rice plants (Umehara et al. 2008). Although the *gs1;2* mutant exhibits a reduced outgrowth of the axillary buds, particularly under ammonium-sufficient conditions, no difference in the *epi*-5DS content between wild-type and *gs1;2* mutant rice plants was found (Ohashi et al. 2015a). Auxin, an inhibitor of axillary bud outgrowth, is synthesized in young leaves and transported basipetally in the polar auxin transport stream mediated by auxin efflux carriers (Leyser 2011). However, no difference in the contents of indole-3-acetic acid (IAA) and amide-linked IAA-amino acid conjugates between wild-type and *gs1;2* mutant rice plants has been found (Ohashi et al. 2017). These results indicate that a lack of GS1;2 in the basal portion decreases the glutamine content required for CK biosynthesis, causing a deficiency in active CK in the axillary bud meristem that is necessary for tillering (Ohashi et al. 2017). Although SLs and auxins are known to reduce the active tiller number, the reduction in the outgrowth of axillary buds observed in *gs1;2* mutant rice was independent of the levels of both phytohormones (Ohashi et al. 2015a, 2017).

Early anatomical analyses have revealed that normal tiller development in rice depends on adequate nitrogen and carbohydrate (sugar and/or starch) levels in the basal portions of rice shoots (Takahashi et al. 1956; Sato 1959). During the vegetative stage, an insufficiency of nitrogen results in a decrease in the number of active tillers (Mae 1997), while sufficient nitrogen promotes tillering (Liu et al. 2011). It is known that shading treatment causes a delay in tillering emergence (Nakano 2000). These facts indicate that the assimilate availability is important in the control of plant axillary bud outgrowth in addition to phytohormones (Domagalska and Leyser 2011; Evers et al. 2011; Waldie et al. 2014; Rameau et al. 2015). Asparagine is synthesized by the transfer of the glutamine-amide group to aspartate by asparagine synthetase (AS) (Lea and Azevedo 2007; Gaufichon et al. 2010; Ohashi et al. 2015b). Rice has two AS isoenzymes (*OsAS1* and *OsAS2*) (Ohashi et al. 2015b). In our previous research, we confirmed that AS (*OsAS1*) is upregulated by glutamine or related metabolites in the basal portions of rice shoots (Ohashi et al. 2018a). In rice roots, AS1 and GS1;2 are important in the primary assimilation of ammonium that is absorbed from the soil (Funayama et al. 2013; Ohashi et al. 2015b). Although it was found that rice mutants lacking AS1 showed an approximately 60% decrease in asparagine content in the basal portions, the outgrowth of axillary buds and tiller numbers were equal to wild-type rice plants (Ohashi et al. 2018a). This indicates that an availability of glutamine, rather than asparagine, in the basal portions of rice shoots may be required for the outgrowth of rice tillers (Ohashi et al. 2018a). Our previous transcriptomic and metabolomic analyses using the basal portions of rice shoots revealed that a lack of GS1;2 reduced sucrose metabolism, which was shown to be caused by the downregulation of *cytosolic fructose 1,6-bisphosphatase2* (*cFBPase2*) in the basal portions of rice shoots (Ohashi et al. 2018b). Rice mutants lacking cFBPase2 were found to have an approximately 30% reduction in total cFBPase activity in the basal portions of their shoots, and the mutants exhibited a decrease in the sucrose content of the basal portions of shoots, but not in leaf blades (Ohashi et 2018b). A mutant lacking cFBPase2 had a lower tiller number compared with early-stage wild-type rice plants (Ohashi et al. 2018b). Based on this, it can be concluded that sufficient glutamine and sucrose are required for the outgrowth of axillary buds for tillering in the basal portions of rice shoots. We have summarized the models of tillering and nitrogen metabolism (Fig. 1).

4 Translocation of Nitrogen: Source to Sink

Most grain nitrogen originates from other organs in rice plants. Nitrogen recycling enables the plant to survive under nitrogen-deficient conditions and helps to maximize grain production. Nitrogen translocation has four stages in rice (Fig. 2):

1. Catabolism of nitrogen compound in senescent organs
2. Conversion of amide to translocated nitrogen form
3. Translocation of nitrogen from source organ to sink organ through phloem
4. Reutilization of translocated nitrogen in sink organ

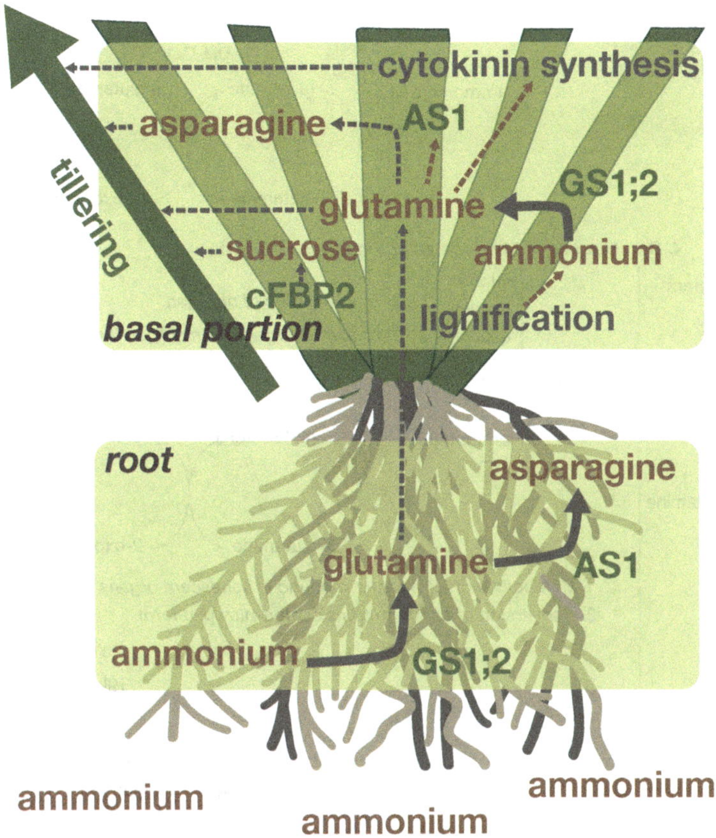

Fig. 1 Tillering and ammonium metabolism. Glutamine, sucrose, and cytokinin stimulate tillering. GS1;2 assimilates ammonium into glutamine in both roots and basal portion. Ammonium is mainly originated from nitrogen fertilizer in roots and from lignification in basal portion. Glutamine stimulates cytokinin and asparagine synthesis in basal portion

4.1 Translocation of Glutamine

Leaf senescence triggers the catabolism of nitrogen compounds, while ribulose 1, 5-bisphosphate carboxylase/oxygenase (Rubisco), other photosynthetic proteins, and chlorophyll are degraded. The resulting small nitrogen compounds leach to the apoplasts, where they are exhausted with the apoplastic fluids from the hydathode. To prevent nitrogen loss, plants have developed machinery to retrieve and convert nitrogen compounds to translocatable forms (Bohner et al. 2015). The translocation of nitrogen from senescent to young organs occurs over the vasculature. A phloem sap analysis of the uppermost internode with brown plant hoppers, *Nilaparvata lugens* Stal, revealed a 40% contribution of glutamine to the total amino acids to phloem sap (Hayashi and Chino 1990). This suggests that glutamine is a major

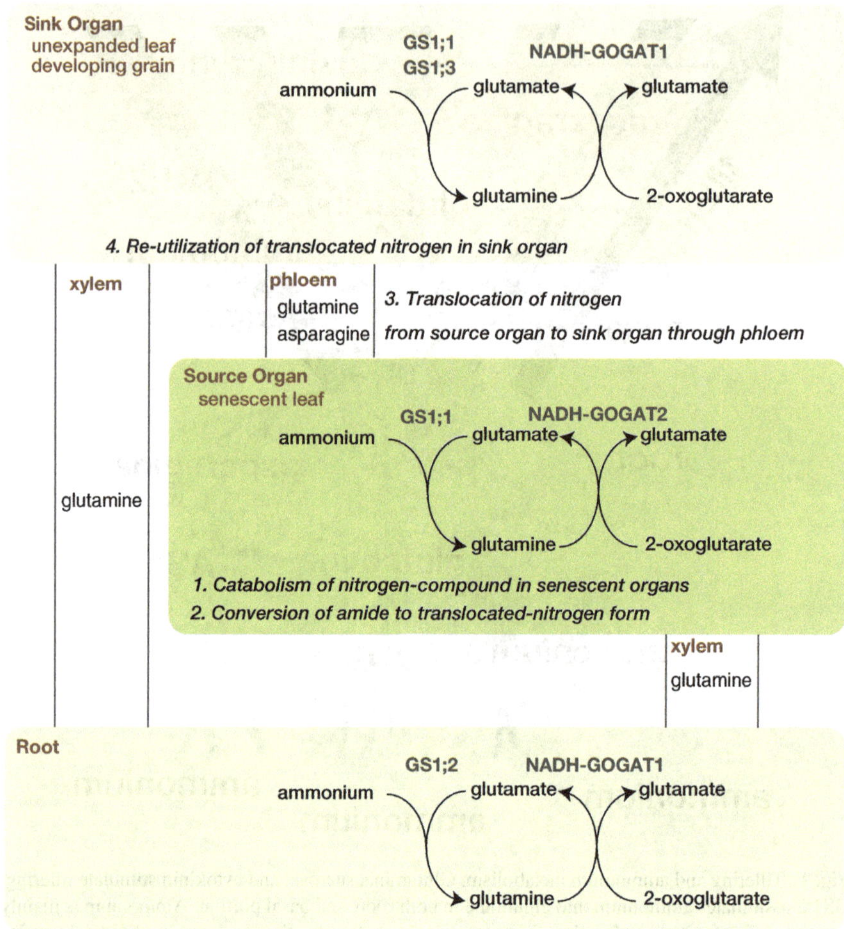

Fig. 2 Nitrogen translocation in rice. Glutamine is translocated from root to both source and sink organ through xylem transport. Glutamine and asparagine are translocated from source organ to sink organ through phloem transport

nitrogen form for translocation in rice. The major nitrogen solute in the xylem is also glutamine, rather than ammonium (Fukumorita and Chino 1982; Kiyomiya et al. 2001).

In glutamine synthesis, the cytosolic GS1 has a key role in senescent leaves. The developmental changes in GS1 protein contents in leaves point to its importance in nitrogen translocation (Kamachi et al. 1991). The amount of GS1 protein is relatively constant during leaf senescent. Conversely, the amount of soluble protein, chlorophyll, GS2, Fd-GOGAT, and Rubisco is decreased during senescence. Immunohistochemistry using a rice GS1-specific antibody has demonstrated the localization of GS1 in parenchyma cells and companion cells of vascular bundles. GS2 was found to be localized in the mesophyll cells (Kamachi et al. 1992a, b). The localization of GS1 is shown to be unchanged during leaf development, although it is increased in older leaves (Sakurai et al. 1996).

The localization of GS1 is well-conserved in different rice cultivars (Obara et al. 2000). Both japonica and indica rice accumulate GS1 in the vascular bundles of senescent leaves. The temporal and spatial distribution of GS1 proteins suggests that GS1 plays a key role in nitrogen translocation from source organs. Real-time PCR analysis has indicated that *GS1;2* is mainly expressed in roots and *GS1;3* is mainly expressed in developing spikelets (Tabuchi et al. 2007). The transcript of *GS1;1* may be a major GS1 isozyme for nitrogen translocation. *GS1;1* is expressed in source organs and sink tissues (Yabuki et al. 2017), and in situ hybridization has shown that GS1;1 is localized in the vascular bundles of developing rice grains. The concentration of glutamate is high in developing young leaves and spikelets, indicating the importance of the transamination of glutamine to form glutamate in sink organs. NADH-GOGAT1 transfers the amide of glutamine to 2-oxoglutarate to form two molecules of glutamate in sink organs. The activity and protein amount of NADH-GOGAT are accumulated in unexpended immature leaves and decrease during leaf growth. In expanded mature leaves and senescent leaves, there are small amounts of NADH-GOGAT protein (Yamaya et al. 1992), whereas there are high levels of Rubisco, Fd-GOGAT, and GS2 proteins. The NADH-GOGAT protein is located in the vascular bundles of unexpanded leaves (Hayakawa et al. 1994), and histochemical analysis indicates the localization of NADH-GOGAT in phloem and parenchyma cells and mestom sheath cells. These biochemical analyses indicate that NADH-GOGAT plays a key role in the metabolization of translocated glutamine in immature leaves. NADH-GOGAT protein is also located in the vascular bundles of developing grains (Hayakawa et al. 1993). The amount of NADH-GOGAT protein is higher than that of Fd-GOGAT during the grain-filling stage, increases after flowering, and reaches a maximum at 10–15 days after flowering. NADH-GOGAT protein is also located in parenchyma cells, nucellar epidermis, nucellar projection, and aleurone cell layers (Hayakawa et al. 1994), all of which take part in nitrogen translocation. Conversely, Fd-GOGAT is mainly located in leaf mesophyll cells and therefore contributes less to nitrogen translocation.

The genomic and cDNA sequences of *NADH-GOGAT* has been previously reported (Goto et al. 1998). Promoter-reporter analysis has shown that tissue- and cell-specific localizations of *NADH-GOGAT* are regulated by the promoter (Kojima et al. 2000). Research has shown that the entire 3.7 kbp of the 5′-upstream region (−2,840 to +886) from the translational start codon of the *NADH-GOGAT* gene is sufficient to express a reporter gene with the same localization as NADH-GOGAT in transgenic rice plants. Analysis of the complete rice genome has revealed that two NADH-GOGAT isozymes and a single Fd-GOGAT are encoded (Hayakawa et al. 2003; Sasaki 2005). The previously reported *NADH-GOGAT* has been renamed *NADH-GOGAT1*, and the newly discovered one is named *NADH-GOGAT2* (Tabuchi et al. 2007). Promoter-reporter analysis indicates that NADH-GOGAT protein localized in developing grains is the product of *NADH-GOGAT1* (Kojima et al. 2000; Tabuchi et al. 2007), whereas NADH-GOGAT2 is localized in the vascular bundles of mature leaves (Tabuchi et al. 2007; Tamura et al. 2011).

Biochemical and molecular approaches indicate that GS1;1 synthesizes glutamine, NADH-GOGAT2 provides glutamate for GS1;1 in source organs, NADH-GOGAT1 synthesizes glutamate, and both GS1;1 and GS1;3 provide glutamine for NADH-GOGAT1 in sink organs.

There is a genetic diversity among GS1 and NADH-GOGAT proteins between japonica and indica rice cultivars (Yamaya et al. 1997). The indica cultivars Chinsurah Boro I and Blue Stick contain higher levels of GS1 than the japonica cultivar Sasanishiki in senescing leaves. Conversely, the japonica cultivar contains higher levels of NADH-GOGAT than the indica cultivars. Transgenic indica cultivars expressing japonica *NADH-GOGAT1* cDNA driven by the *NADH-GOGAT1* promoter show variation in the NADH-GOGAT protein content in unexpanded leaves (Yamaya et al. 2002). Panicle number and spikelet weight are correlated with NADH-GOGAT protein. Notably, transgenic plants containing a higher NADH-GOGAT protein content increase the panicle number and weight of the spikelets. These results suggest a potential for yield increases in indica cultivars by gene modification.

4.2 Translocation of Asparagine

Asparagine is the second most abundant amino acids in phloem sap prepared from flag-leaf sheaths and the most abundant amino acid in endosperm sap (Hayashi and Chino 1990). Asparagine is synthesized through the reaction of glutamine-dependent or ammonium-dependent AS (EC 6.3.5.4) and is catabolized into aspartate and ammonium by the reaction of asparaginase (ASNase, EC 3.5.1.1). AS protein accumulates in high levels in the mature sheaths and spikelets during the ripening process (Nakano et al. 2000). Asparagine concentrations are not parallel to AS protein accumulation in spikelets, while aspartate continuously decreases during ripening (Yabuki et al. 2017). The imbalanced AS protein and asparagine concentration indicate the posttranslational regulation of ASNase. The cellular localization of the AS protein is similar to NADH-GOGAT, suggesting the importance of AS in sink organs. The rice genome encodes two AS genes, the AS1 and AS2 (Sasaki 2005), and real-time PCR indicates that the expression of *AS2* is higher than *AS1*, although both are expressed in developing spikelets (Yabuki et al. 2017). A *Tos*17 insertion line for AS1 has been isolated (Ohashi et al. 2015b), and it has been shown that a loss of AS1 does not change the yield of paddy field-grown rice (Ohashi et al. 2018a), suggesting a limited contribution of AS1 to nitrogen translocation.

4.3 *Amino Acid Permease*

Amino acid transporters are involved in the process of nitrogen translocation (Tegeder 2014). Rice amino acid permease (AAP6) is involved in grain protein contents (GPC) (Peng et al. 2014), and AAP3 (Lu et al. 2018) and AAP5 (Wang et al. 2019) are involved in rice yield. However, it is as yet unclear whether AAPs can transport glutamine and asparagine, and further studies focusing on cell physiology are needed to determine the function of amino acid transport during nitrogen translocation.

5 Yield and Grain Quality

5.1 *Yield of GS/GOGAT Mutants*

A variety of evidence has highlighted the importance of the GS/GOGAT cycle in plant productivity. Quantitative trait locus (QTL) analyses suggest that genotypic differences in productivity between japonica and indica cultivars can partly be explained by *GS1/NADH-GOGAT* loci (Obara et al. 2001, 2004; Yamaya et al. 2002). GS1 has been identified as a QTL for leaf senescence rate and one-spikelet weight, which is indicative of the contribution of GS1 to nitrogen export from senescing leaf blades (Yamaya et al. 2002). The identification of NADH-GOGAT as a QTL for spikelet number and spikelet weight suggests its relationship to grain yield (Yamaya et al. 2002).

Reverse genetic studies have revealed a direct link between GS/GOGAT genes and rice yield (Tabuchi et al. 2005; Tamura et al. 2010, 2011; Funayama et al. 2013). A loss of GS1;1 severely reduces the productivity of rice (Tabuchi et al. 2005). Spikelet number per panicle is reduced by half, fertility is almost reduced to zero, and spikelet weight is reduced by 30% in *gs1;1* mutants (Tabuchi et al. 2005). Due to severe growth retardation in paddy fields, the yield of the *gs1;1* mutant has not been evaluated. The distribution of GS1 transcripts suggests that GS1;1 is the isozyme responsible for nitrogen export from senescing leaves, as *GS1;2* is mainly expressed in roots and *GS1;3* is specifically expressed in spikelets during seedling growth (Tabuchi et al. 2007). Indeed, spikelet weight is decreased in *gs1;1*, which corresponds with the QTL analysis. A loss of GS1;2 leads to a reduction in yield of 70% (Funayama et al. 2013). None of the other yield components, such as the grain number per panicle, grain-filling ratio, and weight of seeds, are influenced; however, the panicle number is reduced by 70% (Funayama et al. 2013). The *gs1;2* phenotype is similar to nitrogen deficiency symptoms in rice (Mae 1997), suggesting a GS1;2-dependent primary ammonium assimilation in roots. As yet, there has been no reverse genetic study of GS1;3. The distribution of GS1 transcripts suggests that GS1;3 does not contribute significantly to productivity, as its transcriptional level is much lower than other GS1s (Tabuchi et al. 2007). An NADH-GOGAT1 knockout mutation causes a decrease in rice yield (Tamura et al. 2010). The panicle number per plant is reduced by 30%, and the spikelet number per panicle is reduced by 15% in *nadh-gogat1* mutants, resulting in decreased panicle production (Tamura et al. 2010). On the other hand, a 1,000-spikelet weight is identical in wild types, and changes in the ratio of filled grains are relatively small. An NADH-GOGAT2 mutation also causes a decrease in rice yield (Tamura et al. 2011), and spikelet numbers per panicle are significantly reduced (by approximately 26–39%) in the *nadh-gogat2* mutant. The proportion of filled grains is also decreased by approximately 5–25%. However, the panicle number and 1,000-spikelet weight are identical to wild types. These data correspond with the QTL analysis. Although the two NADH-GOGAT genes are involved in rice yield, they affect the yield in different ways. As discussed in Chap. 4, the localizations of NADH-GOGAT1 and NADH-GOGAT2 do not overlap; NADH-GOGAT1 mainly localizes in the vascular bundles of immature leaves and developing grains (Hayakawa et al. 1994; Kojima et al. 2000), while NADH-GOGAT2 is mainly localized in senescent leaves (Tabuchi et al. 2007; Tamura et al. 2011).

NADH-GOGAT1 contributes to the reception of translocated glutamine in sink tissues, while NADH-GOGAT2 contributes to the production of glutamine in source tissue with GS1;1 (Yamaya and Kusano 2014). Knockout and overexpression lines indicate the importance of NADH-GOGAT in terms of yield. For example, when the *NADH-GOGAT1* gene from a japonica cultivar was transformed into an indica cultivar, several transgenic lines overexpressing the NADH-GOGAT1 gene showed increased panicle weight per main stem (Yamaya et al. 2002). These findings show promise in terms of breeding rice with a higher sink capacity.

A complete loss-of-function of Fd-GOGAT is lethal to rice (Yang et al. 2016), and several reports have indicated the importance of Fd-GOGAT in rice production (Yang et al. 2016; Bi et al. 2017). A japonica rice cultivar carrying a single-nucleotide mutation in the 22nd exon has been shown to have a reduced yield, reduced spikelet number per panicle of 40%, reduced 1,000 seed weight of 12%, and a severely reduced number of panicles, although the grain-filling ratio was unchanged (Yang et al. 2016). An indica rice cultivar carrying a single-nucleotide mutation in the second exon had a reduction in yield of 30% (Bi et al. 2017).

A mutant that suppresses the growth retardation of *Fd-GOGAT* (*abnormal CK response* 1 or *abc*1) has been previously isolated (Wang et al. 2018). The suppressor mutant *are*1 (*abc*1–1 *repressor* 1) has a single-nucleotide mutation in a gene encoding a chloroplast-localized protein. Surprisingly, the yield of *are*1 was increased by 10–20%, and an increase in the spikelet number per panicle contributed to the higher yield (Wang et al. 2018).

5.2 Grain Protein Content in the Mutants

Human nutrition is dependent on the consumption of cereal grains, and it is therefore important to improve the nutritional value of cereal grains, which can be one of the most important sources of protein in the diet. There is also a potential demand for cultivars with higher grain protein content (GPC) for the production of sweet sake and Shaoxing wine.

Forward genetic studies suggest that GS/GOGAT genes regulate the GPC in durum wheat and QTL analyses of durum wheat have shown genome regions controlling GPC overlap with GS (Gadaleta et al. 2011) and NADH-GOGAT (Nigro et al. 2014). Allelic variants of NADH-GOGAT genes are significantly related to GPC (Nigro et al. 2017). Reverse genetic studies show that a loss of NADH-GOGAT leads to increased GPC in rice (Imagawa et al. 2018). Imagawa et al. (2018) showed the effect of glutamate metabolism on GPC in rice grains with a reverse genetic approach and demonstrated that glutelins, one of the major seed storage proteins in rice grains, were increased in NADH-GOGAT1 mutants in rice, although the total protein concentration was unchanged. SDS-PAGE analysis showed that both the acidic and basic subunits of glutelins increased by 14–19% in the *nadh-gogat1* mutant (Imagawa et al. 2018).

The loss-of-function of NADH-GOGAT2 leads to a significant increase in total GPC. SDS-PAGE analysis has revealed that both glutelins and prolamins, another major seed storage protein in rice, are increased in the *nadh-gogat2* mutant relative to its genetic background "Nipponbare." The contradiction between forward and reverse genetic studies could be explained by considering the grain number phenotype in NADH-GOGAT mutants. A loss of NADH-GOGAT1 and NADH-GOGAT2 severely reduces the grain number, and a marked decrease in the mutant grain number increases the availability of nitrogen for individual grains in NADH-GOGAT mutants and triggers an increase in GPC in the mutant. The next step is to improve the yield of NADH-GOGAT mutants with a higher GPC. The timing and frequency of nitrogen fertilizer topdressing during the panicle formation stage could be a key consideration in further studies (Yoshida 1981).

6 Conclusions and Future Prospects

Though the mechanism of rice nitrogen assimilation has been the subject of some study, many critical questions remain unanswered. For example, the mechanism that orchestrates the signal transduction for ammonium assimilation in rice roots is not yet understood. Protein phosphatase 2a is involved in ammonium-induced NADH-GOGAT1 expression in cultured rice cells (Hirose and Yamaya 1999), and protein kinase regulates the activity of AMT1;2 in rice roots (Beier et al. 2018). These discoveries point to the presence of a signal transduction pathway of phosphorylation and dephosphorylation; however, the details of this pathway are yet to be elucidated. Furthermore, the transcription factors regulating ammonium-induced gene expression are not well-understood. An in-depth understanding of the signal transduction for ammonium assimilation would contribute to the minimization of nitrogen fertilizer use, which is important in the development of sustainable agriculture. Secondly, while AMTs have been studied with reverse genetic approaches in *Arabidopsis* (Loque et al. 2006; Yuan et al. 2007b), there have been few reports in rice (Li et al. 2016), which hinders an understanding of the physiological role of AMTs in rice roots. Further study of AMTs in rice roots is needed as ammonium is a major nitrogen source for paddy field-grown rice and the structure of rice roots is different from those of *Arabidopsis*. In the preparation of knockout mutants, CRISPR/Cas9 may be useful. Thirdly, the localization of amino acid permeases should be studied in detail with reverse genetic studies to determine the nitrogen translocation pathway from source to sink. Finally, GS and GOGAT are expected to improve productivity in rice, as well as in other plants (Thomsen et al. 2014). For example, the expression of an extra copy of native GS1 led to an increase in grain yields and nitrogen use efficiency in barley (Gao et al. 2018). Such findings are promising and should be explored in further detail with other cereal plants. Future research should focus on determining the mechanism behind growth defects and/or retardation found in knockout mutants.

Acknowledgments JSPS KAKENHI Grant Numbers 21688006 and 26450073 to SK and 17H03780 to T.H. Grant-in-Aid for Research Activity Start-up (19K23661 to M.P.B) and Grant-in-Aid for Scientific Research on Innovative Areas (JP18H05490 to T.F) supported this work. The contribution to the reverse genetic work and biochemical work of past and present members of Yamaya laboratory (Tohoku University) is acknowledged.

References

Beier MP, Obara M, Taniai A, Sawa Y, Ishizawa J, Yoshida H, Tomita N, Yamanaka T, Ishizuka Y, Kudo S, Yoshinari A, Takeuchi S, Kojima S, Yamaya T, Hayakawa T (2018) Lack of ACTPK1, an STY kinase, enhances ammonium uptake and use, and promotes growth of rice seedlings under sufficient external ammonium. Plant J 93(6):992–1006. https://doi.org/10.1111/tpj.13824

Bi Z, Zhang Y, Wu W, Zhan X, Yu N, Xu T, Liu Q, Li Z, Shen X, Chen D, Cheng S, Cao L (2017) ES7, encoding a ferredoxin-dependent glutamate synthase, functions in nitrogen metabolism and impacts leaf senescence in rice. Plant Sci 259:24–34. https://doi.org/10.1016/j.plantsci.2017.03.003

Bohner A, Kojima S, Hajirezaei M, Melzer M, von Wiren N (2015) Urea retranslocation from senescing Arabidopsis leaves is promoted by DUR3-mediated urea retrieval from leaf apoplast. Plant J 81(3):377–387

Canfield DE, Glazer AN, Falkowski PG (2010) The evolution and future of earth's nitrogen cycle. Science 330(6001):192–196. https://doi.org/10.1126/science.1186120

Domagalska MA, Leyser O (2011) Signal integration in the control of shoot branching. Nat Rev Mol Cell Biol 12(4):211–221. https://doi.org/10.1038/nrm3088

Evers JB, van der Krol AR, Vos J, Struik PC (2011) Understanding shoot branching by modelling form and function. Trends Plant Sci 16(9):464–467. https://doi.org/10.1016/j.tplants.2011.05.004

Fukumorita T, Chino M (1982) Sugar, amino-acid and inorganic contents in rice phloem sap. Plant Cell Physiol 23(2):273–283

Funayama K, Kojima S, Tabuchi-Kobayashi M, Sawa Y, Nakayama Y, Hayakawa T, Yamaya T (2013) Cytosolic glutamine synthetase1;2 is responsible for the primary assimilation of ammonium in rice roots. Plant Cell Physiol 54:934–943

Gadaleta A, Nigro D, Giancaspro A, Blanco A (2011) The glutamine synthetase (GS2) genes in relation to grain protein content of durum wheat. Funct Integr Genomics 11(4):665–670

Gao Y, de Bang TC, Schjoerring JK (2018) Cisgenic overexpression of cytosolic glutamine synthetase improves nitrogen utilization efficiency in barley and prevents grain protein decline under elevated CO2. Plant Biotechnol J 17:1209. https://doi.org/10.1111/pbi.13046

Gaufichon L, Reisdorf-Cren M, Rothstein SJ, Chardon F, Suzuki A (2010) Biological functions of asparagine synthetase in plants. Plant Sci 179:141–153. https://doi.org/10.1016/j.plantsci.2010.04.010

Good A, Beatty P (2011) Fertilizing nature: a tragedy of excess in the commons. PLoS Biol 9: e1001124

Goto S, Akagawa T, Kojima S, Hayakawa T, Yamaya T (1998) Organization and structure of NADH-dependent glutamate synthase gene from rice plants. BBA-Protein Struct M 1387:298–308

Hayakawa T, Yamaya T, Mae T, Ojima K (1993) Changes in the content of 2 glutamate synthase proteins in spikelets of rice (Oryza sativa) plants during ripening. Plant Physiol 101:1257–1262

Hayakawa T, Nakamura T, Hattori F, Mae T, Ojima K, Yamaya T (1994) Cellular-localization of NADH-dependent glutamate-synthase protein in vascular bundles of unexpanded leaf blades and young grains of rice plants. Planta 193:455–460

Hayakawa T, Hopkins L, Peat LJ, Yamaya T, Tobin AK (1999) Quantitative intercellular localization of NADH-dependent glutamate synthase protein in different types of root cells in rice plants. Plant Physiol 119:409–416

Hayakawa T, Sakai T, Ishiyama K, Hirose N, Nakajima H, Takezawa M, Naito K, Hino-Nakayama-M, Akagawa T, Goto S, Yamaya T (2003) Organization and structure of ferredoxin-dependent glutamate synthase gene and intracellular localization of the enzyme protein in rice plants. Plant Biotechnol 20(1):43–55

Hayashi H, Chino M (1990) Chemical-composition of phloem sap from the uppermost internode of the rice plant. Plant Cell Physiol 31(2):247–251

Hildebrandt TM, Nesi AN, Araujo WL, Braun HP (2015) Amino acid catabolism in plants. Mol Plant 8:1563–1579. https://doi.org/10.1016/j.molp.2015.09.005

Hirose N, Yamaya T (1999) Okadaic acid mimics nitrogen-stimulated transcription of the NADH-glutamate synthase gene in rice cell cultures. Plant Physiol 121(3):805–812

Hirose N, Hayakawa T, Yamaya T (1997) Inducible accumulation of mRNA for NADH-dependent glutamate synthase in rice roots in response to ammonium ions. Plant Cell Physiol 38 (11):1295–1297

Hoshikawa K (1989) The growing rice plant: an anatomical monograph. Nobunkyo, Tokyo

Imagawa F, Minagawa H, Nakayama Y, Kanno K, Hayakawa T, Kojima S (2018) Tos17 insertion in NADH-dependent glutamate synthase genes leads to an increase in grain protein content in rice. J Cereal Sci 84:38–43. https://doi.org/10.1016/j.jcs.2018.09.008

Ireland RJ, Lea PJ (1999) The enzymes of glutamine, glutamate, asparagine, and aspartate metabolism. In: Singh BK (ed) Plant amino acids, biochemistry and biotechnology. Marcel Dekker, New York, pp 49–109

Ishiyama K, Hayakawa T, Yamaya T (1998) Expression of NADH-dependent glutamate synthase protein in the epidermis and exodermis of rice roots in response to the supply of ammonium ions. Planta 204:288–294

Ishiyama K, Inoue E, Tabuchi M, Yamaya T, Takahashi H (2004a) Biochemical background and compartmentalized functions of cytosolic glutamine synthetase for active ammonium assimilation in rice roots. Plant Cell Physiol 45(11):1640–1647

Ishiyama K, Inoue E, Watanabe-Takahashi A, Obara M, Yamaya T, Takahashi H (2004b) Kinetic properties and ammonium-dependent regulation of cytosolic isoenzymes of glutamine synthetase in Arabidopsis. J Biol Chem 279:16598–16605

Izumi M, Hidema K, Makino A, Ishida H (2013) Autophagy contributes to nighttime energy availability for growth in Arabidopsis. Plant Physiol 161:1682–1693

Kamachi K, Yamaya T, Mae T, Ojima K (1991) A role for glutamine-synthetase in the remobilization of leaf nitrogen during natural senescence in rice leaves. Plant Physiol 96:411–417

Kamachi K, Yamaya T, Hayakawa T, Mae T, Ojima K (1992a) Changes in cytosolic glutamine-synthetase polypeptide and its messenger-RNA in a leaf blade of rice plants during natural senescence. Plant Physiol 98:1323–1329

Kamachi K, Yamaya T, Hayakawa T, Mae T, Ojima K (1992b) Vascular bundle-specific localization of cytosolic glutamine-synthetase in rice leaves. Plant Physiol 99:1481–1486

Kiyomiya S, Nakanishi H, Uchida H, Tsuji A, Nishiyama S, Futatsubashi M, Tsukada H, Ishioka NS, Watanabe S, Ito T, Mizuniwa C, Osa A, Matsuhashi S, Hashimoto S, Sekine T, Mori S (2001) Real time visualization of ^{13}N-translocation in rice under different environmental conditions using positron emitting tracer imaging system. Plant Physiol 125(4):1743–1753. https://doi.org/10.1104/pp.125.4.1743

Kojima S, Kimura M, Nozaki Y, Yamaya T (2000) Analysis of a promoter for the NADH-glutamate synthase gene in rice (Oryza sativa): cell type-specific expression in developing organs of transgenic rice plants. Aust J Plant Physiol 27:787–793

Kronzucker HJ, Siddiqi MY, Glass ADM (1997) Conifer root discrimination against soil nitrate and the ecology of forest succession. Nature 385(6611):59–61. https://doi.org/10.1038/385059a0

Kusano M, Tabuchi M, Fukushima A, Funayama K, Diaz C, Kobayashi M, Hayashi N, Tsuchiya YN, Takahashi H, Kamata A, Yamaya T, Saito K (2011) Metabolomics data reveal a crucial role of cytosolic glutamine synthetase 1;1 in coordinating metabolic balance in rice. Plant J 66:456–466

Lea PJ, Azevedo RA (2007) Nitrogen use efficiency. 2. Amino acid metabolism. Ann Appl Biol 151 (3):269–275

Lea PJ, Miflin BJ (1974) Alternative route for nitrogen assimilation in higher-plants. Nature 251 (5476):614–616

Leyser O (2011) Auxin, self-organisation, and the colonial nature of plants. Curr Biol 21(9):R331–R337. https://doi.org/10.1016/j.cub.2011.02.031

Li BZ, Merrick M, Li SM, Li HY, Zhu SW, Shi WM, Su YH (2009) Molecular basis and regulation of ammonium transporter in rice. Rice Sci 16:314–322. https://doi.org/10.1016/S1672-6308(08)60096-7

Li C, Tang Z, Wei J, Qu H, Xie Y, Xu G (2016) The OsAMT1.1 gene functions in ammonium uptake and ammonium–potassium homeostasis over low and high ammonium concentration ranges. J Genet Genomics 43:639–649. https://doi.org/10.1016/j.jgg.2016.11.001

Liu Y, Gu D, Ding Y, Wang Q, Li G, Wang S (2011) The relationship between nitrogen, auxin and cytokinin in the growth regulation of rice (Oryza sativa L.) tiller buds. Aust J Crop Sci 5 (8):1019–1026

Loque D, von Wiren N (2004) Regulatory levels for the transport of ammonium in plant roots. J Exp Bot 55(401):1293–1305. https://doi.org/10.1093/jxb/erh147

Loque D, Yuan L, Kojima S, Gojon A, Wirth J, Gazzarrini S, Ishiyama K, Takahashi H, von Wiren N (2006) Additive contribution of AMT1;1 and AMT1;3 to high-affinity ammonium uptake across the plasma membrane of nitrogen-deficient Arabidopsis roots. Plant J 48:522–534. https://doi.org/10.1111/j.1365-313X.2006.02887.x

Loque D, Lalonde S, Looger LL, von Wiren N, Frommer WB (2007) A cytosolic trans-activation domain essential for ammonium uptake. Nature 446(7132):195–198. https://doi.org/10.1038/nature05579

Lu K, Wu B, Wang J, Zhu W, Nie H, Qian J, Huang W, Fang Z (2018) Blocking amino acid transporter OsAAP3 improves grain yield by promoting outgrowth buds and increasing tiller number in rice. Plant Biotechnol J 16:1710–1722. https://doi.org/10.1111/pbi.12907

Mae T (1997) Physiological nitrogen efficiency in rice: Nitrogen utilization, photosynthesis, and yield potential. In: Ando T, Fujita K, Mae T, Matsumoto H, Mori S, Sekiya J (eds) Plant nutrition for sustainable food production and environment, Developments in plant and soil sciences, vol 78. Springer, Dordrecht, pp 51–60. https://doi.org/10.1007/978-94-009-0047-9_5

Nakano H (2000) Effect of early-stage shading of direct seeded rice on growth and yield components. Jpn J Crop Sci 69(2):182–188. https://doi.org/10.1626/jcs.69.182

Nakano K, Suzuki T, Hayakawa T, Yamaya T (2000) Organ and cellular localization of asparagine synthetase in rice plants. Plant Cell Physiol 41:874–880

Nigro D, Blanco A, Anderson OD, Gadaleta A (2014) Characterization of ferredoxin-dependent glutamine-oxoglutarate amidotransferase (Fd-GOGAT) genes and their relationship with grain protein content QTL in wheat. PLoS One 9(8):e103869

Nigro D, Fortunato S, Giove S, Mangini G, Yacoubi I, Simeone R, Blanco A, Gadaleta A (2017) Allelic variants of glutamine synthetase and glutamate synthase genes in a collection of durum wheat and association with grain protein content. Diversity 9(4):52

Obara M, Sato T, Yamaya T (2000) High content of cytosolic glutamine synthetase does not accompany a high activity of the enzyme in rice (Oryza sativa) leaves of indica cultivars. Physiol Plant 108:11–18. https://doi.org/10.1034/j.1399-3054.2000.108001011.x

Obara M, Kajiura M, Fukuta Y, Yano M, Hayashi M, Yamaya T, Sato T (2001) Mapping of QTLs associated with cytosolic glutamine synthetase and NADH-glutamate synthase in rice (Oryza sativa L.). J Exp Bot 52:1209–1217

Obara M, Sato T, Sasaki S, Kashiba K, Nagano A, Nakamura I, Ebitani T, Yano M, Yamaya T (2004) Identification and characterization of a QTL on chromosome 2 for cytosolic glutamine synthetase content and panicle number in rice. Theor Appl Genet 110:1–11

Ohashi M, Ishiyama K, Kusano M, Fukushima A, Kojima S, Hanada A, Kanno K, Hayakawa T, Seto Y, Kyozuka J, Yamaguchi S, Yamaya T (2015a) Lack of cytosolic glutamine synthetase1;2 in vascular tissues of axillary buds causes severe reduction in their outgrowth and disorder of metabolic balance in rice seedlings. Plant J 81:347–356

Ohashi M, Ishiyama K, Kojima S, Konishi N, Nakano K, Kanno K, Hayakawa T, Yamaya T (2015b) Asparagine synthetase1, but not asparagine synthetase2, is responsible for the biosynthesis of asparagine following the supply of ammonium to rice roots. Plant Cell Physiol 56 (4):769–778

Ohashi M, Ishiyama K, Kojima S, Kojima M, Sakakibara H, Yamaya T, Hayakawa T (2017) Lack of cytosolic glutamine synthetase1;2 activity reduces nitrogen-dependent biosynthesis of cytokinin required for axillary bud outgrowth in rice seedlings. Plant Cell Physiol 58:679–690. https://doi.org/10.1093/pcp/pcx022

Ohashi M, Ishiyama K, Kojima S, Konishi N, Sasaki K, Miyao M, Hayakawa T, Yamaya T (2018a) Outgrowth of rice tillers requires availability of glutamine in the basal portions of shoots. Rice 11:31. https://doi.org/10.1186/s12284-018-0225-2

Ohashi M, Ishiyama K, Kusano M, Fukushima A, Kojima S, Hayakawa T, Yamaya T (2018b) Reduction in sucrose contents by downregulation of fructose-1,6-bisphosphatase 2 causes tiller outgrowth cessation in rice mutants lacking glutamine synthetase1;2. Rice 11(1):65. https://doi.org/10.1186/s12284-018-0261-y

Peng B, Kong H, Li Y, Wang L, Zhong M, Sun L, Gao G, Zhang Q, Luo L, Wang G, Xie W, Chen J, Yao W, Peng Y, Lei L, Lian X, Xiao J, Xu C, Li X, He Y (2014) OsAAP6 functions as an important regulator of grain protein content and nutritional quality in rice. Nat Commun 5:4847. https://doi.org/10.1038/ncomms5847

Rameau C, Bertheloot J, Leduc N, Andrieu B, Foucher F, Sakr S (2015) Multiple pathways regulate shoot branching. Front Plant Sci 5(Jan):741. https://doi.org/10.3389/fpls.2014.00741

Sakamoto T, Matsuoka M (2008) Identifying and exploiting grain yield genes in rice. Curr Opin Plant Biol 11(2):209–214. https://doi.org/10.1016/j.pbi.2008.01.009

Sakurai N, Hayakawa T, Nakamura T, Yamaya T (1996) Changes in the cellular localization of cytosolic glutamine synthetase protein in vascular bundles of rice leaves at various stages of development. Planta 200:306–311

Sakurai N, Katayama Y, Yamaya T (2001) Overlapping expression of cytosolic glutamine synthetase and phenylalanine ammonia-lyase in immature leaf blades of rice. Physiol Plant 113:400–408. https://doi.org/10.1034/j.1399-3054.2001.1130314.x

Sasaki T (2005) The map-based sequence of the rice genome. Nature 436(7052):793–800

Sato K (1959) Studies on starch contained in the tissues of Rice Plant VI. On the elongation of upper lateral buds. Jpn J Crop Sci 28(1):30–32. https://doi.org/10.1626/jcs.28.30

Sonoda Y, Ikeda A, Saiki S, von Wiren N, Yamaya T, Yamaguchi J (2003a) Distinct expression and function of three ammonium transporter genes (OsAMT1;1-1;3) in rice. Plant Cell Physiol 44:726–734. https://doi.org/10.1093/pcp/pcg083

Sonoda Y, Ikeda A, Saiki S, Yamaya T, Yamaguchi J (2003b) Feedback regulation of the ammonium transporter gene family AMT1 by glutamine in rice. Plant Cell Physiol 44:1396–1402. https://doi.org/10.1093/pcp/pcg169

Tabuchi M, Sugiyama K, Ishiyama K, Inoue E, Sato T, Takahashi H, Yamaya T (2005) Severe reduction in growth rate and grain filling of rice mutants lacking OsGS1;1, a cytosolic glutamine synthetase1;1. Plant J 42:641–651. https://doi.org/10.1111/j.1365-313X.2005.02406.x

Tabuchi M, Abiko T, Yamaya T (2007) Assimilation of ammonium ions and reutilization of nitrogen in rice (Oryza sativa L.). J Exp Bot 58:2319–2327. https://doi.org/10.1093/jxb/erm016

Takahashi N, Okajima H, Takagi S, Honda T (1956) The mechanism of tiller development in the rice plant. Jpn J Crop Sci 25:74

Tamura W, Hidaka Y, Tabuchi M, Kojima S, Hayakawa T, Sato T, Obara M, Kojima M, Sakakibara H, Yamaya T (2010) Reverse genetics approach to characterize a function of NADH-glutamate synthase1 in rice plants. Amino Acids 39:1003–1012. https://doi.org/10.1007/s00726-010-0531-5

Tamura W, Kojima S, Toyokawa A, Watanabe H, Tabuchi-Kobayashi M, Hayakawa T, Yamaya T (2011) Disruption of a novel NADH-glutamate synthase2 gene caused marked reduction in spikelet number of rice. Front Plant Sci 2:1–9. https://doi.org/10.3389/fpls.2011.00057

Tegeder M (2014) Transporters involved in source to sink partitioning of amino acids and ureides: opportunities for crop improvement. J Exp Bot 65:1865–1878. https://doi.org/10.1093/jxb/eru012

Thomsen HC, Eriksson D, Moller IS, Schjoerring JK (2014) Cytosolic glutamine synthetase: a target for improvement of crop nitrogen use efficiency? Trends Plant Sci 19:656–663

Tobin AK, Yamaya T (2001) Cellular compartmentation of ammonium assimilation in rice and barley. J Exp Bot 52:591–604

Umehara M, Hanada A, Yoshida S, Akiyama K, Arite T, Takeda-Kamiya N, Magome H, Kamiya Y, Shirasu K, Yoneyama K, Kyozuka J, Yamaguchi S (2008) Inhibition of shoot branching by new terpenoid plant hormones. Nature 455(7210):195–200. https://doi.org/10.1038/nature07272

Waldie T, McCulloch H, Leyser O (2014) Strigolactones and the control of plant development: lessons from shoot branching. Plant J 79(4):607–622

Wallsgrove RM, Turner JC, Hall NP, Kendall AC, Bright SWJ (1987) Barley mutants lacking chloroplast glutamine-synthetase – biochemical and genetic-analysis. Plant Physiol 83(1):155–158. https://doi.org/10.1104/pp.83.1.155

Wang MY, Siddiqi MY, Ruth TJ, Glass ADM (1993a) Ammonium uptake by rice roots (I. Fluxes and subcellular distribution of $^{13}NH_4^+$). Plant Physiol 103(4):1249–1258. https://doi.org/10.1104/pp.103.4.1249

Wang MY, Siddiqi MY, Ruth TJ, Glass ADM (1993b) Ammonium uptake by rice roots (II. Kinetics of $^{13}NH_4^+$ influx across the plasmalemma). Plant Physiol 103(4):1259–1267. https://doi.org/10.1104/pp.103.4.1259

Wang Q, Nian J, Xie X, Yu H, Zhang J, Bai J, Dong G, Hu J, Bai B, Chen L, Xie Q, Feng J, Yang X, Peng J, Chen F, Qian Q, Li J, Zuo J (2018) Genetic variations in ARE1 mediate grain yield by modulating nitrogen utilization in rice. Nat Commun 9(1):735. https://doi.org/10.1038/s41467-017-02781-w

Wang J, Wu B, Lu K, Wei Q, Qian J, Chen Y, Fang Z (2019) The amino acid permease 5 (Osaap5) regulates tiller number and grain yield in rice. Plant Physiol 180(2):1031–1045. https://doi.org/10.1104/pp.19.00034

Yabuki Y, Ohashi M, Imagawa F, Ishiyama K, Beier MP, Konishi N, Umetsu-Ohashi T, Hayakawa T, Yamaya T, Kojima S (2017) A temporal and spatial contribution of asparaginase to asparagine catabolism during development of rice grains. Rice 10:1–10. https://doi.org/10.1186/s12284-017-0143-8

Yamaya T, Kusano M (2014) Evidence supporting distinct functions of three cytosolic glutamine synthetases and two NADH-glutamate synthases in rice. J Exp Bot 65:5519–5525

Yamaya T, Oaks A (2004) Metabolic regulation of ammonium uptake and assimilation. In: Amâncio S, Stulen I (eds) Nitrogen acquisition and assimilation in higher plants, Kluwer handbook series of plant ecophysiology, vol 3. Kluwer Academic, Boston, pp 35–63

Yamaya T, Hayakawa T, Tanasawa K, Kamachi K, Mae T, Ojima K (1992) Tissue distribution of glutamate synthase and glutamine synthetase in rice leaves. Plant Physiol 100:1427–1432

Yamaya T, Tanno H, Hirose N, Watanabe S, Hayakawa T (1995) A supply of nitrogen causes increase in the level of NADH-dependent glutamate synthase protein and in the activity of the enzyme in roots of rice seedlings. Plant Cell Physiol 36(7):1197–1204

Yamaya T, Obara M, Hayakawa T, Sato T (1997) Comparison of contents for cytosolic-glutamine synthetase and NADH-dependent glutamate synthase proteins in leaves of japonica, indica, and javanica rice plants. Soil Sci Plant Nutr 43:1107–1112

Yamaya T, Obara M, Nakajima H, Sasaki S, Hayakawa T, Sato T (2002) Genetic manipulation and quantitative-trait loci mapping for nitrogen recycling in rice. J Exp Bot 53:917–925. https://doi.org/10.1093/jexbot/53.370.917

Yang X, Nian J, Xie Q, Feng J, Zhang F, Jing H, Zhang J, Dong G, Liang Y, Peng J, Wang G, Qian Q, Zuo J (2016) Rice ferredoxin-dependent glutamate synthase regulates nitrogen–carbon metabolomes and is genetically differentiated between japonica and indica subspecies. Mol Plant 9(11):1520–1534. https://doi.org/10.1016/j.molp.2016.09.004

Yoshida S (1981) Fundamentals of rice crop science. International Rice Research Institute, Los Baños

Yuan L, Loque D, Ye F, Frommer WB, von Wiren N (2007a) Nitrogen-dependent posttranscriptional regulation of the ammonium transporter AtAMT1;1. Plant Physiol 143(2):732–744

Yuan L, Loque D, Kojima S, Rauch S, Ishiyama K, Inoue E, Takahashi H, von Wiren N (2007b) The organization of high-affinity ammonium uptake in Arabidopsis roots depends on the spatial arrangement and biochemical properties of AMT1-type transporters. Plant Cell 19(8):2636–2652. https://doi.org/10.1105/tpc.107.052134

How Can We Interpret the Large Number and Diversity of ABA Transporters?

Joohyun Kang, Youngsook Lee, and Enrico Martinoia

Contents

Abstract Abscisic acid (ABA) is generally known as the plant stress hormone. Functioning in a wide range of environmental responses, ABA plays a major role in drought tolerance. In addition to inducing stomatal closure during drought stress, ABA promotes suberization of the exodermis and endodermis, which reduces water loss from the root. Furthermore, ABA increases freezing tolerance and has a complex, but not completely understood, role in plant–pathogen interactions. ABA also functions in plant development; for example, ABA is a central player in maintaining seed dormancy. Whereas the enzymatic steps of ABA biosynthesis have been known for some time, our knowledge of ABA receptors and transporters is quite recent. This is due, at least partially, to redundancy among members of both the ABA receptor and transporter families. Many transporters from different

J. Kang (✉) and E. Martinoia
Division of Integrative Bioscience and Biotechnology, POSTECH, Pohang, Republic of Korea

Institute of Plant and Microbial Biology, University Zurich, Zurich, Switzerland
e-mail: azka@postech.ac.kr

Y. Lee
Division of Integrative Bioscience and Biotechnology, POSTECH, Pohang, Republic of Korea

© Springer Nature Switzerland AG 2020
Progress in Botany (2021) 82: 233–258, https://doi.org/10.1007/124_2020_43,
Published online: 28 December 2020

transporter families cooperate to transport ABA. The weak but distinct phenotypes described for the different loss-of-function mutants indicate that each of these transporters plays a specific role and, at least under a given condition or in a specific tissue, they are not completely redundant. However, for each function described so far, delivery of ABA at the target site requires the activity of several different ABA transporters. This strategy may ensure that ABA is transported to the correct target even if one of the transporters is nonfunctional or that plants can transport ABA under a given condition via several routes.

Keywords Abscisic acid, Drought, Freezing tolerance, Plant pathogen, Seed germination, Transport

Abbreviations

ABA	**Ab**scisic **a**cid
ABA-GE	ABA **g**lucose **e**ster
ABC	**A**TP-**b**inding **c**assette
AIT	**A**BA-**i**mporting **t**ransporters
AtBG1	Arabidopsis thaliana β-glucosidase
AWPM-19	**A**BA-induced **w**heat **p**lasma **m**embrane polypeptide-**19**
Cvi	**C**ape **V**erde **i**slands
DTX/MATE	**D**etoxification **e**fflux **c**arriers/**m**ultidrug and **t**oxic **c**ompound **e**xtrusion
GUS	β-Glucosidase
LATD/NIP	**Lat**eral root **d**efective/**n**umerous **i**nfection threads, **p**olyphenolics
Lr34res	The resistant Lr34 allele
Lr34sus	The susceptible Lr34 allele
NBF	**N**ucleotide **b**inding **f**old
nced3	**N**ine-**c**is-**e**poxycarotenoid **d**ioxygenase 3
NPF	**N**itrate transporter1/**p**eptide transporter **f**amily
OE	**O**ver**e**xpression
ost1	**O**pen **st**omata 1
PM1	**P**lasma **m**embrane protein**1**
PP2C	**P**hos**p**hatase **2C**
PYR/PYL/RCAR	**Py**rabactin **r**esistance1/**PY**R1-**l**ike/**r**egulatory **c**omponents of **A**BA **r**eceptor
TMD	**T**rans**m**embrane **d**omain

1 Introduction

The phytohormone abscisic acid (ABA), a sesquiterpenoid with a 15-carbon ring, has a critical role in a large range of physiological functions in a variety of plant tissues over the entire life span of plants. During seed development, ABA prevents vivipary and contributes to dormancy and desiccation tolerance (Robertson 1955; Meurs et al. 1992; Schwartz et al. 1997; Gubler et al. 2005). ABA also inhibits seedling development (Lopez-Molina et al. 2001). In Arabidopsis the hormone influences root architecture by preserving primary root growth and inhibiting the development of lateral roots at the meristem activation checkpoint (Zhang et al. 2010). In the shoot, ABA's major role is to regulate stomatal aperture and thereby modulate the hydraulic conductance of plants (Kim et al. 2010; Munemasa et al. 2015; Merilo et al. 2018). ABA also plays a role in reproductive organ development, with possible functions in male sterility and fruit development (Peng et al. 2006; Oliver et al. 2007).

Furthermore, ABA is a key mediator of physiological responses to multiple environmental stresses, such as drought, cold, and pathogen attack. In general, the earliest adaptive responses of plants to various stresses are induced by an increase in transcript levels of major ABA biosynthetic genes, which in turn leads to accumulation of cellular ABA and triggers the expression of multiple stress-responsive genes (Inuchi et al. 2001; Tan et al. 2003). These molecular processes have been widely studied in the model plant *Arabidopsis thaliana* (L.) Heynh., where most of the ABA signaling mechanisms have been elucidated.

Abiotic stress conditions, such as water stress, increase ABA levels and thus initiate signaling pathways that induce multiple cascades of molecular and cellular responses, including, in the case of water stress, expression of stress-related genes, cellular ROS accumulation, and stomatal closure. Stomatal closure and ROS accumulation also serve as a mechanism for pathogen defense (Melotto et al. 2006; Desikan et al. 2008), thereby providing a platform for crosstalk between biotic and abiotic stress responses. However, ABA can also directly activate protein kinases, as it does in the activation of SLAC, a guard cell anion channel, leading to stomatal closure (Raghavendra et al. 2010).

For the past several decades, ABA studies have focused on the hormone's biosynthesis and signaling pathways. Almost every ABA biosynthetic enzyme was identified quite some time ago (Nambara and Marion-Poll 2005). However, ABA's receptors (Park et al. 2009; Ma et al. 2009) and transporters (references below) were identified only recently. One of the main reasons for this delay is that, unlike the ABA biosynthetic enzymes, the protein families involved in perception and transport are highly redundant.

Over ten different proteins of the PYR/PYL/RCAR (**py**rabactin **r**esistance1/**PYR**1-like/**r**egulatory **c**omponents of **A**BA **r**eceptor) family act as ABA receptors in different tissues and at various developmental stages (Raghavendra et al. 2010; Gonzalez-Guzman et al. 2012). The ABA transporters that have been identified to date belong to several different protein families – the ABC (**A**TP-**b**inding **c**assette) superfamily, the NPF (**n**itrate transporter1/**p**eptide transporter **f**amily) family, and the DTX/MATE (**d**etoxification **e**fflux **c**arriers/**m**ultidrug and **t**oxic **c**ompound **e**xtrusion) family. ABA transporter mutants often exhibit no apparent phenotypes under normal growth conditions, but show distinct phenotypes under a specific stress condition (e.g., pathogen exposure or drought) or during a specific developmental stage. Thus, ABA transporters may have specific functions and may act in concert with other ABA transporters but also be partially redundant with them.

Why did plants evolve to have such diverse ABA transporters belonging to so many different protein families? Are the physiological functions of these ABA transporters highly redundant or not? The answers to these questions hinge on further studies of mutants with knockouts of multiple ABA transporters. Here, we review the history of ABA transporter studies, the properties of the protein families involved in ABA transport, and the functions of ABA transporters in biological processes. We also propose some hypotheses to explain why plants have evolved so many ABA transporters.

2 Where Does ABA Come from, and Where Is It Going?

In general, ABA meets the definition of a hormone, in that it is synthesized in one place and acts at other places within the plant. However, ABA is synthesized in several places, not just one (Nambara and Marion-Poll 2005), and, as will be described below, there is at least one cell type that relies on cell intrinsic ABA biosynthesis. Furthermore, ABA, like other plant hormones, is often present in conjugated forms. Some of these conjugates can be hydrolyzed under specific conditions (Lee et al. 2006; Xu et al. 2012), releasing active ABA and triggering a signaling cascade within the cell.

Studies of ABA transport and signaling historically focused on how ABA reaches guard cells and induces stomatal closure. Early findings suggested that roots, leaves, and even guard cells can synthesize ABA (Zeevaart and Creelman 1988). Initial attempts to identify the ABA receptor suggested that the receptor was located in the plasma membrane. This implicated that ABA would be perceived at the apoplastic face of the guard cells and had not to be transported into guard cells (Zeevaart and Creelman 1988). Subsequent work revealed that the ABA receptor is a member of the PYR protein family (Park et al. 2009; Ma et al. 2009). These receptors are localized in the cytosol. Therefore this discovery changed the view where ABA is perceived and pointed out that ABA has to cross the plasma membrane to reach its receptor.

As water stress is initially perceived in the root, the ABA required for stomatal closure was hypothesized to be produced in roots and translocated into shoots before being delivered to the guard cells (Zeevaart and Creelman 1988; Jackson 1993). To test this hypothesis, Blackman and Davies (1985) performed split root experiments with maize (*Zea mays* L.). One part of the root system was well watered, while the other was subjected to a reduced water potential. This experimental design allows the shoot to remain unstressed while half of the root is exposed to water stress. The authors reported that plants subjected to this treatment had smaller stomatal apertures than control plants with both parts of their root systems well-watered. However, the authors could not detect any difference in ABA concentrations in the shoot between the treatment and control plants. Further experiments led to the hypothesis that a continuous supply of cytokinins might be the factor allowing maximal stomatal opening.

Other evidence challenges the hypothesis that root-to-shoot transport of ABA or cytokinins is required to control stomata opening in response to drought. For example, such a mechanism cannot account for the fast stomatal reaction observed in tall trees, and an alternative mechanism involving a hydraulic signal has been postulated several times (Christmann et al. 2007). At least for the plants investigated so far, there is no evidence that a chemical signal has to be transported from the root to the shoot to induce stomatal closure. Evidence to the contrary is mainly provided by grafting experiments. For example, Holbrook et al. (2002) grafted tomato (*Solanum lycopersicum* L.) shoots able to produce ABA onto mutant roots unable to do so and showed that stomatal conductance corresponded to the shoot genotype. This result was further confirmed by grafting experiments using split roots or a pressure chamber.

In a detailed study, Christmann et al. (2007) took advantage of a series of Arabidopsis mutants to gain insight into the question of the source of ABA involved in water stress responses in the shoot. These authors also confirmed that ABA is not transported from the root to the shoot; instead, shoot-derived ABA is the signal leading to stomatal closure. They also showed that a sudden change in root water availability results in a rapid change in shoot turgor pressure, which induces rapid synthesis of ABA. These experiments support the hypothesis that a hydraulic signal is the primary cause for the increase in ABA content and the resulting stomatal closure in the shoot.

Stomata close in response not only to water shortage in the soil but also to changes in atmospheric humidity. As mentioned above, guard cell intrinsic ABA synthesis has been postulated for a long time. More recently, Bauer et al. (2013) confirmed this hypothesis and showed that guard cell intrinsic ABA synthesis is important for protecting plants against changes in atmospheric water potential. Although wild-type Arabidopsis plants did not wilt when exposed to dry air, the ABA biosynthesis mutant *aba3-1* did. Rescuing ABA biosynthesis specifically in guard cells of this mutant restored the capacity to close stomata in dry air and the plants did not wilt.

A recent, more comprehensive study compared the corresponding mutants of Arabidopsis, pea (*Pisum sativum*), and tomato and also examined the relative importance of guard cell autonomous and shoot-dependent processes (Merilo et al. 2018). Besides highlighting several important regulatory mechanisms, the authors showed that the pools of ABA synthesized by guard cells and by phloem companion cells are redundant and that their relative importance may be plant specific. Therefore, under certain circumstances, transport of ABA from the leaf tissue may also function in stomatal closure in response to dry air.

Less information is available for other cases that may involve ABA transport. This may be either because the sites of ABA synthesis and action are too close together or because classical tools have not been able to determine whether ABA must be transported or not. One such case concerns seeds and seed germination. Most plants, with the exception of those growing in tropical climates, tightly control seed germination to avoid germination under unfavorable conditions. The mechanisms regulating this process have been studied extensively, and it is now generally accepted that ABA, which represses germination, and gibberellins, which promote germination, act antagonistically to regulate the process (Holdsworth et al. 2008). Dormant seeds typically contain high ABA concentrations, while seeds with impaired ABA production germinate precociously (Robertson 1955). Seed germination is a complex process that varies at least slightly among plants. In Arabidopsis, the embryo is surrounded by an active endosperm and the testa, a dead outer layer of maternal origin. In Arabidopsis, the endosperm is responsible for most or all of the germination-repressive activity (Bethke et al. 2007). Using a seed coat bedding assay, Lee et al. (2010) demonstrated that ABA is released from the endosperm of dormant seeds and is the factor inhibiting embryo development. Hence, export of ABA from the endosperm and uptake of this hormone into the embryo must occur during seed dormancy.

Transport of ABA is likely or may occur in other ABA-dependent processes, but the routes of ABA movement remain largely unknown. For example, under water stress, more suberin is also deposited in the endodermis in an ABA-dependent manner, and lateral root primordia are repressed, but it is not known whether the ABA that regulates these processes is de novo synthesized in situ or transported from elsewhere (Deak and Malamy 2005). ABA also decreases the number of hypodermal passage cells, non-suberized cells of the exodermis (Liu et al. 2019). Finally, in the development of nitrogen-fixing nodules in the Fabaceae, ABA regulates nodule production and probably also bacteroid number (Ding et al. 2008). An ABA transporter involved in this process has been recently identified; however, we still do not have much insight into the source of the ABA delivered to the nodule meristem.

3 Which Proteins Are Involved in ABA Transport?

ABA transport is catalyzed by a large number of transporters belonging to several families (Kuromori et al. 2018). The first class of transporters implicated in ABA transport was the ABC protein family. Subsequently it was demonstrated that some members of the NPF and MATE families are also able to transport ABA. Here, we present some characteristics of the transporter families involved in ABA transport. We have only limited knowledge about the AWPM-19 (**A**BA-induced **w**heat **p**lasma **m**embrane polypeptide-**19**) family; therefore we do not discuss this membrane protein family here.

3.1 The ABC Superfamily

ABC transporters are among the oldest transport proteins and are present in all living organisms from bacteria to humans. These transporters are classified as pumps, since they are directly activated by ATP hydrolysis (Kang et al. 2011; Hwang et al. 2016). The Arabidopsis genome encodes 131 ABC transporters, and a similar or higher number has been reported for all plants analyzed so far. This number is far higher than that for humans (49) and most other organisms.

A functional ABC transporter contains two cytosolic (**n**ucleotide **b**inding **f**old, NBF) and two transmembrane (**t**ransme**m**brane **d**omain, TMD) domains. These domains can be part of one large protein, a so-called full-size ABC transporter, or part of a half-size transporter, which contains only one NBF and TMD, or exist as single proteins corresponding to NBFs and TMDs (bacterial-like). In all cases, however, the proteins have to be assembled to contain two NBFs and two TMDs to act as a functional transporter.

All ABC transporters contain an ABC signature flanked by Walker A and Walker B domains, which are responsible for ATP binding on each NBF. In plants, ABC transporters have been shown to transport a plethora of compounds, such as lipids, secondary compounds, heavy metals, and many hormones. In most cases, a single ABC transporter is able to transport several to many chemically unrelated compounds. All the ABC transporters that have been described as being involved in ABA transport are full- and half-size ABC proteins of the ABCG class. This class has a so-called reverse arrangement, in which the first TMD is located at the N-terminus of the protein, followed by an NBD. So far, AtABCG25, AtABCG30, AtABCG31, and AtABCG40 in Arabidopsis, Lr34 in wheat (*Triticum aestivum* L.), and MtABCG20 in Medicago have been shown to transport ABA (Table 1).

3.2 NPFs (NRT1/PTR Family)

The name NPF (**n**itrate **p**eptide **f**amily) indicates that proteins of this family were originally identified as nitrate transporters (NRTs) and peptide transporters (PTRs) (Longo et al. 2018). The first member of this protein family was isolated in Arabidopsis in a screen for resistance to chlorate, a toxic nitrate analogue. This transporter was subsequently shown to be a nitrate uptake transporter (Tsay et al. 1993). Most NPFs are proton/substrate symporters, but some are facilitators, bidirectional transporters, or proton-coupled potassium antiporters.

This class of transporters has 12 transmembrane α-helices and is part of the large major facilitator superfamily (Léran et al. 2014; Longo et al. 2018). The N-termini and C-termini are both located in the cytosol. Like ABC transporters, NPFs have been identified in all living organisms, from bacteria to fungi to humans, and, also like ABC transporters, the number of NPFs is much higher in plants than in most other organisms. Plant NPFs have been divided into eight subclasses (Léran et al.

Table 1 List and properties of ABA transporters described in the text

Common name	Protein family	Species	Verification of transport activity	Vector	Main function	Major expression tissues	Reference
AtABCG25	ABC protein family	*Arabidopsis thaliana*	Sf9 insect cells	Exporter	ABA accumulation in the apoplastic area around guard cells by exporting biosynthetic ABA from vasculature	Vasculature	Kuromori et al. (2010), Kang et al. (2015)
AtABCG30	ABC protein family	*Arabidopsis thaliana*	Using knockout mutants	Importer	Embryo germination inhibition (dormancy maintenance) by ABA uptake in embryo	Embryo	Kang et al. (2015)
AtABCG31	ABC protein family	*Arabidopsis thaliana*	Using knockout mutants	Exporter	Embryo germination inhibition (dormancy maintenance) by ABA efflux from endosperm	Endosperm	Kang et al. (2015)
AtABCG40	ABC protein family	*Arabidopsis thaliana*	Yeast BY2 cells Arabidopsis protoplast	Importer	Regulation of stomatal aperture under drought condition Embryo germination inhibition (dormancy maintenance) by ABA uptake in embryo	Guard cells Embryo	Kang et al. (2010, 2015)
Lr34	ABC protein family	*Triticum aestivum*	Yeast	Importer	Contribution to fungal resistance	Unknown	Krattinger et al. (2019)
MtABCG20	ABC protein family	*Medicago truncatula*	BY2 cells BY2 cell-derived vesicles	Exporter	Positive regulation of lateral root primordium formation negative effect on the development of nodule primordia Facilitating germination	Vascular bundles and at the sites of lateral root primordium formation Hypocotyl–radicle transition zone of embryos	Pawela et al. (2019)
AIT1/ AtNRT1.2/ AtNPF4.6	NPFs (NRT1/ PTR)	*Arabidopsis*	Yeast	Importer	Regulation of stomatal aperture in inflorescence stems	Vascular tissues in inflorescence stems, leaves, and roots	Kanno et al. (2012)

AtDTX50	MATE	*Arabidopsis*	*E. coli*, Xenopus oocytes, Arabidopsis protoplast	Exporter	Negative regulation of drought tolerance	Guard cells Vascular tissues of leaves, roots, and germinating seeds	Zhang et al. (2014)
OsPM1 (plasma membrane protein1)	AWPM-19 like family	*Oryza sativa*	Yeast	Importer	Regulation of drought tolerance and seed germination	Vasculature, guard cells, root tip, mature embryo	Yao et al. (2018)

2014), but assignment of an NPF to a particular subclass does not necessarily predict its substrate(s). Like ABC transporters, NPFs transport a plethora of substrates such as nitrate, peptides, glucosinolates, auxin, jasmonic acid, and ABA. The ABA importer AIT (**ABA-importing transporters**) was the first and so far only member characterized in detail (Kanno et al. 2012). However, ABA transport activity has been demonstrated or postulated for an additional 11 NPFs (Chiba et al. 2015).

3.3 MATEs

MATE transporters were originally identified as multidrug and toxic compound extrusion proteins conferring resistance against drugs in bacteria (Morita et al. 1998). In Arabidopsis, these transporters are also called DTX proteins. As in the case of ABC transporters and NPFs, they have been identified in most living cells, and plants contain a far higher number of these transporters compared to other organisms. For Arabidopsis, 56 MATEs have been reported, whereas humans have only 2 (Takanashi et al. 2014; Upadhyay et al. 2019).

Most MATEs contain 12 α-helices, with 2 long extensions at the N-terminus and C-terminus, but there are a few exceptions where more or fewer α-helices have been predicted. Furthermore, most MATEs act as proton antiporters, extruding compounds from the cytosol into either the apoplast or the vacuole. MATEs transport a wide variety of compounds, such as toxic abiotic compounds, a large range of plant secondary metabolites (e.g., flavonoids and alkaloids), citrate, salicylic acid, auxin, and ABA. It is also likely that a given MATE can transport multiple chemically unrelated compounds. Recently it was shown that two MATEs act as chloride channels and have a strong impact on stomatal movement (Zhang et al. 2017). AtDTX50 is so far the sole ABA transporter identified in this family (Zhang et al. 2014).

4 What Is the Role of ABA Transporters in Plants?

4.1 Modulation of Hydraulic Conductance in Sessile Plants

Our views on ABA transport have been changed by the results showing that during water stress ABA is not transported from the root to the shoot and subsequently to guard cells but rather is redistributed within a leaf (Christmann et al. 2007). However, even in this case, ABA has to move from the phloem parenchyma cells, the site of biosynthesis, to the guard cells. Since guard cells are not connected to mesophyll or phloem parenchyma cells via plasmodesmata, ABA must cross membranes to make this journey. Older work postulated that an ABA importer on the guard cells is not required, because apoplastic pH was reported to be 5.5 to 6.0. At this pH, a substantial proportion of the ABA (pK_a 4.75) is protonated and can diffuse across membranes. However, it has now been shown that under water stress conditions the pH of the apoplastic space increases to 7.2, and consequently the proportion of protonated, freely diffusible ABA decreases. Under these conditions, only a small fraction of ABA can enter guard cells by simple diffusion, implying that guard cells need a transporter to import ABA from the apoplast to the cytosol. By contrast, for ABA export from the cytosol to the apoplast, a transporter has always been assumed to exist, since at a cytosolic pH of 7.5 only negligible amounts of uncharged, freely diffusible ABA is present.

Two plasma membrane-localized transporters of the ABC protein family were the first ABA transporters to be identified. One, AtABCG25, catalyzes ABA export from the cytosol to the apoplast (Kuromori et al. 2010); the other, AtABCG40 (Kang et al. 2010), catalyzes the uptake of ABA from the apoplast into the cytosol. Both transporters are expressed in roots and shoots. While in shoots AtABCG25 is localized predominantly to the vasculature of leaves, AtABCG40 is mainly expressed in guard cells. ABA transport by these proteins has been studied using membrane vesicles isolated from insect cells overexpressing AtABCG25 and yeast and BY2 cells expressing AtABCG40. Transport was strictly ATP-dependent, and both transporters exhibited high and specific affinity for ABA, with K_Ms of 0.26 and 1 μM for AtABCG25 and AtABCG40, respectively.

Thermal imaging of Arabidopsis plants overexpressing AtABCG25 revealed that these plants were warmer than the wild type, in line with increased ABA release and decreased stomatal aperture. In the case of AtABCG40, thermal imaging revealed that *atabcg40* mutant plants were cooler than the corresponding wild-type plants under osmotic stress conditions or when exogenous ABA was supplied. This result indicated that the stomata of the mutant could not close efficiently in response to elevated ABA. Indeed, measurement of stomatal apertures revealed that *atabcg40* mutant plants were less sensitive to ABA-induced closure. Transformation of the mutant plants with AtABCG40 completely complemented the stomatal defects. These results demonstrate that AtABCG25 acts as an ABA exporter and AtABCG40 as an ABA importer. However, the phenotypes of *atabcg40* mutant plants were quite mild and not nearly as strong as those described for ABA biosynthesis mutants. This raised the question of whether diffusion could play a role or additional ABA transporters are present.

To identify additional ABA transporters, Kanno et al. (2012) devised an elegant screening method. The authors took advantage of the fact that some of the PYR/PYL/RCAR ABA receptors interact with a group of PP2C-type protein phosphatases in the presence of ABA. Using a modified yeast two-hybrid system, they transformed a yeast strain with these two proteins, ABA receptor, PYR1 and ABI1 or HAB1, PP2C-type phosphatase (Park et al. 2009; Ma et al. 2009). In the presence of 0.1 μM ABA, there was no interaction between the partners because the ABA could not enter the yeast cells. The rationale of the screen was that a transporter would allow efficient uptake of ABA into the yeast cells, leading to an interaction between the receptor and the phosphatase. A screen of two cDNA libraries led to the identification of four members of the NRT1/PTR (NPF) family that facilitated interaction between the partners. One of these candidates had previously been characterized as the low-affinity nitrate transporter NRT1.2 (Huang et al. 1999) and exhibited the most pronounced ABA transport activity when expressed in yeast and insect vesicles. This transporter, also called AIT1 by the authors, was characterized in more detail. The protein is localized in the plasma membrane and catalyzes the uptake of ABA with a K_M of 5 μM.

Interestingly, although this transporter is mainly localized to the vasculature of leaves and inflorescences and not in guard cells, thermal imaging of *ait1* mutants showed that inflorescence stems but not leaves were cooler than in the wild type. This result indicates that AIT1 regulates stomatal closing in inflorescence stems but not in leaves. Changes in thermal imaging in line with increased ABA uptake activity were observed in leaves only when AIT1 was overexpressed. The authors hypothesize that AIT1 acts mainly as importer at guard cells in inflorescence stems and in the cells of the vasculature resembles the localization of ABA biosynthetic enzymes to avoid loss of ABA from these cells.

The three other AITs identified in the modified yeast two-hybrid screen were not characterized in detail. In a follow-up paper, the same laboratory looked systemically for NPFs able to transport ABA and extended the method to search for transporters of other hormones, such as gibberellins and jasmonic acid-isoleucine (Chiba et al. 2015). Eleven additional NPFs emerged as candidate ABA transporters from this study, but their physiological roles merit further examination.

A MATE protein has also been identified as an ABA exporter. In an effort to understand the function of MATE/DTX proteins, the laboratory of Professor Luan produced homozygous mutants for a large number of Arabidopsis *MATE* genes. They observed that the *dtx50* mutant was smaller and yellower than the wild type when grown in soil (Zhang et al. 2014). To obtain clues about the substrate(s) of DTX50, the authors screened a multitude of compounds to determine if any would affect the phenotype of the *dtx50* mutant. They observed that the growth of *dtx50* mutants was more severely inhibited than that of the wild type and other mutants in the presence of ABA. This result suggested that *dtx50* mutants might accumulate more ABA and struggle to release it. Indeed, the *dtx50* mutants had increased ABA and increased expression of ABA marker genes.

Using *E. coli* and Xenopus as heterologous expression systems, as well as wild-type and *dtx50* Arabidopsis protoplasts, Zhang et al. (2014) convincingly showed that AtDTX50 is an ABA exporter exhibiting high specificity for the physiological enantiomer of ABA. As expected from the *dtx50* phenotype, DTX50 is localized to the plasma membrane. Expression of DTX50 was mainly observed in veins and guard cells. The guard cell localization was quite surprising, but is consistent with the observation that *dtx50* plants were more drought tolerant and that their stomata closed faster. Apparently, release of ABA from guard cells is important for fine-tuning the regulation of stomatal aperture.

All the work we have described so far used Arabidopsis as a model plant. Much less is known about stomatal regulation in monocotyledons. In a screen for drought-induced genes in rice, Yao et al. (2018) identified OsPM1 (**p**lasma **m**embrane protein**1**), a membrane protein belonging to the AWPM-19 family. It was already known that genes of this small family are strongly induced by abiotic stress (Koike et al. 1997; Rerksiri et al. 2013; Chen et al. 2015). OsPM1 is mainly expressed in mature embryos, vasculature, and guard cells and is highly induced by ABA. By measuring fluorescence resonance energy transfer in yeast expressing OsPM1, the authors showed that OsPM1 exhibits ABA uptake activity. Physiological experiments revealed that OsPM1 plays an important role in tolerating drought stress, since *OsPM1*-RNAi plants had a lower survival rate and OsPM1-overexpression (OE) plants had a higher survival rate compared to the wild type. Measurements of water loss and stomatal aperture (which is very difficult in rice) were in line with the survival rates observed: RNAi plants lost more water, while OE plants retained more water. Furthermore, no differences in stomatal apertures could be detected under control conditions, but significantly more stomata were closed in OE plants compared to the wild type in the presence of ABA. By contrast, RNAi plants had significantly more open stomata. All these results indicate that OsPM1 is an ABA transporter required to regulate stomatal aperture and hence drought resistance in rice.

A study on AtABCG22 is worth mentioning in this context (Kuromori et al. 2011). AtABCG22 is highly expressed in guard cells and is required for stomatal regulation. Mutant plants lacking this transporter lose much more water and are more susceptible to drought. However, AtABCG22 had no detectable ABA transport activity, and furthermore, double mutants of *atabcg22* combined with the biosynthesis mutant *nced3* (**n**ine-**c**is-**e**poxycarotenoid **d**ioxygenase 3) or with *ost1* (**o**pen **st**omata 1) had enhanced ABA-related phenotypes. These results, together with other physiological observations presented in this work, preclude that AtABCG22 acts as an ABA transporter. Instead, the protein may either transport an unidentified factor that regulates stomatal aperture or be involved in an ABA-independent pathway of stomatal regulation.

Merilo et al. (2015) used gas exchange assays to study the roles of different ABA-related genes and ABA transporters in stomatal regulation. Under reduced air humidity, the only mutant that displayed a gas exchange phenotype was *atabcg22*, which, as mentioned above, lacks an ABC protein not involved in ABA transport. In response to exogenous ABA, *atabcg40* and *ait1* had slightly, but not significantly, longer ABA response half-times and a slower ABA-induced decrease in stomatal conductance.

These results can be interpreted in several ways. The large number of ABA importers and exporters and the relatively subtle phenotypes of loss-of-function ABA transporter mutants indicate that there is considerable redundancy among these proteins. Hence, multiple-knockout mutants should be produced to get a clearer picture. Another reason Merilo et al. (2015) might not have detected strong phenotypes is that their experiments were relatively short-term. As shown for the stomatal reaction to changes in relative humidity, the internal ABA stores of guard cells may be sufficient for the initial response. A third point is that, in our experience, plant growth conditions have a strong effect on the reactivity of guard cells. Therefore, different results may be obtained in different laboratories as a function of their growth conditions. This may also explain why some phenotypes have not been reproduced in other laboratories.

4.2 Pathogen Resistance

Numerous reports have dealt with the role of ABA in plant–pathogen interactions (e.g., Cao et al. 2011). Intriguingly, ABA can have opposite effects depending on the situation: it can either promote or inhibit the interaction between the pathogen and the plant. The effect of ABA may depend on the nature of the pathogen, the type of affected tissue, or the developmental stage of the plant, but many questions remain. One of the few genes conferring sustainable pathogen resistance in wheat is the full-size ABCG transporter Lr34res (the resistant Lr34 allele). This allele evolved from the ancestral, susceptible allele (Lr34sus) by two gain-of-function mutations after wheat domestication (Krattinger et al. 2013).

To learn more about the mechanism underlying this important trait, Krattinger et al. (2019) analyzed the transcriptomes of plants expressing the resistant and susceptible forms of the Lr34 transporter. They observed that many ABA-related genes were highly upregulated in plants expressing Lr34res, mimicking to a large extent ABA-dependent stress reactions. To verify that this deregulated expression of ABA-dependent genes has a physiological impact, Krattinger et al. (2019) performed drought stress and stomatal aperture experiments and established that Lr34res plants lost less water and had a reduced stomatal conductance compared to Lr34sus plants. Krattinger et al. (2019) also investigated whether Lr34res could act as an ABA transporter.

Indeed, Lr34res and also, surprisingly, Lr34sus were able to transport ABA in Lr34-expressing yeast cells. However, although immunoblots clearly showed the presence of the Lr34res protein, no Lr34sus protein could be detected. Furthermore, ABA accumulation assays using Lr34-expressing rice seedlings showed that Lr34res-expressing seedlings accumulated more ABA than wild-type and Lr34sus-expressing seedlings. Therefore, although both forms can transport ABA in a heterologous system (yeast), Lr34res is probably the only form that performs this function in plants because it is the only one present in detectable amounts. Nevertheless, exactly how Lr34res-mediated ABA transport confers resistance is an open question. The tissue distribution of Lr34 is still unknown, and the ABA levels in leaves of Lr34res-expressing plants are not changed. Hence, it is tempting to speculate that it is ABA redistribution, not increased levels of the hormone, that leads to pathogen resistance. The ability to visualize ABA redistribution may clarify why ABA renders plants more susceptible to pathogens in some cases and more resistant in others.

Phylogenetic analyses have revealed that there are many homologs of the ABA transporter AtABCG40 in Arabidopsis and other organisms (Hwang et al. 2016). So far, these homologs have been shown to transport of two categories of chemicals: ABA and defense molecules. AtABCG34 and AtABCG36 transport chemicals important in defense against pathogens (Khare et al. 2017; Stein et al. 2006; Lu et al. 2015; Matern et al. 2019), AtABCG30 transports ABA (Kang et al. 2015), and AtABCG31 transports ABA (Kang et al. 2015) and possibly some pathogen defense chemicals (Cho et al. 2019). Indeed, Cho et al. (2019) suggested that many of the full-size ABCGs in plants have co-evolved with the ABCGs of their oomycete pathogens in a long-term arms race.

We can imagine that some of the many AtABCG40 homologs in plants are involved in pathogen defense indirectly via ABA transport, while others evolved to directly transport different substrates for defense against specific pathogens. Future research on the identity of the substrates of these transporters will clarify the many questions on the role of ABA and ABA transporters in plant pathogen defense.

4.3 Regulation of Seed Development and Germination

As mentioned above, seed dormancy depends on a continuous supply of ABA from the endosperm to the embryo. Furthermore, some reports suggest that ABA must also be exported from the embryo to allow seed germination (Holdsworth et al. 2008). All of the studies we described earlier in the section on drought stress and guard cells have also addressed the issue of whether seed germination is affected, but with the exceptions detailed below, none has analyzed this aspect in further detail.

In the case of AtABCG25 and DTX50, the seeds of loss-of-function mutant plants were more sensitive to ABA; their germination and early seed establishment were more severely limited by an exogenous application of ABA than were seeds from wild-type plants (Kuromori et al. 2010; Zhang et al. 2014). By contrast, AIT1 and ABCG40 loss-of-function mutants were less sensitive to exogenously applied ABA compared to the corresponding wild type (Kang et al. 2010; Kanno et al. 2012). These results demonstrate that at least some of the same ABA transporters are involved in drought stress prevention as well as in germination, but the results presented could not explain exactly how these transporters act.

To shed light on this question, Kang et al. (2015) sought to identify ABC transporters that are responsible for the export of ABA from the seed coat and uptake into the embryo. First, the authors analyzed the expression of all 43 ABCG genes from Arabidopsis and established that 10 are highly expressed in seeds. Screening the corresponding mutant lines revealed that four mutants exhibited an altered germination pattern. Second, to establish if these ABCG transporters are more likely involved in ABA release from the seed coat (testa and endosperm) or ABA uptake into the embryo, Kang et al. (2015) separated the seed coat from the embryo and performed a qPCR analysis. Concomitantly the authors generated transgenic plants expressing promoter GUS (β-glucuronidase) reporter gene, *uidA* constructs. The two methods demonstrated that AtABCG25 and AtABCG31 are localized mainly to the endosperm, while AtABCG30 and AtABCG40 are localized mainly to the embryo. These results suggest that AtABCG25, which had already been characterized as an ABA efflux transporter, and AtABCG31 release ABA from the endosperm, whereas AtABCG30 and AtABCG40 import ABA into the embryo (Kang et al. 2015).

To gain insight into the role of these transporters in seed germination, the authors used the seed coat bedding assay established by Lee et al. (2010). When isolated embryos of the *aba2-1* mutant were placed on seed coats of wild-type seeds, germination was completely inhibited. By contrast, when placed on *atabcg31* and *atabcg25* seed coats, *aba2-1* embryos germinated, indicating that AtAGCG31 and AtABCG25 release ABA, and consequently the seed coats of mutants lacking these proteins release less ABA and have an impaired embryo germination inhibitory effect.

To analyze the function of AtABCG30 and AtABCG40, the embryos of mutants lacking these proteins and wild-type embryos were incubated on Cvi (Cape Verde islands) seed coats bedding. These seed coats release more ABA than those of the Columbia ecotype. In line with the expression analysis, germination of wild-type embryos was strongly inhibited, whereas the *atabcg30* and *atabcg40* embryos germinated, indicating that less ABA was taken up by these embryos. These results indicate that AtABCG30 and AtABCG40 import ABA into the embryo.

Finally, transport experiments confirmed that AtABCG25 and AtABCG31 catalyze ABA export from the seed coat (testa and endosperm), whereas AtABCG30 and AtABCG40 act as embryo-localized importers. However, and similar to ABA transport under drought stress, it can be estimated that for both ABA import and export the respective pairs of transporters are responsible for only about 50–60% of the total ABA transport activity. As the NPFs as well as the MATE DTX50 have been shown to also be localized to seeds, they may mediate at least part of the remaining transport activity. A small part of ABA import may also occur by diffusion.

A novel aspect of ABA transport and seed germination was recently reported by Pawela et al. (2019). The authors characterized in detail an ABC transporter, MtABCG20, from *Medicago truncatula* that was strongly induced when the plants were exposed to polyethylene glycol, which mimics water stress, and ABA. They showed that this transporter is localized to the plasma membrane, and, by introducing MtABCG20 into tobacco BY2 cells, they observed that ABA was released much faster by these cells than by those transformed with an empty vector. Furthermore, ABA transport could also be observed in vesicles isolated from these cells. These results indicate that MtABCG20 is an ABA exporter. Using promotor-GUS constructs, Pawela et al. (2019) demonstrated that MtABCG20 is localized to both roots and seeds. *MtABCG20* expression increased strongly after imbibition, and the expression level remained high for more than 3 days if the seeds were kept at 4°C, while the transcript level declined to less than 50% when the seeds were kept at 24°C.

Germination of *mtabcg20* was inhibited to a greater extent by ABA than was the wild type. To establish why this was so, the authors performed a detailed analysis on the site of expression. They observed that in this case the MtABCG20 ABCG transporter was localized to a region that had not previously been reported to contain a transporter, the hypocotyl–radicle transition zone. This is the zone where embryo elongates and is of central importance for germination. The exact mechanism by which embryo-localized ABA inhibits seed germination remains unclear, but apparently reduction of this ABA pool at the specific hypocotyl/radicle elongation zone is a prerequisite for efficient germination. This is in line with the observation of the authors that the ABA concentration in the embryo axis of *mtabcg20* plants is more than double that in the wild type. This result is supported by the observation that much higher transcript levels of a highly ABA-induced PP2C gene (*HAI2*) were detected in *mtabcg20* plants compared to the wild type when ABA was supplied exogenously.

In conclusion, work published so far has shown that ABCG proteins are involved in three steps of seed dormancy and germination: the release of ABA from the endosperm, the uptake of ABA by the embryo, and the release of ABA from the embryo. The first two transport processes contribute to maintaining dormancy and preventing early seed germination. The third process concerns the release of ABA from the embryo and therefore promotes embryo elongation and seed germination. This last observation highlights the notion that the ABA content in the embryo is also an important factor in seed germination, at least in some plants.

4.4 Additional Roles for ABA Transporters

As mentioned in the introduction, ABA functions in many developmental and stress-related processes. While the role of ABA has been clearly demonstrated for most of these processes, it is not always known whether ABA is synthesized directly within the target cells or transported to these cells. For instance, it remains to be established whether ABA, serving as a signal that induces the formation of suberin, which reduces water loss from the roots under drought stress (Hose et al. 2001), is produced directly in the exodermis and endodermis or transported from the root phloem parenchyma cells to these cell layers.

4.4.1 Lateral Root Production

The role of ABA in regulating lateral root formation and elongation varies between plant species. In Arabidopsis, ABA inhibits lateral root development (De Smet et al. 2003). The lateral roots of mutant plants lacking AtABCG40 are smaller when they were exposed to exogenously applied ABA. However, since the detailed localization of AtABCG40 is not known, the function of this transporter in lateral root emergence and elongation remains to be determined (Kang et al. 2010).

Work on Fabaceae indicates that, in this plant family, ABA is involved in the pre-emergence development of lateral roots (Gonzalez et al. 2015). In contrast to Arabidopsis, in Fabaceae, low concentrations of ABA increase the number of lateral root primordia and also the number of total lateral roots (Harris 2015). In line with these results, Pawela et al. (2019) showed that the loss-of-function mutants of MtABCG20 described above produce fewer lateral roots, indicating that this transporter delivers ABA to the sites of lateral root primordia formation and is involved in regulating lateral root number. In Medicago, an NPF (MtNRT1.3) has been shown to regulate lateral root formation and inhibit primary root growth (Pellizzaro et al. 2014). However, since these proteins are involved in nitrate uptake and some members also in ABA transport, it is challenging to distinguish between the effect of nitrate nutrition and ABA.

4.4.2 ABA Transporters in Nitrogen-Fixing Nodules

Most plants rely on soil nitrogen for their growth. However, some plants, such as Fabaceae and a few others, can establish a symbiosis with nitrogen-fixing bacteria, in the case of Fabaceae rhizobia. The symbiosis with nitrogen-fixing bacteria enables the plants to obtain nitrogen independently of the soil nitrogen status (Udvardi and Poole 2013). In a complex crosstalk, plants produce nodules, the organs hosting bacteria, which due to their special environment are also called bacteroids. Plants provide bacteroids mainly with carboxylates, while bacteroids release ammonia through the membrane separating the bacteria from the plant cell, the so-called peribacterial membrane, into the plant cytosol. To establish an equilibrium of these exchanges, the number of nodules has to be tightly controlled, and it has been shown that ABA inhibits nodule formation (Ding et al. 2008). ABA also affects the later stages of nodule formation. Exposure of infected roots to external ABA leads to a decreased amount of leghemoglobin, and hence it can be anticipated that the oxygen concentration will increase in these nodules and reduce nitrogen fixation (González et al. 2001).

Loss-of-function mutants for MtABCG20 produced fewer nodules compared to the wild type (Pawela et al. 2019). The rather small, but significant, changes in nodule number suggest that additional ABA transporters are involved in the regulation of nodule formation. LATD/NIP (**lat**eral root **d**efective/**n**umerous **i**nfection threads, **p**olyphenolics) could be such a candidate, since the expression of this transporter is regulated by ABA (Yendrek et al. 2010). In legumes, *LATD/NIP* function is required for the development of two kinds of lateral root organs, lateral roots and nitrogen-fixing root nodules. However, it remains to be established whether this protein is indeed an ABA transporter.

4.4.3 Internal Transporters Important for ABA Homeostasis

ABA glucose ester (ABA-GE) is the major ABA conjugate and to our knowledge is present in various organs of all plant species analyzed so far (Piotrowska and Bajguz 2011). ABA-GE can be hydrolyzed and serves as a source of biologically active, free ABA. ABA-GE has been reported to be localized to vacuoles (Bray and Zeevaart 1985). However, the first report demonstrating the importance of ABA-GE hydrolysis was the identification of AtBG1, a specific ER-localized ABA-GE β-glucosidase (AtBG1) (Lee et al. 2006). Plants lacking this glucosidase activity exhibit pronounced ABA-deficiency phenotypes, including sensitivity to dehydration, impaired stomatal closure, earlier germination, and lower ABA levels, demonstrating the importance of this conjugated ABA pool.

Two subsequent papers showed that the vacuolar ABA-GE pool also plays an important role in the ABA-related stress response, since two vacuolar β-glucosidases able to hydrolyze vacuolar ABA-GE exhibit a similar, although less pronounced, function as AtBG1 (Wang et al. 2011; Xu et al. 2012). The observation that the single mutants of the vacuolar β-glucosidases exhibit a weaker phenotype than those lacking AtBG1 may be due to a partial redundancy.

These results raise the question of which transporters mediate ABA-GE accumulation in the ER and vacuolar lumen and how free ABA hydrolyzed by glucosidases is exported from these two compartments. Burla et al. (2013) showed that vacuoles can take up ABA-GE via two transport mechanisms. One is directly activated by ATP and exhibits properties consistent with the action of an ABC transporter. Indeed, two vacuolar ABC transporters, AtABCC1 and AtABCC2, are able to transport ABA-GE into vesicles from yeast transgenically expressing these transporters. Vacuolar ABA-GE uptake was also shown to be driven by a proton antiporter. However, in this case the nature of the transporter remains to be elucidated.

The K_M for the uptake of ABA-GE is approximately 1 mM for both transport mechanisms and hence far higher than the cytosolic concentration of this conjugate. Thus, the vacuolar uptake of ABA-GE does not have a regulatory role in ABA-GE storage. Whether ABA released after hydrolysis diffuses across the vacuolar membrane or is delivered to the cytosol by a transporter is unknown. In the acidic vacuole, a large proportion of ABA is present in the protonated, diffusible form and would probably allow efficient delivery to the cytosol. However, the pH of the ER lumen is distinctly less acidic than that of the vacuole, and diffusion of ABA to the cytosol would probably be much less efficient (Martinière et al. 2013). Therefore, a transporter that has yet to be discovered likely mediates the export of ABA from the ER to the cytosol.

5 Concluding Remarks

ABA is involved in many different environmental responses and developmental processes. Not all functions are mentioned in the transport sections; for instance, there is evidence that ABA is also involved in bud dormancy (Cooke et al. 2012; Li et al. 2018), in root hydrotropism and xylem formation, and as a factor regulating the circadian clock and senescence (Dietrich et al. 2017; Ramachandran et al. 2018; Lee et al. 2016). The members of the ABA receptor complex are present as small gene families. In Arabidopsis, there are 14 RCAR/PYR/PYL receptor proteins and 9 clade A PP2Cs (Yoshida et al. 2019). This diversity in ABA receptors greatly expands the number of possible combinations, influencing where and how ABA can act. Indeed, it has been hypothesized that this diversity allows ABA to be perceived in a wide concentration range and in different cells embedded in different tissues (Yoshida

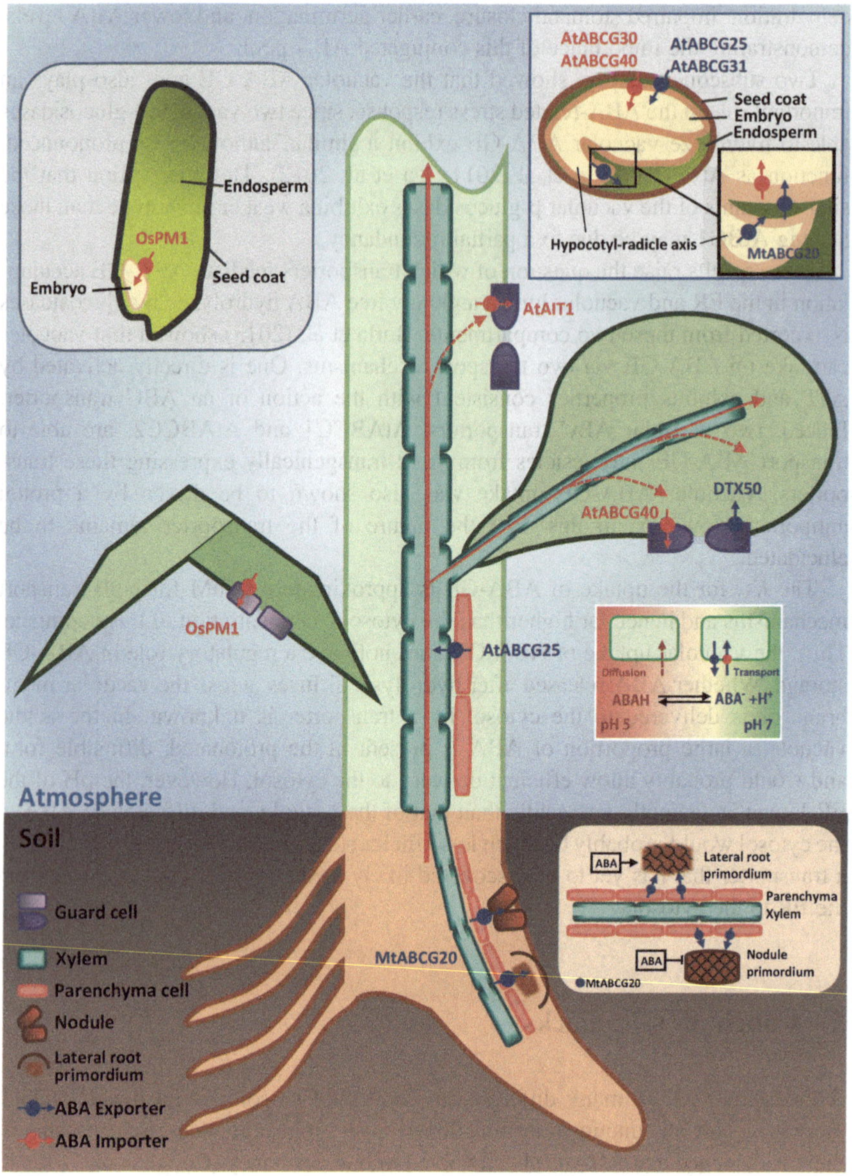

Fig. 1 ABA transporters described to date function in plant development and adaptation to changing environmental conditions. For details, see text

et al. 2019). The large number of potential ABA receptor combinations allows the plant to act in a versatile way to different demands during development and in response to environmental stimuli. Together, the large number of potential receptor combinations implies that ABA transport must be versatile so that the plant can respond to various developmental stages and environmental conditions.

It is striking that the phenotypes of mutants with defects in ABA transporters described so far are rather moderate compared to those lacking enzymes involved in ABA biosynthesis. This suggests that, in all the responses described so far, more than one transporter is involved in delivering ABA and activating a given process. However, if there were complete redundancy among members of the ABA transporter families, mutants lacking an ABA transporter would not have an aberrant phenotype. Hence, more than one transporter is likely required for optimal ABA action.

It is likely that, depending on the growth conditions, the absence of a given ABA transporter leads to a more or less pronounced phenotype. We hypothesize that ABA transport is mediated by several transporters that have overlapping expression in a given tissue or cell type. The presence of several ABA transporters with a limited transport activity may help fine-tune the amount of ABA delivered to a specific site. This is important, since in many of the processes where ABA is involved, ABA is not the only player but participates in a complex phytohormone network and different internal ABA concentrations may have different effects. Furthermore, the presence of several transporters may act as security, since the absence of one transporter would not necessarily lead to plant death.

Acknowledgment The work on ABA transport done in the authors' laboratories was supported by the Basic Science Research Program NRF-2018R1A2A1A05018173 (to Y.L.) through the National Research Foundation of Korea funded by the Ministry of Science and ICT (Information and Communication Technology) and the Swiss National Foundation. J Kang was supported by Ambizione fellowship of the Swiss National Science Foundation (PZ00P3_168041) and Brain Pool Program through the NRF funded by the Ministry of Science and ICT (2019H1D3A2A02102582).

References

Bauer H, Ache P, Lautner S, Fromm J, Hartung W, Al-Rasheid KA, Sonnewald S, Sonnewald U, Kneitz S, Lachmann N, Mendel RR, Bittner F, Hetherington AM, Hedrich R (2013) The stomatal response to reduced relative humidity requires guard cell-autonomous ABA synthesis. Curr Biol 23:53–57

Bethke PC, Libourel IGL, Aoyama N, Chung Y-Y, Still DW, Jones RL (2007) The Arabidopsis aleurone layer responds to nitric oxide, gibberellin, and abscisic acid and is sufficient and necessary for seed dormancy. Plant Physiol 143:1173–1188

Blackman PG, Davies WJ (1985) Root to shoot communication in maize plants of the effects of soil drying. J Exp Bot 36:39–48

Bray EA, Zeevaart JAD (1985) The compartmentation of abscisic acid and β-d-glucopyranosyl abscisate in mesophyll cells. Plant Physiol 79:719–722

Burla B, Pfrunder S, Nagy R, Francisco RM, Lee Y, Martinoia E (2013) Vacuolar transport of abscisic acid glucosyl ester is mediated by ATP-binding cassette and proton-antiport mechanisms in Arabidopsis. Plant Physiol 163:1446–1458

Cao FY, Yshioka K, Desveaux D (2011) The roles of ABA in plant–pathogen interactions. J Plant Res 124:489–499

Chen H, Lan H, Huang P, Zhang Y, Yuan X, Huang X, Huang J, Zhang H (2015) Characterization of OsPM19L1 encoding an AWPM-19-like family protein that is dramatically induced by osmotic stress in rice. Genet Mol Res 14:11994–12005

Chiba Y, Shimizu T, Miyakawa S, Kanno Y, Koshiba T, Kamiya Y, Seo M (2015) Identification of Arabidopsis thaliana NRT1/PTR family (NPF) proteins capable of transporting plant hormones. J Plant Res 128:679–686

Cho CH, Jang S, Choi BY, Hong D, Choi DS, Choi S, Kim H, Han SK, Kim S, Kim MS, Palmgren M, Sohn KH, Yoon HS, Lee Y (2019) Phylogenetic analysis of ABCG subfamily proteins in plants: functional clustering and coevolution with ABCGs of pathogens. Physiol Plant. https://doi.org/10.1111/ppl.13052

Christmann A, Weiler EW, Steudle E, Grill E (2007) A hydraulic signal in root-to-shoot signalling of water shortage. Plant J 52:167–174

Cooke JE, Eriksson ME, Junttila O (2012) The dynamic nature of bud dormancy in trees: environmental control and molecular mechanisms. Plant Cell Environ 35:1707–1728

De Smet I, Signora L, Beeckman T, Inzé D, Foyer CH, Zhang H (2003) An Abscisic acid-sensitive checkpoint in lateral root development of Arabidopsis. Plant J 33:543–555

Deak KI, Malamy J (2005) Osmotic regulation of root system architecture. Plant J 43:17–28

Desikan R, Horák J, Chaban C, Mira-Rodado V, Witthöft J, Elgass K, Grefen C, Cheung MK, Meixner AJ, Hooley R, Neill SJ, Hancock JT, Hauter K (2008) The histidine kinase AHK5 integrates endogenous and environmental signals in Arabidopsis guard cells. PLoS One 3:e2491

Dietrich D, Pang L, Kobayashi A, Fozard JA, Boudolf V, Bhosale R, Antoni R, Nguyen T, Hiratsuka S, Fujii N, Miyazawa Y, Bae T-W, Wells DM, Owen MR, Band LR, Dyson RJ, Jensen OE, King JR, Tracy SR, Sturrock CJ, Mooney SJ, Roberts JA, Bhalerao RP, Dinneny JR, Rodriguez PL, Nagatani A, Hosokawa Y, Baskin TI, Pridmore TP, de Veylder L, Takahashi H, Bennett MJ (2017) Root hydrotropism is controlled via a cortex-specific growth mechanism. Nat Plants 3:17057

Ding Y, Kalo P, Yendrek C, Sun J, Liang Y, Marsh JF, Harris JM, Oldroyd GE (2008) Abscisic acid coordinates nod factor and cytokinin signaling during the regulation of nodulation in Medicago truncatula. Plant Cell 20:2681–2695

Gonzalez AA, Agbévénou K, Herrbach V, Gough C, Bensmihen S (2015) Abscisic acid promotes pre-emergence stages of lateral root development in Medicago truncatula. Plant Signal Behav 10:e977741

González EM, Gálvez L, Arrese-lgor C (2001) Abscisic acid induces a decline in nitrogen fixation that involves leghaemoglobin, but is independent of sucrose synthase activity. J Exp Bot 52:285–293. https://doi.org/10.1093/jexbot/52.355.285

Gonzalez-Guzman M, Pizzio GA, Antoni R, Vera-Sirera F, Merilo F, Bassel GW, Fernández MA, Holdsworth MJ, Perez-Amador MA, Kollist H, Rodriguez PL (2012) Arabidopsis PYR/PYL/RCAR receptors play a major role in quantitative regulation of stomatal aperture and transcriptional response to Abscisic acid. Plant Cell 24:2483–2496

Gubler F, Millar AA, Jacobsen JV (2005) Dormancy release, ABA and pre-harvest sprouting. Curr Opin Plant Biol 8:183–187

Harris JM (2015) Abscisic acid: hidden architect of root system structure. Plants (Basel) 4:548–572

Holbrook NM, Shashidhar VR, James RA, Munns R (2002) Stomatal control in tomato with ABA-deficient roots: response of grafted plants to soil drying. J Exp Bot 53:1503–1514

Holdsworth MJ, Bentsink L, Soppe WJ (2008) Molecular networks regulating Arabidopsis seed maturation, after-ripening, dormancy and germination. New Phytol 179:33–54

Hose E, Clarkson DT, Steudle E, Schreiber L, Hartung W (2001) The exodermis: a variable Apoplastic barrier. J Exp Bot 52:2245–2264

Huang NC, Liu KH, Lo HJ, Tsay YF (1999) Cloning and functional characterization of an Arabidopsis nitrate transporter gene that encodes a constitutive component of low-affinity uptake. Plant Cell 11:1381–1392

Hwang JU, Song W-Y, Hong D, Ko D, Yamaoka Y, Jang S, Yim S, Lee E, Khare D, Kim K, Palmgren M, Yoon HS, Martinoia E, Lee Y (2016) Plant ABC transporters enable many unique aspects of a terrestrial plant's lifestyle. Mol Plant 9:338–355

Inuchi S, Kobayashi M, Taji T, Naramoto M, Seki M, Kato T, Tabata S, Kakubari Y, Yamaguchi-Shinozaki K, Shinozaki K (2001) Regulation of drought tolerance by gene manipulation of 9-cis epoxycarotenoid dioxygenase, a key enzyme in abscisic acid biosynthesis in Arabidopsis. Plant J 27:325–333

Jackson MB (1993) Are plant hormones involved in root to shoot communication? Adv Bot Res 19:103–187

Kang J, Hwang JU, Lee M, Kim YY, Assmann SM, Martinoia E, Lee Y (2010) PDR-type ABC transporter mediates cellular uptake of the phytohormone abscisic acid. Proc Natl Acad Sci U S A 107:2355–2360

Kang J, Park J, Choi H, Burla B, Kretzschmar T, Lee Y, Martinoia E (2011) Plant ABC transporters. Arabidopsis Book 9:e0153

Kang J, Yim S, Choi H, Kim A, Lee KP, Lopez-Molina L, Martinoia E, Lee Y (2015) Abscisic acid transporters cooperate to control seed germination. Nat Commun 6:8113

Kanno Y, Hanada A, Chiba Y, Ichikawa T, Nakazawa M, Matsui M, Koshiba T, Kamiya Y, Seo M (2012) Identification of an abscisic acid transporter by functional screening using the receptor complex as a sensor. Proc Natl Acad Sci U S A 109:9653–9658

Khare D, Choi H, Huh SU, Bassin B, Kim J, Martinoia E, Sohn KH, Paek K-Y, Lee Y (2017) Arabidopsis ABCG34 contributes to defense against necrotrophic pathogens by mediating the secretion of camalexin. Proc Natl Acad Sci U S A 114:E5712–E5720. https://doi.org/10.1073/pnas.1702259114

Kim T-H, Böhmer M, Hu H, Nishimura N, Schroeder JI (2010) Guard cell signal transduction network: advances in understanding abscisic acid, CO_2, and Ca^{2+} signaling. Annu Rev Plant Biol 61:561–591

Koike M, Takezawa D, Arakawa K, Yoshica S (1997) Accumulation of 19-kDa plasma membrane polypeptide during induction of freezing tolerance in wheat suspension-culture cells by abscisic acid. Plant Cell Physiol 38:707–716

Krattinger SG, Jordan DR, Mace ES, Raghavan C, Luo MC, Keller B, Lagudah ES (2013) Recent emergence of the wheat Lr34 multi-pathogen resistance: insights from haplotype analysis in wheat, rice, sorghum and Aegilops tauschii. Theor Appl Genet 126:663–672

Krattinger SG, Kang J, Bräunlich S, Boni R, Chauhan H, Selter LL, Robinson MD, Schmid MW, Wiederhold E, Hensel G, Kumlehn J, Sucher J, Martinoia E, Keller B (2019) Abscisic acid is a substrate of the ABC transporter encoded by the durable wheat disease resistance gene Lr34. New Phytol 223:853–866

Kuromori T, Miyaji T, Yabuuchi H, Shimizu H, Sugimoto E, Kamiya A, Moriyama Y, Shinozaki K (2010) ABC transporter AtABCG25 is involved in abscisic acid transport and responses. Proc Natl Acad Sci U S A 107:2361–2366

Kuromori T, Sugimoto E, Shinozaki K (2011) Arabidopsis mutants of AtABCG22, an ABC transporter gene, increase water transpiration and drought susceptibility. Plant J 67:885–894

Kuromori T, Seo M, Shinozaki K (2018) ABA transport and plant water stress responses. Trends Plant Sci 23:513–522

Lee KH, Piao HL, Kim H-Y, Choi SM, Jiang F, Hartung W, Hwang I, Kwak JM, Lee I-J, Hwang I (2006) Activation of glucosidase via stress-induced polymerization rapidly increases active pools of Abscisic acid. Cell 126:1109–1120

Lee KP, Piskurewicz U, Turecková V, Strnad M, Lopez-Molina L (2010) A seed coat bedding assay shows that RGL2-dependent release of abscisic acid by the endosperm controls embryo growth in Arabidopsis dormant seeds. Proc Natl Acad Sci U S A 107:19108–19113

Lee HG, Mas P, Seo PJ (2016) MYB96 shapes the circadian gating of ABA signaling in Arabidopsis. Sci Rep 6:17754

Léran S, Varala K, Boyer JC, Chiurazzi M, Crawford N, Daniel-Vedele F, David L, Dickstein R, Fernandez E, Forde B, Gassmann W, Geiger D, Gojon A, Gong JM, Halkier BA, Harris JM, Hedrich R, Limami AM, Rentsch D, Seo M, Tsay YF, Zhang M, Coruzzi G, Lacombe B (2014) A unified nomenclature of nitrate transporter 1/PEPTIDE transporter family members in plants. Trends Plant Sci 19:5–9

Li J, Xu Y, Niu Q, He L, Teng Y, Bai S (2018) Abscisic acid (ABA) promotes the induction and maintenance of pear (Pyrus pyrifolia White Pear Group) flower bud endodormancy. Int J Mol Sci 19:E310

Liu G, Stirnemann M, Gübeli C, Egloff S, Courty P-E, Aubry S, Vandenbussche M, Morel P, Reinhardt D, Martinoia E, Borghi L (2019) Strigolactones play an important role in shaping exodermal morphology via a KAI2-dependent pathway. iScience 17:144–154

Longo A, Miles NW, Dickstein R (2018) Genome mining of plant NPFs reveals varying conservation of signature motifs associated with the mechanism of transport. Front Plant Sci 9:1668

Lopez-Molina L, Mongrand S, Chua N-H (2001) A postgermination developmental arrest checkpoint is mediated by abscisic acid and requires the ABI5 transcription factor in Arabidopsis. Proc Natl Acad Sci U S A 98:4782–4787

Lu X, Dittgen J, Piślewska-Bednarek M, Molina A, Schneider B, Svatoš A, Doubský J, Schneeberger K, Weigel D, Bednarek P, Schulze-Lefert P (2015) Mutant allele-specific uncoupling of PENETRATION3 functions reveals engagement of the ATP-binding cassette transporter in distinct tryptophan metabolic pathways. Plant Physiol 168:814–827

Ma Y, Szostkiewicz I, Korte A, Moes D, Yang Y, Christmann A, Grill E (2009) Regulators of PP2C phosphatase activity function as abscisic acid sensors. Science 324:1064–1068

Martinière A, Bassil E, Jublanc E, Alcon C, Reguera M, Sentenac H, Blumwald E, Paris N (2013) In vivo intracellular pH measurements in tobacco and Arabidopsis reveal an unexpected pH gradient in the endomembrane system. Plant Cell 25:4028–4043

Matern A, Böttcher C, Eschen-Lippold L, Westermann B, Smolka U, Döll S, Trempel F, Aryal B, Scheel D, Geisler M, Rosahl S (2019) A substrate of the ABC transporter PEN3 stimulates bacterial flagellin (flg22)-induced callose deposition in Arabidopsis thaliana. J Biol Chem 297:6857–6870

Melotto M, Underwood W, Koczan J, Nomura K, He SY (2006) Plant stomata function in innate immunity against bacterial invasion. Plant Cell 126:969–980

Merilo E, Jalakas P, Kollist H, Brosché M (2015) The role of ABA recycling and transporter proteins in rapid stomatal responses to reduced air humidity, elevated CO_2 and exogenous ABA. Mol Plant 8:657–659

Merilo E, Yarmolinsky D, Jalakas P, Parik H, Tulva I, Rasulov B, Kilk K, Kollist H (2018) Stomatal VPD response: there is more to the story than ABA. Plant Physiol 176:851–864

Meurs C, Basra AS, Karssen CM, van Loon LC (1992) Role of abscisic acid in the induction of desiccation tolerance in developing seeds of Arabidopsis thaliana. Plant Physiol 98:1484–1493

Morita Y, Kodama K, Shiota S, Mine T, Kataoka A, Mizushima T, Tsuchiya T (1998) NorM, a putative multidrug efflux protein, of Vibrio parahaemolyticus and its homolog in Escherichia coli. Antimicrob Agents Chemother 42:1778–1782

Munemasa S, Hauser F, Park J, Waadt R, Brandt B, Schroeder JI (2015) Mechanisms of Abscisic acid-mediated control of stomatal aperture. Curr Opin Plant Biol 28:154–162

Nambara E, Marion-Poll A (2005) Abscisic acid biosynthesis and catabolism. Annu Rev Plant Biol 56:165–185

Oliver SN, Dennis ES, Dolferus R (2007) ABA regulates apoplastic sugar transport and is a potential signal for cold-induced pollen sterility in rice. Plant Cell Physiol 48:1319–1330

Park SY, Fung P, Nishimura N, Jensen DR, Fujii H, Zhao Y, Lumba S, Santiago J, Rodrigues A, Chow TF, Alfred SE, Bonetta D, Finkelstein R, Provart NJ, Desveaux D, Rodriguez PL, McCourt P, Zhu JK, Schroeder JI, Volkman BF, Cutler SR (2009) Abscisic acid inhibits type 2C protein phosphatases via the PYR/PYL family of START proteins. Science 324:1068–1071

Pawela A, Banasiak J, Biała W, Martinoia E, Jasiński M (2019) MtABCG20 is an ABA exporter influencing root morphology and seed germination of Medicago truncatula. Plant J 98:511–523

Pellizzaro A, Clochard T, Cukier C, Bourdin C, Juchaux M, Montrichard F, Thany S, Raymond V, Planchet E, Limami AM, Morère-Le Paven M-C (2014) The nitrate transporter MtNPF6.8 (MtNRT1.3) transports abscisic acid and mediates nitrate regulation of primary root growth in Medicago truncatula. Plant Physiol 166:2152–2165

Peng Y-B, Zou C, Wang D-H, Gong H-Q, Xu Z-H, Bai S-N (2006) Preferential localization of abscisic acid in primordial and nursing cells of reproductive organs of Arabidopsis and cucumber. New Phytol 170:459–466

Piotrowska A, Bajguz A (2011) Conjugates of abscisic acid, brassinosteroids, ethylene, gibberellins, and jasmonates. Phytochemistry 72:2097–2112

Raghavendra AS, Gonugunta VK, Christmann A, Grill E (2010) ABA perception and signaling. Trends Plant Sci 15:395–401

Ramachandran P, Wang G, Augstein F, de Vries J, Carlsbecker A (2018) Continuous root xylem formation and vascular acclimation to water deficit involves endodermal ABA signalling via miR165. Development 145:dev159202

Rerksiri W, Zhang X, Xiong H, Chen X (2013) Expression and promoter analysis of six stress-induced genes in rice. Sci World J 2013:397401

Robertson DS (1955) The genetics of vivipary in maize. Genetics 40:745–760

Schwartz SH, Tan BC, Gage DA, Zeevaart JAD, McCarty DR (1997) Specific oxidative cleavage of carotenoids by VP14 of maize. Science 276:1872–1874

Stein M, Dittgen J, Sánchez-Rodríguez C, Hou B-H, Molina A, Schulze-Lefert P, Lipka V, Somerville S (2006) Arabidopsis PEN3/PDR8, an ATP binding cassette transporter, contributes to nonhost resistance to inappropriate pathogens that enter by direct penetration. Plant Cell 18:731–746

Takanashi K, Shitan N, Yazaki K (2014) The multidrug and toxic compound extrusion (MATE) family in plants. Plant Biotech 31:417–430

Tan BC, Joseph LM, Deng WT, Liu L, Li QB, Cline K, McCarty DR (2003) Molecular characterization of the Arabidopsis 9-cis epoxycarotenoid dioxygenase gene family. Plant J 34:44–56

Tsay YF, Schroeder JI, Feldmann KA, Crawford NM (1993) The herbicide sensitivity gene CHL1 of Arabidopsis encodes a nitrate-inducible nitrate transporter. Cell 72:705–713

Udvardi M, Poole PS (2013) Transport and metabolism in legume-rhizobia symbioses. Annu Rev Plant Biol 64:781–805

Upadhyay N, Kar D, Deepak Mahajan B, Nanda S, Rahiman R, Panchakshari N, Bhagavatula L, Datta S (2019) The multitasking abilities of MATE transporters in plants. J Exp Bot 70:4643–4656

Wang PT, Liu H, Hua HJ, Wang L, Song C-P (2011) A vacuole localized β-glucosidase contributes to drought tolerance in Arabidopsis. Chin Sci Bull 56:3538–3546

Xu Z-Y, Lee KH, Dong T, Jeong JC, Jin JB, Kanno Y, Kim DH, Kim SY, Seo M, Bressan RA, Yun DJ, Hwang I (2012) A vacuolar β-glucosidase homolog that possesses glucose-conjugated abscisic acid hydrolyzing activity plays an important role in osmotic stress responses in Arabidopsis. Plant Cell 24:2184–2199

Yao L, Cheng X, Gu Z, Huang W, Li S, Wang L, Wang Y-F, Xu P, Ma H, Ge X (2018) The AWPM-19 family protein OsPM1 mediates abscisic acid influx and drought response in Rice. Plant Cell 30:1258–1276

Yendrek CR, Lee YC, Morris V, Liang Y, Pislariu CI, Burkart G, Meckfessel MH, Salehin M, Kessler H, Wessler H, Lloyd M, Lutton H, Teillet A, Sherrier DJ, Journet EP, Harris JM, Dickstein R (2010) A putative transporter is essential for integrating nutrient and hormone signaling with lateral root growth and nodule development in Medicago truncatula. Plant J 62:100–112

Yoshida T, Christmann A, Yamaguchi-Shinozaki K, Grill E, Fernie AR (2019) Revisiting the basal role of ABA – Roles outside of stress. Trends Plant Sci 24:625–635. https://doi.org/10.1016/j.tplants.2019.04.008

Zeevaart JAD, Creelman RA (1988) Metabolism and physiology of abscisic acid. Annu Rev Plant Physiol Plant Mol Biol 39:439–473

Zhang H, Han W, De Smet I, Talboys P, Loya R, Hassan A, Rong H, Jürgens G, Paul Knox J, Wang MH (2010) ABA promotes quiescence of the quiescent centre and suppresses stem cell differentiation in the Arabidopsis primary root meristem. Plant J 64:764–774

Zhang H, Zhu H, Pan Y, Yu Y, Luan S, Li L (2014) A DTX/MATE-type transporter facilitates abscisic acid efflux and modulates ABA sensitivity and drought tolerance in Arabidopsis. Mol Plant 7:1522–1532

Zhang H, Zhao F-G, Tang R-J, Yu Y, Song J, Wang Y, Legong L, Luan S (2017) Two tonoplast MATE proteins function as turgor-regulating chloride channels in Arabidopsis. Proc Natl Acad Sci U S A 114:E2036–E2045

Orient in the World with a Single Eye: The Green Algal Eyespot and Phototaxis

Michaela Böhm and Georg Kreimer

Contents

Abstract Motile algae exhibit well-defined movement responses toward or away from a light source, known as positive and negative phototaxis, respectively. To optimize this essential behavior, algae often possess a complex directional optical device, the eyespot. Interest in eyespots has increased considerably during the last decade due to their elaborate ultrastructure and the presence of unique photoreceptors. The latter form an important basis for ongoing developments of optogenetic tools in the field of neurobiology and cell biology. Green algal eyespots, especially that of *Chlamydomonas reinhardtii* P. A. Dangeard, are the best studied with respect to their biophysical function, assembly, positioning, and signaling cascade(s) finally leading to the oriented movement. Here we give a short general introduction to the eyespot and the phototactic orientation of this algal group and summarize recent progress in the diverse eyespot-related areas. We also address emerging novel insights in the homeostatic feedback regulation of phototactic sensitivity by the cells' physiological activities.

M. Böhm and G. Kreimer (✉)
Department of Biology, Friedrich-Alexander-University, Erlangen, Germany
e-mail: michaela.boehm@fau.de; georg.kreimer@fau.de

© Springer Nature Switzerland AG 2020

Progress in Botany (2021) 82: 259–304, https://doi.org/10.1007/124_2020_38,
Published online: 28 December 2020

Abbreviations

ABC1	Activity of BC1 complex
aCRY	Animal-like cryptochrome
ChR	Channelrhodopsin
CHX	Cycloheximide
CK1	Casein kinase 1
CRY	Cryptochrome
DCMU	3-(3,4-Dichlorophenyl)-1,1-dimethylurea
IFT	Intraflagellar transport
PC	Photoreceptor current
pCRY	Plant-type cryptochrome
PHOT	Phototropin
PM	Plasma membrane
PP2C	Protein phosphatase 2C
TRP	Transient receptor potential
UV	Ultraviolet
UVR8	UV resistance locus 8

1 Introduction

The sun has affected life and its evolution from the prebiotic phase until now (Rapf and Vaida 2016). Organisms of the photic zone have developed diverse multilayered mechanisms for detecting, exploiting, and adapting to a light environment with varying intensities and spectral compositions. The basic building blocks used for this are sets of diverse photoreceptors covering the spectral range from UV to the far-red. Photosynthetic organisms use light as an energy source but also as important informational cue to regulate a multitude of developmental and physiological responses. Plant photoreceptors sense even temperature and integrate thereby these central environmental stimuli during the day and night (Jung et al. 2016; Legris et al. 2016; Fuji et al. 2017; Qiu et al. 2019). Regulation of light exposure is essential for all autotrophic organisms to optimize photosynthesis and to avoid stress by either excess or low light levels. For motile organisms, it was thus of utmost importance not only to develop specialized photoreceptors but also to evolve in parallel mechanisms allowing to pick up directional clues and to finally gear toward or away from the light source. This movement along a light gradient, known as positive and negative phototaxis, occurs from cyanobacteria to metazoan. Free-swimming organisms are even capable to follow the light vector in three dimensions (Gehring 2014; Wilde and Mullineaux 2017). Thereby phototaxis can also optimize other essential processes, e.g., increasing the rate of gamete fusion by guiding them to the surface of the water body.

The widespread appearance of phototaxis indicates its great evolutionary advantage. Phylogenetic analyses suggest that algae, fungi, and protists have developed this behavioral response independently at least eight times (Kivic and Walne 1983; Jékely 2009; Gavelis et al. 2017). Concomitantly even single cells evolved diverse complex photoreceptive structures. These represent primordial visual systems, commonly called eyespots. Despite their name and their partially complex ultrastructure, their small size excludes a use for vision as known from higher metazoans. Instead, eyespots are designed to increase the directionality of light perception and thereby the accuracy of the phototactic orientation. For this, algal eyespots include often pigmented or refractive structures, which shield specialized photoreceptors, thereby increasing the front-to-back contrast at the photoreceptor location (reviewed by, e.g., Foster and Smyth 1980; Kreimer 1994; Jékely 2009). These parts of the eyespot are of special interest as even in a small cell, e.g., that of the cyanobacterium *Synechocystis* sp. PCC 6803 or the green alga *Chlamydomonas reinhardtii*, the cell body acts as a focusing lens, which in turn affects the direction of the phototactic orientation (Schuergers et al. 2016; Ueki et al. 2016). Some of these eyespots reached a fascinating complexity on a unicellular level. Some heterotrophic dinoflagellates build camera-type eye-like ocelloids from subcellular components of different endosymbiotic origin, which represent the most complex eyespots of protists (Gavelis et al. 2015; Hayakawa et al. 2015). Loss of complex light focusing eyespots in photosynthetic dinoflagellates correlates with loss of positive phototaxis and directionality (Kreimer 1999; Moldrup et al. 2013).

The presence of light microscopically conspicuous eyespots and the peculiar phototactic behavior led already early to analyses of this light-driven behavior of single cells (Faminzin 1866; Buder 1917). Data gathered from behavioral, ultrastructural, photobiological, biochemical, and cell biological research on different free-swimming algal and protist species has led to the identification of a basic set of three common traits used for a precise phototactic orientation (reviewed by, e.g., Foster and Smyth 1980; Kreimer 1994; Kuhlmann 1998; Jékely 2009; Gavelis et al. 2017). These are (1) a spiral swimming path of a polar cell with axial rotation, (2) precisely placed photoreceptors with downstream signaling cascade(s) affecting flagella beating, and (3) refractive or pigmented structures in close spatial proximity to the photoreceptors that ensure the orientation-dependent periodic excitation of the photoreceptor(s) during movement. Great diversity exists on the other hand by the used photoreceptors and downstream signaling cascades. Known photoreceptors for phototaxis include, e.g., microbial rhodopsins, stentorin, photoactivated adenylate cyclases, and flavoproteins (summarized in Jékely 2009; Gavelis et al. 2017; Govorunova et al. 2017a). Even fusion proteins, e.g., BeGC1, a rhodopsin-guanylyl cyclase fusion from a fungal zoospore, or directly light-gated cation channels like ChR1 and ChR2 from the green alga *C. reinhardtii* accomplish light perception for phototactic responses (Nagel et al. 2002, 2003; Sineshchekov et al. 2002; Avelar et al. 2014). The discovery that both ChRs can depolarize mammalian cells has initiated a growing interest in the molecular identification of the highly diverse algal photoreceptors and their use as optogenetic tools in neurosciences and cell biology (reviewed by Kianianmomeni and Hallmann 2014; Klapoetke et al. 2014; Deisseroth and Hegemann 2017; Govorunova et al. 2017a). Due to the diversity of the involved photoreceptors, also the generated primary signals largely differ. Whereas, e.g., in *Euglena gracilis* G. A. Klebs an increase in cAMP occurs, activation of the ChRs in green algae directly initiates cation fluxes (reviewed by Hegemann 2008).

Phototaxis is the outcome of integrating the light-immanent information (intensity, quality, direction, duration, and polarization) with the organism's actual developmental and physiological state and demands. Hence, also complex cross talks with other signaling cascades will be involved in the overall regulation of the response. As phototaxis often occurs over an enormous range of intensities (e.g., ~10^{15} to >10^{21} photons m^{-2} s^{-1} for *C. reinhardtii*, Berthold et al. 2008), adaptation mechanisms add additional complexity to its regulation. It is thus not surprising that our understanding about the signaling events and their in vivo regulation is still limited and lacks behind the detailed knowledge about structure and properties of primary photoreceptors, e.g., the algal ChRs in ectopic expression systems (recently reviewed by Deisseroth and Hegemann 2017; Govorunova et al. 2017a). Well-established models to understand photo-behavior are the unicellular green alga *C. reinhardtii* and the colonial *Volvox carteri* F. Stein. Their photo-behavior integrates both, fast electrical and slower biochemical networks from different subcellular compartments in a yet not fully understood way. This review focuses on the eyespot and phototactic orientation of green algae – with special emphasis on *Chlamydomonas* – and summarizes basic features as well as recent progress in this field.

2 The Green Algal Eyespot

Underwater, the sun is a diffuse light source. With increasing depth its intensity and spectral composition rapidly decrease. Penetration in clear water is the highest for blue and green light. Turbidity of the water not only affects the depth of penetration but also shifts it more to the green part of the spectrum. Further, surface turbulences and the solar altitude angle affect the amount of light penetrating the water (Kirk 2010). As described below, the eyespot – under close inclusion of the particular movement pattern of the cells – allows precise detection of the average light direction under these constantly changing conditions.

2.1 General Features

The eyespot is typically a singular structure, which is visible in the light microscope as an orange-red spot within the cell. It has always a fixed position at the surface. Most often central positions are observed (Fig. 1a–c), although they can also be located in the anterior or posterior half of the cell. For example, the eyespot of *Spermatozopsis similis* H.R. Preisig & M. Melkonian is always in an anterior

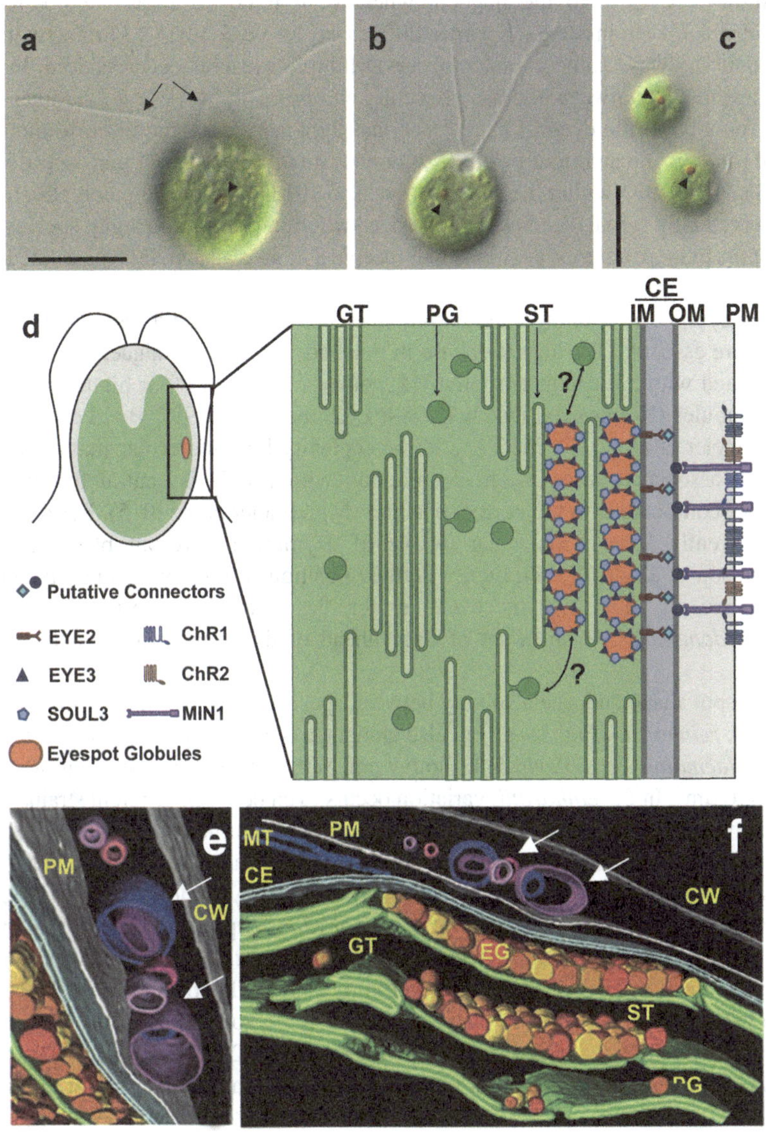

Fig. 1 Appearance and ultrastructure of the eyespot in *Chlamydomonas reinhardtii*. (**a–c**) Differential interference contrast images of unfixed *C. reinhardtii* cells from different strains demonstrating an equatorial eyespot position. (**a**) CC-3403, a cell wall-less strain; (**b**) CC-3403*chr1chr2*, a mutant lacking ChR1 and ChR2, the major photoreceptors for phototaxis (Greiner et al. 2017); and (**c**) CC-125*mat3-1*, a mutant in the cell wall possessing CC-125 background, which exhibits a small cell size phenotype due to a defect in the *MAT3* gene involved in cell cycle control (Greiner et al. 2017). *Arrows*, flagella; *arrowhead*, eyespot; scale bars, 10 μm. (**d**) Schematic drawing (not to scale) of the eyespot demonstrating (i) its multilayered structure involving local specializations from different subcellular compartments and (ii) localization of some known eyespot proteins. For clarity, interactions with the D4 microtubular root are not included, and spacing between the eyespot globules is enlarged (for actual spacing, see f). *CE* chloroplast envelope, composed of the inner and outer chloroplast envelope membrane (*IM* and *OM*), *GT* grana thylakoids, *PG*

position very close to the flagella, whereas all three positions are possible in *C. reinhardtii* strains (e.g., Kreimer 2001; Greiner et al. 2017). Further, within a population always some variance occurs. Multiple eyespots occur seldom. In such species, they are always in close proximity and probably work as a functional unit. The function of the eyespot as a directional light sensor is syntonized to the cell's axial rotation and swimming path (Foster and Smyth 1980; Yoshimura and Kamiya 2001). Its position within the cell will thus also affect capture and modulation of the perceived light signal (see Sect. 2.3). When swimming toward the light, species with eyespots in more anterior positions will receive a more constant, less modulated light signal at the eyespot surface.

For eyespot positioning, the microtubular cytoskeleton and its asymmetric properties are essential. During interphase in most advanced green algae, the eyespot is associated with the daughter rootlet D4, compromised from four highly acetylated microtubules (Holmes and Dutcher 1989; Lechtreck et al. 1997; Boyd et al. 2011a; Mittelmeier et al. 2011). MLT1, a low-complexity high molecular mass phosphoprotein associated with D4, in conjunction with MLT2 is central for directing eyespot components to the correct position (Mittelmeier et al. 2015). For review of the currently known molecular details of asymmetric eyespot placement, see Thompson et al. (2017). Further, a defined position in relation to the flagella and the plane of their beat is essential for the phototactic response; e.g., in *Chlamydomonas* it is an offset of ~45° ahead of this plane (Rüffer and Nultsch 1985).

Eyespot shape and size also vary between species. Most often, they are roundish (e.g., *C. reinhardtii*; Fig. 1a–c), but also elongated or tear-shaped forms are observed (e.g., *Haematococcus pluvialis* J. Flotow em. Wille). Sizes vary between ~0.3 μm^2 and ~10 μm^2. In *C. reinhardtii*, variation occurs even between different strains. Here measured values range between 0.6 and 1.5 μm^2. Often the eyespot size is slightly larger in gametes than that in vegetative cells. Further, light intensity and the duration of exposure regulate the eyespot size dynamically in *Tetraselmis astigmatica* R. E. Norris & Hori and *C. reinhardtii* (Trippens et al. 2012). Flexible size regulation occurs also in colonial species. In *Volvox*, the relative position of the individual cell in the colony is one determining factor. In somatic cells, the size of the eyespot decreases continuously from the anterior to the posterior colony pole. The diameters at the anterior pole are in the range of ~2.5–3.5 μm and those at the posterior pole between ~0.6 and 0.8 μm (Drescher et al. 2010; Ueki et al. 2010).

Fig. 1 (continued) plastoglobule, *PM* plasma membrane, *ST* stroma thylakoid. Double-headed arrows indicate a possible dynamic interconnection between plastoglobules and eyespot globules. (**e**, **f**) Three-dimensional reconstruction of the eyespot from cryoelectron tomography slices (modified from Engel et al. 2015). (**e**) Extracellular unilamellar and bilamellar vesicles (color coded in pink to purple and additionally marked by *white arrows*) between the PM and the cell wall (*CW*) adjacent to the eyespot. (**f**) 3D reconstruction of the whole eyespot, showing the close association of its different parts. Note the association of the eyespot globules (*EG*) with a stroma thylakoid (*ST*) and their dense association forming two tightly packed layers. *MT* microtubules; other abbreviations as in (**d**)

Eyespot size regulation thus is one possibility to regulate the overall photosensitivity of phototaxis. In *C. reinhardtii* the blue light photoreceptor PHOT is essentially involved in this process (Trippens et al. 2012; see Sect. 3.3.2), whereas the factors affecting eyespot size, as a function of the relative position in a colony, are yet still not known. In addition, as the retinal chromophore of the ChRs is orientated almost parallel to the plane of the plasma membrane (Yoshimura 1994), also the geometry of the eyespot surface potentially affects both photosensitivity and directionality. In *C. reinhardtii* the curvature of the eyespot surface appears to be somewhat variable. In chemically fixed cells, strait or slightly convex surfaces dominate, whereas a slight concave surface occurs in a cryofixed sample (Kreimer 2009; Engel et al. 2015). Multicellular volvocine algae possess clear concave surfaces (Arakaki et al. 2013 and references cited therein). In addition, other species have clear central concave depressions, which focus the back reflected light (Kreimer and Melkonian 1990) or protrude beyond the cell surface. Protruding eyespots, elongated parallel to the axis of the cell like in *H. pluvialis*, could facilitate an earlier, longer photoreceptor excitation in conjunction with rotation during the helical swimming behavior. Further, they broaden the scanned area of the environment. Such form and surface specializations occur often in larger species. Although already being early under discussion (e.g., Foster and Smyth 1980; Hegemann and Harz 1998; Kreimer 2001), the significance of these different surface geometries still awaits detailed experimental analyses.

2.2 Ultrastructure

The ultrastructure of the green algal eyespot has been studied and reviewed already in detail (e.g., Melkonian and Robenek 1984; Kreimer 1994, 2001). We here give only an overview and include recent novel findings by cryoelectron microscopy. The functional eyespot is a composed "organelle." It is usually found in close association with the D4 microtubular root, which is important for transport and positioning of the different eyespot components (Thompson et al. 2017). It involves local specialized regions from three different compartments: the PM with the ChRs as primary photoreceptors, the narrow cytosolic space between this PM region and the chloroplast envelope, and the chloroplast. Nonetheless, the different parts are interconnected allowing the isolation of the whole complex from different species (e.g., Kreimer et al. 1991; Eitzinger et al. 2015). Figure 1d–f shows a schematic drawing and 3D reconstructions of the eyespot of *Chlamydomonas* based on cryoelectron tomographs (Engel et al. 2015). The most prominent part is the layers of densely packed carotenoid-rich lipid globules inside the chloroplast. Typically, each globule layer is associated with a thylakoid. The number of layers as well as that of the globules forming them varies between species; single- and double-layered eyespots are most common. The distance from the inner surface of the outer globule layer to the overlying PM (120–200 nm) and the spacing in multilayered eyespots (150–220 nm) are very regular. This and the high carotenoid content of the globules are one basis for the directivity of the whole system (see Sect. 2.3). Interestingly, vestigial osmiophilic eyespot globules are still detectable in cells blocked in the first step of carotenoid biosynthesis (Inwood et al. 2008), indicating that lipophilic components, e.g., triacylglycerides, still do accumulate in the absence of carotenoids. Analysis of the carotenoid biosynthetic pathway in the non-photosynthetic green algal genus *Polytomella*, which is closely related to *Chlamydomonas*, suggests that at least ß-carotene must be formed to build visible eyespots (Asmail and Smith 2016). This agrees well with the dominance of this carotene in eyespot globules (Grung et al. 1994; Satoh et al. 1995; Kreimer 2009). The eyespot globule diameters (80–130 nm) are slightly larger than that of algal plastoglobules. In *C. reinhardtii* these are 64 ± 15 nm (Melkonian and Robenek 1984; Engel et al. 2015), whereas in higher plants they are more variable and highly dynamic ($\sim$30–500 nm; van Wijk and Kessler 2017). Several different lines of evidence – ranging from developmental observations to proteomic analysis – point to an interrelation between these two plastid localized carotene-rich lipid globule types (summarized by Kreimer 2009). Recent proteomic data from plastidic ß-carotene lipid droplets from *Dunaliella* are further supporting their interconnection (Davidi et al. 2014, 2015). Pick et al. (2019) suggested that the hyper-accumulation of plastid ß-carotene globules in response to high light in *Dunaliella bardawil* A. Ben-Amotz & Avron and *D. salina* (Dunal) Teodoresco might be a result of the loss of the capability to organize the eyespot globule layers and a subsequent amplification and remodeling of the former eyespot globules. Besides similarities, however, comparison of the core proteomes reveals also clear differences (Davidi et al. 2015; van Wijk and Kessler 2017). Unfortunately in their cryoelectron tomographic study, Engel et al. (2015) did not analyze whether the eyespot globules are also stabilized by a half-lipid bilayer originating from the stroma side of the thylakoid membranes, as it is the case for higher plant plastoglobules (Austin et al. 2006). Such an analysis would help to understand further the interrelation of these globule types in the green algal chloroplast. Tangential sections and freeze-fractures through the eyespot globule layers have demonstrated their close hexagonal packing with a constant space of ~7–8 nm containing particles between them (Melkonian and Robenek 1984; Kreimer 2009). Protease and lipase treatments demonstrated the protein nature of these particles and their importance for globule stability (Renninger et al. 2001; Schmidt et al. 2007). These two studies also revealed that the interconnections of the globules with the other membrane parts of the functional eyespot are protease sensitive. Although several eyespot-related proteins are known on the molecular level and have been localized by fluorescence microscopy (see Sect. 3.1), to the best of our knowledge, none of them has yet been localized by electron microscopy. Such studies are desirable in the future to substantiate further our current understanding of their localization and function(s) within the eyespot.

A novel addition to the eyespot ultrastructure comes from a cryoelectron tomography approach. Engel et al. (2015) observed a cluster of extracellular uni- and bilamellar vesicles of varying sizes between the ChR-containing PM patch and the cell wall (white arrows in Fig. 1e, f). These vesicles were solely present adjacent to the eyespot, indicating possible functional roles related to the eyespot. Although no functional clue currently exists, it is tempting to suggest that these vesicles might represent micro-vesicles or exosomes formed by fusion of multi-vesicular bodies (MVB) with the PM and subsequent release of their intraluminal vesicles. Wood et al. (2013) reported localized ectosome secretion for the flagella of *Chlamydomonas*. It is well feasible that it can occur also at other places of the cell. Although we currently do not know how the ChRs and other membrane proteins from the PM region of the eyespot are degraded, it is possible that the observed vesicles are a part of their degradation pathway. Pootakham et al. (2010) reported PM protein degradation by proteasome-independent proteolysis in *Chlamydomonas*. In other systems, (mono)ubiquitin-dependent endocytic removal of transporters, receptors, and ion channels is widely used to rapidly downregulate and adapt to physiological changes (e.g., Eguez et al. 2004; Foot et al. 2017). The nature and function of these vesicles, however, await future detailed experimental approaches.

2.3 The Directional Antenna, Signal Generation, and Flagella Responses

Foster and Smyth (1980) were the first to hypothesize that the green algal eyespot is a directional light antenna. They postulated that (1) only light falling from the outside excites the PM-localized photoreceptors as carotenoids in the globules effectively absorb light coming through the cell and (2) that the spacing and structure of the layers in combination with the regular changing refractive indices allow it to act as quarter-wave interference reflector optimized for blue-green light. That is, light of this wavelength range which is falling roughly perpendicular on the eyespot surface and is not absorbed by the ChRs undergoes constructive interference when back reflected. Thereby the intensity at the ChR site and thus the excitation probability increase. On the other hand, negative interference of other wavelengths occurs, and light falling through the cell on the eyespot region will be back scattered. This complex dual function has experimental support from reflection confocal laser microscopy, microspectrometry, and electrophysiology (Kreimer and Melkonian 1990; Sineshchekov 1991; Harz et al. 1992; Kreimer et al. 1992; Schaller and Uhl 1997). The screening properties of the globules are extended to longer wavelengths by enrichment of certain carotenoid species and isomers, which probably also prevents carotenoid crystallization (Kreimer 2009). Intact globule layers are thus important for precision and sign of phototaxis and help to suppress the lens effect of the cells (Kreimer et al. 1992; Matsunaga et al. 2003; Ueki et al. 2016).

How does the cell take advantage of the directionality of the eyespot in relation to its movement and initiation of phototactic steering? In most green algae, the longitudinal cell axis and the direction of movement coincide. This is also true for the biflagellate *C. reinhardtii*. Based on their relative position to the eyespot, the flagella are designated as *cis* and *trans*. They beat usually highly synchronized in a breaststroke-like manner. Analyses of cells with single flagella, however, revealed clear intrinsic differences in the beat frequencies (Fig. 2a, b). Although tightly coupled, the flagella do not beat exactly in one plane, and slight imbalances in the strength and symmetry of their beats occur. Extra beats of the faster beating *trans*-flagellum occasionally interrupt the synchronous breaststroke. In sum, these subtle differences result in a rotation of the cell along its main axis and an overall helical swimming path. When seen from the back, the cell rotates counterclockwise, and one rotation roughly corresponds to a helical turn (Fig. 2a–c; Kamiya and Witman 1984; Rüffer and Nultsch 1985, 1987, 1998; Isogai et al. 2000; Josef et al. 2005; Polin et al. 2009; Wang et al. 2014; Goldstein 2015). Rotation frequencies are ~2 Hz for *Chlamydomonas* and ~0.02–0.3 Hz for the much larger *Volvox* colonies. The phototactic sensitivity of both is optimized to these rotation rates (Yoshimura and Kamiya 2001; Drescher et al. 2010; Goldstein 2015). This spinning coupled to the above described intrinsic eyespot directionality generates periodic signals, which primarily depends on the amount of ChR excitation. The magnitude of ChR activation is a function of the relative alignment of the swimming path with the light direction (Fig. 2d). In a well-aligned cell, only a few ChRs will be excited, whereas a large deviation results in a strong periodic signal. Although in some green algae the eyespot position in relation to the cell axis differs to that observed in *Chlamydomonas*, it is always roughly perpendicular to the direction of movement (Fig. 2e). Thus, also in these cases the generated signal is periodic.

How is this periodic light signal converted into phototactic orientation? Both ChRs are directly light-gated cation ion channels (Nagel et al. 2002, 2003; Sineshchekov et al. 2002). Their excitation initiates rapid depolarization of the plasma membrane in the eyespot region and well-studied PCs, which depend in their amplitude on the light stimulus. In alga, the ChRs mainly conduct Ca^{2+}, H^{+}, and Na^{+}, and the photo-responses strictly depend on extracellular Ca^{2+}. In the eyespot region, intracellular Ca^{2+} concentration changes are central to initiate signaling cascade(s) toward the flagella. Whereas the signaling cascade leading to phototaxis is yet not known in detail, the one leading to the fast (~500 ms) non-oriented photoshock response is based on electrical signal spread and activation of voltage-dependent Ca^{2+}-channels in the flagella (reviewed in Hegemann and Berthold 2009; Sineshchekov et al. 2009; Yoshimura 2011; Wheeler 2017). A transient switch from the breaststroke beat to an undulating waveform beat of the flagella characterizes the photoshock or photophobic reaction. This results in a stop, followed by a short, slow backward swimming phase in a random direction. Only when depolarization by the ChR-dependent PCs exceeds a certain threshold, this response occurs, and Ca^{2+}-dependent currents originating from the flagella accompany it. These currents consist of a fast transient depolarization occurring in the same time as the switch in flagella beat and a sustained slower current. The amplitude of the latter is significantly smaller, its duration is similar as the backward swimming phase, and it is subject to a feedback inhibition by intraflagellar Ca^{2+} (Harz and Hegemann 1991; Harz et al. 1992; Holland et al. 1996, 1997; Ehlenbeck et al. 2002). The voltage-dependent Ca^{2+}-channel CAV2 is one major element in mediating the massive Ca^{2+} influx into the flagella needed for the switch in flagellar beat form and the photoshock response (Hyams and Borisy 1978; Bessen et al. 1980; Matsuda

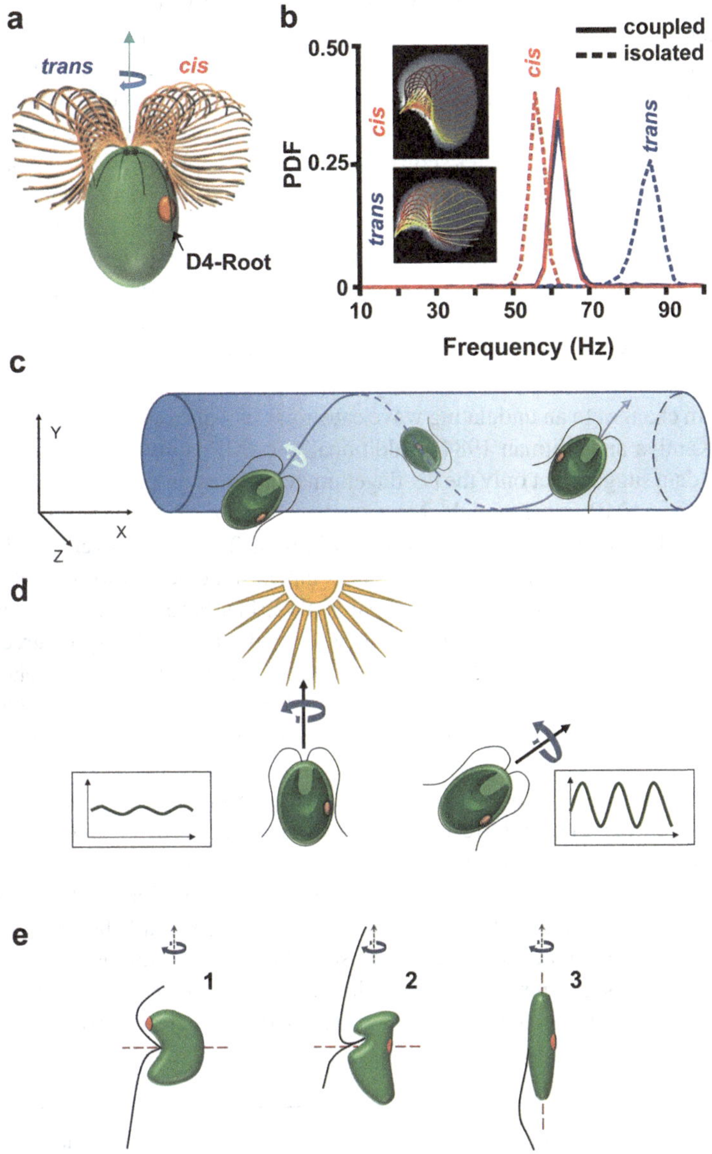

Fig. 2 Signal generation, swimming behavior, and flagella beating. (**a**) Cartoon of a *C. reinhardtii* cell depicting the apical position of the two flagella emerging from the two basal bodies (dark green dots) and their overlaid sequences of the breaststroke beating behavior (modified from Wan et al. 2017). The eyespot is associated with the D4 root emanating from the newly formed daughter basal body. Its slightly longer acetylation is an eyespot-positioning clue (Mittelmeier et al. 2011). The position of the eyespot relative to the flagella defines the *cis*- and *trans*-flagellum, where *cis* is the one closer to the eyespot. Three-dimensional components in the flagella beating induce a rotation of ~2 Hz along the long axis of the cell body (*curved arrow*). *Blue small arrow*: movement direction. (**b**) Removal of either the *cis*- or *trans*-flagellum (*dashed lines*, isolated) revealed intrinsic differences in the in vivo beat frequencies, which are abolished upon regrowth of the removed flagellum to its full length (*continuous line*, coupled; modified from Wang et al. 2014). It is still under debate

et al. 1998; Fujiu et al. 2009). In contrast, to allow positive or negative phototactic orientation, the flagella breaststroke beat must temporarily drop out of synchrony to create different propulsive forces, i.e., one flagellum must dominate until proper alignment of the cell. *Chlamydomonas* flagella respond to both increases and decreases in light intensity. Analyses of flagella from free-swimming cells as well as of cells held on micropipettes showed that the front amplitude of the *cis*-flagellum decreases and that of the *trans*-flagellum increases in positive phototactic cells. For negative phototactic cells, the opposite behavior was recorded. Therefore, the cell will orient toward or away from the light source (Rüffer and Nultsch 1985, 1987, 1991, 1997; Isogai et al. 2000; Josef et al. 2005). The two central second messengers Ca^{2+} and cAMP are involved in flagella dominance control. The *cis*-flagellum dominates at low free Ca^{2+} ($<10^{-8}$ M), and the *trans*-flagellum is more active at 10^{-7}–10^{-6} M Ca^{2+}. Both beat similarly at 10^{-8} M Ca^{2+}, and finally at 10^{-4} M the beat form changes to an undulating wave causing backward swimming (Bessen et al. 1980; Kamiya and Witman 1984). Additionally, cAMP is involved in this control. Current data suggest that only the *cis*-flagellum reduces its propulsive forces upon an increase in cAMP, which might be related to the phosphorylation status of the axonemal dynein IC138 (Saegusa and Yoshimura 2015). However, IC138 is in contrast to other subunits of the I1 dynein complex not essential for phototaxis but for the photoshock response (VanderWaal et al. 2011). Further, the *trans*-flagellum responds faster with respect to beat frequency and stroke velocity to green light stimulation (Josef et al. 2006). The flagellar Ca^{2+}-channels involved in phototaxis still await identification. CAV2 is involved in the photophobic and mechanoresponses, but not in mediating phototaxis (Fujiu et al. 2009; Yoshimura 2011).

In the colonial *Volvox* species, phototactic steering and especially the beat regulation during the photo-responses differ. Here the two flagella of each cell beat in roughly the same direction toward the posterior pole. Direction changes are achieved by changes of the flagellar beating in different portions of the colony. In small *Volvox* species, the flagella of those cells facing toward the light decrease transiently the beat frequency or can even stop. In larger species, however, the beat direction reverses transiently from prevailing toward the posterior to predominantly laterally (Hoops et al. 1999; Drescher et al. 2010; Ueki et al. 2010; Solari et al.

Fig. 2 (continued) how this coupling is achieved. (**c**) *Chlamydomonas* not only rotates around the long axis of the cell but also moves on a helical path due to slight imbalances in the beat of the *cis*- and *trans*-flagella. Thus, the eyespot scans the environment for the light source and is thereby alternately illuminated and shaded. Thus, as shown in (**d**), the magnitude of the deviation of the swimming path from the light source is one major determining factor for the signal magnitude generated by ChR excitation. (**e**) The eyespot position in relation to the cell axis differs but is always roughly perpendicular to the direction of movement, in flagellate green algae. Shown examples are *Nephroselmis olivacea* (1), *Mesostigma viride* (2), or *Pediomonas spec.* (3). The typical situation for *Chlamydomonas*-type cells, where movement direction and cell axis are identical, is depicted in (a). *Dashed red line*, longitudinal cell axis; *dashed black arrow*, movement direction. Redrawn with modifications from Kreimer (2001)

2011). These coordinated changes do not need cytoplasmic connections between the individual cells. Recent analyses of *V. rousseletii* G. S. West cell models revealed that the Ca^{2+} responsiveness of the flagella exhibits a clear anterior to posterior gradient: high at the anterior pole and almost zero at the posterior one in this large species. Unfortunately, cAMP was not included in the analyses. Beat reversal is involved in both photoshock and phototaxis. It is this graded response along the anterior-posterior axis together with the variation in eyespot size, which appears to be important for phototaxis in *Volvox* (Ueki et al. 2010; Ueki and Wakabayashi 2018). Many cell biological and molecular issues on how the different Volvox species regulate phototaxis are still unclear, just to name a few: How does the relative position of the individual somatic cell along the anterior/posterior axis of the colony determine eyespot size and the tilt toward the anterior pole? How can the relative position determine the axonemal Ca^{2+}-sensitivity gradient? What is the molecular and structural basis of the different flagellar responses observed between small and large *Volvox* species? Finally, detailed studies of flagella and phototactic behavior for other multicellular species from the Chlamydomonadales would be helpful to understand the mechanisms allowing oriented movement responses after the transition from uni- to multicellularity. Already the four-celled *Tetrabaena socialis* (Dujardin) H. Nozaki & M. Itoh colony exhibits rotational asymmetry in arrangement of microtubular rootlets and separation of basal bodies like in *Volvox* (Arakaki et al. 2013).

It follows from the above that the correct differential flagella responses initiating the phototactic response must be phased precisely with ChR sensitivity, signal perception, and the transduction cascades from the eyespot to the flagella. Hence, many mutants affected in flagella beating exhibit also changes in phototactic behavior. For instance, mutation in Bug22, a protein that controls the planar beat, causes slow and three-dimensional beating of the flagella and results in a loss of phototaxis (Meng et al. 2014). The reason for this is that in *Chlamydomonas* ChR sensitivity is optimized to changes in the light intensity within the frequency range of cell body rotation (Yoshimura and Kamiya 2001). Consequently, the PCs are faster in *Chlamydomonas* than in *Volvox* (Braun and Hegemann 1999; Hegemann and Berthold 2009). Also fast mechanisms adapting the ChR sensitivity under continuous light, e.g., formation and ratio of the different conducting ChR states (Hegemann et al. 2005; Bruun et al. 2015; Kuhne et al. 2019), will probably be optimized to fit the rotational speed. In addition, the absorption properties of the ChRs must fit the optical properties (absorption, reflection, refraction) of the eyespot, especially those changing with the varying angle of light incidence caused by the helical swimming. It is obvious that a complex interplay of separate regulatory mechanisms at the eyespot, signal transduction, and flagella level is indispensable to make phototaxis such a robust behavior. Whereas already a great deal of knowledge has been accumulated for the flagella, still very limited information is available for the eyespot and the signaling processes toward the flagella. Here detailed future research is desirable for a better understanding of the overall integration of this photo-behavior.

3 Protein Composition and Expression Profiles

Currently over 100 algal whole-genome sequences are available. Among them are 25 genomes from members of the Chlorophyta, including *C. reinhardtii* and *V. carteri* (Merchant et al. 2007; Prochnik et al. 2010; Blaby-Haas and Merchant 2019). Thus, the basis for transcriptomic, proteomic, and mutational approaches is constantly growing and allows diverse comparative analyses. Analyses of eyespot- and phototaxis-related signaling processes will also further benefit from these still growing datasets. Once ample molecular data are gathered for the unicellular *C. reinhardtii* and *V. carteri f. nagariensis* (~2,000 cells; spherical), analyses, e.g., related to necessary changes and specializations linked to adaptation of phototactic steering during the process of becoming multicellular, can be tackled. Here the genomes of *T. socialis* (4 cells, square) and *Gonium pectorale* O. F. Müller (~16 cells, planar) will be of value. Until now, the genome data of *Chlamydomonas* and *Volvox* have already been indispensable for diverse proteomic and transcriptomic approaches.

3.1 Eyespot Proteins: An Overview

To unravel eyespot-related functions, knowledge about its protein composition is essential. First trials relied on the identification of photoreceptors from eyespot preparations by biochemical approaches. Here the alternative splicing products COP1 and COP2 were the first reported retinal binding receptors. Immuno-localization and enrichment in eyespot preparations confirmed their presence in the eyespot region. However, both are not involved in phototaxis (Deininger et al. 1995; Fuhrmann et al. 2001; Trippens et al. 2017), and their cellular functions are still unknown. COP2 is chloroplast-localized and co-purifies with and stabilizes the Ycf4 complex, which is important for photosystem I assembly, and might be involved in the regulation of this important process (Ozawa et al. 2009; Kreimer 2009). Recent disruption of the gene coding for COP1/2 (Greiner et al. 2017) has now paved the way to unravel the in vivo function(s) for both proteins. Based on expression studies in *Xenopus* oocytes, Tian et al. (2018) have challenged recently their function as photoreceptors and members of the opsin family. Thus, further analyses on COP1/2 function(s) are highly desirable. The major light receptors involved in phototaxis of *C. reinhardtii*, ChR1 and ChR2, were identified later by searching EST libraries and genome databases (Nagel et al. 2002, 2003; Sineshchekov et al. 2002; Suzuki et al. 2003). Currently, a diversity of algal ChRs acting either as cation or as anion channels are identified by transcriptome sequencing, thereby broadening the optogenetic tool kit. Natural anion ChRs are, however, currently only known from the Cryptophytes (e.g., Klapoetke et al. 2014; Govorunova et al. 2015, 2017a, b). We discuss the ChRs in Sect. 3.3.1 in more detail together with other photoreceptors related to the eyespot and phototactic behavior.

Identification of other eyespot related proteins relied mainly on two approaches: forward genetic screens and isolation of eyespots or parts of them followed by biochemical approaches. In the genetic approaches, first screens usually take advantage of an altered phototactic behavior, followed by a second focused on aberrant eyespot features. To narrow down the mutants to eyespot-related proteins, dual screens are mandatory as phototaxis is affected by a multiplicity of factors, ranging, e.g., from the photosynthetic efficiency to the function of the BBSome in the flagella (Pazour et al. 1995; Lechtreck et al. 2009; Kim et al. 2016; Liu and Lechtreck 2018). However, even then mutants missing a visual eyespot can end up in the identification of mutated genes for general pathways, e.g., the carotenoid biosynthetic pathway, which are also important for normal eyespot development (Ueki et al. 2016). As discussed already above (see Sect. 2.3), such mutants are useful in deciphering the optical properties of the eyespot and cell. Such screens identified also loci central for eyespot assembly and positioning (*EYE2*, *EYE3*, *MIN1*, and *MLT1*; Lamb et al. 1999). Whereas MLT1 localizes on the D4 rootlet and is central for the formation of a single eyespot after cytokinesis, the other proteins are found within the eyespot (see Fig. 1d; Roberts et al. 2001; Mittelmeier et al. 2008, 2015; Boyd et al. 2011b). Thompson et al. (2017) summarized the currently known details about these proteins. A mutation in *MLT1* was also identified independently within the frame of an insertional mutagenesis screen for non-phototactic mutants. Here it was the only gene out of seven phototaxis-related genes, which affected the eyespot visually. The others were involved in signaling toward and responses of the flagella (Pazour et al. 1995).

Biochemical approaches mainly relied on the development of methods for the enrichment of intact eyespots and different eyespot fractions (e.g., Kreimer et al. 1991; Deininger et al. 1995; Renninger et al. 2001; Schmidt et al. 2006; Eitzinger et al. 2015). Using such preparations proteins such as COP1/2 and GAP1, a not further characterized putatively globule-associated protein, or protease-protected subclasses of eyespot globule-associated proteins were identified (Deininger et al. 1995; Renninger et al. 2001; Schmidt et al. 2007). Large-scale proteomic approaches to highly purified eyespot preparations identified later 202 proteins with $\geq$2 different peptides. The 72 proteins identified with $\geq$5 peptides probably represent the core eyespot proteome (Schmidt et al. 2006; Eitzinger et al. 2015). It includes, besides all known eyespot proteins, for example, also putative structural proteins, such as many plastid lipid-associated protein/fibrillin (PAP-fibrillin) domain-possessing proteins, enzymes involved in carotene and retinal biosynthesis, and potentially signaling-related proteins. The latter represent ~16% of the core proteome and include photoreceptors, kinases, and members of the PP2C family and Ca^{2+}-binding proteins. Use of these proteomic datasets in combination with insertional mutant libraries (e.g., Dent et al. 2015; Li et al. 2016, 2019; Cheng et al. 2017) will accelerate our understanding of their functional relevance for photo-orientation and the eyespot.

Currently, only for a few candidates from this list, experimental validation for a role in eyespot development or phototaxis exists. The blue light receptors PHOT and SOUL3, a globule-localized hemin-binding protein (Fig. 1d), are involved in eyespot size regulation. SOUL3 additionally affects eyespot positioning (Trippens et al. 2012; Schulze et al. 2013). PP2Cs are involved in ChR1 dephosphorylation, and CK1 has, among others, effects on the circadian control of phototaxis (Schmidt et al. 2006; Böhm et al. 2019). Eitzinger et al. (2015) confirmed with largely intact eyespots the core proteome and put more potential candidates on the list. Gain in intactness, however, results also in an increase of contaminants. Several reasons complicate the decision what to regard as a contaminant in such preparations. Major points are (1) the functional eyespot is an "organelle" composed of different subcellular compartments, (2) the actual protein composition of the eyespot might change upon changing conditions, and (3) protein moonlighting may also occur within the eyespot. The following examples will illustrate these difficulties. Recent evidence points to dynamic, light-dependent redistributions of rhodopsins between the flagella and eyespot in *Chlamydomonas* (Awasthi et al. 2016). The functional significance is yet not clear. Restriction of cycling of phospholipase D, a protein important for intracellular signaling, between the cell body and the flagella impairs phototaxis (Lechtreck et al. 2009; Liu and Lechtreck 2018). Thus, ChR redistributions might as well have functional significance. Although only a small proportion of the cellular PHOT is located to the eyespot, this blue light receptor affects phototaxis in several ways (see Sect. 3.3; Huang et al. 2004; Trippens et al. 2012). Further, indirect evidence points to potential moonlighting functions of the α- and ß-subunits of the chloroplast ATP synthase within the eyespot (Schmidt et al. 2007). The Ca^{2+}-sensing receptor CAS represents another instructive example. In *Chlamydomonas* it is present in the thylakoid and eyespot proteome (Allmer et al. 2006; Schmidt et al. 2006). It is involved in regulating photo-acclimation, cyclic photosynthetic electron flow under anoxia, and the CO_2-concentrating mechanism (Petroutsos et al. 2011; Terashima et al. 2012; Wang et al. 2016). CAS is additionally interrelated to adapting the photo-behavior under low light (Trippens et al. 2017). Interestingly the ambient CO_2 level affects its subcellular distribution. Under high CO_2, it forms clusters in all thylakoid membranes. Low CO_2 led to a restriction to thylakoids running through the pyrenoid and to the eyespot (Yamano et al. 2018). This example highlights also the need of high-resolution localization of (potentially) eyespot-related proteins. Such data are currently restricted to analyses by fluorescence microscopy (e.g., Deininger et al. 1995; Kreimer 2009; Mittelmeier et al. 2011; Luck et al. 2012; Awasthi et al. 2016; Yamano et al. 2018). High-resolution light microscopy localized EYE2 to the chloroplast envelope and SOUL3 to the globules (Boyd et al. 2011b; Schulze et al. 2013). In the future electron microscopy approaches will help to unravel their exact localization within the eyespot. In combination with high-resolution live cell imaging, it will also help to understand protein dynamics during assembly, steady state, and disassembly of the eyespot.

3.2 *Diurnal Phasing of Expression Levels of Eyespot-Related Genes*

Eyespot distribution to the progenies occurs in the Chlorophyta by either division, a semiconservative fashion, where one is kept and the other is newly formed, or de novo (Holmes and Dutcher 1989; Suda 2003; Kreimer 2009). In *C. reinhardtii*, eyespot de novo synthesis occurs. Its disassembly starts shortly prior to beginning of mitosis. Cell division takes place early in the dark phase, and after its completion, both eyespots of the daughter cells are assembled (Holmes and Dutcher 1989; Wood et al. 2012). Zones et al. (2015) have published the diurnal transcriptome of highly synchronized *C. reinhardtii* cultures. In *Chlamydomonas* the majority of the genome is differentially expressed (Zones et al. 2015; Strenkert et al. 2019). We were interested in the expression patterns of selected genes of photoreceptors, proteins important for structure, development, and eyespot placement, and some proteins identified by proteomic approaches. In Fig. 3, we plotted the expression patterns of 22 genes falling in the above categories from the dataset of Zones et al. (2015). As the maximum reads per kilobase per million mapped of the different transcripts greatly differ (e.g., *COP1/2*, 454, and *MLT1*, 4.4; Table 1), each dataset is normalized to its maximum to allow a better comparison. The expression levels of the majority of the selected genes correlate well with cell division and eyespot assembly and peak early in the dark phase. Examples are transcripts in which mutations lead to loss of a visible eyespot (*EYE2, EYE3*) or those of globule-associated proteins like *SOUL3* and *PLAP7* (Fig. 3c–e). EYE2 possess a LysM domain, and a significant proportion of proteins with this domain in *C. reinhardtii* (20%) are part of the eyespot proteome (Boyd et al. 2011b; Thompson et al. 2017). Transcripts for other eyespot proteins with this domain (MIN1, PLAP8, LMR1) also follow this pattern. These proteins have a hydrophobic character. MIN1 possesses additionally a lipid bilayer interacting C2 domain (Schmidt et al. 2006; Mittelmeier et al. 2008; Eitzinger et al. 2015). Thus, these expression patterns support the suggestion that these proteins might be involved in facilitating adhesion of the different eyespot components (Thompson et al. 2017).

A continuous increase upon onset of the dark phase is also seen for the transcript levels of the two primary photoreceptors for phototaxis, ChR1 and ChR2 (COP3 and COP4). Maximal increases occur during the last 2 h of the night phase, i.e., when eyespot formation is already finished. Upon onset of illumination, their transcript levels rapidly decrease within 1 h. The transcript decrease is more dramatic for ChR1 than for ChR2. After that, the levels increase slightly again before they drop 4 h before the night phase again to a minimum. The pattern of the transcript level of Cr2c-Cyclop1, a member of a new opsin family in green algae with a light-inhibited guanylyl cyclase activity (Tian et al. 2018), is similar to that of ChR2 (Fig. 3a, b). A similar behavior occurs for almost all selected transcripts. The levels of *COP1/2* and that of the two enzyme rhodopsins (*COP5* and *COP8* alias *HRK1* and *HKR4*, respectively) increase then again in the light, with the *COP1/2* exhibiting a sharp peak 4–5 h after onset of the light phase (Fig. 3a). None of these photoreceptors is

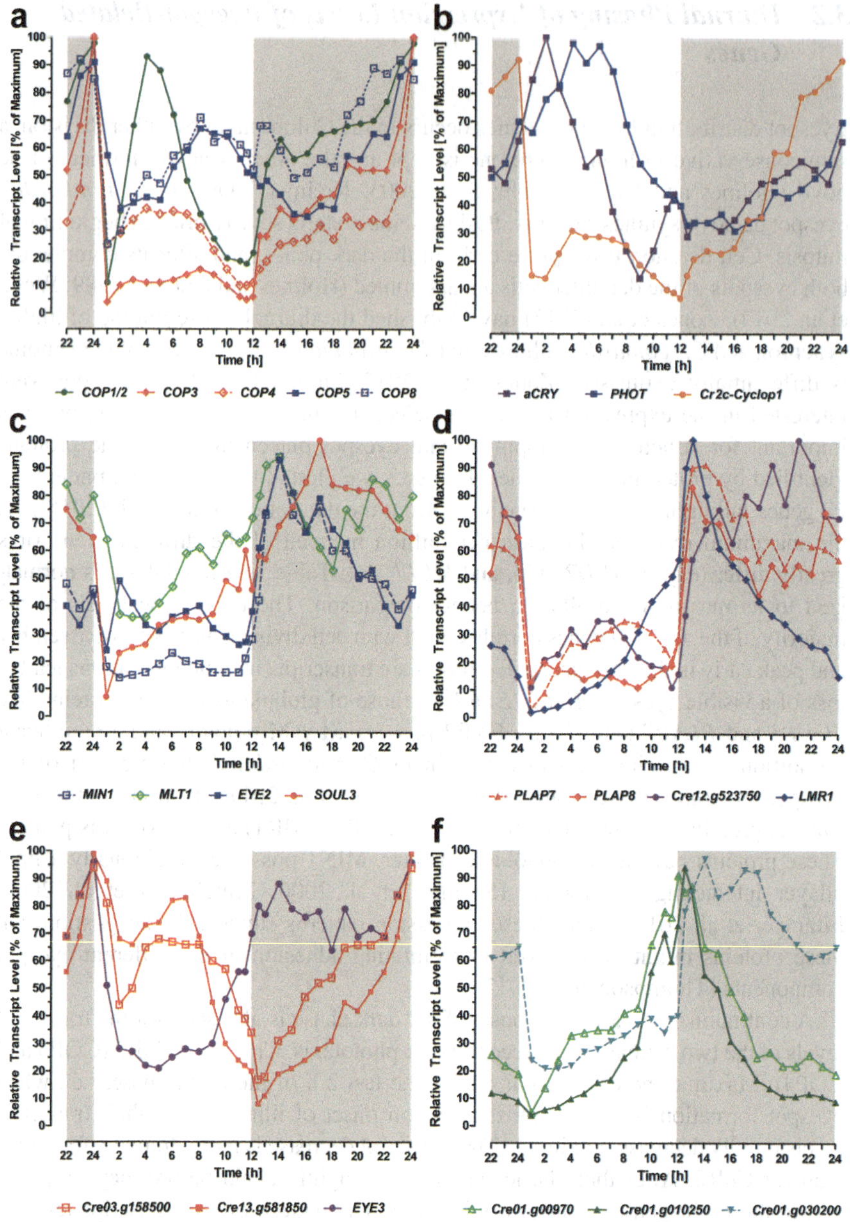

Fig. 3 Diurnal phasing of expression levels of selected genes coding for known photoreceptors and proteins known or assumed to be involved in eyespot development and structure, function, and signaling. The means of two independent replicates from synchronized cultures grown at a 12:12 h LD cycle normalized to the peak expression (selected from supplemental dataset 1 of Zones et al. 2015) are shown. As the actual maximal reads per kilobase per million mapped diverge by a factor of ~100 (Zones et al. 2015; Table 1), we plotted the data as percentage of the corresponding maximum to allow an easier comparison. For better visualization, the last three data points from the dark phase were plotted twice. *Light gray shading* marks the dark phase. The cells divide early in

Table 1 Maximum reads per kilobase per million mapped (mean of two replicates) for the genes plotted in Fig. 3 over a 12 h light: 12 h dark cycle observed by Zones et al. (2015)

Functional classification	Gene alias or phytozome Cr#	Maximal RPKM
Photoreceptors	*COP1/2*	454
	COP3 (ChR1)	284
	COP4 (ChR2)	65.6
	COP5 (HKR1)	55.7
	COP8 (HKR4)	9.6
	Cr2c-Cyclop1	22.8
	aCry	12.4
	Phot	33.7
Structural and positioning	*Min1*	28.7
	MLT1	4.4
	EYE2	15.6
	SOUL3	46.3
	PLAP7	15.6
	PLAP8	20.7
	Cre12.g523750	8.3
	LMR1	154.6
ABC1 kinase family	*Cre03.g158500*	29.3
	Cre13.g581850	53.3
	EYE3	65.7
PP2c family	*Cre01.g00970*	5.7
	Cre01.g010250	12.8
	Cre01.g030200	155.6

RPKM maximum reads per kilobase per million mapped

directly involved in the movement responses, although immunofluorescence localized them all to the eyespot (Deininger et al. 1995; Luck et al. 2012; Awasthi et al. 2016). The levels of *aCRY* and *PHOT* also peak in the first half of the light phase. PHOT affects the ChR1 protein level in vegetative cells (Trippens et al. 2012). Preliminary data indicate that also aCRY in gametes (Zou 2016) and vegetative cells (Böhm and Kreimer, unpublished data) affects ChR1 (see Sect. 3.3.3). The transcript levels of both ABC1-like kinases of the core eyespot proteome peak in a similar time window as that of PHOT (Fig. 3b, e). Their transcript levels are minimal when cell division starts and then rise in the night phase to a maximum at the end of the night

Fig. 3 (continued) the dark phase (~13–17 h; Wood et al. 2012; Zones et al. 2015). (**a, b**) Transcript levels for photoreceptors: (a) retinal-based and (b) *aCRY* and *PHOT*. *COP3* = *ChR1*; *COP4* = *ChR2*; *COP5* = *HRK1*; *COP8* = *HRK4*. (**c, d**) Transcript levels of structural proteins: (c) eyespot size, formation and positioning, (d) putative core globule/eyespot proteins with PAP/fibrillin and/or LysM domains. (**e**) Transcripts levels of kinases belonging to the ABC1 kinase family and (**f**) of phosphatases of the PP2c family identified by proteomic approaches to the eyespot

phase. They are homologs of the ABC1-like kinases 1 and 3 of *Arabidopsis.* Both are thought to regulate the tocopherol metabolism in *Arabidopsis*, probably via affecting the tocopherol cyclase stability and other enzymes including the ABC1-like kinase 6 (e.g., Lundquist et al. 2013; Martinis et al. 2014; Lohscheider et al. 2016; van Wijk and Kessler 2017). By analogy, these two eyespot-associated kinases might have a role in eyespot globuli stabilization. Notably, the transcript level of the eyespot globule localized EYE3, a homolog of the *Arabidopsis* ABC1-like kinase 6 and essential for globule formation (Boyd et al. 2011b), has a constant high level in the night phase (Fig. 3e). It will be interesting to test if these two ABC1-like kinases are involved in EYE3 regulation. Alternatively, as first suggested for the plastoglobule proteins of higher plants (Lohscheider et al. 2016), these kinases in conjunction with EYE3 might be important for the globule layer assembly and stability (Thompson et al. 2017). Interestingly, some proteins with potential impact on eyespot structure are phosphoproteins (Wagner et al. 2008). ChR1 and ChR2 are also phosphoproteins, and pharmacological evidence suggests involvement of members of the PP2c family in ChR1 dephosphorylation (Wagner et al. 2008; Wang et al. 2014; Böhm et al. 2019). Transcripts for the three PP2Cs identified in the eyespot proteomes (Schmidt et al. 2006; Eitzinger et al. 2015) all peak within the first 2 h of the dark phase (Fig. 3f). The transcript levels of the dominant two PP2Cs rapidly decrease in the night phase, whereas that of the PP2C-like phosphatase (*Cre01.g030200*) remains high until the beginning of the light phase.

In summary, most of the analyzed eyespot relevant genes in *Chlamydomonas* are expressed in a highly coordinated and similar pattern. Some, but not all, eyespot genes cluster within linkage groups and are co-expressed (Thompson et al. 2017). Upon accumulation of more knowledge about the different eyespot components and signaling cascades initiated by ChR excitation, the high-resolution dataset of Zones et al. (2015) will be useful. As protein levels not necessarily match transcript levels, quantitative analyses of the eyespot proteins of interest should be included whenever possible. Only the combination of multiomic approaches like the recent one by Strenkert et al. (2019) with a focus on photosynthesis will help to decipher further the networks underlying the regulation of eyespot formation, photomovement responses, and the embedment in the physiological needs of the cells.

Also for *Volvox* transcriptional analyses, including a very detailed analysis of its different photoreceptors, are available (Kianianmomeni et al. 2009; Kianianmomeni and Hallmann 2015; Klein et al. 2017). These analyses demonstrated that the two *Volvox* channelrhodopsins (*VChR1 and 2*) are highly expressed in somatic cells but not in the reproductive cells. Low light stimulates expression of both. Other environmental and developmental factors, however, led to more selective effects on the transcript levels of the two *VChRs*. Notably, UV-A, green, yellow, and red light affect the *VChR1* levels, and they respond highly sensitive to environmental cues. For *VChR2* such changes were not seen. *V. carteri* encodes not only ChRs but also opsins of the histidine-kinase group (VcHKR1-4). Three of them possess adenylate/guanylate cyclase domains (Kianianmomeni and Hallmann 2015). Their expression levels show cell-type specificity and changes during the sexual life cycle, indicating specific roles in sexual development; e.g., *VcHKR4* transcript levels increase in response to the sexual inducer by ~300% (Kianianmomeni and Hallmann 2015; Tian et al. 2018). Recently, Tian et al. (2018) renamed them as two-component cyclase opsins (2c-Cyclops). Expression levels of Vc2c-Cyclop1 are, except during embryogenesis, quite low and do not differ for different stress conditions, except for light stress (moderate upregulation) and UV-A/yellow light (downregulation; Tian et al. 2018). However, as for *Chlamydomonas*, currently no evidence for involvement of Vc2c-Cyclop1 in regulating phototaxis is available. It could have an indirect effect via modulating the intracellular cGMP level. The basal levels of cGMP and cAMP differ between *Chlamydomonas* strains with positive and negative phototactic behavior. Those exhibiting positive taxis had higher levels (Boonyareth et al. 2009). However, these authors saw only significant changes in cAMP – but not cGMP – levels upon illumination with 514 nm light. As the guanylyl cyclase activity of both Cyclop1 members is light inhibited with the strongest inhibition observed at ~540 nm (Tian et al. 2018), even an indirect involvement of Cyclop1 appears thus not likely. For more details on the rhodopsin-like and other photoreceptors in *Volvox*, the reader is referred to the in-depth analysis of Kianianmomeni and Hallmann (2015).

3.3 Eyespot-Related Photoreceptors

As already briefly mentioned in Sect. 2.1, there is burgeoning evidence that some aspects of the eyespot and phototactic behavior are under the control of a photoreceptor network. Current data points to receptors absorbing in the blue/UV range: namely, the ChRs, PHOT, and both pCRY and aCRY. Except for the CRYs, proteomic approaches demonstrated their presence in the eyespot (Schmidt et al. 2006; Wagner et al. 2008; Eitzinger et al. 2015). Although *C. reinhardtii* responds to UV light with negative phototaxis (e.g., Nultsch et al. 1971; Trippens et al. 2012), the UV-B receptor UVR8 was not identified in the eyespot proteome. Low UV-B exposure inducing adaptational responses does not significantly affect, e.g., the transcript level of the CHRs but induces increased transcript levels of PHOT as well as aCRY (Tilbrook et al. 2016). Thus, UVR8 might still be indirectly involved in such a network. Further, it is well known for *C. reinhardtii* that both phototactic and chemotactic movements are not only light-controlled but also affected by the circadian clock. Highest phototactic sensitivity is observed during the day. Important environmental clues like temperature and the photoperiod synchronize the clock. Different wavelengths ranging from violet to red are effective in the clock entrainment. Although phytochromes from eukaryotic algae sense a broad range of wavelengths ranging from blue, green, orange to far-red (Rockwell et al. 2014), its involvement can be excluded as the *Chlamydomonas* genome lacks this photoreceptor. Instead current evidence supports pCRY, which is also known as CPH1, and a red/violet signaling pathway with a not yet unequivocally identified photoreceptor as candidates for clock setting (Johnson et al. 1991; Kondo et al. 1991; Niwa et al. 2013; Forbes-Stovall et al. 2014; Kinoshita et al. 2017). Potential eyespot signaling inputs into the clock are not yet understood at all. Silencing of CK1, a part of the clock in many organisms, has – among others – clear effects on the period of circadian phototaxis. Further, this kinase is present in the proteomes of the eyespot and flagella (Pazour et al. 2005; Schmidt et al. 2006). We refer readers interested in details about the algal circadian clock to reviews by, e.g., Schulze et al. (2010), Matsuo and Ishiura (2010, 2011), Noordally and Millar (2015), and Ryo et al. (2016). Below we will concentrate on the discussion of ChRs, PHOT, and CRYs.

3.3.1 Channelrhodopsins

Green algal ChRs are directly light-gated cation channels conducting in algae mainly Ca^{2+}, H^+, and Na^+, whereas ectopically expressed they conduct mainly H^+. With their planar all-*trans,6s-trans* retinal chromophore and seven-transmembrane domain structure, they belong to the microbial-type rhodopsins. Due to N- and C-terminal extensions, they are larger than other rhodopsins (*Chlamydomonas*, ChR1 ~76.5 kDa, ChR2 ~77.2 kDa; *Volvox*, ChR1 ~86.6 kDa, ChR2 ~77.8 kDa). The functional ChRs are homodimers. Further details on, e.g., structure, ion permeation pathway, and photocycles are summarized in diverse reviews (Hegemann and Berthold 2009; Kianianmomeni and Hallmann 2014; Schneider et al. 2015; Deisseroth and Hegemann 2017; Govorunova et al. 2017a; Kuhne et al. 2019). Both ChRs are photochromic, and ChR1 is additionally protochromic, with a peak absorbance shift from ~470 to ~500 nm at more acidic conditions. This shift is accompanied by significant increases in the current amplitude at wavelengths above 525 nm (Berthold et al. 2008). Light-induced H^+-extrusion across the chloroplast envelope membranes is a long known phenomenon in plants and algae (Höhner et al. 2016). Thus, acidification of the small cytosolic space below the CHR patch allows indirect coupling of photosynthesis and ChR activity via expansion of the effective wavelength range and current modulation. This might be an early part of the system controlling the phototactic sign. In general, the ChR properties are well adapted to the optics of the eyespot and the movement pattern. For instance, their dichroic orientation within the membrane increases the eyespot directionality, and their sensitivity is optimized to the frequency range of cell body rotation (Yoshimura 1994; Yoshimura and Kamiya 2001). Further, eyespot reflectivity changes with the angle of light incidence from ~480 nm for a perpendicular incidence toward longer wavelength with decreasing angles (Foster and Smyth 1980). Such shifts would favor the accumulation of less conductive ChR sub-states according to the most recent photocycle model for light adaptation and temporal evolution of cation conductance under continuous illumination (Kuhne et al. 2019). However, it should be kept in mind that a great diversity regarding spectral sensitivity, kinetics, and inactivation of the currents as well the pH dependence exists even within the genus *Chlamydomonas* (Hou et al. 2012; Klapoetke et al. 2014).

In algae, ChR excitation evokes two PCs, an early, fast and a late PC. The latter saturates already at low light intensities, whereas the fast PC reaches a plateau only at saturating intensities. For *Chlamydomonas* it is established that both ChRs contribute to the generation of the PCs and phototaxis. Their relative contribution, however, varies between strains, vegetative cells, and gametes. In *Volvox* both ChRs are additionally involved in developmental processes (Sineshchekov and Spudich 2005; Berthold et al. 2008; Hegemann and Berthold 2009; Kianianmomeni et al. 2009; Sineshchekov et al. 2009; Greiner et al. 2017). Some evidence (e.g., PC amplitude changes and other data from RNA_i mutants depleted by either ChR1 or ChR2) indicates that besides their direct channel activity, other mechanisms are involved in initiating phototaxis in alga at low light (summarized by Sineshchekov et al. 2009). Under these conditions, the ChR ion channel activities alone are not sufficient for significant plasma membrane depolarization, and signal amplification via, e.g., secondary ion channels or changes in second messenger molecules is supposed. At intermediate and higher light intensities, a coexistence of direct ChR channel activity and signal amplification is proposed (Kateriya et al. 2004; Sineshchekov et al. 2009). However, the mechanisms and relative contributions of ChR1 and ChR2 to the phototactic response are still unclear. Interestingly, high light reduces ChR2 promoter activity (Fuhrmann et al. 2004). The functional significance of their different concentrations, both on the cellular and strain level, is not yet understood. However, their protein levels are clearly interconnected. A massive increase in ChR1 occurred in a ChR2-RNAi strain (Govorunova et al. 2004). Vice versa, in ChR1 deletion strains, ChR2 increases. The magnitude of the increase, however, depends on the strain (Böhm and Kreimer, unpublished results). Eyespot size regulation shows also some yet not understood strain dependencies. For instance, in 302cw deletion of ChR1 has no significant effect, whereas in CC-3403 deletion of either ChR1 or ChR2 leads to a significant size reduction (Trippens et al. 2012; Greiner et al. 2017). Nevertheless, it should be kept in mind that the small eyespot size of strain 302cw might have hampered the detection of such subtle size differences.

Light affects the cellular ChR1 and ChR2 content at two levels. The transcript levels of ChR1 decrease at onset of the day phase within 1 h by 96%. In contrast, ChR2 is less affected, and only high light reduces ChR2 promoter activity (Fig. 3a; Fuhrmann et al. 2004). At the protein level, the ChR1 concentration changes in a light intensity- and wavelength-dependent manner, both in the absence and presence of the translation inhibitor CHX. ChR1 degradation is highest in the blue-violet range, whereas red and green light is less effective (Fig. 4a–c). The decrease in the ChR2 level is much slower (Böhm and Kreimer, unpublished results). These data point to specific light-dependent regulatory mechanisms in the turnover of both ChRs. One of the involved receptors in ChR1, but not in ChR2, degradation is PHOT (Trippens et al. 2012; see Sect. 3.3.2). Both ChRs are phosphoproteins, and the phosphorylation status of ChR1 exhibits dynamic and rapid responses to key physiological stimuli, but PHOT is not involved in its phosphorylation (Wagner et al. 2008; Wang et al. 2014; Böhm et al. 2019; for details to ChR1 phosphorylation, see Sect. 4). Further, induction of ChR1 hyper-phosphorylation does not lead to an increased degradation (Boness and Kreimer, unpublished data). Analyses of different PHOT knockout strains showed that there are further light-dependent components (Trippens et al. 2012; Böhm and Kreimer, unpublished results). As will be

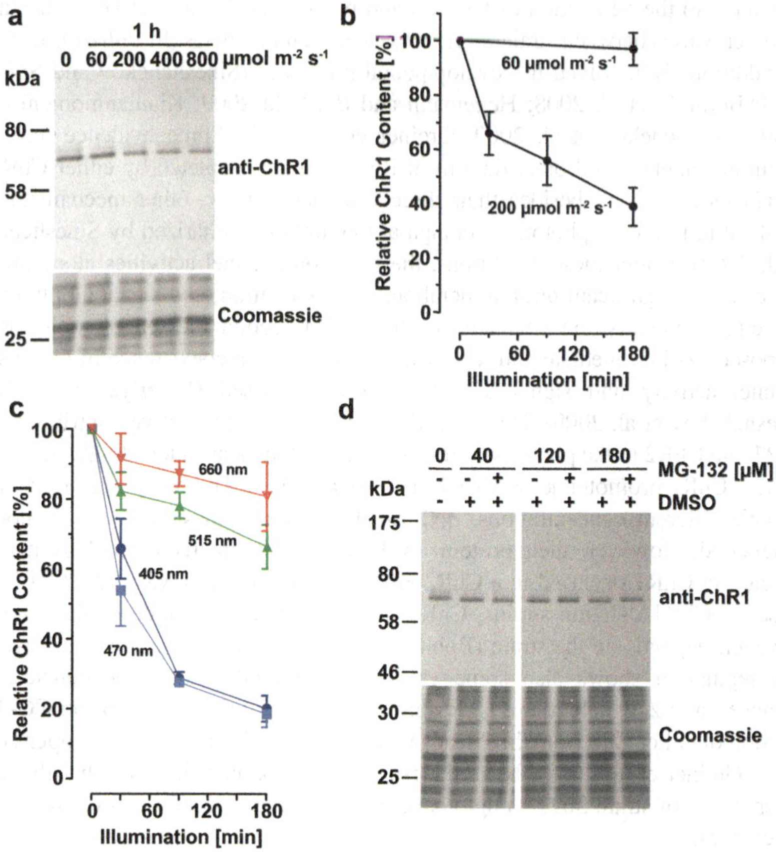

Fig. 4 Light affects the level of ChR1 in an intensity-dependent manner. (**a**) *Chlamydomonas* cultures (strain cw15) were grown in TAP medium under a 14 h light (60 µmol m^{-2} s^{-1}), 10 h dark cycle for 5 days. Cells from the dark phase were then either kept in darkness or were illuminated for 1 h with the indicated intensities of white light. Cells were then directly injected into MetOH:Chl (2:1), and the precipitated proteins were separated by SDS-PAGE prior to immunoblot analysis with anti-ChR1(1:5000). (**b**) To analyze whether the decrease in the ChR1 level with increasing light intensity is a consequence of enhanced degradation, cultures at the end of the night phase were preincubated for 1 h in the dark with CHX, which blocks protein synthesis at the 80S ribosomes. Cells were then illuminated for the indicated times and intensities in the presence of the inhibitor. Proteins were extracted as in (**a**), and following Western blotting the ChR1 band was quantified. The content prior to illumination was set as 100%. Data plotted are the mean ± SD of blots from 3 independent cultures. (**c**) Spectral dependence of the light-induced ChR1degradation. Cultures of strain cw15, grown as described under (**a**), were preincubated at the end of the night phase with CHX for 1 h in the dark, transferred into fresh TAP medium (pH 8) supplemented with CHX, and illuminated with the indicated wavelengths (100 µmol m^{-2} s^{-1}; LEDs) from the top. Proteins were extracted as in (**a**), and following Western blotting the ChR1 band was quantified. The amounts prior to illumination were set as 100%. Data plotted are the mean ± SD of blots from 3 independent cultures. (**d**) Effect of the proteasome inhibitor MG-132 on light-induced ChR1 degradation. Cultures of cw15 were grown and treated with CHX as described in (**c**). Prior to illumination with white light (200 µmol m^{-2} s^{-1}), the cultures were either supplemented with the indicated

discussed in Sect. 3.3.3, preliminary data suggest a possible role of aCRY. To the best of our knowledge, almost nothing is yet known about the degradation pathway (s). We therefore started recently with pharmacological approaches. As shown in Fig. 4d, the cell-permeable proteasome inhibitor MG-132 has no effect in CHX-treated illuminated cells. Involvement of this pathway like, e.g., in the blue/red light-induced degradation of pCRY and aCRY or the circadian clock component ROC15 in *Chlamydomonas*, is unlikely (Reisdorph and Small 2004; Niwa et al. 2013; Müller et al. 2017; Zou et al. 2017).

To understand the in vivo regulation of phototaxis in more detail, knowledge about ChR regulation under constant illumination and potential interaction partners is important. ChRs form dimers. Berthold et al. (2008) showed that ChR1 but not ChR2 forms dimers in vivo via disulfide bridges. In our hands, however, also ChR2 forms multimers (Böhm and Kreimer, in preparation). Our observations are in accordance with data from spin distance measurements, cryoelectron microscopy, and the crystal structure from ectopic expressed ChR2. It forms stable dimers mainly via interaction between helices 3 and 4 and two extracellular cysteine residues (Cys34, Cys36). Even when these residues are mutated, the dimers are stable (Müller et al. 2011; Krause et al. 2013; Volkov et al. 2017). Only recently, Awasthi et al. (2018) were the first to show that ChR1 and ChR2 do physically interact and that their extended C-termini are involved. Thus, ChRs probably also form heteromeric dimers. Neither the amounts, the physiological significance, nor under which conditions these are formed is yet known. However, as the functional ChRs are homodimers, it may have potential regulatory roles. To understand better the in vivo regulation and function of ChRs, knowledge of additional potential interaction partners is important. Immunoprecipitation using solubilized crude extracts points to ChR1 interaction with some components of the IFT machinery (Awasthi et al. 2016). This substantiates diverse immunofluorescence data, which support a directed transport of ChR1 to the eyespot along cytoskeletal elements and the D4 rootlet (summarized by Thompson et al. 2017). Similar evidence also exists for ChR2 (Awasthi et al. 2018). Additionally, also several proteins involved in vesicle formation, trafficking, and signaling were coprecipitated besides proteins with general housekeeping functions in the study of Awasthi et al. (2016). Among the few proteins specifically pointed out by them is a putative voltage-gated calcium channel. As earlier suggested such a channel could be involved in generation of photocurrents under low light (Kateriya et al. 2004; Sineshchekov et al. 2009). Knockout strains of this channel, however, were not disturbed in the strictly Ca^{2+}-dependent ChR1 phosphorylation (Böhm et al. 2019), indicating that it is most likely not involved in signal amplification. We have several independent in vivo

Fig. 4 (continued) concentrations of MG-132 dissolved in DMSO or DMSO alone. As an additional control, CHX-treated cultures without any addition were used. Proteins were extracted and separated by SDS-PAGE prior to immunoblot analysis as described in (**a**). A typical Western blot from several independent repetitions is shown

experimental evidences that ChR1 forms complexes of higher molecular mass (Böhm, Erhard and Kreimer; unpublished data). Currently, one of our aims lies in the identification of the composition of these complexes and to understand under which conditions they are formed.

3.3.2 Phototropin

PHOTs are present in all major land plant lineages and green algae, where they regulate diverse key physiological responses under blue light control. In most land plants and the Zygnematales, two PHOTs occur, whereas in liverworts, hornworts, and all other green algae, it is a single-copy gene (Christie 2007; Kianianmomeni and Hallmann 2014; Li et al. 2015). In *Chlamydomonas* PHOT is present in different subcellular compartments including flagella and eyespot (Huang et al. 2004; Schmidt et al. 2006). Im et al. (2006) reported a major role of PHOT in regulating expression changes in genes encoding enzymes involved in carotenoid and chlorophyll biosynthesis in this alga. Further, it is involved in feedback regulatory processes of photosynthesis (Petroutsos et al. 2016) and the control of the sexual life cycle (Huang and Beck 2003). With respect to motility-related responses, PHOT participates in both the chemotactic and phototactic responses (Ermilova et al. 2004; Trippens et al. 2012; Ermilova and Zalutskaya 2014). Regarding phototaxis it is involved in adaptive responses at different levels and seems to be important for desensitization of the response in a blue light-dependent manner. The suggested fine-tuning role of PHOT for photo-behavior upon increasing blue light intensities by Trippens et al. (2012) would also be beneficial for an increased gamete fusion at the surface of the water body. PHOT is part of the regulatory pathway affecting the ChR1 protein content but has no influence on that of ChR2. PHOT seems to be involved in regulating the ChR1 starting level before onset of illumination as well as in the light intensity-dependent reduction during the first hours of illumination. Further, it is also involved in the light-dependent size regulation of the eyespot and the sign regulation of the phototactic behavior (Trippens et al. 2012). Evidence presented by these authors additionally points to overlapping and independent signaling functions of the C-terminal kinase and the N-terminal LOV domains of PHOT. Monte Carlo simulations (Peter et al. 2014) also support this conclusion by Trippens et al. (2012). In this context, formation of a C-terminal truncated PHOT in a light intensity-dependent manner in addition to the light-activated PHOT might allow a synergistic signaling and might help to respond more subtle to the actual light environment. However, independent PHOT deletion mutants in different backgrounds still exhibit a light intensity-dependent ChR1 decrease, indicating that it is not the only photoreceptor involved (Trippens et al. 2012; Böhm and Kreimer, unpublished results). Further details about these regulatory pathways are still unknown.

PHOT functions as a molecular link between photoreception, photosynthesis, and photoprotection (Petroutsos et al. 2016). As phototaxis is an efficient and direct way to cope with high light and otherwise unfavorable light conditions, PHOT-mediated fine-tuning of the phototactic response and desensitization of the eyespot signaling under increasing blue light intensities underlines this role (Trippens et al. 2012). Besides PHOT's eyespot-related functions, its partial axonemal localization is suggestive for an involvement in other processes of motility regulation and flagella function. For example, reversible flagella adhesion of *Chlamydomonas* on surfaces is a blue light-dependent process (Huang et al. 2004; Kreis et al. 2017). However, neither its effects on flagella beating nor on IFT-mediated surface gliding (Shih et al. 2013) have, to the best of our knowledge, yet been published. Other functions of PHOT in accordance with its flagella localization include the regulation of initial steps of the sexual life cycle. Gamete mating depends on the activity state of the agglutinins on the flagella surface, which is under the control of PHOT as well as the initial gamete formation itself (Huang and Beck 2003; Huang et al. 2004).

In *Volvox* the eyespot size of the individual cells depend on their relative position in the colony (Drescher et al. 2010; Ueki et al. 2010). Thus, it is conceivable that PHOT might also be involved in this process. The PHOT expression level is high in somatic cells compared to gonidia and embryos (Kianianmomeni and Hallmann 2015). It will also be interesting to see whether PHOT and ChR1 levels are interconnected in a similar way as in *Chlamydomonas*. As recently CRISPR/Cas9 mutagenesis has successfully been applied also in *Volvox* (Ortega-Escalante et al. 2019), future analysis of PHOT knockout strains and strains overexpressing the PHOT-kinase domain will help to analyze its role in this colonial green alga.

3.3.3 Cryptochromes: aCRY and pCRY

Cryptochromes are blue light photoreceptors evolutionarily related to the light-activated DNA repair enzymes known as photolyases and are found in bacteria, fungi, algae, plants, and animals (Chaves et al. 2011). For a recent phylogenetic analysis of the green algal cryptochrome photolyase family, see Kottke et al. (2017). *Chlamydomonas* encodes one aCRY, one pCRY (also known as CPH1), one PHR2, and two CRY-DASH proteins (Reisdorph and Small 2004; Merchant et al. 2007; Beel et al. 2012; Kottke et al. 2017). In the *V. carteri* genome in addition to one aCRY and one pCRY, five genes encoding CRY-DASHs are present (Prochnik et al. 2010; Kottke et al. 2017). Currently only aCRY and pCRY are related to some extent to phototaxis by functional analysis. We therefore restrict our discussion to these cryptochromes.

The *Chlamydomonas* aCRY has interesting properties. It exhibits a broad spectral response and acts as both a blue and red light-absorbing photoreceptor. This is accomplished via the unusual stability of the blue light formed neutral radical state (Beel et al. 2012, 2013; Spexard et al. 2014; Oldemeyer et al. 2016; Franz et al. 2018). As reported for other aCRYs from unicellular algae (Coesel et al. 2009; Heijde et al. 2010), it is bifunctional acting in UV-induced DNA repair and as a sensory photoreceptor. Similar to *Ostreococcus tauri* C. Courties & M.-J. Chrétiennot-Dinet, where it is involved in the maintenance of the circadian clock, it is also involved in *C. reinhardtii* in the control of transcript levels of important clock components and diverse other transcripts. Additionally it is part of the network controlling the sexual life cycle (Heijde et al. 2010; Beel et al. 2012; Zou et al. 2017; Franz et al. 2018). Further, analysis of aCRY knockdown mutants points to a contribution of this receptor to the photoaccumulation behavior in both vegetative cells and gametes of *Chlamydomonas* (Zou et al. 2017). Preliminary evidence also connects aCRY to the control of the protein levels of ChR1 in gametes. ChR1 levels in gametes of knockdown mutants are reduced to a lower extent than in the wild-type cells (Zou 2016). aCRY is found mainly in the soluble fraction, where it appears to be present in complexes. Details about the composition of the complexes in vivo are not yet known. Homomer formation via its C-terminal extension has been demonstrated in vitro (Oldemeyer et al. 2016). Recently, Franz-Badur et al. (2019) showed that the C-terminus binds to the photolyase homology domain. Light affects the subcellular distribution of aCRY. During the night, it is mainly present in the cell body, whereas it accumulates in the nucleus during the daytime (Zou et al. 2017). However, as for PHOT, a proportion of aCRY in *Chlamydomonas* is also associated peripherally with membranes. The amounts of these forms change during the sexual cycle. The highest amounts of the membrane-associated form are reported for pregametes and gametes, whereas the soluble form is degraded in these cell types (Kottke et al. 2017; Zou et al. 2017). In vegetative cells aCRY is relatively stable in the light (Zou et al. 2017). In contrast, pCRY is rapidly degraded in the light and accumulates in complexes in the night. It is involved in the control of gametogenesis and zygote germination but also affects the circadian phototactic behavior (period and phase; Forbes-Stovall et al. 2014; Müller et al. 2017).

Although our knowledge about the function of green algal CRYs is rapidly increasing (reviewed by Kottke et al. 2017), our understanding of their integration into the regulatory networks of the cell is still limited. This holds especially true for the integration of aCRY and pCRY into the control of phototactic movements. The abovementioned preliminary data on ChR1 levels in aCRY mutants as well as the increased transcript levels of both PHOT and aCRY upon low UV exposure (Tilbrook et al. 2016) make it conceivable that similar to PHOT (Trippens et al. 2012) also aCRY might help to adapt the overall phototactic sensitivity. This suggestion, however, needs future experimental proof.

4 Adaptation and Homeostatic Regulation of Phototactic Sensitivity

Phototaxis is the outcome of integrating a fluctuating light-immanent information over a wide intensity range with the actual physiological and developmental states of the cell, which themselves are often affected by light-driven processes. Additionally, phototaxis is under the regime of a circadian rhythm, and the photo-response is affected by other tactic movement-inducing stimuli. Hence, generation of the particular movement responses involves not only the directional light information but also complex cross talks and integration with other signaling cascades as well as adaptational mechanisms. It is long known that photosynthetic microalgae play an important role as primary producers and thus also for the global carbon cycle (Field et al. 1998). Thus, a better understanding of the processes how the different inputs are integrated into an optimal orientation in the light is highly desirable, especially under the currently rapid changing environmental conditions. Unfortunately, knowledge about the molecular basis of such sensory integrating pathways is even for the well-studied photo-behavior of *C. reinhardtii* still very limited.

What do we currently know about adaptive and other mechanisms regulating phototactic sensitivity in this model alga, and on what time scale do they occur? The different emerging mechanisms range from pico- and milliseconds up to days. The fastest adaptive mechanism is the gradual inactivation of the ChR-dependent PCs in continuous light. Here light-induced conversion between the *all*-trans and *13-cis* retinal isomers and concomitant changes in the relative contribution of two receptor cycling states (*syn* and *anti*) are involved. These states differ in conductance and ion selectivity (Hegemann et al. 2005; Bruun et al. 2015). PC desensitization seems not to be linked to the level of ChR bleaching but to the magnitude of the plasma membrane depolarization following ChR activation. PC dark recovery involves K^+ currents, indicating that repolarization of the plasma and flagella membrane are important components in the overall sensitivity regulation in this alga (Govorunova et al. 1997; Berthold et al. 2008). Some evidence pinpoints a subset of TRP ion channels as interesting candidates for regenerating resting membrane potentials adapted to the current physiological state of the cell (Arias-Darraz et al. 2015a, b). Several TRP members are encoded by the *Chlamydomonas* genome. Recently, a first structure – with previously unknown and unique properties – of one of its TRP channels was published. Further, TRPs respond to diverse physical and chemical stimuli and are, e.g., important for mechano- and thermoreception in this alga (Wheeler and Brownlee 2008; Fujiu et al. 2011; Arias-Darraz et al. 2015a, b; McGoldrick et al. 2019).

Elements affecting the overall excitability of the cell are interesting candidates for central roles in the sensory integrating pathway(s) needed for a fine-tuned homeostatic regulation of the phototactic movements. For example, one possibility how the photosynthetic electron transport might exert its long-known effects on the sign of phototaxis might be via affecting the resting membrane potential. However, other more indirect links, e.g., effects on the cytosolic free Ca^{2+} concentration and the redox poise, must also be considered (Takahashi and Watanabe 1993; Wakabayashi et al. 2011; Trippens et al. 2017). The existence of such links gets further support by the identical time scale observed for the dynamics of the adaptation of phototaxis and photosynthesis to a series of light on-off cycles (Arrieta et al. 2017). Also, the finding reported by Kim et al. (2016) that a strong relationship between the photosystem II operating efficiency and the rate of negative photo-orientation exists points to intricate relationships between these two processes. The reasons for this correlation could be, however, manifold ranging from photokinetic effects due to a better ATP supply to faster adaptation processes of the photosynthetic machinery to high light or even a sensitized phototactic signaling. Besides photosynthesis, the levels of cyclic nucleotides (cAMP and cGMP) are in the discussion to be key determinants of the sign of phototaxis. Their levels in the dark differ in strains exhibiting initial positive or negative phototaxis. Additionally, pharmacological manipulation of the cAMP level by, e.g., IBMX changes the intensity-response curve of phototaxis (Boonyareth et al. 2009). The molecular basis for this effect is not yet clear. Effects on the resting membrane potential are one possibility. Ion channels related to cyclic nucleotide-gated channels are present in the *C. reinhardtii* genome (Verret et al. 2010). In general, elements contributing to the resting plasma membrane potential and its repolarization and/or affecting the speed and sensitivity of it following ChR-induced depolarization are interesting candidates for such integrating pathways. Here one of the several actions of PHOT, albeit more indirect, in regulating the overall phototactic sensitivity might be seen. Mimicking high light conditions by overexpressing either the PHOT kinase domain or the LOV domains without the kinase, which resembles the C-terminal truncated PHOT formed in a light intensity-dependent manner, resulted in light-adapted cells with negative phototactic response (Trippens et al. 2012). The underlying molecular mechanisms are also here still unknown. Toward this end, it will be interesting to see whether plasma membrane-localized ion channels/pumps appear among the direct or indirect targets of PHOT in *Chlamydomonas*. Alternatively, PHOT targets in the flagella would be candidates of prime importance, as a differential regulation of the flagella beat finally determines the phototactic behavior (see Sect. 2.3). Additionally, the importance of the resting membrane potential for a normal response becomes also clear from mutant analyses. For example, ptx3, a phototaxis mutant isolated by Pazour et al. (1995), has probably an abnormal resting potential.

Direct targeting of the primary photoreceptors also represents an effective way to integrate rapidly different physiological signals. Here reversible posttranslational modifications are of special interest, as, e.g., phosphorylation is widely used for receptor regulation. Both ChRs are phosphoproteins (Wagner et al. 2008; Wang et al. 2014). Recently, Böhm et al. (2019) demonstrated rapid and dynamic responses of the in vivo ChR1 phosphorylation status to changes in light intensity and different other key physiological stimuli. These changes exhibit a strict dependency on elevated cytosolic Ca^{2+} levels. Further manipulation of the cellular redox state, IBMX treatments, and inhibition of the linear photosynthetic electron flow by DCMU affected the phosphorylation of ChR1 and led to a change in phototactic behavior (Takahashi and Watanabe 1993; Boonyareth et al. 2009; Wakabayashi et al. 2011; Trippens et al. 2017; Böhm et al. 2019). Most interestingly, ChR1 phosphorylation changes with the phototactic behavior and responds to plasma membrane hyper- and depolarization in a light-independent way (Böhm et al. 2019). Currently it is not known how phosphorylation affects ChR activity. It might affect potential interactions of ChR1 with other, yet unidentified proteins. Positions of some of the identified phosphosites close to the ion-conducting pore are suggestive for possible effects on channel properties. Analyses of ectopically expressed ChRs, where phosphomimetic exchanges at one of these positions affected kinetics and resulted in a higher Ca^{2+} conductance, support this assumption (Kato et al. 2012; Plazzo et al. 2012; Böhm et al. 2019). The currently available data indicate that a high level of phosphorylated ChR1 correlates with an adapted, less-sensitive photoreceptive state and that a lower level correlates with a more sensitive photoreceptive state. As changes in the ChR1 phosphorylation status also occur in the dark, this mechanism would allow a fine-tuned sensitivity adjustment already before the break of dawn. The light intensity-dependent rapid adaptation of this fast posttranslational ChR1 modification would additionally allow also fine-tuned regulation under fluctuating continuous light. Thus, reversible ChR1 phosphorylation occurring on a time scale of a few seconds to minutes is most likely a further element of desensitization in *Chlamydomonas* vision (Böhm et al. 2019). In vertebrate and invertebrate visual systems, rhodopsin phosphorylation is a key element in this process (Arshavsky 2002; Lee et al. 2004). The kinase(s) involved in ChR1 phosphorylation still await their identification. Two of the eyespot-localized kinases, PHOT and CK1, can be excluded (Böhm et al. 2019). Pharmacological evidence indicates that the involved protein phosphatase(s) belong to the PP2C group. The two dominating phosphatases in the eyespot belong to this group (Schmidt et al. 2006; Böhm et al. 2019). PP2Cs and PP2C-like proteins in higher plants are part of Ca^{2+} signaling cascades and mediate Ca^{2+}-specificity (Kudla et al. 2018). The regulation of these enzymes counteracting ChR1 phosphorylation in the eyespot area is still unknown. As the phosphatases are in vivo active in the dark and in the light (Böhm et al. 2019), their embedment in the overall sensitivity regulation must involve light-independent mechanisms.

The decreases in the ChR transcript and protein levels (Figs. 3a and 4a–c) represent somewhat slower directly light-dependent mechanisms regulating the phototactic sensitivity. They occur on time scales ranging from minutes to hours, as discussed already in Sects. 3.2, 3.3.1, and 3.3.2. Regulation of the ChR1 protein levels might involve not only PHOT but also aCRY (Trippens et al. 2012; Zou 2016; Böhm and Kreimer, unpublished results). In addition, the much slower eyespot size regulation involves PHOT. In both cases, PHOT acts as a negative regulator, desensitizing the eyespot with increasing blue light intensities. Light intensity-dependent formation of a C-terminal truncated PHOT resulting in an activated PHOT without kinase activity in addition to the full-length activated PHOT might allow a synergistic signaling and an even more subtle adjustment to the actual light environment (Trippens et al. 2012). PHOT turned recently out to be one central photoreceptor involved in the acclimation to fluctuating light conditions via linking photosynthesis and photo-protection. Here it finally regulates the thermal dissipation of excess light via controlling expression of light-harvesting complex stress-related protein 3 and thereby affects high-energy quenching. This pathway involves Ca^{2+} signaling via CAS, a Ca^{2+}-sensing receptor, and probably cyclic nucleotides (Petroutsos et al. 2011, 2016; Allorent and Petroutsos 2017). Notably, Ca^{2+} is also the central messenger in ChR1 phosphorylation changes, and CAS overexpression affects the sign of phototaxis at low light intensities on a longer time scale (Trippens et al. 2017; Böhm et al. 2019). These findings further underline (1) that one of the links used to integrate photo-orientation with other signaling mechanisms most likely involves Ca^{2+} signaling and (2) that it affects rapid as well as slower signaling processes. Also in *C. reinhardtii*, Ca^{2+}-dependent signaling processes are manifold, highly dynamic, and central to diverse responses. However, our knowledge, compared to higher plants, about this important second messenger is still limited. Wheeler (2017) recently reviewed the Ca^{2+}-dependent signaling processes in *Chlamydomonas*. To understand better its function in phototactic signaling, future correlative measurements of its local concentration changes, phototaxis, and physiological data, e.g., ChR phosphorylation and PCs under steady-state illumination on short and longer time scales, would be highly desirable. Here, although requiring highest possible spatiotemporal resolution, those changes in the eyespot region and within the flagella would be most informative. Although imaging of cytosolic Ca^{2+} in green algae is still being at its beginnings, dynamic Ca^{2+} measurements within the cytosol and flagella have already been published (Wheeler et al. 2008; Collingridge et al. 2013; Bickerton et al. 2016).

Al-Hijab et al. (2019) recently published another example for cross talk of different stimuli affecting the positioning of *C. reinhardtii* in the water column. They analyzed the effect of the phytohormone abscisic acid on the alga. Most of the components of the abscisic acid signaling machinery are also present in green algae (Wang et al. 2015), although its functions in alga are still poorly understood. In higher plants, some of the abscisic acid signaling is Ca^{2+}-dependent, and PP2Cs mediate Ca^{2+} specificity (Brandt et al. 2015; Kudla et al. 2018). In *Chlamydomonas* external abscisic acid affects the gravitactic behavior, and high light suppresses this response. Interestingly, low light reverts the sign of the tactic movement to a point source of this phytohormone (Al-Hijab et al. 2019). It is still completely unclear how this modulation by light and abscisic acid is embedded in the signaling cascade (s) leading to the movement responses. Gravitactic orientation in *Chlamydomonas* involves membrane excitability. Further, despite some overlapping elements, its downstream signaling is at least partially independent from that of phototaxis (Yoshimura et al. 2003). Integration of phototaxis and chemosensory behavior is well known from *Chlamydomonas*. Chemotactic behavior to, e.g., bicarbonate and ammonium is most sensitive in the night phase (Byrne et al. 1992; Govorunova and Sineshchekov 2003, 2007; Ermilova and Zalutskaya 2014; Choi et al. 2016). Phototaxis, on the other hand, peaks even in space during the day (Mergenhagen and Mergenhagen 1987). Only a few studies have tried to analyze the molecular basis for the integration of these signaling pathways. However, the data point to clear interactions. For example, some chemoattractants suppress photoaccumulation and PCs rapidly after application. During gametogenesis, PHOT is involved in inhibiting chemotaxis toward nitrite (Ermilova et al. 1993; Govorunova and Sineshchekov 2003; Ermilova and Zalutskaya 2014).

Chlamydomonas also exhibits thermotaxis (Sekiguchi et al. 2018). Interestingly, PHOT and other plant photoreceptors also act as thermosensors (Jung et al. 2016; Fuji et al. 2017; Qiu et al. 2019). In mammalian sperms, opsins are involved in thermotaxis using probably two light-independent signaling pathways (Pérez-Cerezales et al. 2015). The molecular basis and mechanisms of thermotaxis in *Chlamydomonas* still await detailed analysis. Thus, it will be interesting to see whether the ChRs or the other opsins are involved in temperature sensing in green algae. Knowledge about such interconnections as well as with chemotaxis will become more relevant as temperature rises and nutrient pollution is increasingly a severe problem in many aqueous environments.

As indicated in this section, we are at the very beginning to understand how integration of the different mechanisms might influence the final optimal positioning of a single cell. The above-discussed exemplified aspects represent only a minute fraction of the necessary interconnections to be analyzed. Intense future work is needed to unravel the underlying molecular mechanisms and signaling networks. We are sure that during these studies many novel and surprising aspects of the complexity of motile cells with a single "eye" will be discovered. Some of these future results might even help to illuminate further general principles of how cells translate and use light signals and how proteins involved in such pathways function. Good examples coming from unicellular systems are the algal ChRs and other microbial opsins, viz., bacterio- and halorhodopsin. They had and still have an ongoing profound impact on our understanding of how light-gated proteins, involved in regulating ion fluxes across membranes, work at the molecular and atomic scale (reviewed in, e.g., Schneider et al. 2015; Deisseroth and Hegemann 2017; Govorunova et al. 2017a; Engelhard et al. 2018; Wickstrand et al. 2019). Further, ChR2 from *Chlamydomonas* was even the founding molecule for the emerging of a new technology, optogenetics (Boyden et al. 2005). This technique uses the expression of light-sensitive proteins in cells, tissues, or organisms to modulate diverse cellular functions, e.g., the membrane potential, cell signaling, or protein-protein interaction by well-controlled temporal and spatial illumination. Already 5 years after the application of ChR2 to depolarize neurons by Boyden and co-workers, optogenetics was chosen as the method of the year by Nature Methods (Pastrana 2010) and listed in an article describing breakthroughs of the decade in Science (News Staff 2010). This still rapidly expanding technology is now using diverse light-sensitive proteins or protein domains and allows new insights into complex biological functions in diverse intact systems, not only in animals but also in plants (e.g., Deisseroth 2011, Cohen 2016, Rost et al. 2017; Andres et al. 2019 and literature cited therein). Such progress was enabled and advanced by cooperations of researchers from diverse disciplines ranging from, e.g., phycology, physiology, cell and molecular biology, biophysics, structural biology to neurobiology. Deepening of our understanding of the different aspects of photo-orientation in green algae and its regulation will be only possible by continuing and increasing such interdisciplinary joint efforts in the future.

Acknowledgments G.K. wants to thank all colleagues and especially the members of his laboratory, past and present, who share his fascination for unicellular algae and their eyes during the last three decades. Without their stimulating discussions, help, patience, and engagement, a lot of the work in the group would have never been done.

References

Al-Hijab L, Gregg A, Davies R, Macdonald H, Ladomery M, Wilson I (2019) Abscisic acid induced a negative geotropic response in dark-incubated *Chlamydomonas reinhardtii*. Sci Rep 9:12063. https://doi.org/10.1038/s41598-019-48632-0

Allmer J, Naumann B, Markert C, Zhang M, Hippler M (2006) Mass spectrometric genomic data mining: novel insights into bioenergetics pathways in *Chlamydomonas reinhardtii*. Proteomics 6:6207–6220

Allorent G, Petroutsos D (2017) Photoreceptor-dependent regulation of photoprotection. Curr Opin Plant Biol 37:102–108

Andres J, Blomeier T, Zurbriggen MD (2019) Synthetic switches and regulatory circuits in plants. Plant Physiol 179:862–884

Arakaki Y, Kawai-Toyooka H, Hamamura Y, Higashiyama T, Noga A, Hirono M, Olson BJSC, Nozaki H (2013) The simplest integrated multicellular organism unveiled. PLoS One 8:e81641. https://doi.org/10.1371/journal.pone.0081641

Arias-Darraz L, Cabezas D, Colenso CK, Alegrí-Acros M, Bravo-Moraga F, Varas-Concha I, Almonacid DE, Madrid R, Brauchi S (2015a) A transient receptor potential ion channel in *Chlamydomonas* shares key features with sensory transduction-associated TRP channels in mammals. Plant Cell 27:177–188

Arias-Darraz L, Colenso CK, Veliz LA, Vivar JP, Cardenas S, Brauchi S (2015b) A TRP conductance modulates repolarization after sensory depolarization in *Chlamydomonas reinhardtii*. Plant Sign Behav 10(8):e1052924. https://doi.org/10.1080/15592324.2015.1052924

Arrieta J, Chioccioli M, Polin M, Tuval I (2017) Phototaxis beyond turning: persistent accumulation and response acclimation of the microalga *Chlamydomonas reinhardtii*. Sci Rep 7:3447. https://doi.org/10.1038/s41598-017-03618-8

Arshavsky VY (2002) Rhodopsin phosphorylation: from terminating single photon responses to photoreceptor dark adaptation. Trends Neurosci 25:124–126

Asmail SR, Smith DR (2016) Retention, erosion, and loss of the carotenoid biosynthetic pathway in the nonphotosynthetic green algal genus *Polytomella*. New Phytol 209:899–903

Austin JR 2nd, Frost E, Vidi P-A, Kessler F, Staehlin LA (2006) Plastoglobules are lipoprotein subcompartments of the chloroplast that are permanently coupled to thylakoid membranes and contain biosynthetic enzymes. Plant Cell 18:1693–1703

Avelar GM, Schumacher RI, Zaini PA, Leonard G, Richards TA, Gomes SL (2014) A rhodopsin-guanylyl cyclase gene fusion functions in visual perception in a fungus. Curr Biol 24:1234–1240

Awasthi M, Ranjan P, Sharma K, Veetil SK, Kateriya S (2016) The trafficking of bacterial type rhodopsins into the *Chlamydomonas* eyespot and flagella is IFT mediated. Sci Rep 6:34646. https://doi.org/10.1038/srep34646

Awasthi M, Ranjan P, Kateriya S (2018) Cytoplasmic extensions of the channelrhodopsins 1 and 2 interacts in *Chlamydomonas reinhardtii*. J Appl Biotechnol Bioeng 5:84–90

Beel B, Prager K, Spexard M, Sasso S, Weiss D, Müller N, Heinnickel M, Dewez D, Ikoma D, Grossman AR, Kottke T, Mittag M (2012) A flavin binding cryptochrome photoreceptor responds to both blue and red light in *Chlamydomonas reinhardtii*. Plant Cell 24:2992–3008

Beel B, Müller N, Kottke T, Mittag M (2013) News about cryptochrome photoreceptors in algae. Plant Signal Behav 8(2):e22870

Berthold P, Tsunoda SP, Ernst OP, Mages W, Gradmann D, Hegemann P (2008) Channelrhodopsin-1 initiates phototaxis and photophobic responses in *Chlamydomonas* by immediate light-induced depolarization. Plant Cell 20:1665–1677

Bessen M, Fay RB, Witman GB (1980) Calcium control of waveform in isolated flagellar axonemes of *Chlamydomonas*. J Cell Biol 86:446–455

Bickerton P, Sello S, Brownlee C, Pittman JK, Wheeler GL (2016) Spatial and temporal specificity of Ca^{2+} signaling in *Chlamydomonas reinhardtii* in response to osmotic stress. New Phytol 212:920–933

Blaby-Haas CE, Merchant SS (2019) Comparative and functional algal genomics. Annu Rev Plant Biol 70:605–638

Böhm M, Boness D, Fantisch E, Erhard H, Frauenholz J, Kowalzyk Z, Marcinkowski N, Kateriya S, Hegemann P, Kreimer G (2019) Channelrhodopsin-1 phosphorylation changes with the phototactic behavior and responds to physiological stimuli in *Chlamydomonas*. Plant Cell 31:886–910

Boonyareth M, Saranak J, Pinthong D, Sanvarinda Y, Foster KW (2009) Roles of cyclic AMP in regulation of phototaxis in *Chlamydomonas reinhardtii*. Biologia 64:1058–1065

Boyd JS, Gray MM, Thompson MD, Horst CI, Dieckmann C (2011a) The daughter four-membered microtubule rootlet determines anterior-posterior positioning of the eyespot in *Chlamydomonas*. Cytoskeleton 68:459–469

Boyd JS, Mittelmeier TM, Lamb MR, Dieckmann CL (2011b) Thioredoxin-family protein EYE2 and Ser/Thr kinase EYE3 play interdependent roles in eyespot assembly. Mol Biol Cell 22:1421–1429

Boyden ES, Zhang F, Bamberg E, Nagel G, Deisseroth K (2005) Millisecond- timescale, genetically targeted optical control of neural activity. Nat Neurosci 8:1263–1268

Brandt B, Munemasa S, Wang C, Nguyen D, Yong T, Yang PG, Poretsky E, Belnap TF, Waadt R, Alemán F, Schroeder JI (2015) Calcium specificity signaling mechanisms in abscisic acid signal transduction in *Arabidopsis* guard cells. elife 4:e03599

Braun FJ, Hegemann P (1999) Two light-activated conductances in the eye of the green alga *Volvox carteri*. Biophys J 76:1668–1678

Bruun S, Stoeppler D, Keidel A, Kuhlmann U, Luck M, Diehl A, Geiger MA, Woodmansee D, Trauner D, Hegemann P, Oschkinat H, Hildebrandt P, Stehfest K (2015) Light-dark adaptation of channelrhodopsin involves photoconversion between the all-*trans* and 13-*cis* retinal isomer. Biochemistry 54:5389–5400

Buder J (1917) Zur Kenntnis der phototaktischen Richtungsbewegungen. Jahrb Wiss Bot 58:105–220

Byrne TE, Wells MR, Johnson CH (1992) Circadian rhythms of chemotaxis to ammonium and methylammonium uptake in *Chlamydomonas*. Plant Physiol 98:879–886

Chaves I, Pokorny R, Byrdin M, Hoang N, Ritz T, Brettel K, Essen LO, van der Horst GT, Batschauer A, Ahmad M (2011) The cryptochromes: blue light photoreceptors in plants and animals. Annu Rev Plant Biol 62:335–364

Cheng X, Liu G, Ke W, Zhao L, Lv B, Ma X, Xu N, Xia X, Deng X, Zheng C, Huang K (2017) Building a multipurpose insertional mutant library for forward and reverse genetics in *Chlamydomonas*. Plant Methods 13:36. https://doi.org/10.1186/s13007-017-0183-5

Choi HI, Kim JYH, Kwak HS, Sung YJ, Sim SJ (2016) Quantitative analysis of the chemotaxis of a green alga, *Chlamydomonas reinhardtii*, to bicarbonate using diffusion-based microfluidic device. Biomicrofluidics 10:014121. https://doi.org/10.1063/1.4942756

Christie JM (2007) Phototropin blue-light receptors. Annu Rev Plant Biol 58:21–45

Coesel S, Mangogna M, Ishikawa T, Heijde M, Rogato A, Finazzi G, Todo T, Bowler C, Falciatore A (2009) Diatom PtCPF1 is a new cryptochrome/photolyase family member with DNA repair and transcription regulation activity. EMBO Rep 10:655–661

Cohen AE (2016) Optogenetics: tuning the microscope on its head. Biophys J 110:997–1003

Collingridge P, Brownlee C, Wheeler GL (2013) Compartmentalized calcium signaling in cilia regulates intraflagellar transport. Curr Biol 23:2311–2318

Davidi L, Shimoni E, Khozin-Goldberg I, Zamir A, Pick U (2014) Origin of ß-carotene- rich plastoglobuli in *Dunaliella bardawil*. Plant Physiol 164:2139–2156

Davidi L, Levin Y, Ben-Dor S, Pick U (2015) Proteome analysis of cytoplasmatic and plastidic β-carotene lipid droplets in *Dunaliella bardawil*. Plant Physiol 167:60–79

Deininger W, Kröger P, Hegemann U, Lottspeich F, Hegemann P (1995) Chlamyrhodopsin represents a new type of sensory photoreceptor. EMBO J 14:5849–5858
Deisseroth K (2011) Optogenetics. Nat Methods 8:26–29
Deisseroth K, Hegemann P (2017) The form and function of channelrhodopsin. Science 357: eaan5544. https://doi.org/10.1126/science.aan5544
Dent RM, Sharifi MN, Malnoë A, Haglund C, Calderon RH, Wakao S, Niyogi KK (2015) Large-scale insertional mutagenesis of *Chlamydomonas* supports phylogenomic functional prediction of photosynthetic genes and analysis of classical acetate-requiring mutants. Plant J 82:337–351
Drescher K, Goldstein RE, Tuval I (2010) Fidelity of adaptive phototaxis. Proc Natl Acad Sci U S A 107:11171–11176
Eguez L, Chung YS, Kuchibhatla A, Paidhungat M, Garrett S (2004) Yeast Mn^{2+} transporter, Smf1p, is regulated by ubiquitin-dependent vacuolar protein sorting. Genetics 167:107–117
Ehlenbeck S, Gradmann D, Braun F-J, Hegemann P (2002) Evidence for a light- induced H^+ conductance in the eye of the green alga *Chlamydomonas reinhardtii*. Biophys J 82:740–751
Eitzinger N, Wagner V, Weisheit W, Geimer S, Boness D, Kreimer G, Mittag M (2015) Proteomic analysis of a fraction with intact eyespots of *Chlamydomonas reinhardtii* and assignment of protein methylation. Front Plant Sci 6:1085. https://doi.org/10.3389/fpls.2015.01085
Engel BD, Schaffer M, Kuhn Cuellar L, Villa E, Plitzko JM, Baumeister W (2015) Native architecture of the *Chlamydomonas* chloroplast revealed by in situ cryo- electron tomography. elife 4:e04889
Engelhard C, Chizhov I, Siebert F, Engelhard M (2018) Microbial halorhodopsins: light-driven chloride pumps. Chem Rev 118:10629–10645
Ermilova E, Zalutskaya Z (2014) Regulation by light of chemotaxis to nitrite during the sexual life cycle in *Chlamydomonas reinhardtii*. Plan Theory 3:113–127
Ermilova EV, Zalutskaya ZM, Gromov BV (1993) Chemotaxis towards sugars in *Chlamydomonas reinhardtii*. Curr Microbiol 27:47–50
Ermilova EV, Zalutskaya ZM, Huang K, Beck CF (2004) Phototropin plays a crucial role in controlling changes in chemotaxis during the initial phase of the sexual life cycle in *Chlamydomonas*. Planta 219:420–427
Faminzin A (1866) Die Wirkung des Lichtes auf die Bewegung der Chlamidomonas pulvisculus Ehr., Euglena viridis Ehr. und Oscillatoria insignis Tw. Melanges Biologiques tires du Bulletin de l'Ácademie Imperial des Sciences de St-Petersbourg, pp 73–93
Field CB, Behrenfeld MJ, Randerson JT, Falkowski P (1998) Primary production of the biosphere: integrating terrestrial and oceanic components. Science 281:237–240
Foot N, Henshall T, Kumar S (2017) Ubiquitination and the regulation of membrane proteins. Physiol Rev 97:253–281
Forbes-Stovall J, Howton J, Young M, Davis G, Chandler T, Kessler B, Rinehart CA, Jacobshagen S (2014) *Chlamydomonas reinhardtii* strain CC-124 is highly sensitive to blue light in addition to green and red light in resetting its circadian clock, with the blue-light photoreceptor plant cryptochrome likely acting as negative modulator. Plant Physiol Biochem 75:14–23
Foster KW, Smyth RD (1980) Light antennas in phototactic algae. Microbio Rev 44:572–630
Franz S, Ignatz E, Wenzel S, Zielosko H, Putu EPGN, Maestre-Reyna M, Tsai MD, Yamamoto J, Mittag M, Essen L-O (2018) Structure of the bifunctional cryptochrome aCRY from *Chlamydomonas reinhardtii*. Nucleic Acid Res 46:8010–8022
Franz-Badur S, Penner A, Straß S, von Horsten S, Linne U, Essen L-O (2019) Structural changes within the bifunctional cryptochrome/photolyase CraCRY upon blue light excitation. Sci Rep 9:9896. https://doi.org/10.1038/s41598-019-45885-7
Fuhrmann M, Stahlberg A, Govorunova E, Rank S, Hegemann P (2001) The abundant retinal protein of the *Chlamydomonas* eye is not the photoreceptor for phototaxis and photophobic responses. J Cell Sci 114:3857–3863
Fuhrmann M, Hausherr A, Ferbitz L, Schödl T, Heitzer M, Hegemann P (2004) Monitoring dynamic expression of nuclear genes in *Chlamydomonas reinhardtii* by using a synthetic luciferase reporter gene. Plant Mol Biol 55:869–881

Fuji Y, Tanaka H, Konno N, Ogasawara Y, Hamashima N, Tamura S, Hasegawa S, Hayasaki Y, Okajima K, Kodamaa Y (2017) Phototropin perceives temperature based on the lifetime of its photoactivated state. Proc Natl Acad Sci U S A 114:9206–9211

Fujiu K, Nakayama Y, Yanagisawa A, Sokabe M, Yoshimura K (2009) *Chlamydomonas* CAV2 encodes a voltage-dependent calcium channel required for the flagellar waveform conversion. Curr Biol 19:133–139

Fujiu K, Nakayama Y, Iida H, Sokabe M, Yoshimura K (2011) Mechanoreception in motile flagella of *Chlamydomonas*. Nat Cell Biol 13:630–632

Gavelis G, Hayakawa S, White R III, Gojobori T, Suttle CA, Keeling PJ, Leander BS (2015) Eye-like ocelloids are built from different endosymbiotically acquired components. Nature 523:204–207

Gavelis GS, Keeling PJ, Leander BS (2017) How exaptations facilitated photosensory evolution: seeing the light by accident. BioEssays 39:1600266. https://doi.org/10.1002/bies.201600266

Gehring WJ (2014) The evolution of vision. WIREs Dev Biol 3:1–40. https://doi.org/10.1002/wdev.96

Goldstein RE (2015) Green algae as model organisms for biological fluid dynamics. Annu Rev Fluid Mech 47:343–375

Govorunova EG, Sineshchekov OA (2003) Integration of photo- and chemosensory signaling pathways in Chlamydomonas. Planta 216:535–540

Govorunova EG, Sineshchekov OA (2007) Chemotaxis in the green flagellate alga *Chlamydomonas*. Biochem Mosc 70:717–725

Govorunova EG, Sineshchekov OA, Hegemann P (1997) Desensitization and dark recovery of the photoreceptor current in *Chlamydomonas reinhardtii*. Plant Physiol 115:633–642

Govorunova EG, Jung K-H, Sineshchekov OA, Spudich JL (2004) *Chlamydomonas* sensory rhodopsins A and B: cellular content and role in photophobic responses. Biophys J 86:2342–2349

Govorunova EA, Sineshchekov OA, Janz R, Liu X, Spudich JL (2015) Natural light- gated anion channels: a family of microbial rhodopsins for advanced optogenetics. Science 349:647–650

Govorunova EG, Sineshchekov OA, Li H, Spudich JL (2017a) Microbial rhodopsins: diversity, mechanisms, and optogenetic applications. Annu Rev Biochem 86:845–872

Govorunova EG, Sineshchekov OA, Rodarte EM, Janz R, Morelle O, Melkonian M, Wong GK-S, Spudich JL (2017b) The expanding family of natural anion channelrhodopsins reveals large variations in kinetics, conductance, and spectral sensitivity. Sci Rep 7:43358. https://doi.org/10.1038/srep43358

Greiner A, Kelterborn S, Evers H, Kreimer G, Sizova I, Hegemann P (2017) Targeting of photoreceptor genes via zinc-finger nucleases and CRISPR/Cas9 in *Chlamydomonas reinhardtii*. Plant Cell 29:2498–2518

Grung M, Kreimer G, Calenberg M, Melkonian M, Liaaen-Jensen S (1994) Carotenoids in the eyespot apparatus of the flagellate green alga *Spermatozopsis similis*: adaptation to the retinal-based photoreceptor. Planta 193:38–43

Harz H, Hegemann P (1991) Rhodopsin-regulated calcium currents in *Chlamydomonas*. Nature 351:489–491

Harz H, Nonnengäßer C, Hegemann P (1992) The photoreceptor current of the green alga *Chlamydomonas*. Phil Trans R Soc London Ser B 338:39–52

Hayakawa S, Takaku Y, Hwang JS, Horiguchi T, Suga H, Gehring W, Ikeo K, Gojobori T (2015) Function and evolutionary origin of unicellular camera-type eye structure. PLoS One 10: e0118415. https://doi.org/10.1371/journal.pone.0118415

Hegemann P (2008) Algal sensory photoreceptors. Annu Rev Plant Biol 59:167–189

Hegemann P, Berthold P (2009) Sensory photoreceptors and light control of flagellar activity. In: Witman GB (ed) The Chlamydomonas source-book, vol 3, 2nd edn. Academic Press, Cambridge, pp 395–429

Hegemann P, Harz H (1998) How microalgae see the light. In: Caddick MX, Baumberg S, Hodgson DA, Phillip-Jones MK (eds) Society for general microbiology symposium, vol 56. Cambridge University Press, London, pp 95–105

Hegemann P, Ehlenbeck S, Gradmann D (2005) Multiple photocycles of channelrhodopsin. Biophys J 89:3911–3918

Heijde M, Zabulon G, Corellou F, Ishikawa T, Brazard J, Usman A, Sanchez F, Plaza P, Martin M, Falciatore A, Todo T, Bouget F-Y, Bowler C (2010) Characterization of two members of the cryptochrome/photolyase family from *Ostreococcus tauri* provides insights into the origin and evolution of cryptochromes. Plant Cell Environ 33:1614–1626

Höhner R, Aboukila A, Kunz H-H, Venema K (2016) Proton gradients and proton- dependent transport processes in the chloroplast. Front Plant Sci 7:218. https://doi.org/10.3389/fpls.2016.00218

Holland EM, Braun FJ, Nonnengässer C, Harz H, Hegemann P (1996) Nature of rhodopsin-triggered photocurrents in *Chlamydomonas*. 1. Kinetics and influence of divalent ions. Biophys J 70:924–931

Holland EM, Harz H, Uhl R, Hegemann P (1997) Control of phobic behavioral responses by rhodopsin-induce photocurrents in *Chlamydomonas*. Biophys J 73:1395–1401

Holmes JA, Dutcher SK (1989) Cellular asymmetry in *Chlamydomonas reinhardtii*. J Cell Sci 94:273–285

Hoops HJ, Brighton MC, Stickles SM, Clement PR (1999) A test of two possible mechanisms for phototactic steering in *Volvox carteri* (Chlorophyceae). J Phycol 35:539–547

Hou S-Y, Govorunova EG, Ntefidou M, Lane CE, Spudich EN, Sineshchekov OA, Spudich JL (2012) Diversity of *Chlamydomonas* channelrhodopsins. Photochem Photobiol 88:119–128

Huang K, Beck CF (2003) Phototropin is the blue-light receptor that controls multiple steps in the sexual life cycle of the green alga *Chlamydomonas reinhardtii*. Proc Natl Acad Sci U S A 100:6269–6274

Huang K, Kunkel T, Beck CF (2004) Localization of the blue-light receptor phototropin to the flagella of the green alga *Chlamydomonas reinhardtii*. Mol Biol Cell 15:3605–3614

Hyams JS, Borisy GG (1978) Isolated flagellar apparatus of *Chlamydomonas*: characterization of forward swimming and alteration of waveform and reversal of motion by calcium ions *in vitro*. J Cell Sci 33:235–253

Im CS, Eberhard S, Huang K, Beck CF, Grossman AR (2006) Phototropin involvement in the expression of genes encoding chlorophyll and carotenoid biosynthesis enzymes and LHC apoproteins in *Chlamydomonas reinhardtii*. Plant J 48:1–16

Inwood W, Yoshihara C, Zalpuri R, Kim K-S, Kustu S (2008) The ultrastructure of a *Chlamydomonas reinhardtii* mutant strain lacking phytoene synthase resembles that of a colorless alga. Mol Plant 1:925–937

Isogai N, Kamiya R, Yoshimura K (2000) Dominance between the two flagella during phototactic turning in *Chlamydomonas*. Zool Sci 17:1261–1266

Jékely G (2009) Evolution of phototaxis. Philos Trans R Soc B 364:2795–2808

Johnson CH, Kondo T, Hastings JW (1991) Action spectrum for resetting the circadian phototaxis rhythm in the CW15 strains of *Chlamydomonas*. II. Illuminated cells. Plant Physiol 97:1122–1129

Josef K, Saranak J, Foster KW (2005) An electro-optic monitor of the behavior of *Chlamydomonas reinhardtii* cilia. Cell Motil Cytoskeleton 61:83–96

Josef K, Saranak J, Foster KW (2006) Linear systems analysis of the ciliary steering behavior associated with negative phototaxis in *Chlamydomonas reinhardtii*. Cell Motil Cytoskeleton 63:758–777

Jung J-H, Domijan M, Klose C, Biswas S, Ezer D, Gao M, Khattak AK, Box MS, Charoensawan V, Cortijo S, Kumar M, Grant A, Locke JCW, Schäfer E, Jaeger KE, Wigge PA (2016) Phytochromes function as thermosensors in *Arabidopsis*. Science 354:886–889

Kamiya R, Witman GB (1984) Submicromolar levels of calcium control the balance of beating between the two flagella in demembranated models of *Chlamydomonas*. J Cell Biol 98:97–107

Kateriya S, Nagel G, Bamberg E, Hegemann P (2004) "Vision" in single-celled algae. News Physiol Sci 19:133–137

Kato HE, Zhang F, Yizhar O, Ramakrishnan C, Nishizawa T, Hirata K, Ito J, Aita Y, Tsukazaki T, Hayashi S, Hegemann P, Maturana AD, Ishitani R, Deiseroth K, Nureki O (2012) Crystal structure of the channelrhodopsin light-gated cation- channel. Nature 482:369–375

Kianianmomeni A, Hallmann A (2014) Algal photoreceptors: *in vivo* functions and potential applications. Planta 239:1–26

Kianianmomeni A, Hallmann A (2015) Transcriptional analysis of *Volvox* photoreceptors suggests the existence of different cell-type specific light- signaling pathways. Curr Genet 61:3–18

Kianianmomeni A, Stehfest K, Nematollahi G, Hegemann P, Hallmann A (2009) Channelrhodopsins of *Volvox carteri* are photochromic proteins that are specifically expressed in somatic cells under control of light, temperature, and the sex inducer. Plant Physiol 151:347–366

Kim JYH, Kwak HS, Sung YJ, Choi HI, Hong ME, Lim HS, Lee J-H, Lee SY, Sim SJ (2016) Microfluidic high-throughput selection of microalgal strains with superior photosynthetic productivity using competitive phototaxis. Sci Rep 6:21155. https://doi.org/10.1038/srep21155

Kinoshita A, Niwa Y, Onai K, Yamano T, Fukuzawa H, Ishiura M, Matsuo T (2017) CSL encodes a leucine-rich-repeat protein implicated in red/violet light signaling to the circadian clock in *Chlamydomonas*. PLoS Genet 13(3):e1006645. https://doi.org/10.1371/journal.pgen.1006645

Kirk JTO (2010) Light and photosynthesis in aquatic ecosystems, 3rd edn. Cambridge University Press, New York

Kivic PA, Walne PL (1983) Algal photosensory apparatus probably represent multiple parallel evolutions. Biosystems 16:31–38

Klapoetke NC, Murata Y, Kim SS, Pulver SR, Birdsey-Benson A, Cho YK, Morimoto TK, Chuong AS, Carpenter EJ, Tian Z, Wang J, Xie Y, Yan Z, Zhang Y, Chow BY, Surek B, Melkonian M, Jayaraman V, Constantine-Paton M, Wong GK, Boyden ES (2014) Independent optical excitation of distinct neural populations. Nat Methods 11:338–346. https://doi.org/10.1038/nmeth.2836

Klein B, Wibberg D, Hallmann A (2017) Whole transcriptome RNA-Seq analysis reveals extensive cell type-specific compartmentalization in *Volvox carteri*. BMC Biol 15:111. https://doi.org/10.1186/s12915-017-0450-y

Kondo T, Johnson CH, Hastings JW (1991) Action spectrum for resetting the circadian phototaxis rhythm in the cw 15 strains of *Chlamydomonas*. I. Cells in darkness. Plant Physiol 95:197–205

Kottke T, Oldemeyer S, Wenzel S, Zou Y, Mittag M (2017) Cryptochrome photoreceptors in green algae: unexpected versatility of mechanisms and functions. J Plant Physiol 217:4–14

Krause N, Engelhard C, Heberele J, Schlesinger R, Bittl R (2013) Structural differences between the closed and open states of channelrhodopsin-2 as observed by EPR spectroscopy. FEBS Lett 587:3309–3313

Kreimer G (1994) Cell biology of phototaxis in flagellated algae. Int Rev Cytol 148:229–310

Kreimer G (1999) Reflective properties of different eyespot types in dinoflagellates. Protist 150:311–323

Kreimer G (2001) Light reception and signal modulation during photoorientation of flagellate algae. In: Lebert M, Häder D-P (eds) Comprehensive series in photosciences, vol 1. Elsevier, Amsterdam, pp 193–227

Kreimer G (2009) The green algal eyespot apparatus: a primordial visual system and more? Curr Genet 55:19–43

Kreimer G, Melkonian M (1990) Reflection confocal laser scanning microscopy of eyespots in flagellate green alga. Eur J Cell Biol 53:101–111

Kreimer G, Brohsonn U, Melkonian M (1991) Isolation and partial characterization of the photoreceptive organelle for phototaxis of a flagellate green alga. Eur J Cell Biol 55:318–327

Kreimer G, Overländer C, Sineshchekov OA, Stolzis H, Nultsch W, Melkonian M (1992) Functional analysis of the eyespot in *Chlamydomonas reinhardtii* mutant ey 627, mt^{-}. Planta 188:513–521

Kreis CT, Le Blay M, Linne C, Makowski MM, Bäumchen O (2017) Adhesion of *Chlamydomonas* microalgae to surfaces is switchable by light. Nat Phys 14:45–49

Kudla J, Becker D, Grill E, Hedrich R, Hippler M, Kummer U, Parniske M, Romeis T, Schumacher K (2018) Advances and current challenges in calcium signaling. New Phytol 218:414–431

Kuhlmann H-W (1998) Photomovements in ciliated protozoa. Naturwissenschaften 85:143–154

Kuhne J, Vierock J, Tennigkeit SA, Dreier M-A, Wietek J, Petersen D, Gavriljuk K, El-Mashtoly SF, Hegemann P, Gerwert K (2019) Unifying photocycle model for light adaptation and temporal evolution of cation conductance in channelrhodopsin-2. Proc Natl Acad Sci U S A 116:9380–9389

Lamb MR, Dutcher SK, Worley CK, Dieckmann CL (1999) Eyespot assembly mutants in *Chlamydomonas reinhardtii*. Genetics 153:721–729

Lechtreck KF, Reize IB, Melkonian M (1997) The cytoskeleton of the naked green fagellate *Spermatozopsis similis* (Chlorophyta): flagellar and basal body developmental cycle. J Phycol 33:254–265

Lechtreck KF, Johnson EC, Sakai T, Cochran D, Ballif BA, Rush J, Pazour GJ, Ikebe M, Witman GB (2009) The *Chlamydomonas reinhardtii* BBSome is an IFT cargo required for export of specific signaling proteins from flagella. J Cell Biol 187:1117–1132

Lee S-J, Xu H, Montell C (2004) Rhodopsin kinase activity modulates the amplitude of the visual response in *Drosophila*. Proc Natl Acad Sci U S A 101:11874–11879

Legris M, Klose C, Burgie ES, Rojas CC, Neme M, Hiltbrunner A, Wigge PA, Schäfer E, Vierstra RD, Casal JJ (2016) Phytochrome B integrates light and temperature signals in *Arabidopsis*. Science 354:897–900

Li F-W, Rothfels CJ, Melkonian M, Villarreal JC, Stevenson DW, Graham SW, Wong GK-S, Mathews S, Pryer KM (2015) The origin and evolution of phototropins. Front Plant Sci 6:637. https://doi.org/10.3389/fpls.2015.00637

Li X, Zhang R, Patena W, Gang SS, Blum SR, Ivanova N, Yue R, Robertson JM, Lefebvre P, Fitz-Gibbon ST, Grossman AR, Jonikas MC (2016) An indexed, mapped mutant library enables reverse genetics studies of biological processes in *Chlamydomonas reinhardtii*. Plant Cell 28:367–387

Li X, Patena W, Fauser F, Jinkerson RE, Saroussi S, Meyer MT, Ivanova N, Robertson JM, Yue R, Zhang R, Vilarrasa-Blasi J, Wittkopp TM, Ramundo S, Blum SR, Goh A, Laudon M, Srikumar T, Lefebvre PA, Grossman AR, Jonikas MC (2019) A genome-wide algal mutant library reveals a global view of genes required for eukaryotic photosynthesis. Nat Genet. https://doi.org/10.1038/s41588-019-0370-6

Liu P, Lechtreck KF (2018) The Bardet–Biedl syndrome protein complex is an adapter expanding the cargo range of intraflagellar transport trains for ciliary export. Proc Natl Acad Sci U S A 115. https://doi.org/10.1073/pnas.1713226115

Lohscheider JN, Friso G, van Wijk KJ (2016) Phosphorylation of plastoglobular proteins in *Arabidopsis thaliana*. J Exp Bot 67:3975–3984

Luck M, Mathes T, Bruun S, Fudim R, Hagedorn R, Nguyen TMT, Kateriya S, Kennis JTM, Hildebrandt P, Hegemann P (2012) A photochromic histidine kinase rhodopsin (HKR1) that is bimodally switched by ultraviolet and blue light. J Biol Chem 287:40083–40090

Lundquist PA, Poliakov A, Giacomelli L, Friso G, Appel M, McQuinn RP, Krasnoff SB, Rowland E, Ponnala L, Sun Q, van Wijk KJ (2013) Loss of plastoglobule kinases ABC1K1 and ABC1K3 causes conditional degreening, modified prenyl-lipids, and recruitment of the jasmonic acid pathway. Plant Cell 25:1818–1839

Martinis J, Glauser G, Valimareanu S, Stettler M, Zeeman SC, Yamamoto H, Shikanai T, Kessler F (2014) ABC1K1/PGR6 kinase: a regulatory link between photosynthetic activity and chloroplast metabolism. Plant J 77:269–283

Matsuda A, Yoshimura K, Sineshchekov OA, Hirono M, Kamiya R (1998) Isolation and characterization of novel *Chlamydomonas* mutants that display phototaxis but not photophobic response. Cell Motil Cytoskeleton 41:353–362

Matsunaga S, Watanabe S, Sakaushi S, Miyamura S, Hori T (2003) Screening effect diverts the swimming directions from diaphototactic to positive phototactic in a disk-shaped green flagellate *Mesostigma viride*. Photochem Photobiol 77:324–332

Matsuo T, Ishiura M (2010) New Insights into the circadian clock in *Chlamydomonas*. Int Rev Cell Mol Biol 280:281–314

Matsuo T, Ishiura M (2011) *Chlamydomonas reinhardtii* as a new model system for studying the molecular basis of the circadian clock. FEBS Lett 585:1495–1502

McGoldrick LL, Singh AK, Demirkhanyan L, Lin T-Y, Casner RG, Zakharian E, Sobolevsky AI (2019) Structure of the thermo-sensitive TRP channel TRP1 from the alga *Chlamydomonas reinhardtii*. Nat Commun 10:1–12. https://doi.org/10.1038/s41467-019-12121-9

Melkonian M, Robenek H (1984) The eyespot apparatus of flagellated green alga: a critical review. Prog Phycol Res 3:193–268

Meng D, Cao M, Oda T, Pan J (2014) The conserved ciliary protein Bug22 controls planar beating of *Chlamydomonas* flagella. J Cell Sci 127:281–287

Merchant SS, Prochnik SE, Vallon O, Harris EH, Karpowicz SJ, Witman GB, Terry A, Salamov A, Fritz-Laylin LK, Maréchal-Drouard L, Marshall WF, Qu L, Nelson DR, Sanderfoot AA, Spalding MH, Kapitonov VV, Ren Q, Ferris P, Lindquist E, Shapiro H, Lucas SM, Grimwood J, Schmutz J, Cardol P, Cerutti H, Chanfreau G, Chen C, Cognat V, Croft MT, Dent R, Dutcher S, Fernández E, Ferris P, Fukuzawa H, González-Ballester D, González-Halphen D, Hallmann A, Hanikenne M, Hippler M, Inwood M, Jabbari K, Kalanon M, Kuras R, Lefebvre PA, Lemaire SD, Lobanov AV, Lohr M, Manuell A, Meier I, Mets L, Mittag M, Mittelmeier T, Mornoney JV, Moseley J, Napoli C, Nedelcu AM, Niyogi K, Novoselov SV, Paulsen IT, Pazour G, Purton S, Ral J, Riaño-Pachón DM, Riekhof W, Rymarquis L, Schroda M, Stern D, Umen J, Willows R, Wilson N, Zimmer SL, Allmer J, Balk J, Bisova K, Chen C, Elias M, Gendler K, Hauser C, Lamb MR, Ledford H, Long JC, Minagawa J, Page MD, Pan J, Pootakham W, Roje S, Rose A, Stahlberg E, Terauchi AM, Yang P, Ball S, Bowler C, Dieckmann CL, Gladyshev VN, Green P, Jorgensen R, Mayfield S, Mueller-Roeber B, Rajamani S, Sayre RT, Brokstein P, Dubchak I, Goodstein D, Hornick L, Huang YW, Jhaveri J, Luo Y, Martínez D, Ngau WCA, Otillar B, Poliakov A, Porter A, Szajkowski L, Werner G, Zhou K, Gregoriev IV, Rokhsar DS, Grossman AR (2007) The evolution of key animal and plant functions is revealed by analysis of the *Chlamydomonas* genome. Science 318:245–251

Mergenhagen D, Mergenhagen E (1987) The biological clock of *Chlamydomonas reinhardtii* in space. Eur J Cell Biol 43:203–207

Mittelmeier TM, Berthold P, Danon A, Lamb MR, Levitan A, Rice ME, Dieckmann CL (2008) The C2 domain protein MIN1 promotes eyespot organization in *Chlamydomonas*. Eukaryot Cell 7:2100–2112

Mittelmeier TM, Dieckmann C, Boyd JS, Lamb MR (2011) Asymmetric properties of the *Chlamydomonas reinhardtii* cytoskeleton direct rhodopsin photoreceptor localization. J Cell Biol 193:741–753

Mittelmeier TM, Thompson MD, Lamb MR, Lin H, Dieckmann CL (2015) MLT1 links cytoskeletal asymmetry to organelle placement in *Chlamydomonas*. Cytoskeleton 72:113–123

Moldrup M, Moestrup Ø, Hansen PJ (2013) Loss of phototaxis and degeneration of an eyespot in long-term algal cultures: evidence from ultrastructure and behaviour in the dinoflagellate *Kryptoperidinium foliaceum*. J Euk Microbiol 60:327–334

Müller M, Bamann C, Bamberg E, Kühlbrandt W (2011) Projection structure of Channelrhodopsin-2 at 6 Å resolution by electron crystallography. J Mol Biol 414:86–95

Müller N, Wenzel S, Zou Y, Künzel S, Sasso S, Weiß D, Prager K, Grossman A, Kottke T, Mittag M (2017) A plant cryptochrome controls key features of the *Chlamydomonas* circadian clock and its life cycle. Plant Physiol 174:185–201

Nagel G, Ollig D, Fuhrmann M, Kateriya S, Musti AM, Bamberg E, Hegemann P (2002) Channelrhodopsin-1: a light-gated proton channel in green algae. Science 296:2395–2398

Nagel G, Szellas T, Huhn W, Kateriya S, Adeishvili N, Berthold P, Ollig D, Hegemann P, Bamberg E (2003) Channelrhodopsin-2, a directly light-gated cation- selective membrane channel. Proc Natl Acad Sci U S A 100:13940–13945

News Staff (2010) Insights of the decade: stepping away from the trees for a look at the forest. Science 330:1612–1613

Niwa Y, Matsuo T, Onai K, Kato D, Tachikawa M, Ishiura M (2013) Phase-resetting mechanism of the circadian clock in *Chlamydomonas reinhardtii*. Proc Natl Acad Sci U S A 110:13666–13671

Noordally ZB, Millar AJ (2015) Clocks in algae. Biochemistry 54:171–183

Nultsch W, Throm G, von Rimscha I (1971) Phototaktische untersuchungen an *Chlamydomonas reinhardtii* dangeard in homokontinuierlicher Kultur. Arch Mikrobiol 80:351–369

Oldemeyer S, Franz S, Wenzel S, Essen L-O, Mittag M, Kottke T (2016) Essential role of an unusually long-lived tyrosyl radical in the response to red light of the animal-like cryptochrome aCRY. J Biol Chem 291:14062–14071

Ortega-Escalante JA, Jasper R, Miller SM (2019) CRISPR/Cas9 mutagenesis in *Volvox carteri*. Plant J 97:661–672

Ozawa S, Nield J, Terao A, Stauber EJ, Hippler M, Koike H, Rochaix J-D, Takahashi Y (2009) Biochemical and structural studies of the large Ycf4-photosystem I assembly complex of the green alga *Chlamydomonas reinhardtii*. Plant Cell 21:2424–2442

Pastrana E (2010) Optogenetics: controlling cell function with light. Nat Methods 8:24–25

Pazour GJ, Sineshchekov OA, Witman GB (1995) Mutational analysis of the phototransduction pathway of *Chlamydomonas reinhardtii*. J Cell Biol 131:427–440

Pazour GJ, Agrin N, Leszyk J, Witman GB (2005) Proteomic analysis of a eukaryotic cilium. J Cell Biol 170:103–113

Pérez-Cerezales S, Boryshpolets S, Afanzar O, Brandis A, Nevo R, Kiss V, Eisenbach M (2015) Involvement of opsins in mammalian sperm thermotaxis. Sci Rep 5:16146. https://doi.org/10.1038/srep16146

Peter E, Dick B, Stambolic I, Baeurle SA (2014) Exploring the multiscale signaling behavior of phototropin1 from *Chlamydomonas reinhardtii* using a full-residue space kinetic Monte Carlo molecular dynamics technique. Proteins 82:2018–2040

Petroutsos D, Busch A, Janssen I, Trompelt K, Bergner SV, Weinl S, Holtkamp M, Karst U, Kudla J, Hippler M (2011) The chloroplast calcium sensor CAS is required for photoacclimation in *Chlamydomonas reinhardtii*. Plant Cell 23:2950–2963

Petroutsos D, Tokutsu R, Maruyama S, Flori S, Greiner A, Magneschi L, Cusant L, Kottke T, Mittag M, Hegemann P, Finazzi G, Minagawa J (2016) A blue- light photoreceptor mediates the feedback regulation of photosynthesis. Nature 537:563–566

Pick U, Zarka A, Boussiba S, Davidi L (2019) A hypothesis about the origin of carotenoid lipid droplets in the green algae *Dunaliella* and *Haematococcus*. Planta 249:31. https://doi.org/10.1007/s00425-018-3050-3

Plazzo AP, de Franceschi N, da Broi F, Zonta F, Sanasi MF, Filippini F, Mongillo M (2012) Bioinformatic and mutational analysis of channelrhodopsin-2 protein cation-conducting pathway. J Biol Chem 287:4818–4825

Polin M, Tuval I, Drescher K, Gollub JP, Goldstein RE (2009) *Chlamydomonas* swims with two "gears" in a eukaryotic version of run-and-tumble locomotion. Science 325:487–490

Pootakham W, Gonzalez-Ballester D, Grossman AR (2010) Identification and regulation of plasma membrane sulfate transporters in *Chlamydomonas*. Plant Physiol 153:1653–1668

Prochnik SE, Umen J, Nedelcu AM, Hallmann A, Miller SM, Nishii I, Ferris P, Kuo A, Mitros T, Fritz-Laylin LK, Hellsten U, Chapman J, Simakov O, Rensing SA, Terry A, Pangilinan J, Kapitonov V, Jurka J, Salamov A, Shapiro H, Schmutz J, Grimwood J, Lindquist E, Lucas S, Grigoriev IV, Schmitt R, Kirk D, Rokhsar DS (2010) Genomic analysis of organismal complexity in the multicellular green alga *Volvox carteri*. Science 329:223–226

Qiu Y, Li M, Kim RJ-A, Moore CM, Chen M (2019) Daytime temperature is sensed by phytochrome B in *Arabidopsis* through a transcriptional activator HEMERA. Nat Commun 10 (1):1–13. https://doi.org/10.1038/s41467-018-08059-z

Rapf RJ, Vaida V (2016) Sunlight as an energetic driver in the synthesis of molecules necessary for life. Phys Chem Chem Phys 18:20067–20084

Reisdorph NA, Small GD (2004) The CPH1 gene of *Chlamydomonas reinhardtii* encodes two forms of cryptochrome whose levels are controlled by light- induced proteolysis. Plant Physiol 134:1546–1554

Renninger S, Backendorf E, Kreimer G (2001) Subfractionation of eyespot apparatuses from the green alga *Spermatozopsis similis*: isolation and characterization of eyespot globules. Planta 213:51–63

Roberts DGW, Lamb MR, Dieckmann CL (2001) Characterization of the eye2 gene required for eyespot assembly in *Chlamydomonas reinhardtii*. Genetics 158:1037–1049

Rockwell NC, Duanmu D, Martin SS, Bachy C, Price DC, Bhattacharya D, Worden AZ, Lagarias JC (2014) Eukaryotic algal phytochromes span the visible spectrum. Proc Natl Acad Sci U S A 111:3871–3876

Rost BR, Schneider-Warme F, Schmitz D, Hegemann P (2017) Optogenetic tools for subcellular applications in neuroscience. Neuron 96:572–603

Rüffer U, Nultsch W (1985) High-speed cinematographic analysis of the movement of *Chlamydomonas*. Cell Motil 5:251–263

Rüffer U, Nultsch W (1987) Comparison of the beating of *cis*-flagella and *trans*- flagella of *Chlamydomonas* cells held on micropipettes. Cell Motil Cytoskeleton 7:87–93

Rüffer U, Nultsch W (1991) Flagellar photoresponses of *Chlamydomonas* cells held on micropipettes: II. Change in flagellar beat pattern. Cell Motil Cytoskeleton 18:269–278

Rüffer U, Nultsch W (1997) Flagellar photoresponses of ptx1, an nonphototactic mutant of *Chlamydomonas*. Cell Motil Cytoskeleton 37:111–119

Rüffer U, Nultsch W (1998) Flagellar coordination in *Chlamydomonas* cells held on micropipettes. Cell Motil Cytoskeleton 41:297–307

Ryo M, Matsuo T, Yamashino T, Ichinose M, Sugita M, Aoki S (2016) Diversity of plant circadian clocks: insights from studies of *Chlamydomonas reinhardtii* and *Physcomitrella patens*. Plant Signal Behav 11:e1116661. https://doi.org/10.1080/15592324.2015.1116661

Saegusa Y, Yoshimura K (2015) cAMP controls the balance of the propulsive forces generated by the two flagella of *Chlamydomonas*. Cytoskeleton 72:412–421

Satoh M, Hori T, Tsujimoto K, Sasa T (1995) Isolation of eyespots of green algae and analyses of pigments. Bot Mar 38:467–474

Schaller K, Uhl R (1997) A microspectrometric study of the shielding properties of eyespot and cell body in *Chlamydomonas*. Biophys J 73:1573–1578

Schmidt M, Gessner G, Luff M, Heiland I, Wagner V, Kaminski M, Geimer S, Eitzinger N, Reißenweber T, Voytsekh O, Fiedler M, Mittag M, Kreimer G (2006) Proteomic analysis of the eyespot of Chlamydomonas reinhardtii provides novel insights into its components and tactic movements. Plant Cell 18:1908–1930

Schmidt M, Luff M, Mollwo A, Kaminski M, Mittag M, Kreimer G (2007) Evidence for a specialized localization of the chloroplast ATP-synthase subunits α, β and γ in the eyespot apparatus of *Chlamydomonas reinhardtii* (Chlorophyceae). J Phycol 43:284–294

Schneider F, Grimm C, Hegemann P (2015) Biophysics of channelrhodopsin. Annu Rev Biophys 44:167–186

Schuergers N, Lenn T, Kampmann R, Meissner MV, Esteves T, Temerinac-Ott M, Korvink JG, Lowe AR, Mullineaux CW, Wilde A (2016) Cyanobacteria use micro-optics to sense light direction. elife 5:e12620

Schulze T, Prager K, Dathe H, Kelm J, Kießling P, Mittag M (2010) How the green alga *Chlamydomonas reinhardtii* keeps time. Protoplasma 244:3–14

Schulze T, Schreiber S, Iliev D, Boesger J, Trippens J, Kreimer G, Mittag M (2013) The heme-binding protein SOUL3 of *Chlamydomonas reinhardtii* influences size and position of the eyespot. Mol Plant 6:931–944

Sekiguchi M, Kameda S, Kurosawa S, Yoshida M, Yoshimura K (2018) Thermotaxis in *Chlamydomonas* is brought about by membrane excitation and controlled by redox conditions. Sci Rep 8:16114. https://doi.org/10.1038/s41598-018-34487-4

Shih SM, Engel BD, Kocabas F, Bilyard T, Gennerich A, Marshall WF, Yildiz A (2013) Intraflagellar transport drives flagellar surface motility. elife 2:e00744

Sineshchekov OA (1991) Photoreception in unicellular flagellates: bioelectric phenomena in phototaxis. In: Douglas RH (ed) Light in biology and medicine, vol 2. Plenum, New York, pp 523–532

Sineshchekov OA, Spudich JL (2005) Sensory rhodopsin signaling in green flagellate algae. In: Briggs WR, Spudich JL (eds) Handbook of photosensory receptors. Wiley, Weinheim, pp 25–42

Sineshchekov OA, Jung KH, Spudich JL (2002) Two rhodopsins mediate phototaxis to low- and high-intensity light in *Chlamydomonas reinhardtii*. Proc Natl Acad Sci U S A 99:8689–8694

Sineshchekov OA, Govorunova EG, Spudich JL (2009) Photosensory functions of channelrhodopsins in native algal cells. Photochem Photobiol 85:556–563

Solari CA, Drescher K, Goldstein RE (2011) The flagellar photoresponse in *Volvox* species (Volvocaceae, Chlorophyceae). J Phycol 47:580–583

Spexard M, Thöing C, Beel B, Mittag M, Kottke T (2014) Response of the sensory animal-like cryptochrome aCRY to blue and red light as revealed by infrared difference spectroscopy. Biochemistry 53:1041–1050

Strenkert D, Schmollinger S, Gallaher SD, Salomé PA, Purvine SO, Nicora CD, Mettler-Altmann T, Soubeyrand E, Weber APM, Lipton MS, Basset GJ, Merchant SS (2019) Multiomics resolution of molecular events during a day in the life of *Chlamydomonas*. Proc Natl Acad Sci U S A 116:2374–2383

Suda S (2003) Light microscopy and electron microscopy of *Nephroselmis spinosa* sp nov (Prasinophyceae, Chlorophyta). J Phycol 39:590–599

Suzuki T, Yamasaki K, Fujita S, Oda K, Iseki M, Yoshida K, Watanabe M, Daiyasu H, Toh H, Asamizu E, Tabata S, Miura K, Fukuzawa H, Nakamura S, Takahashi T (2003) Archaeal-type rhodopsins in *Chlamydomonas*: model structure and intracellular localization. Biochem Biophys Res Commun 301:711–717

Takahashi T, Watanabe M (1993) Photosynthesis modulates the sign of phototaxis of wild-type Chlamydomonas reinhardtii. Effects of red background illumination and 3-(39,49-dichlorophenyl)-1,1-dimethylurea. FEBS Lett 336:516–520

Terashima M, Petroutsos D, Hudig M, Tolstygina I, Trompelt K, Gabelein P, Fufezan C, Kudla J, Weinl S, Finazzi G, Hippler M (2012) Calcium-dependent regulation of cyclic photosynthetic electron transfer by a CAS, ANR1, and PGRL1 complex. Proc Natl Acad Sci U S A 109:17717–17722

Thompson MD, Mittelmeier TM, Dieckmann CL (2017) Chlamydomonas: the eyespot. In: Hippler M (ed) Chlamydomonas: molecular genetics and physiology. Microbiology monographs, vol 30. Springer, Berlin, pp 257–281

Tian Y, Gao S, von der Heyde EL, Hallmann A, Nagel G (2018) Two-component cyclase opsin of green algae are ATP-dependent and light-inhibited guanylyl cyclases. BMC Biol 16:144–161

Tilbrook K, Dubois M, Crocco CD, Yin R, Chappuis R, Allorent G, Schmid-Siegert E, Goldschmidt-Clermont M, Ulm T (2016) UV-B perception and acclimation in *Chlamydomonas reinhardtii*. Plant Cell 28:966–983

Trippens J, Greiner A, Schellwat J, Neukam M, Rottmann T, Lu Y, Kateriya S, Hegemann P, Kreimer G (2012) Phototropin influence on eyespot development and regulation of phototactic behavior in *Chlamydomonas reinhardtii*. Plant Cell 24:4687–4702

Trippens J, Reißenweber T, Kreimer G (2017) The chloroplast calcium sensor protein CAS affects phototactic behaviour in *Chlamydomonas reinhardtii* (Chlorophyceae) at low light intensities. Phycologia 56:261–270

Ueki N, Wakabayashi KI (2018) Detergent-extracted *Volvox* model exhibits an anterior-posterior gradient in flagellar Ca^{2+} sensitivity. Proc Natl Acad Sci U S A 115:E1061–E1068. https://doi.org/10.1073/PNAS.1715489115

Ueki N, Matsunaga S, Inouye I, Hallmann A (2010) How 5000 independent rowers coordinate their strokes in order to row into the sunlight: phototaxis in the multicellular green alga *Volvox*. BMC Biol 8:103

Ueki N, Ide T, Mochiji S, Kobayashi Y, Tokutsu R, Ohnishi N, Yamaguchi K, Shigenobu S, Tanaka K, Minagawa J, Hisabori T, Hirono M, Wakabayashi KI (2016) Eyespot-dependent determination of the phototactic sign in *Chlamydomonas reinhardtii*. Proc Natl Acad Sci U S A 113:5299–5304

van Wijk KJ, Kessler F (2017) Plastoglobuli: plastid microcompartments with integrated functions in metabolism, plastid developmental transitions, and environmental adaptation. Annu Rev Plant Biol 68:253–289

VanderWaal KE, Yamamoto R, Wakabayashi K-I, Fox L, Kamiya R, Dutcher SK, Bayly PV, Sale WS, Porter ME (2011) Bop5 mutations reveal new roles for the IC138 phosphoprotein in the regulation of flagellar motility and asymmetric waveforms. Mol Biol Cell 22:2862–2874

Verret F, Wheeler G, Taylor AR, Farnham G, Brownlee C (2010) Calcium channels in photosynthetic eukaryotes: implications for evolution of calcium-based signalling. New Phytol 187:23–43

Volkov O, Kovalev K, Polovinkin V, Borshchevskiy V, Bamann C, Astashkin R, Marin E, Popov A, Balandin T, Wilbold D, Büldt G, Bamberg E, Gordeliy V (2017) Structural insights into ion conduction by channelrhodopsin 2. Science 358:eaan8862. https://doi.org/10.1126/science.aan8862

Wagner V, Ullmann K, Mollwo A, Kaminski M, Mittag M, Kreimer G (2008) 1182 the phosphoproteome of a *Chlamydomonas reinhardtii* eyespot fraction includes key proteins of the light signaling pathway. Plant Physiol 146:772–788

Wakabayashi KI, Misawa Y, Mochiji S, Kamiya R (2011) Reduction-oxidation poise regulates the sign of phototaxis in *Chlamydomonas reinhardtii*. Proc Natl Acad Sci U S A 108:11280–11284

Wan KY, Leptos KC, Goldstein RE (2017) Lag, lock, sync, slip: the many 'phases' of coupled flagella. J R Soc Interface 11:20131160

Wang H, Gau B, Slade WO, Juergens M, Li P, Hicks LM (2014) The global phosphoproteome of *Chlamydomonas reinhardtii* reveals complex organellar phosphorylation in the flagella and thylakoid membrane. Mol Cell Proteomics 13:2337–2353

Wang C, Yang L, Li S-S, Han G-Z (2015) Insights into the origin and evolution of the plant hormone signalling machinery. Plant Physiol 167:872–886

Wang L, Yamano T, Takane S, Niikawa Y, Toyokawa C, Ozawa S, Tokutsu R, Takahashi Y, Minagawa J, Kanesaki Y, Yoshikawa H, Fukuzawa H (2016) Chloroplast-mediated regulation of CO_2-concentrating mechanism by Ca^{2+}-binding protein CAS in the green alga *Chlamydomonas reinhardtii*. Proc Natl Acad Sci U S A 113:12586–12591

Wheeler GL (2017) Calcium-dependent signalling processes in Chlamydomonas. In: Hippler M (ed) Chlamydomonas: molecular genetics and physiology. Microbiology monographs, vol 30. Springer, Berlin, pp 257–281

Wheeler GL, Brownlee C (2008) Ca^{2+} signalling in plants and green algae - changing channels. Trends Plant Sci 13:506–514

Wheeler GL, Joint I, Brownlee C (2008) Rapid spatiotemporal pattering of cytosolic Ca^{2+} underlies flagellar excision in *Chlamydomonas reinhardtii*. Plant J 53:401–413

Wickstrand C, Nogly P, Nago E, Iwata S, Standfuss J, Neutze R (2019) Bacteriorhodopsin: structural insights revealed using x-ray lasers and synchrotron radiation. Annu Rev Biochem 88:59–83

Wilde A, Mullineaux CW (2017) Light-controlled motility in prokaryotes and the problem of directional light perception. FEMS Microbiol Rev 41:900–922

Wood CR, Wang Z, Diener D, Zones JM, Rosenbaum J, Umen JG (2012) IFT proteins accumulate during cell division and localize to the cleavage furrow in *Chlamydomonas*. PLoS One 7: e30729. https://doi.org/10.1371/journal.pone.0030729

Wood CR, Huang K, Diener DR, Rosenbaum JL (2013) The cilium secretes bioactive ectosomes. Curr Biol 23:906–911

Yamano T, Toyokawa C, Fukuzawa H (2018) High-resolution suborganellar localization of Ca^{2+}-binding protein CAS, a novel regulator of CO_2- concentrating mechanism. Protoplasma 255:1015–1022

Yoshimura K (1994) Chromophore orientation in the photoreceptor of *Chlamydomonas* as probed by stimulation with polarized light. Photochem Photobiol 60:594–597

Yoshimura K (2011) Stimulus perception and membrane excitation in unicellular alga *Chlamydomonas*. In: Luan S (ed) Coding and decoding of calcium signals in plants, vol 10. Springer, Berlin, pp 79–92

Yoshimura K, Kamiya R (2001) The sensitivity of *Chlamydomonas* photoreceptor is optimized for the frequency of cell body rotation. Plant Cell Physiol 42:665–672

Yoshimura K, Matsuo Y, Kamiya R (2003) Gravitaxis in *Chlamydomonas reinhardtii* studied with novel mutants. Plant Cell Physiol 44:1112–1118

Zones JM, Blaby IK, Merchant SS, Umen JG (2015) High-resolution profiling of a synchronized diurnal transcriptome from *Chlamydomonas reinhardtii* reveals continuous cell and metabolic differentiation. Plant Cell 27:2743–2769

Zou Y (2016) The role of cryptochromes in the sexual life cycle of *Chlamydomonas reinhardtii*. PhD thesis, Faculty of Biology and Pharmacy, Friedrich Schiller University Jena, Germany

Zou Y, Wenzel S, Müller N, Prager K, Jung E-M, Kothe E, Kottke T, Mittag M (2017) An animal-like cryptochrome controls the *Chlamydomonas* sexual cycle. Plant Physiol 174:1334–1347

Bidirectional Lateral Transport Barriers in Serving Plant Organs and Integral Plant Functioning: Localized Lignification, Suberinization, and Cutinization

Ulrich Lüttge

Contents

Abstract In their organs – roots, leaves, and stems – higher plants possess bidirectional barriers blocking apoplastic transport by diffusion of water and solutes. These are in the roots, (1) an endodermis (EN), (2) an exodermis (EX), and (3) EN-type cell layers in N_2-fixing root nodules; in the leaves (4) envelopes of glands, gland hairs, and trichomes, (5) bundle sheaths, and (6) EN-type cell layers around the haustoria of rust fungi; (7) an EN-type cell layer in stems; and (8) a structurally invisible transport barrier in the root tip. Cell wall encrustations by lignin and adcrustations by suberin lamellae and by cutin prevent transport within the free spaces of the cell walls and transport across the cell walls. The latter is mediated by plasmodesmata.

U. Lüttge (✉)
Department of Biology, Technical University of Darmstadt, Darmstadt, Germany
e-mail: luettge@bio.tu-darmstadt.de

© Springer Nature Switzerland AG 2019
Progress in Botany (2021) 82: 305–336, https://doi.org/10.1007/124_2019_36,
Published online: 21 September 2019

Individually these transport barriers at strategic locations in the plant organs enforce switches between apoplastic and symplastic transport. These switches involve transport across membranes, and this ensures metabolic control over the solutes transported. Collectively the complement of the bidirectional transport barriers regulates the integrated harmonious functioning of the plants as whole entire organisms.

The cell wall structures of the apoplastic barriers in some cases, e.g., the root EN, have been shown not to be static anatomical elements but to be dynamic in response to internal development and environmental cues. Recent studies identify molecular signaling modules of developmental integrity of the radial cell wall encrustations in the Casparian strips in the root EN. Molecular studies also demonstrate hormonal control of reversible suberinization in response to environmental factors. The *Scr* gene and SCARECROW transcription factor regulate some of the barriers at distant locations within the plants. This opens a fascinating outlook for more work of plant molecular biology unraveling the integrated functioning of the transport-barrier complement.

Abbreviations

ABA	Abscisic acid
AFS	Apparent free space
BS	Bundle sheath
CASP	CASPRIAN STRIP MEMBRANE DOMAIN PROTEINS
CIF	CASPARIAN STRIP INTEGRITY FACTORS
CL	Cellulose layers
CS	Casparian strip
DFS	Donnan free space
EN	Endodermis
EX	Exodermis
M	Mesophyll
PD	Plasmodesmos, plasmodesmata
PEP	Phosphoenolpyruvate
PEPC	Phosphoenolpyruvate carboxylase
PM	Plasma membrane
PW	Primary cell wall
ROS	Reactive oxygen species
RuBISCO	Ribulosebisphosphate carboxylase/oxygenase
SGN	SCHENGEN RECEPTOR LIKE KINASE
SL	Suberin lamellae
WFS	Water free space

1 Introduction

Roots are the gateways for the exchange of stationary terrestrial plants with their pedospheric environment. In primary roots the endodermis (EN) is a bidirectional lateral transport barrier restricting, by localized lignification and suberinization, uncontrolled entry and exit of solutes, especially inorganic ions, into and from the plant as a whole via the apoplastic spaces (Geldner 2013). At a critical location, the EN demands control and enforces selective transport by membranes and the living cytoplasm. Therefore, there is a voluminous literature on structure and function of the EN. Currently, understanding of the formation and regulation of this barrier is making great progress by investigations at the molecular level using specific mutants. Fascinating studies reveal the functional and even also structural versatility of the root EN in development and in environmental adaptations.

However, while the root EN is the most prominent example, regulation of lateral transport in plant organs by barriers ensuring control is a general principle. In roots the EN in the inner cortex is assisted at the outer cortex by an exodermis (EX) below the rhizodermis. An EN is found in roots of legumes at nitrogen-fixing nodules. In leaves of angiosperms, bidirectional lateral transport barriers control transport by glands, gland hairs, and epidermal trichomes. In the lamina of leaves, such barriers occur in bundle sheaths of the veins. In one of these cases, they play an essential role in sustaining a globally important mode of primary photosynthetic production, viz., C_4 photosynthesis. Even in pathogenic interactions, EN-type structures are developed in coevolution. In the needles of conifers, the bundle sheaths possess an EN, which in structure and functions resembles that in roots. EN also occur in shoots.

The voluminous literature on roots fails to consider the entire complement of bidirectional transport barriers in context. To elaborate their functional interdependence appears fascinating, therefore, not only as a descriptive enumeration of similarities but also in view of the recent detection of molecular links of genetic programs, which regulate the identity of EN in roots and leaves within the same plants. This is the aim of the present essay. Structural details of the lateral transport barriers are surveyed. Established observations are recalled and related to the current progress. Can the progress in molecular analysis in unraveling dynamic signaling and functioning of the root EN elaborate general principles, which may govern whole-plant performance?

2 Switches Between Apoplast and Symplast

In plants we basically distinguish three processes of transport with different spatial reaches and different types of pathways and mechanisms, namely, short, medium, and long-distance transport. In an anthropomorphic analogy, they correspond to pedestrian, tramway, and high-speed express train (Lüttge 2017). Short-distance transport in plants is across biomembranes with a thickness of several nanometers,

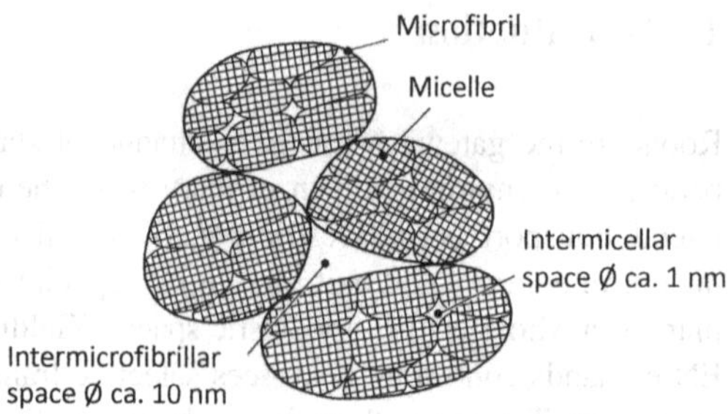

Fig. 1 Micelles and microfibrils in cell walls and the water free spaces between them after Frey-Wyssling (1959) and Robards (1970)

i.e., the plasma membrane (PM) between cells and their outside, as well as the membranes of compartments within cells. Long-distance transport occurs in the tracheids and vessels of the xylem and the sieve tubes of the phloem in the vascular bundles and can span many centimeters up to over 100 m in tall trees. Medium-distance transport links short and long-distance transports within tissues. It extends over several millimeters and represents the essential means of lateral transport within plant organs. Its pathways can structurally be very different, namely, the apoplast and the symplast, respectively.

The apoplast is the entire space outside the living cytoplasm, i.e., outside the PM, although it is a matter of opinion if one might include the vacuolar cell sap. Leaving this question aside, all of the space of cell walls but also of dead cells, such as the transporting elements of the xylem, builds up the apoplast. In the cell walls, the intermicellar and interfibrillar spaces between the micelles and fibrils of cellulose constitute the transport pathway (Fig. 1). With diameters of about 1 and 10 nm, respectively, these two spaces appear to be freely accessible to water, mineral ions, and small solute molecules. They therefore were termed apparent free space (AFS). Particular diffusing particles are water molecules in the water free space (WFS), while electric surface charges affect the transport of ions in what is called Donnan free space (DFS) (Walker and Pitman 1976; Läuchli 1976; Lüttge and Higinbotham 1979). Either way, medium-distance transport in the AFS is passive driven by physical forces and their gradients. For realizing the necessity of lateral apoplastic transport barriers – the central topic of this essay – it is indispensable to understand that the apoplastic transports are not subject to direct metabolic control. This can only occur in the living symplasm and is typically located in membranes. Therefore, switches between apoplastic and symplastic transport are crucial and need to be structurally enforced. The enforcement is imposed by apoplastic diffusion barriers. If there were continuous AFS pathways across roots, for example, metabolic control of transport into and out of the whole plant would be impossible.

Transport in the symplast can occur by diffusion in the cytoplasm and via the bridges of the plasmodesmata (PD) between cells (Fig. 2). Acceleration is enabled by cytoplasmic streaming, which is an active process under the control of actin and myosin filaments of the cytoskeleton. Symplastic transport was demonstrated and first studied in great detail by Arisz using the long narrow leaves of the water plant *Vallisneria* (Arisz 1956, 1960, 1969; Arisz and Wiersema 1966).

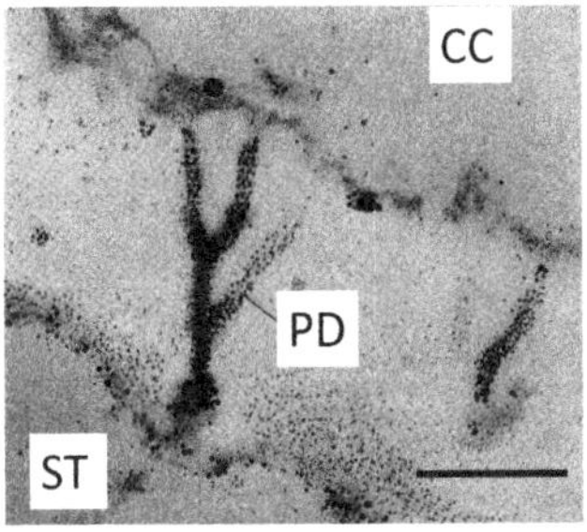

Fig. 2 Transport of Cl^- via plasmodesmata (PD) between a sieve tube (ST) and a companion cell (CC) in the phloem of leaves of *Limonium vulgare* (Mill.). The labeling for electron microscopy was obtained by precipitating the Cl^- with Ag^+ obtaining the electron dense particles (Ziegler and Lüttge 1967)

Switches between apoplastic and symplastic transport located at membranes which separate the two pathways are essential for exerting metabolic control over the transported solutes. The first switch occurs at PMs in the external AFS, e.g., in the case of roots in their outer medium. These membranes harbor ATPases pumping protons and various metal ions at the expense of ATP and ion channels and aquaporins, through which transport is driven by physicochemical gradients built up by the ATPases. Further switches are required at strategic lateral locations, enforced by apoplastic diffusion barriers as described in Sect. 3.

3 Apoplastic Barriers at Strategic Locations in Lateral Transport Functions

3.1 Encrustations and Adcrustations: Lignin, Suberin, and Cutin

Apoplastic transport barriers are established by modifications of the cell wall matrix by encrustations and adcrustations (Frey-Wyssling 1959). When cell walls are lignified, the low molecular building stones of lignin, i.e., the phenyl-propane derivatives of the monolignoles, imbibe the free space of the walls before they polymerize to the hydrophobic macromolecular structure of lignins. These, metaphorically speaking, are interior crusts or, in scientific terms, encrustations. These block the apoplastic transport within the free spaces of the cell wall matrix. Conversely hydrophobic suberin lamellae are layered on top of cell walls adjacent to the cytoplasm (Fig. 3). In such a way, adcrustations are formed, preventing apoplastic transport across cell walls. At the periphery of plant organs, precursors of cutin

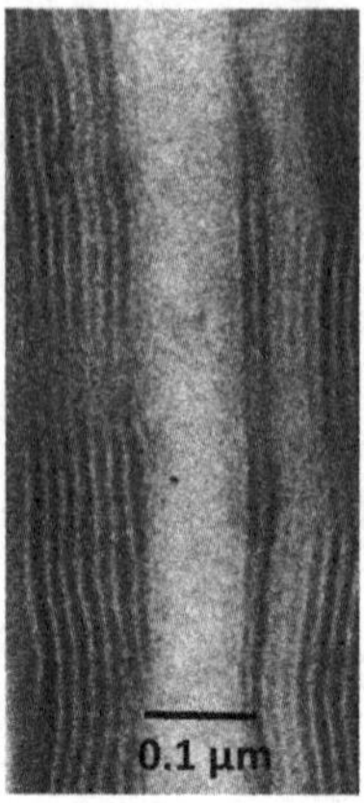

Fig. 3 Suberin lamellae in the cell walls of a wound periderm of the potato tuber. (Electron micrograph by H. Falk from Läuchli 1976)

diffuse through the outer cell walls of epidermis cells and polymerize to hydrophobic adcrustations of cuticles, which also block traversing transport. All three of them – lignin, suberin, and cutin – can be involved in constituting apoplastic barriers for lateral transport functions at strategic locations in plant organs.

3.2 Lateral Transport in Roots

3.2.1 Endodermis (EN)

In roots the morphogenetically innermost cell layer of the cortex forms a particular sheath of the thickness of one cell around the vascular central cylinder. This sheath forms the so-called endodermis (EN) which develops to an effective bidirectional barrier of apoplastic transport across the root. Mostly three developmental states are distinguished; some authors discern a fourth state (Fig. 4; van Fleet 1961; Läuchli 1976; Lüttge and Higinbotham 1979; Mertz and Brutnell 2014).

State I is the primary EN where the radial primary cell walls are locally encrusted by lignin (Fig. 4a). A complete continuous band is formed like a banderole around the radial walls of all EN cells and thus the entire EN, constituting the so-called Casparian band or strip (CS). It consists of lignin and suberin accompanied by large amounts of cell wall carbohydrate and protein (Zeier and Schreiber 1999; Schreiber et al. 1999; Zeier et al. 1999a, b). The CS is tightly associated with the PM as shown under plasmolysis where PM remains attached to the strip ("band plasmolysis"; Enstone and Peterson 1997). The apoplastic transport is effectively blocked between the cortex and the stele (Fig. 5).

In state II, the secondary endodermis, suberin lamellae (SL) are layered onto the walls all around the EN cells (Fig. 4b). In addition to suberin with poly-aliphatic and poly-phenolic residues, they contain lignin and cell wall proteins (Zeier and Schreiber 1999; Ranathunge et al. 2011). Subsequently tertiary cellulose walls are layered on top of the SL (Fig. 4c). They are often strongly lignified. This constitutes

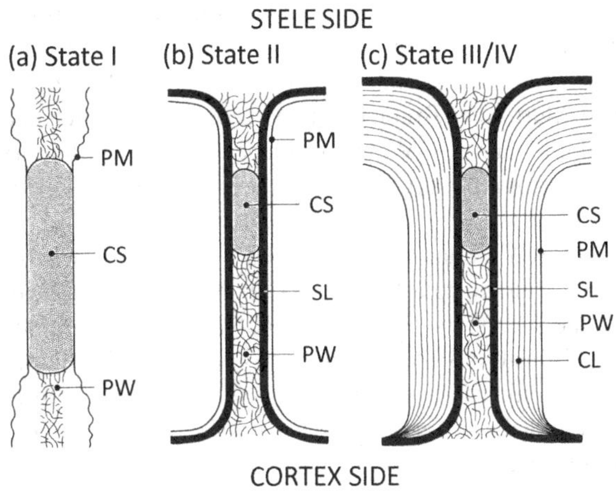

Fig. 4 Scheme of adjacent cell walls of the primary (**a**), secondary (**b**), and tertiary (**c**) endodermis. The lignification of the Casparian strip (CS) blocks radial transport in the cell walls, suberin lamellae (SL) block transport across living cells of the endodermis, and thick cellulose layers strengthen the dead cells of the tertiary endodermis, where the entire cell wall may become lignified. For annotations see list of abbreviations (Läuchli 1976; Lüttge and Higinbotham 1979). Note that (**b**) and (**c**) are at a different scale than (**a**)

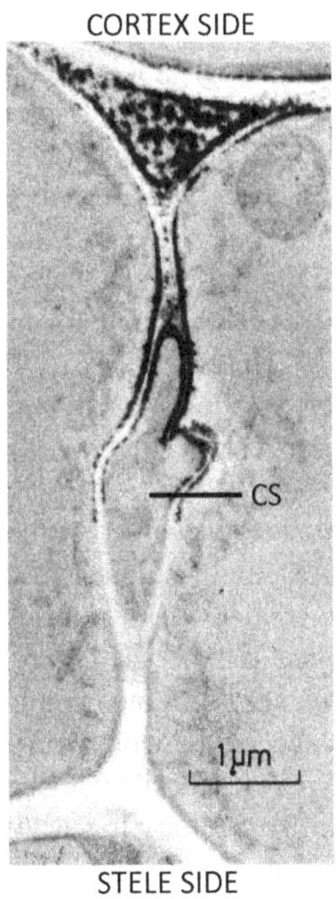

Fig. 5 Cell wall between two adjacent cells of the endodermis of a barley root where the Casparian strip blocks radial transport between cortex and stele as shown by the labeling with $La(NO_3)_3$, the heavy metal La forming electron dense particles. (Robards and Robb 1974, from Läuchli 1976)

state III, the tertiary endodermis. In monocotyledons the deposits of cellulose walls are particularly expressed on the radial and the inner tangential walls of the EN, so that U-shaped structures are formed (Ziegler et al. 1963; Meyer et al. 2009). Van Fleet (1961) and Mertz and Brutnell (2014) separate the latter phenomenon as a particular and additional state IV. The U-shaped cells in the tertiary EN are often completely isolated from the surrounding tissues so that eventually they die (Ziegler et al. 1963).

In states II and III (IV), transport across the EN would be blocked not only via the apoplast but also via the symplast. However, there are a few cells in the EN, which do not undergo the secondary and tertiary modifications. These so-called passage cells (Peterson and Enstone 1996) maintain a symplastic continuum between cortex and stele. The apoplastic route remains completely blocked as the CS extends over the passage cells and the CS structure is maintained in the entire secondary and tertiary endodermis (Ziegler et al. 1963). However, the symplastic route is still open locally via the passage cells.

3.2.2 Exodermis (EX)

In very many plant species, the hypodermal cell layer of the outer root cortex beneath the root epidermis or rhizodermis differentiates forming a structure very similar to the EN. It develops in more basal parts above the root tip to a mature cell layer called exodermis (EX) with CS and lignin in the radial cell walls (state I), suberinization of all cell walls (state II), and state III/IV modifications and also possesses passage cells (Peterson 1988; Perumalla et al. 1990; Peterson and Perumalla 1990; Enstone and Peterson 1992; Peterson and Enstone 1996; Hose et al. 2001; Wu et al. 2005). The EX separates the cortex from the root external medium.

At the root periphery, the EX represents an additional lateral transport barrier in support of the EN (Peterson 1987, 1988; Zimmermann and Steudle 1998; Zeier et al. 1999a, b; Lehmann et al. 2000; Hose et al. 2001; Schröder et al. 2002). The EX enforces membrane transport and symplastic movement right at the surface of roots where apoplastic tracer transport can occur only via wounds, e.g., of penetrating lateral roots (Enstone and Peterson 1992). Comparative studies with maize roots cultivated in conditions that either or not lead to EX formation showed the presence of EX to reduce the hydraulic conductivity and, hence, affect water transport across the roots. The EX appeared, however, to be a lesser barrier to transport of charged solutes such as nutrient ions as compared to the EN (Zimmermann and Steudle 1998).

A spectacular example of the need for potential radial transport control by an EX is given by some aerial roots, especially of epiphytic orchids. The aerial roots form a dense multicellular layer of hypodermal cells which die and are then filled with air, the so-called *velamen radicum*. By capillary forces this velamen absorbs water from rain or dew with dissolved minerals. The EX as the outermost living cell layer of the cortex inside the velamen can ensure membrane and symplast control (Benzing 1989; Goh and Kluge 1989).

Particular anatomical observations are those of the occurrence of a multilayered ("multiseriate") EX in the roots of several plant species, e.g., *Carex arenaria* (L.) (Robards et al. 1979), *Agave deserti* (Engelm.) (North and Nobel 1995), *Iris germanica* (L.), *Iris pumila* (L.), and *Iris pallida* (Lam.) but not so in other species of *Iris* (Meyer et al. 2009, 2011), *Typha* spec., and *Phragmites australis* ((Cav.) Trin. ex Steud.) (see Table 2 of Meyer et al. 2009 with further species listed in the older literature). The three to five layers of the EX of *C. arenaria* provide an extremely effective permeability barrier to water and ion flows (Robards et al. 1979). The number of the EX cell layers and their suberinization in *A. deserti* increases during soil drying reducing root hydraulic conductivity (North and Nobel 1995). The multilayered EX of adventitious roots of *I. germanica* was studied in detail by Meyer et al. (2009, 2011). It consists of up to four concentric layers below the rhizodermis. The anticlinal radial walls have encrustations typical of CS. However, these encrustations also extend into the tangential walls of the adjoining EX cell layers. Meyer et al. (2009) termed this "continuous circumferential Casparian band," which in its three-dimensional architecture does not look like a band, though. There are suberin lamellae around the entire cells deposited concurrently in development with the CS material, and eventually a state III with lignified tertiary cellulosic walls is also developed. There are no passage cells. However, the symplast is available to radial transport. Inward-directed symplastic transport occurs outside the EX layers across some of the rhizodermal cells which remain alive. EX cells also remain alive, and there are symplastic connections via plasmodesmata between rhizodermis, EX, and cortex cells. Solutes unable to traverse cell membranes must be efficiently repelled by the cell wall modifications of the multilayered EX blocking the apoplastic pathway. Comparisons with root zones where the multilayered EX was not yet developed showed that the hydraulic conductivity is tremendously reduced in the presence of an EX and permeation of NaCl drops below the detection limit. The EX is an effective bilateral apoplastic transport barrier in the addition to the EN preventing loss of water to dry soil and supporting resistance to salinity (see below Sect. 6.3).

3.2.3 The Apoplasmic Permeability of the Root Apex

As we have seen in Sect. 3.2.1, the morphogenetically innermost cell layer of the root cortex, the EN, is a readily visible structure that unfolds its barrier functions. However, it is astonishing that more recent investigations remain rather unaware of the fact that there also is a structurally invisible lateral transport barrier in the root apices. This barrier is located at the border between the periblem, from which later the cortex is formed, and the plerome generating the vascular cylinder, i.e., exactly at the position where the EN is differentiating. Lüttge and Weigl (1962) have demonstrated the blockade of apoplastic transport at this location with micro-autoradiographs of root tips after short-term local peripheral application of radioactively labeled ions ($^{45}Ca^{2+}$, $^{35}SO_4^{2-}$; Fig. 6). It was confirmed decades later (Peterson and Emanuel 1981). Enstone and Peterson (1992) have used the fluorescent dye

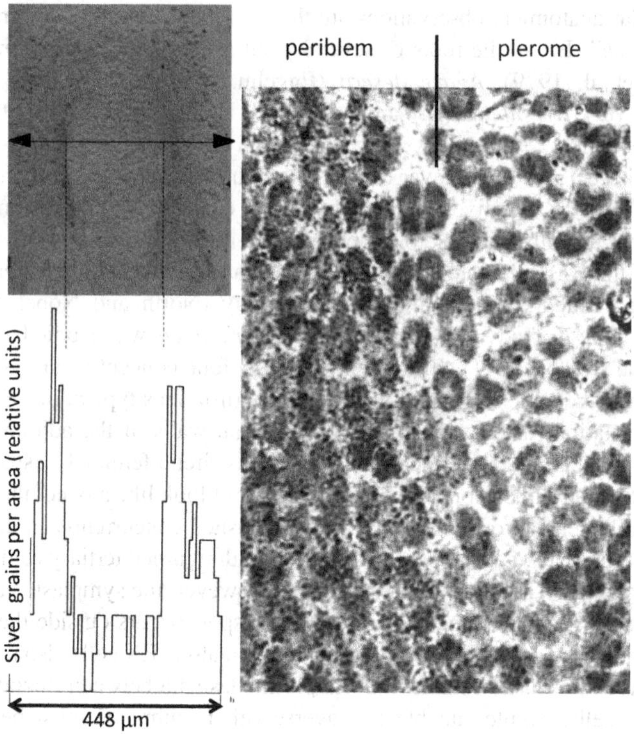

Fig. 6 Micro-autoradiographs of cross sections of a pea root 1.6 mm above the apical meristem of the root tip to which radioactively labelled $^{35}SO_4^{2-}$ had been applied for 5 min. The developed silver grains per unit area across the section as indicated in the upper left have been counted under a microscope. The transport barrier at the interface of periblem and plerome is seen in the magnification (Lüttge and Weigl 1962)

berberine precipitated for microscopy by thiocyanate as apoplastic tracer. So they also demonstrated that in the root apices, the interior tissue is apoplastically isolated, although suberized structures are missing.

Both labeling techniques have shown that passive apoplastic diffusion is blocked at the periblem–plerome border in the non-differentiated root tip tissues. What could be the structural or functional basis of this barrier? Both studies do not answer this question. Since nothing is seen microscopically, it must reside at the molecular level. As the tracers, divalent ions and berberine, respectively, are that much different, a role of ion-exchange matrices seems to be unlikely. Enstone and Peterson (1992) discuss that interfibrillar spaces (Fig. 1) might not be large enough in unexpanded cell walls where cells did not yet undergo extension growth. Therefore, the AFS might not be sufficiently developed beyond the barrier in the immature apical root zone. For understanding the functioning of the protection of the meristematic tissues of the root apex from uncontrolled influx of material into the symplast by the obvious lateral transport barrier, molecular studies appear to be urgent and indispensable.

3.2.4 Symbiotic Interactions: Root Nodules

Root nodules of leguminous plants (*Fabaceae*) are fixing atmospheric nitrogen in symbiosis with bacteria of the *Rhizobiaceae*. This requires massive lateral transport in two directions between the nodule tissue with the N_2-fixing bacteroids (modified bacterial cells) and the plant's vascular system (Fig. 7). At the periphery of the root nodules, there are several elements of the vascular network. These bundles in their phloem deliver photosynthetic products, mainly sucrose, from the shoot and leaves to the nodules, which are needed for energy metabolism. They also provide the precursors for the synthesis of glutamate via the glutamate synthase cycle as acceptor of NH_3 produced by N_2 reduction. In the opposite direction, the bundles in their xylem export the N_2 fixation products, mainly asparagine and aspartic aid, glutamine and glutamic acid, the amino acid alanine, and also the derivatives allantoine and allantoate of the purine base xanthin (Pate et al. 1969; Pate 1976).

The substantial lateral transport is checked at EN-like bidirectional barriers of apoplastic transport, which control diffusion into and out of the nodules. The entire nodule is covered by an EN with CS. For the exchange between the nodules and the rest of the plants via the vascular bundles, it is essential that each of the bundles is also surrounded by an EN with CS, blocking apoplastic transport. In many although not in all nodulated leguminous species, a pericycle exists inside the EN of the nodule vascular bundles with cells which possess a highly enlarged extracytoplasmic apoplast space due to the formation of a labyrinth of cell walls (Fig. 7, Pate et al. 1969; Pate 1976). Such types of cells are often found in plants at locations of massive

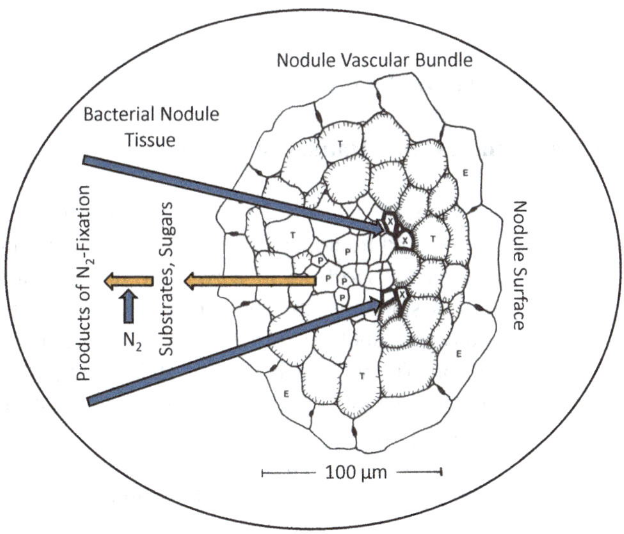

Fig. 7 Vascular bundle of a root nodule of *Pisum sativum* (L.) with EN and CS (E), pericycle transfer cells (T), phloem (P), and Xylem (X) and the massive lateral transport of substrates and N_2-fixation products indicated. (Drawing of the vascular bundle from Pate 1976)

transport between the apoplast and the symplast, and they are therefore named transfer cells. These cells facilitate apoplastic connections. However, the bidirectional lateral transport of sugars and N compounds across the EN (Fig. 7) due to the CS can only be symplastic and involves plasmodesmata (Pate et al. 1969; Pate 1976).

3.2.5 Phi-Cell Layers

Some roots develop cell layers with thickenings in their radial cell walls much like the CS, but where the thickenings are considerably inflated, so that in cross sections of roots, they look like an upper case Greek letter Phi (ϕ). Layers with ϕ-cells in roots are observed both in the inner cortex just adjacent to the EN and also as a hypodermis at the periphery of the cortex (Haas et al. 1976; Mackenzie 1979; Peterson et al. 1981). The ϕ-strips are made of cellulosic wall material and contain lignin but no suberin. The similarity of the ϕ-cell layers to EN with CS and to EX has suggested that they also function as lateral transport barriers. It was found, however, that dyes marking apoplastic transport routes penetrated the ϕ-strips, while in the same roots, dye movement was blocked by the CS. Therefore, the ϕ-cell layers rather have a mechanical support function than being a barrier for the apoplastic transport of small molecules (Peterson et al. 1981). Nevertheless, the anatomical similarity with CS remains intriguing. Regarding their developmental biology, it would be very interesting to learn if they share molecular links and signaling with EN and bundle sheaths such as the *SCARECROW* system (Sect. 3.3.2.2) and the CIF 1/2 – SGN3 – SGN1 module (Sect. 4.2).

3.3 Lateral Transport in Leaves

3.3.1 Epidermal Glands, Gland Hairs, and Trichomes

Lateral transport of solutes out of and into leaves is mediated by the activity of peripheral glands, gland hairs, and trichomes. With respect to the origin and nature of solutes eliminated by such transport out of leaves, we distinguish secretion of substances originating from assimilatory metabolism, e.g., nectar sugars, excretion of products from dissimilative metabolism, and recretion of solutes not modified by metabolism, e.g., mineral ions (Frey-Wyssling 1935, 1972; Lüttge 2019b). All of these elimination processes require metabolic control by switches between apoplastic and symplastic transport.

In perfect analogy to transport across roots (Sect. 3.2), specifically localized blockades of the apoplastic pathways are built up. In addition to lignification and suberinization, cutinization is observed (Frey-Wyssling 1935; Lüttge 1971; Gunning and Hughes 1976; Hill and Hill 1976; Lüttge and Schnepf 1976). Some examples are depicted in Fig. 8. In salt hairs (Fig. 8a) and in simpler salt glands sunken in the epidermis of leaves (Fig. 8b, c), there are cuticular envelopes in the cell walls around

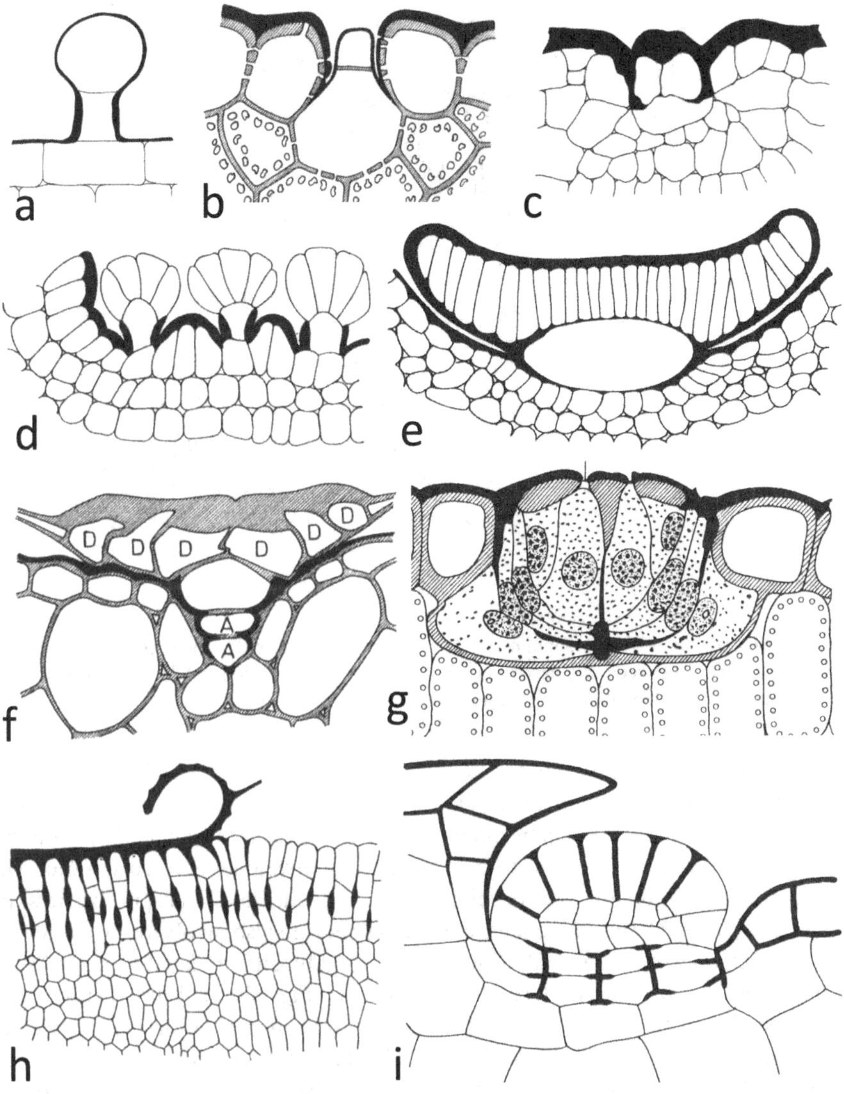

Fig. 8 Blockade of apoplastic transport by cutinization (cuticular envelopes) and by CS-like cell wall incrustations (indicated in black) in epidermal hairs and glands. (**a**) Salt hair of the salt bush *Atriplex*, (**b**) salt gland of the grass *Spartina*, (**c**) salt gland of the mangrove *Aegiceras*, (**d**) extrafloral nectary trichomes of *Syringa sargentiana* (C. K. Schneid.), (**e**) extrafloral nectary gland of *Glaziova*, (**f**) scale of the bromeliad *Tillandsia usneoides* (L.) (D – dead cells of the disk like surface of the scale, A – living absorption cells), (**g**) salt gland of *Limonium vulgare* (Mill.), (**h**) extrafloral nectary glands of *Hevea*, and (**i**) digestive pitcher gland of the carnivorous plant *Nepenthes compacta* (Hort. ex Gentil). (After Hill and Hill (1976) and Lüttge and Higinbotham (1979))

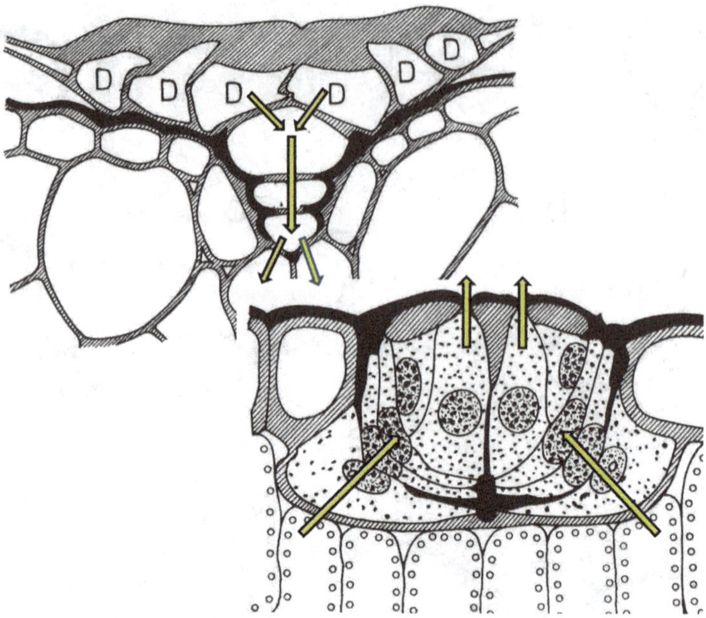

Fig. 9 Magnification of the diagrams (**f**) and (**g**) in Fig. 8 to show where cell wall blockade enforces switches between apoplastic and symplastic transport in ion uptake by bromeliad trichomes and salt recretion by *L. vulgare*, respectively (yellow arrows)

them, which make purely apoplastic recretion impossible. The solution with the recreted ions often accumulates at the top of the gland cells until the hydrostatic pressure thus built up opens pores in the cuticle for release (Feng et al. 2014; Lüttge 2019b). Similarly in some nectary glands, cuticular envelopes enforce secretion via the symplast (Fig. 8d, e). Such envelopes are peripheral lateral transport barriers. In more complex epidermal structures, internal barriers are also involved, such as those in the trichomes of bromeliads (Fig. 8f), the highly differentiated salt glands of *Limonium* (Fig. 8g), the nectary glands of *Hevea* (Fig. 8h), and the glands in the pitchers of the carnivorous plants of the genus *Nepenthes* (Fig. 8i). In the nectar-secreting glands of *Hevea*, cell walls perpendicular to the leaf surface are encrusted. The glands of the pitcher wall tissue of *Nepenthes* serve both the release and uptake of solutes to and from the pitcher fluid, respectively. Digestive enzymes are secreted and organic molecules obtained from the prey are absorbed. Here a cuticular envelope as well as perpendicular cell wall incrustations is involved. In Fig. 9 the schemes of trichomes of bromeliads and salt glands of *Limonium* are magnified to demonstrate how barriers prevent exclusive apoplastic transport and enforce transport across membranes and symplastic movement.

In the phytothelmata or tanks of bromeliads formed by the overlapping basal parts of the leaves of rosettes, the trichomes or leaf scales (Fig. 8f) serve the uptake of solutes originating from decomposing litter and debris (Benzing et al. 1976). These are primarily mineral ions. Thus, the functional analogy of the apoplastic transport barriers with the CS in roots is evident. Passive uptake of water and solutes begins with diffusion into the dead cells of the disk-like upper part of the trichomes (D in Figs. 8f and 9), and then controlled uptake into the absorption cells (A in Figs. 8f and 9) occurs, followed by symplastic transport inward. Organic molecules from the putrefying debris may also be absorbed including compounds from small animals trapped in the tanks, so that some bromeliads have been considered proto-carnivorous (Givnish et al. 1984).

The salt glands of *Limonium* are complex structures consisting of 20 cells. At the basis there are large so-called "collecting" cells, where the cell wall encrustations and the cuticular envelope block apoplastic transport, while symplastic transport delivers the salt to the recretion cells and export via pores in the peripheral cuticle (Figs. 8g and 9). The glands of *Limonium* are perhaps the best investigated salt glands (Hill and Hill 1976), including recent studies at the level of molecular biology (Sect. 5).

3.3.2 Mesophyll and Bundle Sheath

General Functions of Lateral Transport in the Lamina of Leaves

The most prominent features of lateral transport in the lamina of leaves are unloading of minerals from the transpirational water stream in the xylem and loading of photosynthetic metabolites, mainly sucrose, into the phloem of the vascular bundles.

In some grass leaves, e.g., wheat, cell layers are arranged around the veins, so-called mestome sheaths, which carry suberin lamellae and can only be crossed by symplastic transport, i.e., constitute effective lateral transport barriers in the apoplast (Kuo et al. 1974). Endodermal cell layers similar to those in the roots are also developed in some cases around the vascular bundles in leaves.

In needles of *Pinus bungeana* Zucc. ex Engl., the bundle sheath cells have CS-like structures typical of state I of the root EN. States II and III are not developed. The needle EN controls the outflow of water from the bundles into the adjacent peripheral arm palisade parenchyma. Here, the bundles are embedded in a so-called transfusion parenchyma. This parenchyma comprises dead tracheids for water transport from the bundle xylem through the EN to the arm palisade parenchyma and, eventually, stomata, together with live cells for assimilate transport from the palisade cells across the EN to the bundle phloem. However, the CS in the needles is a less effective barrier than in roots (Wu et al. 2005; Chen et al. 2011). It lacks participation of lignin in contrast to the root CS of the same species. The transport of the apoplastic tracer calcofluor which tightly binds to cellulose was blocked. However, berberine could pass the CS in needles other than in roots indicating a higher porosity of the CS in the needles. This implies that some apoplastic transport between bundles and mesophyll is maintained. Control over mineral ions unloaded from the transpiration stream and supplied to the mesophyll in the needles may occur to some extent, while plants are relying on the primary control of xylem loading by the EN in the roots.

For loading of the phloem with photosynthetic assimilates, it is noteworthy that there are two possible pathways, namely, a symplastic and an apoplastic one. Historically in investigations apoplastic loading was considered first and regarded as the only possible mechanism (Giaquinta 1983; Delrot 1987). However, in considerable efforts of research especially by Aart van Bel and collaborators (van Bel 1993, further refs. below), evidence for symplastic loading was also accumulated. The anatomical requirements for both pathways are realized in leaves (van Bel et al. 1988). Symplastic loading occurs when photosynthetic products move directly within the symplast from the green mesophyll cells to the companion cells of the sieve tube elements and into the sieve tubes. Plasmodesmata between the cells of the symplastic continuum are of dominant importance (van Bel 1987, 1993; van Bel et al. 1988, 1992). Apoplastic loading involves release of the photosynthetic products via the PM into the apoplast of the mesophyll, apoplastic transport to the companion cells, and uptake into the symplast of the companion cell-sieve tube complex by active membrane transport. In our context of lateral apoplastic diffusion barriers, it is essential to note that apoplastic loading requires the absence of such barriers in the pathway. Symplastic and apoplastic loading are attributes of different plant species. However, there are also very many species where both may occur in parallel. Therefore, we can distinguish three modes, (1) symplastic, (2) apoplastic, and (3) mixed symplastic/apoplastic loading (van Bel 1987, 1993; van Bel and Gamalei 1992; van Bel et al. 1994). This much reminds to the composite model of transport across roots (Sect. 6.2).

Support of a Major Mode of Inorganic Carbon Acquisition: C_4-Photosynthesis

Above we have noticed the suberized cell layers around the veins of the C_3 species wheat (see above). A physiologically preeminent role is played by such cell wall modifications between the green cells of the mesophyll (M) and the bundle sheaths (BS) in plants performing C_4 photosynthesis. Restricting apoplastic transport supports control of the CO_2-concentrating mechanism of this mode of photosynthesis. The C_4 mode of photosynthesis involves a cooperation of green cells of the mesophyll and the bundle sheath. In brief, primary fixation of atmospheric CO_2 is via phosphoenolpyruvate carboxylase (PEPC) in the mesophyll. The C_4 compounds formed, especially the dicarboxylate malate and also the amino acid aspartate, are transported to the bundle sheath where they are decarboxylated. This leads to a large concentrating effect of CO_2 so that ribulose-bis-phosphate carboxylase/oxygenase (RuBISCO) can operate close to its thermodynamic optimum, i.e., substrate saturation for CO_2 fixation and assimilation in the Calvin cycle. Photorespiration, i.e., the reaction with oxygen, is minimized. The C_3 compounds pyruvate and alanine resulting from decarboxylation are recycled to the mesophyll for regenerating PEP as the CO_2 acceptor. This metabolic cycle involves a requirement of massive metabolite transport across the leaves of C_4 plants between BS and M.

In the early analysis of C_4 photosynthesis, C. Barry Osmond was the first to recognize the strong analogy to the root model with the movement of ions across the

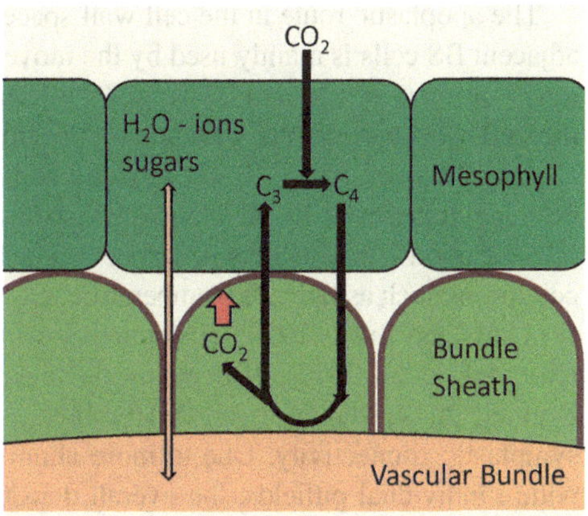

Fig. 10 Apoplastic barrier between mesophyll and BS cells of C_4 plants by suberinization of the outer tangential and the radial walls of the BS cells (brown contours) restricting leakage of CO_2 (red arrow). (Simplified sketch after the scheme of Fig. 2 in Mertz and Brutnell 2014)

cortex and into the stele (Osmond 1971). A typical endodermis is missing in the leaves of C_4 plants. However, since the M and BS cells are arranged in concentric rings around the veins, the BS cell layer can be viewed analogous to an EN around vascular bundles (Slewinski et al. 2012). Esau (1953) already has considered the BS of angiosperm leaves to be an EN. A similar apoplastic transport barrier as in roots is structurally differentiated at the interface of M and BS cells with both secondary cell wall thickening and suberinization (Fig. 10; Osmond 1971; Danila et al. 2016). The suberinization in different C_4 species may be localized to the outer tangential walls and the radial walls of the BS cells or also surround the entire symplast (Osmond 1971; Mertz and Brutnell 2014).

Most likely the analogy is more than just superficial. Slewinski et al. (2012) have explored homologies between the development of root EN and leaf BS at the molecular level, asking if both share elements of cell identity. They demonstrated that the auxin efflux protein PIN is present in both developing EN cells in roots and BS cells in leaves of the C_4 species maize. This marker suggests that there is expression of similar genetic programs in root EN and leaf BS. The authors then found that a close molecular link is given by a common homologue of the *Arabidopsis* SCARECROW (AtSCR) transcription factor. The *Scr* gene regulates endodermal cell identity in roots (Di Laurenzio et al. 1996) and is specifically expressed in the EN of maize roots (Lim et al. 2000) as well as in leaf vascular tissue (Lim et al. 2005) and BS of maize (Li et al. 2010). Thus, a conserved genetic pathway is possibly involved in the root endodermal CS formation and the leaf BS suberinization in the C_4 plant maize (Slewinski et al. 2012; Mertz and Brutnell 2014).

Notwithstanding these strong similarities, the functional analogy of the BS as a specific cell layer with the EN in the roots is not perfect. The suberin lamellae in the radial walls of the BS cells do not fuse, so that a small free apoplastic space remains in-between them (Mertz and Brutnell 2014). For lateral symplastic transport, there are no particular cells like the passage cells in states II and III of the root EN. However, there are plasmodesmata of pit fields in the suberized cell walls of BS (Danila et al. 2016).

The apoplastic route in the cell wall space in-between the suberized cell walls of adjacent BS cells is mainly used by the movement of water, ions, and sugars in both directions between M and BS (Fig. 10; Mertz and Brutnell 2014). However, due to the cell wall barrier, the transport of the metabolites (malate/pyruvate, aspartate/alanine) between M and BS is predominantly symplastic and restricted to plasmodesmata. It is driven by the metabolite gradients of the different pool sizes in M and BS (Leegood 1985). BS conductance can be regulated in response to environmental conditions such as light and temperature, e.g., by opening/closing of plasmodesmata (Robards and Lucas 1990; Bilska and Sowinski 2010; Bellasio and Griffiths 2013; Kromdijk et al. 2014). It depends on the density of plasmodesmata (Mertz and Brutnell 2014; Danila et al. 2016). In C_4 grasses, such as maize, there is high symplastic connectivity. Due to more abundant pitfields and higher PD frequency within individual pitfields, the overall density of plasmodesmata is nine times that observed in C_3 grasses. The high density is generally expressed in the leaves, i.e., not only between M and BS cells but also between adjacent M cells (Danila et al. 2016).

The most essential function of the cell wall suberinization of BS cells is the control of CO_2 leaking (Fig. 10). Because of the high concentration of CO_2 in the BS, which is the very physiological achievement and eco-physiological advantage of C_4 photosynthesis, there is a strong diffusion gradient potentially driving backflow of CO_2 and leakiness (Furbank and Hatch 1987; Kromdijk et al. 2014). Measurements of leakiness have shown that it is under very complex metabolic control and also regulated in response to growth conditions, particularly light (Bellasio and Griffiths 2013). However, the blockade of the apoplastic route of diffusive leakage is prerequisite for the effectiveness of the metabolic control mechanisms (Kromdijk et al. 2014).

3.3.3 Pathogenic Interactions: Rust Fungi

When infecting leaves, the intercellular mycelium of rust fungi in susceptible hosts develops haustoria into the plant cells. Around the cells of the necks of such haustoria, a band is formed which is osmiophilic in electron microscopy and closely mimics the CS of EN. The band has been studied in *Vigna sinensis* (Torner) Savi/*Uromyces phaseoli* (Pers.) Wint. (i.e., cowpea/cowpea rust) and *Zea mays* L./*Puccinia sorghi* Schw. interactions (Heath 1976). By means of markers, the haustorial neckband was shown to block apoplastic transport across the interface between the two organisms of host and parasite. From the parasite perspective, this means symplastic control is warranted over essential substances obtained from the host. The similarity between the haustorial band and the EN is a fascinating analogy or even homology. Is it a convergence due to independent parallel evolution or are there even common molecular links, where we may again think of the *SCARECROW* system (see above) and the CIF 1/2 – SGN3 – SGN1 module (Sect. 4.2)?

3.4 Shoots with Endodermis (EN) and Casparian Strips (CS)

In some stems, e.g., in the epicotyls of pea (*Pisum sativum* L.), there are also EN cell layers with CS, very similar to those of roots. Their development in the shoots – but not in the roots of the same plants – is regulated in complex ways by light (Sack 1987; Karahara and Shibaoka 1994).

4 Developmental Signaling Functions at the Root Endodermis (EN)

4.1 Systemic Functions at a Strategic Point of Whole-Plant Performance

Regulating the interactions of whole plants with their external root environment in the soil, the EN exerts fundamental systemic functions at a strategic point of the whole-plant system (Sect. 6). Recent progress at the molecular level has shown that signaling mechanisms are involved in integrated performance. The lignification producing the CS of the primary EN, which needs to be maintained up to the expression of the tertiary EN (i.e., state III), when this is formed, and the suberinization leading to the secondary EN (state II) are processes independent of each other (Barberon et al. 2016). Both processes are regulated in different ways. The effectiveness of the blockade of apoplastic transport by the CS in separating cortex and stele is controlled by a sensitive molecular signaling process (Sect. 4.2, Fig. 11). The blockade between the symplasts of cortex and stele by suberinization of the EN cells is reversibly modified in response to environmental cues under the antagonistic influence of the phytohormones ABA and ethylene (Sect. 4.3, Fig. 12).

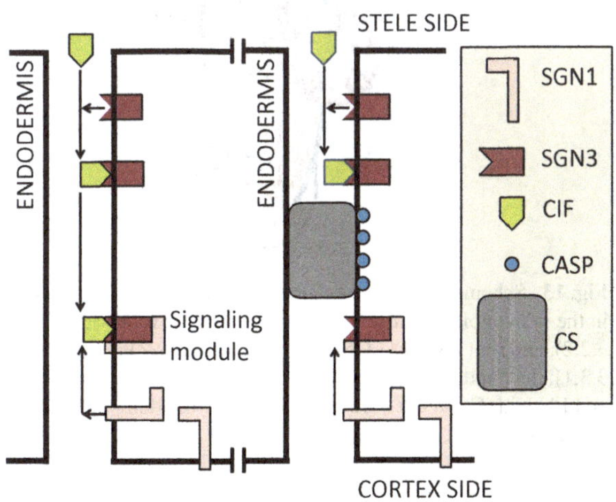

Fig. 11 Dynamics of buildup and maintenance of the Casparian strip. For details see text and list of abbreviations. (Simplified sketch after the scheme of Fig. 4a in Doblas et al. 2017a)

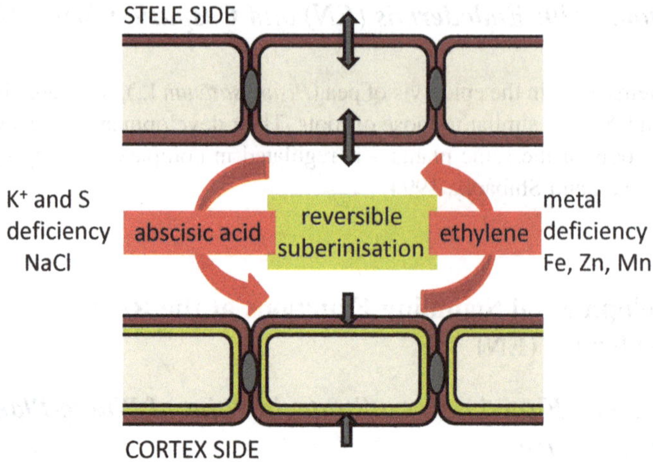

Fig. 12 Dynamic regulation of secondary endodermis suberinization under hormonal control in response to a wide range of nutrient stresses

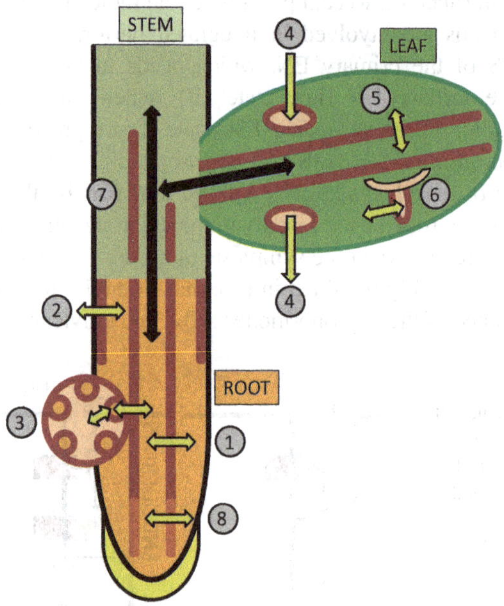

Fig. 13 Schematic view of the lateral transport barriers (heavy brown contours and yellow arrows) in the system of the entire plant as a whole: The root with the EN (1, Sect. 3.2.1), the EX (2, Sect. 3.2.2), and root nodules (3, Sect. 3.2.4); the leaves with glands, gland hairs, and trichomes (4, Sect. 3.3.1); the bundle sheaths (5, Sect. 3.3.2); the EN around the haustoria of rust fungi (6, Sect. 3.3.3); and EN with CS in the shoots (7, Sect. 3.4). In addition there is the anatomically not differentiated barrier in the root tip (8, Sect. 3.2.3)

The stelar elements of long-distance transport, phloem and xylem, connect roots and shoots (Fig. 13). Thus the hormonal control of the EN in the root is an important contribution to whole-plant performance (Lüttge 2013, 2019a).

4.2 *Casparian Strip (CS) Integrity*

Molecular elements of a composed complex signaling module regulating the buildup and the maintenance of the CS are identified with a remarkable set of specific mutants of *Arabidopsis* (Alassimone et al. 2016; Doblas et al. 2017a). Some of these molecular modules have been named SCHENGEN reminding political border regulations in Europe (Fig. 11). Two small peptides of CASPARIAN STRIP INTEGRITY FACTORS (CIF 1/2) are produced in the stele. They can move apoplastically into the cell wall between adjacent EN cells and bind to a SCHENGEN RECEPTOR-LIKE KINASE (SGN3) in the plasma membrane. For assembly of a functional signaling module, a second ligand SGN1 is required. However, the latter is exclusively produced at the cortex side of the EN. As long as there is no CS, SGN1 can reach the CIF-SGN3 complex via the apoplastic path, and so the effective CIF 1/2 – SGN3 – SGN1 signaling module is assembled. This then affects the localization of CASPARIAN STRIP MEMBRANE DOMAIN PROTEINS (CASPs), which are not produced in the EN itself (Doblas et al. 2017b). The CASPs organize enzymes involved in lignin biosynthesis such as localized endodermal peroxidases and NADPH oxidase (Lee et al. 2013) with further downstream ROS-dependent enzymes (Lee et al. 2013; Doblas et al. 2017b). However, as soon as the CS is formed and becomes tight, the buildup and turnover of the CIF 1/2 – SGN3 – SGN1 signaling module stops because CIFs cannot diffuse to the SGN3-SGN1 complex anymore. The apoplastic path is now blocked by the CS, so that further CS formation is stalled. In case that the CS gets damaged and, thus, the path opens again, CIF 1/2 – SGN3 – SGN1 is reconstructed, and repair can set in. The CIF 1/2 – SGN3 – SGN1 module is signaling specific CS requirements in response to environmental stress such as salinity (Doblas et al. 2017b; Köster et al. 2019, Sect. 6.3) and dynamic necessities of maintenance. That kind of signaling endodermal barrier integrity is possible because the essential elements CIF and SGN1 are originating from the stele and cortex, respectively, so that the opening/closing of the apoplastic path can be sensed.

4.3 *Dynamic Regulation of Secondary Endodermis (EN) Suberinization*

Suberinization of the endodermal cells leading to state II EN is under hormonal control in response to a wide range of nutrient stresses with ABA and ethylene as antagonists (Barberon et al. 2016, Fig. 12), where Ca^{2+} is a regulator upstream of ABA (Köster et al. 2019). ABA is a general regulator of suberinization, and most genes involved in it are dependent of this stress phytohormone (Hose et al. 2001; Doblas et al. 2017b). Under deficiency of K^+ and sulfur and under the stress of NaCl salinity (see also Sect. 6.3), ABA stimulates the formation of the CS. Antagonistically ethylene elicits the removal of the suberinization under metal deficiencies of Fe, Zn, and Mn (Doblas et al. 2017b). Ethylene can cause disappearance of suberin lamellae by upregulating a broad array of enzymes involved in the degradation of suberin (Naseer et al. 2012; Barberon et al. 2016). The dynamics of the nutrient-induced plasticity with formation and removal of suberinization, respectively, are astonishing. The dynamically balanced expression of the specific state II structure of EN reduces loss of K^+ and S compounds and entry of Na^+ from the stele and facilitates metal uptake into the stele, respectively, in response to different types of stresses (Sect. 6.3).

5 Molecular Control of Salt Gland Development

Given the obvious functional analogy of apoplastic transport barriers, one might expect similar developmental controls in the gland systems as in the roots (Sects. 4.2 and 4.3). Molecular studies on salt glands advanced recently. This work largely concentrates on the identification of transporter molecules of membranes of salt glands (Yuan et al. 2016a, b; Lüttge 2019b). However, investigations of the development of salt glands also reveal molecular control of features relevant for medium-distance transport. Salt glands are the first differentiated epidermal structures in emergent leaves of *Limonium bicolor* (Bunge) O. Kuntze. A transcriptome survey showed that 26 genes were involved (Yuan et al. 2015). These included genes important for the expression of structural properties of pathways of medium-distance transport, such as the organization of plasmodesmata for symplastic transport.

Mutants of *L. bicolor* have shown that the molecular composition of the apoplastic transport barrier, the cuticular envelope, is modified (Yuan et al. 2013). The cuticle of the salt glands of *L. bicolor* and the mangrove *Aegialitis rotundifolia* Roxb. contains ferulic acid as shown by blue autofluorescence under UV radiation at 330–380 nm (Yuan et al. 2013; Deng et al. 2015). Ferulic acid decreases cuticle extensibility and provides rigidity. As a consequence, with the recreted fluid accumulating at the top of glands below the cuticle, a higher hydrostatic pressure is required to build up before cuticular pores open for release. Yuan et al. (2013) isolated mutants of *L. bicolor* in which autofluoresecence and hence ferulic acid in the cuticle were increased (*fii*) or decreased (*fid*). The volume of the recreted fluid in *fii* was higher than in *fid* and wild type. In *fii* the increased volume enhanced the hydrostatic pressure.

6 Integration of Whole-Plant Relations

6.1 *Self-Organization of Root-Shoot Integration in the Stationary Life of Plants in Contrasting Root and Shoot Environments*

Life of plants is stationary at the location where they germinated and got established. Their fixation brings about that they live simultaneously in two physically contrasting milieus, the solid mineral ground and the gaseous atmosphere. One may ask why plants could not pass their lives by moving around as in fact do many algae floating and swimming in their aqueous medium. However, in contrast to aquatic plants, higher plants on land cannot gain minerals from water surrounding them but must mobilize them from the lithosphere and soil of their pedosphere. Sessility is required, therefore, as mineral acquisition from the ground is a time-demanding process and requires staying in place. In fact, only by simultaneously residing in the two different above and belowground environments plants as completely autotrophic organisms can provide their essential ecosystem, biome, and biosphere services. These are making reduced carbon from photosynthesis of atmospheric CO_2 and reduced nitrogen and sulfur from oxidized forms in the inorganic environment available to the entire food web of life. Living in their two contrasting environments, plants demonstrate complex self-organization by integrating roots and foliated shoots as their main modules into the emerging level of individual organisms (Lüttge 2013, 2019a).

In the organization of whole-plant systems, the lateral transport barriers play prominent roles of structure and function. Often the blockade of apoplastic transport is not complete, so that permeability may remain to some extent, also depending on the nature of transported solutes (Hose et al. 2001; White 2001; Ranathunge et al. 2005; Steudle 2011). There is different resistance, for example, to H_2O, NaCl, and O_2 (Ranathunge et al. 2011). Resistance also depends on the chemical composition of suberin (Zimmermann et al. 2000; Schreiber et al. 2005). However, hindrance of apoplastic transport by the barriers in general is very effective. Figure 13 summarizes the various apoplastic barriers described in this essay, depicting their integration in the whole-plant system. Their most advanced regulatory function is observed in the roots by means of EN and EX. The root is the gateway between the entire plants, and their inorganic pedospheric milieu along which regulation of bidirectional transport is indispensable.

6.2 *The Composite Root Model of Ernst Steudle*

Water and ions can move inward across the rhizodermis in the apoplast and enter the symplast sooner or later before reaching the EN. Passage of the EX with CS, where present, needs transport via the cytoplasm, but from here apoplastic transport is

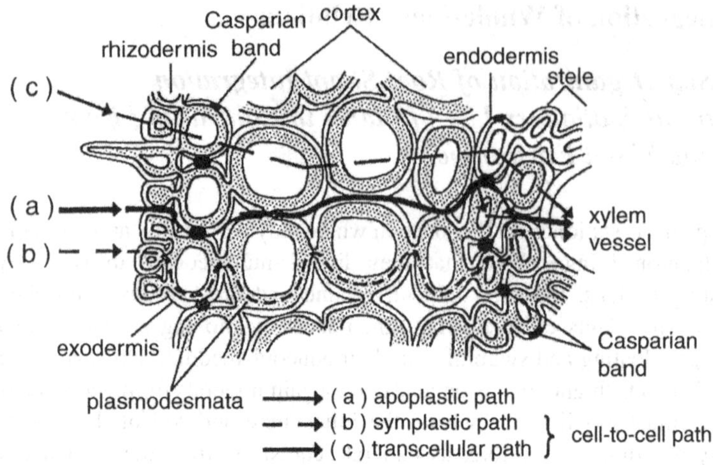

Fig. 14 Scheme of the composite model with the three pathways of transport across roots. From Steudle (2011)

possible up to the EN with CS. Then the cytoplasm must be crossed, but apoplastic transport can occur again up to loading the xylem elements. This can be called (a) the apoplastic pathway. After entering the symplast in the EX or somewhere in the outer cortex, transport may continue symplastically via the EN and the pericycle of the central cylinder up to the xylem parenchyma cells from where xylem loading may take place. This is a cell-to-cell pathway, which may be called (b) the symplastic pathway. A complication arises when along the cells of the symplastic pathway transport also passes through the vacuoles into and out across their tonoplasts. This is still a cell-to-cell pathway but called (c) the transcellular pathway (Fig. 14).

Parts of these pathways can be used interactively in transport traversing the root. Ernst Steudle together with his colleagues has dedicated much work to understanding the integration of these three pathways in the transport of water and mineral ions as sustained by the lateral transport barriers of EN and EX. They developed the concept of a composite root model (Steudle 1994, 2000, 2011; Steudle and Peterson 1998; Hose et al. 2001; Barberon et al. 2016).

Since in all pathway combinations the EN and EX with their CS can only be passed via the cytoplasm of their cells, plasmodesmata frequency is crucial. This was studied in onion roots (*Allium cepa* L.) by Ma and Peterson (2001). Plasmodesmata are most frequent at the interfaces EX/central cortex, central cortex/EN, and EN/pericycle while being rarer at epidermis/EX and pericycle/stelar parenchyma transitions.

Regulatory effects of root EN and EX at the whole-plant level involve signaling by transport of the phytohormones ABA and cytokinins in the transpiration stream. This is modulated by apoplastic water flow across roots. The release of the phytohormones from root cells is positively correlated with increased water flow (Hose et al. 2001).

6.3 Regulation of EN and EX Resistance in Response to Environmental Conditions

For a consideration of the regulatory functions of EN and EX, it is supportive that they are not just static anatomical structures but highly dynamic. Their resistance is variable in response to environmental conditions, such as:

- Nutritional deficiencies
- Drought, with the problem of water loss from the roots to dry soil
- Salinity (see below)
- Toxic metals
- Water logging (see below)

(Hose et al. 2001; Doblas et al. 2017a, b; Barberon et al. 2016).

Under water-logging EN and EX control diffusive radial loss of oxygen via the roots as coming from the shoot (De Simone et al. 2003; Chen et al. 2011; Ranathunge et al. 2011). Ranathunge et al. (2011) compared roots of rice grown in a stagnant aqueous medium, which was simulating the effects of water logging during flooding and in an aerated medium. Both EN and EX in the stagnant medium developed to tight barriers against radial oxygen loss. The absolute amounts of suberin, lignin, and esterified aromatics (ferulic and coumaric acid) were significantly higher than in the aerated roots.

Especially well-studied are responses to salinity (Hose et al. 2001; Karahara et al. 2004; Chen et al. 2011; Köster et al. 2019). Upon stress impact, suberinization and lignification are increased, which may be (partially) reversible (Sect. 4). Salinity induces EN and accelerates EN formation advancing closer toward the root tip in cotton (Reinhardt and Rost 1995), maize (Karahara et al. 2004), and rice (Krishnamurthy et al. 2009, 2011). In maize conditioning under moderate salt stress of 100 mM NaCl caused deposition of additional hydrophobic aliphatic suberin. Subsequent stronger salt stress of 200 mM NaCl led to lower Na^+ levels in the apoplast, reduced Na^+ accumulation, and improved survival of the plants (Krishnamurthy et al. 2011). In maize the radial extension of the CS within the endodermal cell walls was significantly increased under salt stress, which presumably enforces its function as apoplastic barrier (Karahara et al. 2004).

7 Outlook

Philosophically borders and borderlines are required anywhere for separating different states, entities, or conditions whether they are geographical and administrative areas or even issues of conceptual thinking, belief, and debate. They are important for sustainment of diversity. For this, however, structured and controlled exchange across borders is indispensable. Good borders are required for organizational integration. Conversely, bad nonnegotiable borders lead to narcissistic self-containment, stagnancy and sterility, and eventual death of the encased entities.

The full complement of lateral transport barriers of higher plants surveyed in this essay is a vital example of the function of solid, well-defined, and evolutionarily generated borders in the emergent self-organization of the plants as whole entire organisms from integration of their parts as modules. The appropriate function of the borders is underlined by the notion that they are "bidirectional" transport barriers and by the recognition of the various ways allowing exchange across them, as discussed in the essay.

The most recent studies demonstrating that the anatomical structures of these barriers in important cases are not only static but can be very dynamic responses, open the way to advanced exploration. The already documented regulation of some of these barriers at distant locations within the plants by the same SCARECROW factor and the *Scr* gene opens fascinating outlooks to further research at the level of molecular biology of plants. The comprehensive scope of the entire complement of lateral transport barriers in whole-plant functioning and their role in plant evolution and development comes into reach, truly representing an "evo-devo" perspective.

Acknowledgment I thank C. Barry Osmond for his help with the literature on the mesophyll/bundle sheath interface in the leaves of C_4 plants. I am grateful to Rainer Matyssek for reading a draft of the essay and making valuable comments and to Francisco Cánovas for editing the essay.

References

Alassimone J, Fujita S, Doblas VG, Van Dop M, Barberon M, Kalmbach L, Vermeer JEM, Rojas-Murcia N, Santuari L, Hardtke SC, Geldner N (2016) Polarly localized kinase SGN1 is required for Casparian strip integrity and positioning. Nature Plants 2:16113

Arisz WH (1956) Significance of the symplasm theory for transport across the root. Protoplamsa 46:5–62

Arisz WH (1960) Symplasmatischer Transport in *Vallisneria*-Blättern. Protoplasma 52:309–343

Arisz WH (1969) Intercellular polar transport and the role of the plasmodesmata in coleoptiles and *Vallisneria* leaves. Acta Bot Neerl 18:14–38

Arisz WH, Wiersema EP (1966) Symplasmatic long-distance transport in *Vallisneria* plants investigated by means of autoradiograms. Proc Kon Ned Akad Wetenschap Ser C 69:223–241

Barberon M, Vermeer JEM, de Bellis D, Wang P, Naseer S, Andersen TG, Humbel BM, Nawrath C, Takano J, Salt DE, Geldner N (2016) Adaptation of root function by nutrient-induced plasticity of endodermal differentiation. Cell 164:447–459

Bellasio C, Griffiths H (2013) Acclimation to low light by C_4 maize: implications for bundle sheath leakiness. Plant Cell Environ 37:1046–1058

Benzing DH (1989) The mineral nutrition of epiphytes. In: Lüttge U (ed) Vascular plants as epiphytes. Evolution and ecology. Springer, Berlin, pp 167–199

Benzing DH, Henderson K, Kessel B, Sulak J (1976) The absorptive capacities of bromeliad trichomes. Am J Bot 63:1009–1014

Bilska A, Sowinski P (2010) Closure of plasmodesmata in maize (*Zea mays*) at low temperature: a new mechanism for inhibition of photosynthesis. Ann Bot 106:675–686

Chen T, Cai X, Wu X, Karahara I, Schreiber L, Lin J (2011) Casparian strip development and its potential function in salt tolerance. Plant Signaling and Behavior 6:1499–1502

Danila FR, Quick WP, White RG, Furbank RT (2016) The metabolite pathway between bundle sheath and mesophyll: quantification of plasmodesmata in leaves of C_3 and C_4 monocots. Plant Cell 18:1461–1471

De Simone O, Haase K, Müller E, Junk WJ, Hartmann K, Schreiber L, Schmidt W (2003) Apoplastic barriers and oxygen transport properties of hypodermal cell walls in roots from four Amazonian tree species. Plant Physiol 132:206–217

Delrot S (1987) Phloem loading: apoplastic or symplastic? Plant Physiol Biochem 25:676–676

Deng Y, Feng Z, Yuan F, Guo J, Suo S, Wang B (2015) Identification and functional analysis of the autofluorescent substance in *Limonium bicolor* salt glands. Plant Physiol Biochem 97:20–27

Di Laurenzio L, Wysocka-Diller J, Malamy JE, Pysh L, Helariutta Y, Freshour G et al (1996) The *SCARECROW* gene regulates an asymmetric cell division that is essential for generating the radial organization of the Arabidopsis root. Cell 86:423–433

Doblas VG, Smakowska-Luzan E, Fujita S, Alassimone J, Barberon M, Madalinski M, Belkhadir Y, Geldner N (2017a) Root diffusion barrier control by a vasculature-derived peptide binding to the SGN3 receptor. Science 355:280–284

Doblas VG, Geldner N, Barberon M (2017b) The endodermis, a tightly controlled barrier for nutrients. Curr Op Plant Biol 39:136–143

Enstone DE, Peterson CA (1992) The apoplastic permeability of root apices. Can J Bot 70:1502–1512

Enstone DE, Peterson CA (1997) Suberin position and band plasmolysis in the maize (*Zea mays* L.) root exodermis. Can J Bot 75:1188–1199

Esau K (1953) Plant anatomy. Wiley, New York

Feng Z, Sun Q, Deng Y, Sun S, Zhang J, Wang B (2014) Study on pathway and characteristics of ion secretion of salt glands of *Limonium bicolor*. Acta Physiol Plant. https://doi.org/10.1007/s11738-014-1644-3

Frey-Wyssling A (1935) Die Stoffausscheidungen der höheren Pflanzen. Springer, Berlin

Frey-Wyssling A (1959) Die pflanzliche Zellwand. Springer, Berlin

Frey-Wyssling A (1972) Elimination processes in higher plants. Saussurea 3:79–90

Furbank RT, Hatch MD (1987) Mechanism of C_4 photosynthesis. Plant Physiol 85:958–964

Geldner N (2013) The endodermis. Annu Rev Plant Biol 64:531–558

Giaquinta RT (1983) Phloem loading of sucrose. Ann Rev Plant Phys 54:892–898

Givnish TJ, Burkhardt EL, Happel R, Weintraub J (1984) Carnivory in the bromeliad *Brocchinia reducta*, with a cost-benefit model for the general restriction of carnivorous plants to sunny, moist, nutrient-poor habitats. Am Nat 124:479–497

Goh CJ, Kluge M (1989) Gas exchange and water relations in epiphytic orchids. In: Lüttge U (ed) Vascular plants as epiphytes. Evolution and ecology. Springer, Berlin, pp 139–166

Gunning BES, Hughes JE (1976) Quantitative assessment of symplastic transport of pre-nectar into the trichomes of *Abutilon* nectaries. Aust J Plant Physiol 3:619–637

Haas DL, Carothers ZB, Robins RR (1976) Observations of the phi-thickenings and casparian strips in *Pelargonium* roots. Am J Bot 63:863–867

Heath MC (1976) Ultrastructural and functional similarity of the haustorial neck-band of rust fungi and the Casparian strip of vascular plants. Can J Bot 54:2484–2489

Hill AE, Hill BS (1976) Elimination processes by glands. Mineral ions. In: Lüttge U, Pitman MG (eds) Transport in plants II. Part B tissues and organs. Encyclopedia of plant physiology, New Series vol 2. Springer, Berlin, pp 225–243

Hose E, Clarkson DT, Steudle E, Schreiber L, Hartung W (2001) The exodermis: a variable apoplastic barrier. J Exp Bot 52:2245–2264

Karahara I, Shibaoka H (1994) The Casparian strip in pea epicotyls: effects of light on its development. Planta 192:269–275

Karahara I, Ikeda A, Kondo T, Uetake Y (2004) Development of the Casparian strip in primary roots of maize under salt stress. Planta 219:41–47

Köster P, Wallrad L, Edel KH, Faisal M, Alatar AA, Kudla J (2019) The battle of two ions: Ca^{2+} signalling against Na^{+} stress. Plant Biol 21(Suppl 1):39–48

Krishnamurthy P, Ranathunge K, Franke R, Prakash HS, Schreiber L, Mathew MK (2009) The role of root apoplastic transport barriers in salt tolerance of rice (*Oryza sativa* L). Planta 230:119–134

Krishnamurthy P, Ranathunge K, Nayak S, Schreiber L, Mathew MK (2011) Root apoplastic barriers block Na^+ transport to shoots in rice (*Oryza sativa* L.). J Exp Bot 62:4215–4228

Kromdijk J, Ubierna N, Cousins AB, Griffiths H (2014) Bundle-sheath leakiness in C_4 photosynthesis: a careful balancing act between CO_2 concentration and assimilation. J Exp Bot 65:3443–3457

Kuo J, O'Brien TB, Canny MJ (1974) Pit-field distribution, plasmodesmatal frequency, and assimilate flux in the mestome sheath cells of wheat leaves. Planta 121:97–118

Läuchli A (1976) Apoplasmic transport in tissues. In: Lüttge U, Pitman MG (eds) Transport in plants II. Part B tissues and organs. Encyclopedia of plant physiology, New Series vol 2. Springer, Berlin, pp 3–34

Lee Y, Rubio MC, Alassimone J, Geldner N (2013) A mechanism for localized lignin deposition in the endodermis. Cell 1523:402–412

Leegood RC (1985) The intercellular compartmentation of metabolites in leaves of *Zea mays* L. Planta 164:163–171

Lehmann H, Stelzer R, Holzamer S, Kunz U, Gierth M (2000) Analytical electron microscopical investigations on the apoplastic pathways of lanthanum transport in barley roots. Planta 211:816–822

Li P, Ponnala L, Gandotra N, Wang L, Si Y, Tausta L et al (2010) The developmental dynamics of the maize leaf transcriptome. Nat Genet 42:1060–1067

Lim J, Helariutta Y, Specht CD, Jung J, Sims L, Bruce WB, Diehn S et al (2000) Molecular analysis of the *SCARECROW* gene in maize reveals a common basis for radial patterning in diverse meristems. Plant Cell 12:1307–1318

Lim J, Jung JW, Lim CE, Lee M-H, Kim BJ, Kim M et al (2005) Conservation and diversification of *SCARECROW* in maize. Plant Mol Biol 59:619–630

Lüttge U (1971) Structure and function of plant glands. Ann Rev Plant Physiol 22:23–44

Lüttge U (2013) Whole-plant physiology: synergistic emergence rather than modularity. Progr Bot 74:165–190

Lüttge U (2017) Faszination Pflanzen. Springer, Heidelberg

Lüttge U (2019a) Plants: unitary organisms emerging from integration and self-organization of modules. In: Wegner LH, Lüttge U (eds) Emergence and modularity in life sciences. Springer, Heidelberg, pp 171–193

Lüttge U (2019b) Elimination of salt by recretion: salt glands and gland-supported bladders in recretohalophytes. In: Hasanuzzaman M, Shabala S, Fujita M (eds) Halophytes and climate change: adaptive mechanisms and potential uses. CAB International, Wallingford, pp 223–239

Lüttge U, Higinbotham N (1979) Transport in plants. Springer, New York

Lüttge U, Schnepf E (1976) Elimination processes by glands. Organic substances. In: Lüttge U, Pitman MG (eds) Transport in plants II. Part B tissues and organs. Encyclopedia of plant physiology, New Series vol 2. Springer, Berlin, pp 245–277

Lüttge U, Weigl J (1962) Mikroautoradiographische Untersuchungen der Aufnahme und des Transportes von $^{35}SO_4^{--}$ und $^{45}Ca^{++}$ in Keimwurzeln von *Zea mays* L. und *Pisum sativum* L. Planta 58:113–126

Ma F, Peterson CA (2001) Frequencies of plasmodesmata in *Allium cepa* L. roots: implications for solute transport pathways. J Exp Bot 52:1051–1061

Mackenzie KAD (1979) The development of the endodermis and phi layer of apple roots. Protoplasma 100:21–32

Mertz RA, Brutnell TP (2014) Bundle sheath suberization in grass leaves: multiple barriers to characterization. J Exp Bot 65:3371–3380

Meyer CJ, Seago JL, Peterson CA (2009) Environmental effects on the maturation of the endodermis and multiseriate exodermis of *Iris germanica* roots. Ann Bot 103:687–702

Meyer CJ, Peterson CA, Steudle E (2011) Permeability of *Iris germanica*'s multiseriate exodermis to water, NaCl, and ethanol. J Exp Bot 62:1911–1926
Naseer S, Lee Y, Lapierre C, Franke R, Nawrath C, Geldner N (2012) Casparian strip diffusion barrier in Arabidopsis is made of a lignin polymer without suberin. Proc Natl Acad Sci U S A 209:10101–10106
North GB, Nobel PS (1995) Hydraulic conductivity of concentric root tissues of *Agave deserti* Engelm. under wet and drying conditions. New Phytol 130:47–57
Osmond CB (1971) Metabolite transport in C_4 photosynthesis. Aust J Biol Sci 24:159–163
Pate JS (1976) Transport in symbiotic systems fixing nitrogen. In: Lüttge U, Pitman MG (eds) Transport in plants II. Part B tissues and organs. Encyclopedia of plant physiology, New Series vol 2. Springer, Berlin, pp 278–303
Pate JS, Gunning BES, Briarty LG (1969) Ultrastructure and functioning of the transport system of the leguminous root nodule. Planta 85:11–34
Perumalla CJ, Peterson CA, Enstone DE (1990) A survey of angiosperm species to detect hypodermal Casparian bands. I. Roots with a uniseriate hypodermis and epidermis. Bot J Lin Soc 103:93–112
Peterson CA (1987) The exodermal Casparian band of onion roots blocks the apoplastic movement of sulphate ions. J Exp Bot 38:2068–2081
Peterson CA (1988) Exodermal Casparian bands: their significance for ion uptake by roots. Physiol Plant 72:204–208
Peterson CA, Emanuel ME (1981) Pathway of movement of apoplastic fluorescent dye tracers through the endodermis at the site of secondary root formation in corn (*Zea mays*) and broad bean (*Vicia faba*). Can J Bot 59:618–625
Peterson CA, Enstone DE (1996) Functions of passage cells in the endodermis and exodermis of roots. Physiol Plant 97:592–598
Peterson CA, Perumalla CJ (1990) A survey of angiosperm species to detect hypodermal Casparian bands. II. Roots with a multiseriate hypodermis or epidermis. Bot J Lin Soc 103:113–125
Peterson CA, Emanuel ME, Weerdenburg CA (1981) The permeability of phi thickenings in apple (*Pyrus malus*) and geranium (*Pelargonium hortorum*) roots to an apoplastic fluorescent dye tracer. Can J Bot 59:1107–1110
Ranathunge K, Steudle E, Lafitte R (2005) A new precipitation technique provides evidence for the permeability of the Casparian band to ions in young roots of corn (*Zea mays* L.) and rice (*Oryza sativa* L.). Plant Cell Environ 28:1450–1462
Ranathunge J, Lin J, Steudle E, Schreiber L (2011) Stagnant deoxygenated growth enhances root suberization and lignifications, but differentially affects water and NaCl permeabilities in rice (*Oryza sativa* L.) roots. Plant Cell Environ 34:1223–1240
Reinhardt DH, Rost TL (1995) Salinity accelerates endodermal development and induces an exodermis in cotton seedling roots. Environ Exp Bot 35:563–574
Robards AW (1970) Electron microscopy and plant ultrastructure. McGraw-Hill, London
Robards AW, Lucas WJ (1990) Plasmodesmata. Ann Rev Plant Physiol Plant Mol Biol 41:369–419
Robards AW, Robb ME (1974) The entry of ions and molecules into roots: an investigation using electron-opaque tracers. Planta 120:1–12
Robards AW, Clarkson DT, Sanderson J (1979) Structure and permeability of the epidermal/hypodermal layers of the sand sedge (*Carex arenaria* L.). Protoplasma 101:331–347
Sack FD (1987) The structure of the stem endodermis in etiolated pea seedlings. Can J Bot 65:1514–1519
Schreiber L, Hartmann K, Skrabs M, Zeier L (1999) Apoplastic barriers in roots: chemical composition of endodermal and hypodermal cell walls. J Exp Bot 50:1267–1280
Schreiber L, Franke R, Hartmann KD, Ranathunge K, Steudle E (2005) The chemical composition of suberin in apoplastic barriers affects radial hydraulic conductivity differently in the roots of rice (*Oryza sativa* L. cv. IR64) and corn (*Zea mays* L. cv. Helix). J Exp Bot 56:1427–1436

Schröder M, Kunz U, Stelzer R, Lehmann H (2002) On the evidence of a diffusion barrier in the outer cortex apoplast of cress roots (*Lepidium sativum*), demonstrated by analytical electron microscopy. J Plant Phys 159:1197–1204
Slewinski TL, Anderson AA, Zhang CK, Turgeon R (2012) Scarecrow plays a role in establishing Kranz anatomy in maize leaves. Plant Cell Phys 53:2030–2037
Steudle E (1994) Water transport across roots. Plant Soil 167:79–90
Steudle E (2000) Water uptake by plant roots: an integration of views. Plant Soil 226:45–56
Steudle E (2011) Hydraulic architecture of vascular plants. In: Lüttge U, Beck E, Bartels D (eds) Plant desiccation tolerance. Ecological studies, vol 215. Springer, Berlin, pp 185–207
Steudle E, Peterson CA (1998) How does water get through roots? J Exp Bot 49:775–788
van Bel AJE (1987) The apoplast concept of phloem loading has no universal validity. Plant Physiol Biochem 25:677–686
van Bel AJE (1993) Strategies of phloem loading. Ann Rev Plant Physiol Plant Mol Biol 44:253–281
van Bel AJE, Gamalei YV (1992) Ecophysiology of phloem loading in source leaves. Plant Cell Environ 15:265–270
van Bel AJE, van Kesteren WJP, Papenhuijzen C (1988) Ultrastructural indications for coexistence of symplastic and apoplastic phloem loading in *Commelina benghalensis* leaves. Difference in ontogenic development, spatial arrangement and symplastic connections of the two sieve tubes in the minor vein. Planta 176:159–172
van Bel AJE, Gamalei YV, Ammerlaan A, Bik LPM (1992) Dissimilar phloem loading in leaves with symplasmic or apoplasmic minor-vein configurations. Planta 186:518–525
van Bel AJE, Ammerlaan A, van Dijk AA (1994) A three-step screening procedure to identify the mode of phloem loading in intact leaves. Evidence for symplasmic and apoplasmic phloem loading associated with the type of companion cell. Planta 192:31–39
van Fleet DS (1961) Histochemistry and function of the endodermis. Bot Rev 27:165–220
Walker NA, Pitman MG (1976) Mesurement of fluxes across membranes. In: Lüttge U, Pitman MG (eds) Transport in plants II. Part A cells. Encyclopedia of plant physiology, New Series vol 2. Springer, Berlin, pp 93–126
White PJ (2001) The pathways of calcium movement to the xylem. J Exp Bot 52:891–899
Wu X, Lin JX, Lin QQ, Wang J, Schreiber L (2005) Casparian strips in needles are more solute-permeable than endodermal transport barriers in roots of *Pinus bungeana*. Plant Cell Physiol 46:1799–1808
Yuan F, Chen M, Leng BY, Wang BS (2013) An efficient autofluorescence method for screening *Limonium bicolor* mutants for abnormal salt gland density and salt secretion. S Afr J Bot 88:110–117
Yuan F, Amy Lu M-J, Leng B-Y, Zheng GY, Feng Z-T, Li P-H, Zhu X-G, Wang BS (2015) Comparative transcriptome analysis of developmental stages of the *Limonium bicolor* leaf generates insights into salt gland differentiation. Plant Cell Environ 38:1637–1657
Yuan F, Leng B, Wang B (2016a) Progress in studying salt secretion from the salt glands in recretohalophytes: how do plants secrete salt? Frontiers Plant Sci 7:1–12
Yuan F, Amy Lyu MJ, Leng B-Y, Zhu X-G, Wang B-S (2016b) The transcriptome of NaCl-treated *Limonium bicolor* leaves reveals the genes controlling salt secretion of salt gland. Plant Mol Biol 91:241–256
Zeier J, Schreiber L (1999) Fourier transform infrared-spectroscopic characterization of isolated endodermal cell walls from plant roots: chemical nature in relation to anatomical development. Planta 209:537–542
Zeier J, Goll A, Yokoyama M, Karahara I, Schreiber L (1999a) Structure and chemical compostition of endodermal and rhizodermal/hypodermal walls of several species. Plant Cell Environ 22:271–279
Zeier J, Ruel K, Ryser U, Schreiber L (1999b) Chemical analysis and immunolocalistion of lignin and suberin in endodermal and hypodermal/rhizodermal cell walls of developing maize (*Zea mays* L.) primary roots. Planta 209:1–12

Ziegler H, Lüttge U (1967) Die Salzdrüsen von *Limonium vulgare*. II Mitteilung Die Lokalisierung des Chlorids. Planta 74:1–17

Ziegler H, Weigl J, Lüttge U (1963) Mikroautoradiographischer Nachweis der Wanderung von $^{35}SO_4^{--}$ durch die Tertiärendodermis der *Iris*-Wurzel. Protoplasma 56:362–370

Zimmermann HM, Steudle E (1998) Apoplastic transport across young maize roots: effect of the exodermis. Planta 206:7–19

Zimmermann HM, Hartmann K, Schreiber L, Steudle E (2000) Chemical composition of apoplastic transport barriers in relation to radial hydraulic conductivity of maize roots (*Zea mays* L.). Planta 210:302–311

The Haustorium of Phytopathogenic Fungi: A Short Overview of a Specialized Cell of Obligate Biotrophic Plant Parasites

Álvaro Polonio, Alejandro Pérez-García, Jesús Martínez-Cruz, Dolores Fernández-Ortuño, and Antonio de Vicente

Contents

Abstract Among all biotic stresses to which plants are subjected, the biotrophic fungal pathogens of rust and powdery mildew are the most economically relevant. They are characterized by their ability to develop specialized infective structures called haustoria. The fungal haustorium has been recognized as a fungal structure with a key role in disease establishment and has been implicated in essential processes, such as nutrient uptake and effector delivery. However, despite the early description of this fungal structure, many details of its composition, development or effector functions remain unsolved. In this work, we provide an overview of the current knowledge of the fungal haustorium, including the most recent isolation techniques and expression studies. We can conclude that the fungal haustorium is a complex structure, with a high level of expression of genes associated with nutrient uptake and pathogenesis and with a high level of protein synthesis, which seems to

Á. Polonio, A. Pérez-García, J. Martínez-Cruz, D. Fernández-Ortuño, and A. de Vicente (✉)
Departamento de Microbiología, Universidad de Málaga, Málaga, Spain

Instituto de Hortofruticultura Subtropical y Mediterránea “La Mayora”, Universidad de Málaga, Consejo Superior de Investigaciones Científicas (IHSM–UMA–CSIC), Málaga, Spain
e-mail: adevicente@uma.es

© Springer Nature Switzerland AG 2020

Progress in Botany (2021) 82: 337–356, https://doi.org/10.1007/124_2020_45,
Published online: 16 December 2020

be related to the release of secreted proteins. Although recent molecular studies have significantly advanced the knowledge of this structure, many questions remain unsolved. We hope that the development of novel techniques of genetic manipulation based on the capability of the fungal haustorium to uptake dsRNA, siRNA or T-DNA will allow us to answer these questions in the near future.

1 Introduction

Plants are constantly exposed to challenges from their environment and from biotic stresses. These biotic stresses include several organisms, such as viroids, viruses, bacteria, oomycetes, fungi and nematodes (Chaudhari et al. 2014). Biotrophic fungal pathogens are undoubtedly among the most fascinating of these organisms, as they develop in plants very differently from other fungal counterparts and require live cells for growth and development. The relationships between biotrophic fungi and their plant hosts are quite diverse, from symbiosis to parasitism. Thus, whereas the endomycorrhizal fungi maintain a symbiotic relationship with plant roots, rust and powdery mildew fungi have a parasitic relationship with their hosts. In the case of endomycorrhizae, fungi mainly obtain sugars from plant cells while helping plants obtain other nutrients, such as phosphorus, sulphur, nitrogen and several minerals from soil. In contrast, in the case of parasitic fungi, they compete with the plant for the use of different nutrients, such as sugars and amino acids (Szabo and Bushnell 2001), and are responsible for serious diseases in many crop plants. For the development of these diseases, rust and powdery mildew fungi need to overcome two layers of plant immune systems, PAMP-triggered immunity (PTI), or the recognition of pathogen-associated molecular patterns (PAMPs) (van Loon et al. 2006), and effector-triggered immunity (ETI) (Jones and Dangl 2006), which is the recognition of fungal effectors mainly through resistance (R) proteins (van Loon et al. 2006).

The powdery mildew and rust are the most important biotrophic fungi. Powdery mildew fungi (Phylum *Ascomycota*, Order *Erysiphales*) comprise 16 genera with approximately 900 species that infect nearly 10,000 species of both mono- and dicotyledonous plants (Braun and Cook 2012). Rust fungi (Phylum *Basidiomycota*, Order *Uredinales*) comprise approximately 100 genera with approximately 7,000 species (Berbee and Taylor 1993; Hahn 2000; Cummins and Hiratsuka 2003; Swann et al. 2011). They belong to the largest group of plant pathogenic fungi, affecting many economically important crops and causing significant yield losses (Hückelhoven 2005; Jakupović et al. 2006; Micali et al. 2008; Yin et al. 2009). Although the life cycles of powdery mildew and rust are quite different, both share the presence of a typical infection structure, the haustorium. This structure, however, is not exclusive to fungal pathogens and is also present in other biotrophic plant parasites, such as oomycetes (Grenville-Briggs and van West 2005). This structure has attracted great attention from plant pathologists since its first description in the mid-nineteenth century (von Mohl 1853). Currently, it is widely accepted that the fungal haustorium plays a key role in disease establishment and development by two main functions: the acquisition of nutrients to sustain parasitic growth and the release of effector proteins with different functions in pathogenesis, such as the manipulation of plant immunity (Jones and Dangl 2006; Oliva et al. 2010; Sánchez-Vallet et al. 2013; Martínez-Cruz et al. 2014; Lo Presti et al. 2015).

Despite the early description of the fungal haustorium, many questions remain unanswered and come to mind when we talk about this structure. The main unsolved issues of haustorial biology are related to its establishment into the host cell, its composition, its ability to avoid host recognition, its manner to uptake nutrients from host cells and the specific functions of its arsenal of effectors secreted into host plant cells.

2 Haustorial Cell Biology

2.1 *Haustorial Development and Establishment*

Fungal haustoria exist in a variety of morphologies to adapt to the architecture of the host cell. The wide morphological spectrum of haustoria is exemplified in the case of rust fungi, which have different stages, monokaryotic and dikaryotic, each associated with different haustorial morphologies (Mendgen et al. 2000). The monokaryotic stage of rust fungi produces M-haustoria, which appears as a hyphal extension without morphological differentiation (Gold and Mendgen 1991), whereas the dikaryotic haustoria are developed from external haustorial mother cells and present marked structural modifications compared to hyphal cells (Heath and Skalamera 1997). For powdery mildew fungi, two main morphologies have been described: the highly branched multidigitate haustoria typical of barley powdery mildew *Blumeria graminis* f. sp. *hordei* (Lambertucci et al. 2019) and globular haustoria that are highly decorated with tubular lobes typical of the *Arabidopsis* powdery mildew *Golovinomyces orontii* and cucurbit powdery mildew *Podosphaera xanthii* (Micali et al. 2011; Martínez-Cruz et al. 2014). In this work, we will explain the process of haustorial development focused on powdery mildew haustoria, which have been studied less than rust haustoria.

First, prior to haustorium formation, it is necessary for the fungus to penetrate through the plant cell wall, which occurs by the release of lytic enzymes (Eichmann and Hückelhoven 2008) and by the formation of a fungal structure termed the appressorium. From the appressorium, the so-called penetration peg emerges, which is a specialized hypha that penetrates the cuticle and plant cell wall and finally enlarges to produce the haustorial body (Bushnell and Bergquist 1974). Later, the haustorial body completely develops with prolongations emerging from it, known as the haustorial lobes (Fig. 1a, b) (Gil and Gay 1977; Mackie et al. 1991; Martínez-Cruz et al. 2014). The haustorium is not truly an intracellular structure, since the process of its formation causes the invagination of the host plasma membrane, which remains surrounding the haustorium, giving rise to the so-called extrahaustorial

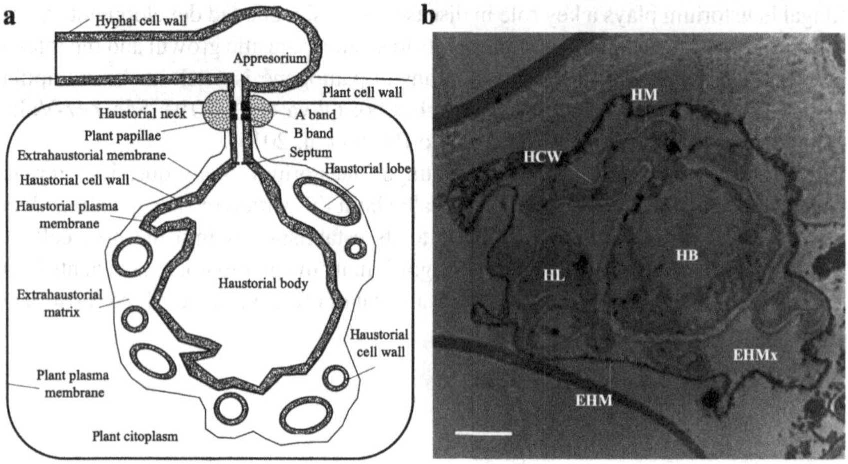

Fig. 1 The powdery mildew haustorial complex. (**a**) Schematic representation of a transverse section of a powdery mildew haustorial complex and its association with a host plant epidermal cell. (**b**) A cross section of a *P. xanthii* haustorial complex examined by transmission electron microscopy (TEM). Abbreviations are *HCW* haustorial cell wall, *HM* haustorial membrane, *HL* haustorial lobe, *EHMx* extrahaustorial matrix, *EHM* extrahaustorial membrane. Bar, 2 μm. Picture "**b**" was taken from Martínez-Cruz et al. (2014)

membrane (EHM) (Fig. 1a, b) (Gil and Gay 1977; Mackie et al. 1991; Martínez-Cruz et al. 2014).

The appressorium and haustorium are separated by the haustorial neck, where the EHM and the haustorial membrane (HM) are sealed by two neckband regions called the A band and B band (Chong and Harder 1980; Stumpf and Gay 1990; Mendgen et al. 2000), whereas a septum with a pore, coming from the papilla rupture (Eichmann and Hückelhoven 2008), separates the haustorial cytoplasm and the appressorium cytoplasm (Fig. 1a) (Gil and Gay 1977; Mackie et al. 1991; Wang et al. 2009; Micali et al. 2011). All of this leaves a space covered by a complex amorphous matter between the EHM and the HM termed the extrahaustorial matrix (EHMx) (Gil and Gay 1977; Koh et al. 2005), which can be considered a symplastic compartment (Voegele and Mendgen 2003) and seems ideal for nutrient uptake from the host (Gil and Gay 1977; Micali et al. 2011) (Fig. 1a, b). Collectively, the haustorial body, haustorial lobes, HM, EHM and EHMx are called the haustorial complex (HC) (Mackie et al. 1991; Micali et al. 2011; Martínez-Cruz et al. 2014). In fact, when we refer colloquially to haustoria, we are truly referring to haustorial complexes.

During the haustorial establishment process, the cytoskeleton of host plant cells suffers a polarized reorganization (Kobayashi et al. 1997; Takemoto and Hardham 2004) related to the plant immune response (Thomma et al. 2011; Henty-Ridilla et al. 2013). On the one hand, changes in the organization of microtubules occur to mediate the functions of certain proteins by directly binding to microtubules or by the indirect transport of these proteins (Schmidt and Panstruga 2011). On the other hand, the reorganization of actin filaments plays a role in physical alteration and is rapidly initiated in response to the attempted penetration of fungus to promote the traffic of vesicles with papilla formation components to this penetration site (Takemoto and Hardham 2004). Moreover, actin filaments accumulate around the nucleus, repositioning it closer to the penetration site to allow a faster immune response by causing changes in the expression of several genes (Eichmann et al. 2004). It is believed that fungal effectors are directed towards the plant cytoskeleton manipulation, whereas some peptides secreted from the plants may be directed to the fungal cytoskeleton. Therefore, the cytoskeleton seems to be a mutual target during fungus-plant combat, with effectors from both sides attacking their opponents (Schmidt and Panstruga 2007).

2.2 Haustorial Composition

As previously indicated, the origin of the haustorium is a hypha, so it seems obvious to think that its composition is the same. Nevertheless, several unique features have been described in the haustorium (Mackie et al. 1991; Micali et al. 2011; Martínez-Cruz et al. 2014). First, the EHM is a modification of the plant plasma membrane because its composition and structure are different (Hückelhoven and Panstruga 2011). The EHM is more convoluted and presents a greater thickness than the plant plasma membrane, exhibiting a substantial amount of associated polysaccharides and lacking intramembrane particles (Gil and Gay 1977; Mackie et al. 1991). Moreover, in contrast to the plant plasma membrane, in the EHM, there is an absence of ATPase activity (Spencer-Phillips and Gay 1981; Mackie et al. 1991). Nevertheless, the resistance protein RPW8.2, which is involved in the accumulation of hydrogen peroxide, is localized in the EHM and is essential to haustorium reception (Wang et al. 2009; Kim et al. 2014). The EHM has been the focus of several studies, and although its origin remains unknown, two hypotheses have been formulated in this regard. The first indicates that the origin of EHM is the consequence of severe modifications of the plant plasma membrane as a consequence of fungal activity over time. The second hypothesis suggests that its origin is due to de novo formation during haustorium formation (Mackie et al. 1991; Wang et al. 2009; Kim et al. 2014). A recent study has demonstrated that the EHM shares common features with the endoplasmic reticulum membrane, but it does not depend on conventional secretion, suggesting the possibility of a nonconventional secretory pathway from endoplasmic reticulum that may provide the necessary material to this membrane (Kwaaitaal et al. 2017).

On the other hand, the composition of the bands of the haustorial neck is different from that of the fungal cell wall. They are composed mainly of iron, phosphorus, β-glucans and lipid and proteinaceous compounds (Chong and Harder 1980; Stumpf and Gay 1990; Mendgen et al. 2000). More specifically, the A band seems to be rich in β-1,3-glucans, and its lipid compounds are attached to chitin and β-1,3-glucans, whereas the B band seems to be rich in β-1,4-glucans, and its lipid compounds are bound to proteinaceous components (Stumpf and Gay 1990).

Other evidence that suggests the molecular differentiation of the haustorium during its development is the presence of an N-linked glycoprotein with a size of 62 kDa, which has been described as unique in the haustorial membrane of pea powdery mildew, suggesting that the external composition of the haustorium differs from the composition of the rest of the fungal membranes (Mackie et al. 1991). Similarly, the differences seem to also be extensible to the haustorial cell wall. This is the case for three unknown carbohydrate epitopes present in the haustorial cell wall, but not in other fungal structures, which have been described in the flax rust fungus *Melampsora lini* by the production of monoclonal antibodies in mice immunized with isolated haustoria (Murdoch et al. 1998). Moreover, although molecular patterns associated with the fungal cell wall, such as chitin and β-1,3-glucans, are present on the surface of the haustorium, chitosan or any other modification of haustorial chitin does not seem to occur as they occur in the hyphal cell wall (Micali et al. 2011). The chitin modification to chitosan has been described as one of the strategies employed by fungi to avoid chitin recognition by the host, since it presents a lower power of elicitation than chitin (Mochizuki et al. 2011; Sánchez-Vallet et al. 2013). Therefore, the mechanism employed by the haustorium to suppress chitin recognition remains unknown.

Another striking point regarding the haustorium is the presence of a substantial amount of vesicles located mostly in the haustorial lobes (Martínez-Cruz et al. 2014) (Fig. 2) and in the EHMx (Micali et al. 2011), suggesting the existence of a specific exosome-mediated secretion pathway in the haustorium.

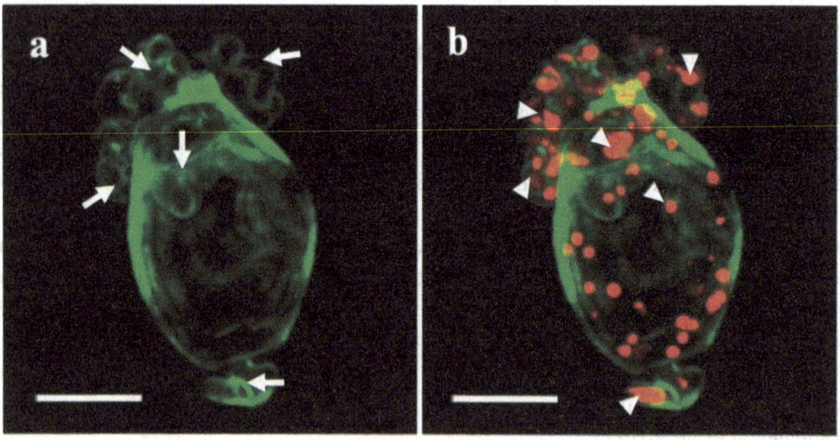

Fig. 2 Detection of vesicles in an isolated immature haustorium of *P. xanthii* examined by confocal laser scanning microscopy. (**a**) Specific staining of the haustorial cell wall with WGA-Alexa Fluor 488 conjugate (green). (**b**) Overlay of cell wall haustorial staining with WGA-Alexa Fluor 488 and specific staining of membrane vesicles with FM4-64 (red). The haustorial lobes are indicated with arrows, whereas the vesicles are indicated with arrowheads. Bars, 4 μm. Pictures were taken from Martínez-Cruz et al. (2014)

3 Isolation of Haustoria

To perform specific analyses of haustorial cells (e.g. gene expression analyses), it is necessary to properly isolate these fungal cells to accurately evaluate the genes being expressed and to deduce the metabolic activities within haustorial cells without the interference of similar activities from plant or hyphal cells. The isolation of fungal haustoria has been performed by different approaches: affinity chromatography with lectin concanavalin A columns (Hahn and Mendgen 1992), gradient centrifugation using sucrose (Tiburzy et al. 1992) and isopycnic centrifugation with Percoll (Pain et al. 1994). Affinity chromatography and isopycnic centrifugation are the most commonly used isolation strategies. Although ConA and Percoll approaches have been slightly modified (Godfrey et al. 2010; Micali et al. 2011; Link et al. 2014; Martínez-Cruz et al. 2014), the haustorial preparations obtained by both isolation techniques still present a high amount of plant contaminants. Recently, a novel haustorial isolation method based on fluorescence-activated cell sorting (FACS) was described (Polonio et al. 2019), which seems to overcome this limitation

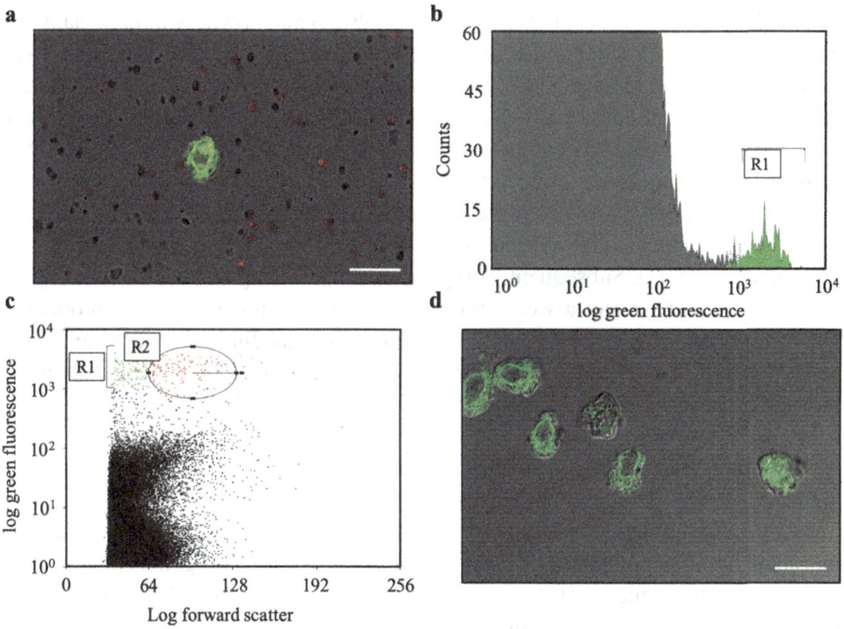

Fig. 3 Haustorial isolation of *P. xanthii* using flow cytometry. (**a**) Homogenate before fluorescence-activated cell sorting, where it is possible to see a haustorial cell stained with WGA-Alexa Fluor 488 (green fluorescence) and plant contaminants such as chloroplasts (red). (**b**) The R1 population, those structures with the highest green fluorescence, corresponds with fungal structures. (**c**) The R2 population, corresponding with haustorial cells, represents those structures that combined the highest green fluorescence and the relative haustorial size (log forward scatter). (**d**) Final homogenate after flow cytometry showing the haustorial preparation virtually free of contaminants. Bars, 25 μm. The figure was taken from Polonio et al. (2019)

(Fig. 3). Using this approach, after the initial homogenization of powdery mildew-infected plant tissues, a treatment with a WGA-Alexa Fluor 488 conjugate of the homogenate was applied to label haustorial cells (Fig. 3a). Then, a selection of the desired cell population (haustoria) was performed, which combined high green fluorescence and forward scatter properties that corresponded to the haustorial size (Fig. 3b, c). By this approach, it was possible to obtain a haustorial preparation virtually free of visible contaminants (Fig. 3d), solving the presence of plant contaminants in the haustorial preparations obtained by previous techniques (Godfrey et al. 2010; Weßling et al. 2012; Link et al. 2014).

4 Haustorial Molecular Physiology

4.1 Haustorial Gene Expression

Despite the different approaches used for the isolation of haustoria, only a few transcriptomic studies of rust (Hahn and Mendgen 1997; Jakupović et al. 2006; Yin et al. 2009; Garnica et al. 2013; Link et al. 2014) and powdery mildew haustoria (Godfrey et al. 2010; Weßling et al. 2012; Polonio et al. 2019; Sharma et al. 2019) have been reported to date. This is surely due to the difficulty of haustorial isolation and the fact that all haustorial isolation techniques are time consuming, and hence, the gene expression status can be altered (Weßling et al. 2012). The first study on haustorial gene expression, which was conducted in the common bean rust fungus *Uromyces appendiculatus*, identified a metallothionein gene and genes related to thiamine biosynthesis and nutrient uptake, such as amino acid transporters (Hahn and Mendgen 1997). Subsequent works showed that the highest proportion of genes expressed in rust haustoria were involved in protein synthesis, energy production and metabolism, with relatively abundant transcripts encoding secreted proteins (Yin et al. 2009) and nutrient transporters (Jakupović et al. 2006; Yin et al. 2009). However, these works only allowed the study of a small portion of haustorial-expressed genes.

The emergence of high-throughput sequencing technologies has allowed us to increase the understanding of gene expression in haustorial cells. In this way, transcriptomic studies of rust haustoria showed that they are especially active in the acquisition of amino acids, sugars, phosphorus and nitrogen compounds, in response to stress and in protein synthesis (Garnica et al. 2013; Link et al. 2014). High-throughput sequencing studies of powdery mildew haustoria suggested that gene expression in rust and powdery mildew haustoria is quite similar (Weßling et al. 2012; Polonio et al. 2019). These transcriptomic analyses showed that powdery mildew haustoria are also active in protein metabolism, primary metabolism, energy production, response to stress, reactive oxygen species (ROS) scavenging, vesicle-mediated transport and pathogenesis. Moreover, a recent study on the *P. xanthii* haustorial transcriptome showed a high expression of ncRNAs, suggesting a putative regulation of gene expression mediated by these ncRNAs. Furthermore, the RNase function was highly represented among the top 50 highly expressed ncRNAs (Polonio et al. 2019). In both rust and powdery mildew haustoria, a substantial proportion of transcripts corresponded with genes encoding secreted proteins, among which a high amount corresponded to effector candidates, that is, those putatively secreted proteins of small size without homology to other proteins in databases (Weßling et al. 2012; Garnica et al. 2013; Link et al. 2014). In summary, the gene expression analyses of haustoria indicate that these cells present high protein synthesis and are directly involved in nutrient uptake, energy production, plant defence suppression, ROS scavenging, pathogenesis and probably the release of effector proteins.

4.2 Nutrient Uptake in the Haustorium

One vital aspect of the lifestyle of obligate biotrophic fungi is the acquisition and mobilization of nutrients from the host cells. With this regard, it is interesting to focus on the arsenal of protein transporters and enzymes required for this purpose. In haustoria-forming fungi, the haustorium seems to be the main structure responsible for nutrient uptake, and thus, the specific expression in the haustorium of several nutrient transporters was proposed long ago. Thus, the putative hexose transporter HXT1 and the amino acid transporters AAT1, AAT2 and ATT3 have been widely described in several studies on rust haustoria (Mendgen et al. 2000; Struck et al. 2002, 2004; Voegele et al. 2002; Garnica et al. 2013). Biochemical studies of AAT1 and AAT3 showed that they act as amino acid transporters with specificity for L-histidine/L-lysine and L-leucine/L-methionine/L-cysteine, respectively (Struck et al. 2002, 2004). Moreover, HXT1 and AAT2 transporters were immunolocalized exclusively in the haustorial plasma membrane (Mendgen et al. 2000; Voegele et al. 2002).

The main ion pumps in fungi and plants are the H^+-ATPases present in the plasma membrane, which seem to play a key role in active nutrient uptake (Sondergaard et al. 2004). This is the case for PMA1, a plasma membrane ATPase essential for the growth of *Saccharomyces cerevisiae* (Serrano et al. 1986), whose hydrolytic activity was severalfold higher in microsomal vesicles of isolated haustoria of the faba bean rust fungus *Uromyces viciae-fabae* than in ungerminated urediniospores and germ tubes (Struck et al. 1996). This fact, together with the molecular characterization of this enzyme and its suggested autoregulation (Struck et al. 1998), supports the function of PMA1 in nutrient uptake from host cells.

In the case of powdery mildew fungi, although nine predicted sugar transporters and six putative amino acid transporters were identified in *Arabidopsis* powdery mildew haustoria, none of them were described among the top 50 expressed genes in these haustorial cells (Weßling et al. 2012), as expected according to previous results obtained in rust haustoria, where those transporters were highly expressed (Hahn and Mendgen 1997; Jakupović et al. 2006; Duplessis et al. 2011). This fact is noteworthy, since in the case of powdery mildew fungi, glucose seems to be the main carbon source obtained from host leaves (Sutton et al. 1999; Fotopoulos et al. 2003). Although an MFL maltose transporter and PMA1 were induced in late stages of fungal development (Weßling et al. 2012) and despite that the transcript abundance does not determine the protein levels, the lack of identification of hexose transporters in a proteomic study of haustoria from barley powdery mildew *B. graminis* (Godfrey et al. 2009) indicates that powdery mildews probably use few sugar transporters. This fact could suggest that powdery mildew haustoria, in contrast to rust haustoria, may use other alternatives for carbohydrate uptake. However, a more recent transcriptomics study of *P. xanthii* haustoria described an MFS sugar transporter among the top 50 expressed genes in this structure (Polonio et al. 2019), suggesting different strategies for carbohydrate uptake among powdery mildew fungi.

Although hexose and amino acid transporters have been extensively studied, other nutrient transporters have been described in haustoria. This is the case for the sulphate transporter described in wheat stripe rust haustoria (Yin et al. 2009) or the inorganic phosphate transporter found among the top 50 expressed genes in the haustoria of *Arabidopsis* powdery mildew (Weßling et al. 2012), suggesting the implication of powdery mildew haustoria in sulphur and phosphorus acquisition.

4.3 Haustorial Effectors

The presence of a high amount of transcripts encoding putative secreted proteins in both rust and powdery mildew haustoria (Weßling et al. 2012; Garnica et al. 2013; Link et al. 2014; Polonio et al. 2019; Sharma et al. 2019) suggests the participation of the haustorium in the release of effectors directly inside the host cells and highlights the importance of this structure not only in the acquisition of nutrients but also in pathogenesis, for example, in the manipulation of host plant defence responses.

Rust transferred protein 1 (RTP1) of *U. viciae-fabae* was the first protein that was shown to be specifically expressed in the haustorium. This protein is translocated to the host cell during the rust infection (Hahn and Mendgen 1997; Kemen et al. 2005) and acts as a protease inhibitor in the supernatants of yeast cultures, suggesting a putative role as an inhibitor of host protease activity associated with plant defence (Pretsch et al. 2013). Moreover, it was also found forming filamentous structures and aggregates in EHMx, and hence, a putative function in the accommodation of haustorium into the host cell was also postulated (Kemen et al. 2013). Other proteins have been identified to be secreted by rust haustoria. This is the case for four Avr proteins of *M. lini*, which have been confirmed to be translocated into flax cells (Rafiqi et al. 2010).

Recent transcriptomics studies have identified many putative secreted proteins in rust haustoria. Thus, the haustorial transcriptome of *Puccinia striiformis* f. sp. *tritici* has revealed the abundance of transcripts of cysteine-rich proteins among the haustorial effector candidates and identified that most of the secreted proteins expressed in haustoria were differentially expressed compared to those expressed in germinated spores (Garnica et al. 2013), which is consistent with the major expression of effector candidate genes in haustoria, where they can directly act by modifying the functions of the host cell. The same transcriptomics study identified a considerably high number of extracellular cell wall-modifying enzymes, such as chitinases, in the haustorial transcriptome of *U. appendiculatus*, where these proteins and those related to the response to biotic stimulus were considerably enriched (Link et al. 2014). Extracellular cell wall-modifying enzymes were also identified in the haustorial transcriptomes of *G. orontii*, *P. xanthii* and *Erysiphe pisi*. In the case of *E. pisi*, two glycosylase/hydrolase/chitin-binding proteins were overexpressed by >twofold in the haustorial samples, whereas in the case of *P. xanthii*, a putative lytic polysaccharide monooxygenase of chitin was expressed specifically in haustoria (Weßling et al. 2012; Polonio et al. 2019; Sharma et al. 2019). This suggests the importance of these enzymes in haustorial physiology. The latest powdery mildew haustorial transcriptomics studies have also identified other secreted proteins exclusively or mostly expressed in haustoria, such as a putative and an annotated acid phosphatase, a SnodProt1 protein, a dodecin protein and several putative adhesion proteins, which could be involved in phosphorus acquisition, protection against reactive oxygen species and haustorial accommodation, respectively (Polonio et al. 2019), and several ribonuclease/ribotoxin domain-containing proteins, two putative virulence factors, a heat-shock protein 70 and a cutinase were also identified (Sharma et al. 2019). Furthermore, genes encoding some of these proteins were silenced, resulting in compromised powdery mildew infections (Sharma et al. 2019). These results further support the attributed importance of the haustorium and their secreted proteins in disease establishment and pathogenesis.

In contrast to rust haustorial secreted proteins, where no characteristic sequence motifs have been identified among all predicted haustorial secreted proteins (Garnica et al. 2013; Link et al. 2014), the N-terminal Y/F/WxC-motif is conserved among the effector candidates identified in *B. graminis* haustoria, which is located around the first 24 amino acids after the signal peptide cleavage site (Godfrey et al. 2010), and in *E. pisi*, where this motif was the only found significantly enriched among the effector candidates (Sharma et al. 2019). This motif seems to be involved in the translocation of proteins to host cells or in allowing correct folding of the effector proteins (Hacquard et al. 2012). To date, these functions have not been conclusively demonstrated; however, the fact that approximately 20% of the *B. graminis* haustorial transcriptome encodes Y/F/WxC proteins suggests that this domain may have a key function in the physiology of powdery mildew fungi (Godfrey et al. 2010).

5 The Haustorium as a Gateway for the Introduction of Genetic Material in Biotrophic Fungi

A major problem in the study of biotrophic fungi is their intrinsic recalcitrance to genetic manipulation (Micali et al. 2008). Over time, approaches such as particle bombardment (Bhairi and Staples 1992; Li et al. 1993; Christiansen et al. 1995; Webb et al. 2006; Djulic et al. 2011) and electroporation (Vela-Corcía et al. 2018) have been used to transform both rust and powdery mildew fungi. However, these methods are poorly reproducible and unstable and have limited applicability.

In recent years, a novel system called host-induced gene silencing (HIGS) has emerged to overcome the limitations of previous approaches and is currently widely used in the functional analysis of genes in obligate biotrophs. This method is based on the capability of haustorium-forming pathogens to take genetic material, in this case double-stranded RNA (dsRNA) or small interfering RNA (siRNA), through the haustorium, which leads to subsequent gene silencing of the gene of interest (Nowara et al. 2010; Pliego et al. 2013). Although the precise mechanism of RNA uptake by haustoria remains unknown, this tool has been successfully used in the functional analysis of powdery mildew and rust effectors (Nowara et al. 2010; Pliego et al. 2013; (Panwar et al. 2013). A variant of the HIGS system used for gene silencing in *Puccinia triticina* (Yin et al. 2010) is the so-called virus-induced gene silencing (VIGS), which is based on the ability of eukaryotic organisms to fight against infections caused by viruses (Waterhouse et al. 2001; Unver and Budak 2009; Kirigia et al. 2014). During viral infections, a dsRNA replication intermediate is produced, recognized and cleaved by Dicer, a ribonuclease that processes the dsRNA to produce siRNA. Later, this siRNA guides the RISC complex for the degradation of mRNA corresponding to the target gene (Velásquez et al. 2009). To exploit this procedure, viral vectors such as tobacco mosaic virus (TMV), tobacco rattle virus (TRV), potato virus X (PVX) and barley strip mosaic virus have been modified to carry the inserts from the fungal genes to be silenced (Lu et al. 2003).

Recently, the use of *Agrobacterium tumefaciens* has also been incorporated in powdery mildew research. In particular, *Agrobacterium* has been used for two purposes: gene silencing and genetic transformation. Thus, a method designated *A. tumefaciens*-mediated host-induced gene silencing (ATM-HIGS) was recently used to analyse the function of several effector candidates of *P. xanthii* (Martínez-Cruz et al. 2018a). The method is based on the transformation of melon cotyledon cells with silencing constructs via *Agrobacterium*. Thus, plant cells produce the interfering RNA (RNAi) that will be taken up by haustoria by unknown mechanisms, leading to gene silencing (Fig. 4a, b). *Agrobacterium* has also been used to transform *P. xanthii*. Two methods have been proposed. The first is a variant of the typical *Agrobacterium*-mediated transformation (AMT) system; in this case, *P. xanthii* conidia are directly exposed to *Agrobacterium* and transformed by the bacterium (Martínez-Cruz et al. 2017). The second method has been designated "transformation by growth onto agroinfiltrated tissues" (TGAT). Using this method, the leaf tissue is first transformed by *Agrobacterium* and then inoculated with

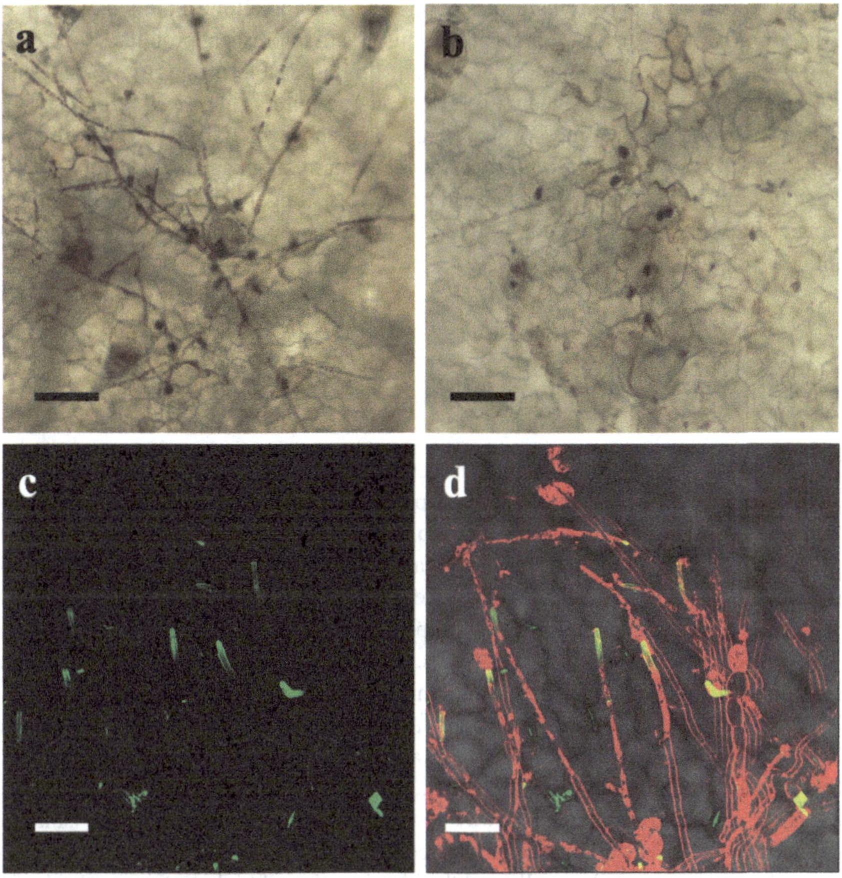

Fig. 4 Pictures illustrating the ATM-HIGS and TGAT systems for gene silencing and transient transformation of powdery mildew fungi, respectively. (**a**, **b**) *A. tumefaciens*-mediated host-induced gene silencing (ATM-HIGS) in *P. xanthii*. (**a**) Empty vector negative control. (**b**) Reduction in fungal growth in terms of penetration points (black spots, corresponding with haustoria) after silencing of the *P. xanthii* effector gene *PEC034*. (**c**, **d**) Transformation by growth onto agro-infiltrated tissues (TGAT) in *P. xanthii*. (**c**) Localization of the effector fusion PEC2-GFP (green) in hyphal tips. (**d**) Overlay of PEC2-GFP and fungal membranes stained with FM4–64. Bars are 100 μm (pictures "**a**" and "**b**"); 50 μm (pictures "**c**" and "**d**"). Pictures "**a**" and "**b**" were taken from Martínez-Cruz et al. 2018a), and "**c**" and "**d**" were taken from Martínez-Cruz et al. (2018b)

P. xanthii (Martínez-Cruz et al. 2018b). In this case, according to the authors' hypothesis, the T-DNA is transferred to *P. xanthii* via haustoria. This was supported by the co-localization of translational fluorescent fusions of the Rab5 protein from *P. xanthii* and the VirD2 protein from *A. tumefaciens* into small vesicles of haustoria, suggesting that endocytosis is the mechanism by which haustoria acquire the T-DNA (Martínez-Cruz et al. 2018b). The VirD2 protein was shown to migrate from the haustorium to the nucleus of hyphal cells, suggesting the possibility that hyphal cells may also be transformed. Thus, this method allowed the localization of effector proteins by means of translational fusions to fluorescent proteins (Fig. 4c, d). However, both methods show an important limitation: the instability of the transformation. It seems that when the selection pressure is removed, the T-DNAs disappear from the genome. However, these methods of transient transformation of *P. xanthii* by *Agrobacterium* are considered an extraordinary step forward for powdery mildew research. The haustorium, the key fungal cell for gene silencing in fungal biotrophs, could also be the key for the genetic transformation of powdery mildew fungi.

6 Conclusions and Perspectives

Rusts and powdery mildews are the largest groups of plant pathogenic fungi. They affect many economically important crops, inflicting significant yield losses every season. These fungi parasitize plants to use them as living substrates. For this purpose, pathogenic biotrophic fungi have developed a specialized structure of parasitism called the haustorium. Despite its implication in disease establishment, this structure remains largely unknown, perhaps because its isolation and study are very difficult. However, in recent years, it has been possible to obtain valuable molecular information that has shed some light to the most unknown aspects of haustorial physiology. In this way, many aspects of the haustorial composition have been determined, such as the presence of a substantial number of vesicles located in the haustorial lobes and in the EHMx, suggesting the existence of a specific exosome-mediated secretion pathway in the haustorium. Moreover, new isolation techniques, such as fluorescence-activated cell sorting, in combination with high-throughput sequencing, have allowed the identification of high protein synthesis, for example, of many nutrient transporters, and the high expression of candidate secreted proteins that are exclusive to this structure, supporting the hypothesis of the secretion of specific effectors by the haustorium. Furthermore, these gene expression studies have provided extensive information on many genes expressed in haustoria, such as genes involved in energy production, plant defence suppression, ROS scavenging, pathogenesis and adhesion, deepening the knowledge of this structure. Finally, the recent advances in the genetic manipulation of these fungi, due to the ability of the haustorium to uptake dsRNA, siRNA and T-DNA, make the haustorium the key to the development of methods for the genetic transformation of biotrophic fungi and opens an avenue of possibilities that will allow us to analyse in detail many processes and functions in the future and, therefore, answer many of the unsolved questions that remain to date.

Acknowledgements This work was supported by a grant from the "Agencia Estatal de Investigación (AEI)" (AGL2016-76216-C2-1-R), cofinanced by FEDER funds (European Union). A.P. was supported by a PhD fellowship (BES-2014-068602) from the former "Ministerio de Economía y Competitividad (MINECO)".

References

Berbee ML, Taylor JW (1993) Dating the evolutionary radiations of the true fungi. Can J Bot 71:1114–1127

Bhairi S, Staples RC (1992) Transient expression of the β-glucuronidase gene introduced into *Uromyces appendiculatus* uredospores by particle bombardment. Phytopathology 82:986–989

Braun U, Cook RTA (2012) Taxonomic manual of the Erysiphales (powdery mildews). CBS-KNAW Fungal Biodiversity Centre, Utrecht

Bushnell WR, Bergquist SE (1974) Aggregation of host cytoplasm and the formation of papillae and haustoria in powdery mildew of barley. Phytopathology 65:310–318

Chaudhari P, Ahmed B, Joly DL, Germain H (2014) Effector biology during biotrophic invasion of plant cells. Virulence 5:703–709

Chong J, Harder DE (1980) Ultrastructure of haustorium development in *Puccinia coronata avenae*. I. Cytochemistry and electron probe X-ray analysis of the haustorial neck ring. Can J Bot 58:2496–2505

Christiansen SK, Knudsen S, Giese H (1995) Biolistic transformation of the obligate plant pathogenic fungus, *Erysiphe graminis* f. sp. *hordei*. Curr Genet 29:100–102

Cummins GB, Hiratsuka Y (2003) Illustrated genera of rust fungi, 3rd edn. American Phytopathological Society, St. Paul

Djulic A, Schmid A, Lenz H, Sharma P, Koch C, Wirsel SGR, Voegele RT (2011) Transient transformation of the obligate biotrophic rust fungus *Uromyces fabae* using biolistics. Fungal Biol 115:633–642

Duplessis S, Cuomo CA, Lin Y et al (2011) Obligate biotrophy features unraveled by the genomic analysis of rust fungi. Proc Natl Acad Sci U S A 108:9166–9171

Eichmann R, Hückelhoven R (2008) Accommodation of powdery mildew fungi in intact plant cells. J Plant Physiol 165:5–18

Eichmann R, Schultheiss H, Kogel KH, Huckelhoven R (2004) The barley apoptosis suppressor homologue BAX inhibitor-1 compromises nonhost penetration resistance of barley to the inappropriate pathogen *Blumeria graminis* f. sp. *tritici*. Mol Plant Microbe Interact 17:484–490

Fotopoulos V, Gilbert MJ, Pittman JK, Marvier AC, Buchanan AJ, Sauer N, Hall JL, Williams LE (2003) The monosaccharide transporter gene, AtSTP4, and the cell-wall invertase, AtBfruct1, are induced in *Arabidopsis* during infection with the fungal biotroph *Erysiphe cichoracearum*. Plant Physiol 132:821–829

Garnica DP, Upadhyaya NM, Dodds PN, Rathjen JP (2013) Strategies for wheat stripe rust pathogenicity identified by transcriptome sequencing. PLoS One 8:e67150

Gil F, Gay JL (1977) Ultrastructural and physiological properties of the host interfacial components of haustoria of *Erysiphe pisi in vivo* and *in vitro*. Physiol Plant Pathol 10:1–12

Godfrey D, Zhang Z, Saalbach G, Thordal-Christensen H (2009) A proteomics study of barley powdery mildew haustoria. Proteomics 9:3222–3232

Godfrey D, Böhlenius H, Pedersen C, Zhang Z, Emmersen J, Thordal-Christensen H (2010) Powdery mildew fungal effector candidates share N-terminal Y/F/WxC-motif. BMC Genomics 11:317

Gold RE, Mendgen K (1991) Rust basidiospore germlings and disease initiation. In: Cole GT, Hoch HC (eds) The fungal spore and disease initiation in plants and animals. Plenum Press, New York, pp 67–99

Grenville-Briggs LJ, van West P (2005) The biotrophic stages of oomycete–plant interactions. Academic Press, San Diego, pp 217–243

Hacquard S, Joly DL, Lin Y-C et al (2012) A comprehensive analysis of genes encoding small secreted proteins identifies candidate effectors in *Melampsora larici-populina* (poplar leaf rust). Mol Plant Microbe Interact 25:279–293

Hahn M (2000) The rust fungi. Cytology, physiology and molecular biology of infection. In: Kronstad JW (ed) Fungal pathology. Kluwer Academic Publishers, Dordrecht, pp 267–306

Hahn M, Mendgen K (1992) Isolation by ConA binding of haustoria from different rust fungi and comparison of their surface qualities. Protoplasma 170:95–103

Hahn M, Mendgen K (1997) Characterization of *in planta*-induced rust genes isolated from a haustorium-specific cDNA library. Mol Plant Microbe Interact 10:427–437

Heath MC, Skalamera D (1997) Cellular interactions between plants and biotrophic fungal parasites (JH Andrews, IC Tommerup, and JABT-A in BR Callow, Eds.). Adv Bot Res 24:195–225

Henty-Ridilla JL, Shimono M, Li J, Chang JH, Day B, Staiger CJ (2013) The plant actin cytoskeleton responds to signals from microbe-associated molecular patterns. PLoS Pathog 9: e1003290

Hückelhoven R (2005) Powdery mildew susceptibility and biotrophic infection strategies. FEMS Microbiol Lett 245:9–17

Hückelhoven R, Panstruga R (2011) Cell biology of the plant-powdery mildew interaction. Curr Opin Plant Biol 14:738–746

Jakupović M, Heintz M, Reichmann P, Mendgen K, Hahn M (2006) Microarray analysis of expressed sequence tags from haustoria of the rust fungus *Uromyces fabae*. Fungal Genet Biol 43:8–19

Jones JD, Dangl JL (2006) The plant immune system. Nature 444:323–329

Kemen E, Kemen AC, Rafiqi M, Hempel U, Mendgen K, Hahn M, Voegele RT (2005) Identification of a protein from rust fungi transferred from haustoria into infected plant cells. Mol Plant Microbe Interact 18:1130–1139

Kemen E, Kemen A, Ehlers A, Voegele R, Mendgen K (2013) A novel structural effector from rust fungi is capable of fibril formation. Plant J 75:767–780

Kim H, O'Connell R, Maekawa-Yoshikawa M, Uemura T, Neumann U, Schulze-Lefert P (2014) The powdery mildew resistance protein RPW8.2 is carried on VAMP721/722 vesicles to the extrahaustorial membrane of haustorial complexes. Plant J 79:835–847

Kirigia D, Runo S, Alakonya A (2014) A virus-induced gene silencing (VIGS) system for functional genomics in the parasitic plant *Striga hermonthica*. Plant Methods 10:1–8

Kobayashi Y, Kobayashi I, Funaki Y, Fujimoto S, Takemoto T, Kunoh H (1997) Dynamic reorganization of microfilaments and microtubules is necessary for the expression of non-host resistance in barley coleoptile cells. Plant J 11:525–537

Koh S, André A, Edwards H, Ehrhardt D, Somerville S (2005) *Arabidopsis thaliana* subcellular responses to compatible *Erysiphe cichoracearum* infections. Plant J 44:516–529

Kwaaitaal M, Nielsen ME, Böhlenius H, Thordal-Christensen H (2017) The plant membrane surrounding powdery mildew haustoria shares properties with the endoplasmic reticulum membrane. J Exp Bot 68:5731–5743

Lambertucci S, Orman KM, Das Gupta S, Fisher JP, Gazal S, Williamson RJ, Cramer R, Bindschedler LV (2019) Analysis of barley leaf epidermis and extrahaustorial proteomes during powdery mildew infection reveals that the PR5 thaumatin-like protein TLP5 is required for susceptibility towards *Blumeria graminis* f. sp. *hordei*. Front Plant Sci 10:1138

Li A, Altosaar I, Heath MC, Horgen PA (1993) Transient expression of the β-glucuronidase gene delivered into urediniospores of *Uromyces appendiculatus* by particle bombardment. Can J Plant Pathol 15:1–6

Link TI, Lang P, Scheffler BE et al (2014) The haustorial transcriptomes of *Uromyces appendiculatus* and *Phakopsora pachyrhizi* and their candidate effector families. Mol Plant Pathol 15:379–393

Lo Presti L, Lanver D, Schweizer G, Tanaka S, Liang L, Tollot M, Zuccaro A, Reissmann S, Kahmann R (2015) Fungal effectors and plant susceptibility. Annu Rev Plant Biol 66:513–545

Lu R, Martin-Hernandez AM, Peart JR, Malcuit I, Baulcombe DC (2003) Virus-induced gene silencing in plants. Methods 30:296–303

Mackie AJ, Roberts AM, Callow JA, Green JR (1991) Molecular differentiation in pea powdery-mildew haustoria. Planta 183:399–408

Martínez-Cruz J, Romero D, Dávila JC, Pérez-García A (2014) The *Podosphaera xanthii* haustorium, the fungal Trojan horse of cucurbit-powdery mildew interactions. Fungal Genet Biol 71:21–31

Martínez-Cruz J, Romero D, de Vicente A, Pérez-García A (2017) Transformation of the cucurbit powdery mildew pathogen *Podosphaera xanthii* by *Agrobacterium tumefaciens*. New Phytol 213(4):1961–1973

Martínez-Cruz J, Romero D, de la Torre FN, Fernández-Ortuño D, Torés JA, de Vicente A, Pérez-García A (2018a) The functional characterization of *Podosphaera xanthii* candidate effector genes reveals novel target functions for fungal pathogenicity. Mol Plant-Microbe Interact 31:914–931

Martínez-Cruz J, Romero D, De Vicente A, Pérez-García A (2018b) Transformation by growth onto agro-infiltrated tissues (TGAT), a simple and efficient alternative for transient transformation of the cucurbit powdery mildew pathogen *Podosphaera xanthii*. Mol Plant Pathol 19:2502–2515

Mendgen K, Struck C, Voegele RT, Hahn M (2000) Biotrophy and rust haustoria. Physiol Mol Plant Pathol 56:141–145

Micali C, Göllner K, Humphry M, Consonni C, Panstruga R (2008) The powdery mildew disease of *Arabidopsis*: A paradigm for the interaction between plants and biotrophic fungi. Arabidopsis Book 6:e0115

Micali CO, Neumann U, Grunewald D, Panstruga R, O'Connell R (2011) Biogenesis of a specialized plant-fungal interface during host cell internalization of *Golovinomyces orontii* haustoria. Cell Microbiol 13:210–226

Mochizuki S, Saitoh K, Minami E, Nishizawa Y (2011) Localization of probe-accessible chitin and characterization of genes encoding chitin-binding domains during rice-*Magnaporthe oryzae* interactions. J Gen Plant Pathol 77:163–173

Murdoch LJ, Kobayashi I, Hardham AR (1998) Production and characterisation of monoclonal antibodies to cell wall components of the flax rust fungus. Eur J Plant Pathol 104:331–346

Nowara D, Gay A, Lacomme C, Shaw J, Ridout C, Douchkov D, Hensel G, Kumlehn J, Schweizer P (2010) HIGS: host-induced gene silencing in the obligate biotrophic fungal pathogen *Blumeria graminis*. Plant Cell 22:3130–3141

Oliva R, Win J, Raffaele S et al (2010) Recent developments in effector biology of filamentous plant pathogens. Cell Microbiol 12:705–715

Pain NA, Green JR, Gammie F et al (1994) Immunomagnetic isolation of viable intracellular hyphae of *Colletotrichum lindemuthianum* (Sacc. & Magn.) Briosi & Cav. from infected bean leaves using a monoclonal antibody. New Phytol 127:223–232

Panwar V, McCallum B, Bakkeren G (2013) Host-induced gene silencing of wheat leaf rust fungus *Puccinia triticina* pathogenicity genes mediated by the barley stripe mosaic virus. Plant Mol Biol 81:595–608

Pliego C, Nowara D, Bonciani G et al (2013) Host-induced gene silencing in barley powdery mildew reveals a class of ribonuclease-like effectors. Mol Plant Microbe Interact 26:633–642

Polonio Á, Seoane P, Claros MG, Pérez-Garciá A (2019) The haustorial transcriptome of the cucurbit pathogen *Podosphaera xanthii* reveals new insights into the biotrophy and pathogenesis of powdery mildew fungi. BMC Genomics 20:543

Pretsch K, Kemen A, Kemen E, Geiger M, Mendgen K, Voegele R (2013) The rust transferred proteins—a new family of effector proteins exhibiting protease inhibitor function. Mol Plant Pathol 14:96–107

Rafiqi M, Gan PHP, Ravensdale M, Lawrence GJ, Ellis JG, Jones DA, Hardham AR, Dodds PN (2010) Internalization of flax rust avirulence proteins into flax and tobacco cells can occur in the absence of the pathogen. Plant Cell 22:2017–2032

Sánchez-Vallet A, Saleem-Batcha R, Kombrink A, Hansen G, Valkenburg D-J, Thomma BP, Mesters JR (2013) Fungal effector Ecp6 outcompetes host immune receptor for chitin binding through intrachain LysM dimerization. eLife 2:e00790

Schmidt SM, Panstruga R (2007) Cytoskeleton functions in plant-microbe interactions. Physiol Mol Plant Pathol 71:135–148

Schmidt SM, Panstruga R (2011) Pathogenomics of fungal plant parasites: what have we learnt about pathogenesis? Curr Opin Plant Biol 14:392–399

Serrano R, Kielland-Brandt MC, Fink GR (1986) Yeast plasma membrane ATPase is essential for growth and has homology with (Na+ + K+), K+- and Ca2+-ATPases. Nature 319:689–693

Sharma G, Aminedi R, Saxena D, Gupta A, Banerjee P, Jain D, Chandran D (2019) Effector mining from the *Erysiphe pisi* haustorial transcriptome identifies novel candidates involved in pea powdery mildew pathogenesis. Mol Plant Pathol 20:1506–1522

Sondergaard TE, Schulz A, Palmgreen M (2004) Energization of transport processes in plants. Roles of the plasma membrane H+-ATPase. Plant Physiol 136:2475–2482

Spencer-Phillips PTN, Gay JL (1981) Domains of ATPase in plasma membranes and transport through infected plant cells. New Phytol 89:393–400

Struck C, Hahn M, Mendgen K (1996) Plasma membrane H+-ATPase activity in spores, germ tubes, and haustoria of the rust fungus *Uromyces viciae-fabae*. Fungal Genet Biol 20:30–35

Struck C, Siebels C, Rommel O, Wernitz M, Hahn M (1998) The plasma membrane H + -ATPase from the biotrophic rust fungus *Uromyces fabae*: molecular characterization of the gene *PMA1* and functional expression of the enzyme in yeast. Mol Plant-Microbe Interact 11:458–465

Struck C, Ernst M, Hahn M (2002) Characterization of a developmentally regulated amino acid transporter (AAT1p) of the rust fungus *Uromyces fabae*. Mol Plant Pathol 3:23–30

Struck C, Mueller E, Martin H, Lohaus G (2004) The *Uromyces fabae* UfAAT3 gene encodes a general amino acid permease that prefers uptake of in planta scarce amino acids. Mol Plant Pathol 5:183–189

Stumpf M, Gay J (1990) The composition of *Erysiphe pisi* haustorial complexes with special reference to the neckbands. Physiol Mol Plant Pathol 37:125–143

Sutton PN, Henry MJ, Hall JL (1999) Glucose, and not sucrose, is transported from wheat to wheat powdery mildew. Planta 208:426–430

Swann EC, Frieders EM, McLaughlin DJ (2011) Urediniomycetes. In: McLaughlin DJ, Blackwell M, Spatafora JW (eds) Systematics and evolution. Springer, Berlin, pp 37–56

Szabo LJ, Bushnell WR (2001) Hidden robbers: the role of fungal haustoria in parasitism of plants. Proc Natl Acad Sci U S A 98:7654–7655

Takemoto D, Hardham AR (2004) The cytoskeleton as a regulator and target of biotic interactions in plants. Plant Physiol 136:3864–3876

Thomma BP, Nurnberger T, Joosten MH (2011) Of PAMPs and effectors: the blurred PTI–ETI dichotomy. Plant Cell 23:4–15

Tiburzy R, Martins E, Reisener H (1992) Isolation of haustoria of *Puccinia graminis* f. sp. tritici from wheat leaves. Exp Mycol 328:324–328

Unver T, Budak H (2009) Virus-induced gene silencing, a post transcriptional gene silencing method. Int J Plant Genomics 2009:198680

van Loon LC, Geraats BPJ, Linthorst HJM (2006) Ethylene as a modulator of disease resistance in plants. Trends Plant Sci 11:184–191

Vela-Corcía D, Romero D, De Vicente A, Pérez-García A (2018) Analysis of β-tubulin-carbendazim interaction reveals that binding site for MBC fungicides does not include residues involved in fungicide resistance. Sci Rep 8:7161

Velásquez AC, Chakravarthy S, Martin GB (2009) Virus-induced gene silencing (VIGS) in *Nicotiana benthamiana* and tomato. J Vis Exp 28:1292

Voegele RT, Mendgen K (2003) Rust haustoria : uptake and beyond. New Phytol 159:93–100

Voegele RT, Struck C, Hahn M, Mendgen K (2002) The role of haustoria in sugar supply during infection of broad bean by the rust fungus *Uromyces fabae*. Proc Natl Acad Sci U S A 98:8133–8138

von Mohl H (1853) Ueber die Traubenkrankheit. Bot Z 11:585–590

Wang W, Wen Y, Berkey R, Xiao S (2009) Specific targeting of the *Arabidopsis* resistance protein RPW8.2 to the interfacial membrane encasing the fungal haustorium renders broad-spectrum resistance to powdery mildew. Plant Cell 21:2898–2913

Waterhouse PM, Wang M, Lough T (2001) Gene silencing as an adaptive defence against viruses. Nature 411:834–842

Webb CA, Szabo LJ, Bakkeren G, Garry C, Staples RC, Eversmeyer M, Fellers JP (2006) Transient expression and insertional mutagenesis of *Puccinia triticina* using biolistics. Funct Integr Genomics 6:250–260

Weßling R, Schmidt SM, Micali CO, Knaust F, Reinhardt R, Neumann U, Ver Loren van Themaat E, Panstruga R (2012) Transcriptome analysis of enriched *Golovinomyces orontii* haustoria by deep 454 pyrosequencing. Fungal Genet Biol 49:470–482

Yin C, Chen X, Wang X, Han Q, Kang Z, Hulbert SH (2009) Generation and analysis of expression sequence tags from haustoria of the wheat stripe rust fungus *Puccinia striiformis* f sp tritici. BMC Genomics 10:626

Yin C, Jurgenson JE, Hulbert SH (2010) Development of a host-induced RNAi system in the wheat stripe rust fungus *Puccinia striiformis* f. sp. *tritici*. Mol Plant-Microbe Interact 24:554–561

Transmission of Phloem-Limited Viruses to the Host Plants by Their Aphid Vectors

Jaime Jiménez, Aránzazu Moreno, and Alberto Fereres

Contents

Abstract Plant viruses produce important economic losses in crops worldwide. The study of their transmission by insects has been key in order to develop new strategies to interfere with their spread. Studies on the monitoring of aphid probing behavior by

J. Jiménez (✉)
Instituto de Ciencias Agrarias – Consejo Superior de Investigaciones Científicas (ICA-CSIC), Madrid, Spain

Department of Entomology and Nematology, University of Florida, Gainesville, FL, USA
e-mail: jaime.jimenez@ufl.edu

A. Moreno and A. Fereres
Instituto de Ciencias Agrarias – Consejo Superior de Investigaciones Científicas (ICA-CSIC), Madrid, Spain

© Springer Nature Switzerland AG 2020
Progress in Botany (2021) 82: 357–382, https://doi.org/10.1007/124_2020_47,
Published online: 14 December 2020

using the electrical penetration graph (EPG) technique allowed to study the different aphid stylet activities in plant tissues associated with the transmission of plant viruses. Aphids produce intermittent intracellular punctures (commonly named as "potential drops": pds) along the stylet pathway, ultimately reaching the phloem tissues. The phloem-pd has been recently described as the key stylet activity in plant cells associated with the transmission of phloem-limited viruses by aphids. This behavioral pattern represents the first report of a brief intracellular puncture produced by aphids in phloem tissues associated to the transmission of a phloem-limited virus. A single brief phloem-pd (3–5 s) was mandatory to transmit both semipersistently and persistently transmitted, phloem-limited viruses by aphids. Stylet penetration of both sieve elements and companion cells of the host plant during the occurrence of the phloem-pd by *M. persicae* were confirmed by using confocal laser-scanning microscopy together with CT-microtomography. In this chapter, we revisit the main approaches recently carried out in the field of aphid probing and feeding behavior in association with the transmission of phloem-limited viruses transmitted in either a semipersistent or persistent manner.

1 Introduction

Pests have reduced the productivity of crops since the dawn of agriculture and farmers have been looking for ways of protecting their crops from these organisms (Oerke 2006). Among biotic factors, the losses concerning global food production caused by plant diseases are estimated in 10%, with viruses as the second biotic factor, just after phytopathogenic fungi (Strange and Scott 2005; Oerke 2006). Even though weeds represent the most important pest in many crops; however, the efficacy of their control reaches almost the 75%, compared to 39% for animal pests, 32% for pathogens, and just 5% for plant viruses (Oerke 2006). The poor success in their control makes viruses such an important yield-limiting factor in crops which require challenging methods for their control. The use of synthetic insecticides has been a long tradition to control arthropod pests and therefore interfering with virus transmission. However, the increased use of pesticides often resulted in the emergence of resistance of arthropods to active ingredients and not resulting in a significant decrease of crop losses (Oerke 2006; Foster et al. 2014). Moreover, chemical methods applied for disease control can result in environmental contamination and presence of chemical residues in the food, in addition to social and economic problems (Song et al. 2017). Plant viruses, as strict intracellular pathogens, are not easily controlled chemically; therefore new strategies interfering with virus transmission by insect vectors need to be developed (van Hemden 2018).

In this chapter, we will review the main approaches and findings recently achieved in the field of aphid-transmitted, phloem-limited viruses. Even though this type of virus transmission began to be studied decades ago (Sylvester 1956; Power et al. 1991), the development of techniques and devices to study aphid probing behavior in plant (McLean and Kinsey 1964; Tjallingii 1978) has allowed to elucidate the specific aphid stylet activities involved in the transmission of many plant viruses (Hodges and McLean 1969; Powell 1991; Prado and Tjallingii 1994; Martín et al. 1997). New advances have been recently achieved in the mechanisms of transmission of phloem-limited viruses either transmitted in a semipersistent (SP) or persistent (P) manner by aphids (Jiménez et al. 2018, 2020a, b). Also, the EPG technique combined with modern techniques of cryofixation with liquid nitrogen and aphid track analysis by confocal laser-scanning microtomography was essential to determine the exact localization of the aphid stylets during the production of the specific stylet activities involved in the inoculation of semipersistently transmitted, phloem-limited viruses (Jiménez et al. 2020a).

2 Aphids as Virus Vectors

Arthropods are likely the most successful animal phylum, with Insecta as the largest class of this phylum. Within Insecta, the order Hemiptera comprises small sap-sucking insects such as aphids (Hemiptera: Aphididae) and whiteflies (Hemiptera: Aleyrodidae) with a needlelike stylet bundle consisting of two mandibular and two maxillary stylets. Concretely, 300 Hemipteran species are responsible for the transmission of more than 70% of all known insect-borne viruses (Fereres and Raccah 2015). Particularly, the family Aphididae includes the greatest proportion of insect vectors, transmitting about one third of all plant viruses studied (Hogenhout et al. 2008).

There are two major types of aphid life cycles based on how the insect utilizes host plants: host alternation (heteroecious) and nonhost alternating (monoecious or autoecious). Host-alternating aphids live on one plant species in winter (primary host – usually woody plants), migrate to an unrelated plant species (secondary host – usually herbaceous plants) in spring/summer, and migrate back to the primary host in autumn. After reproducing parthenogenetically during the summer, sexual forms appear in the fall. Sexual reproduction gives rise to eggs that are laid on the primary host after males and sexual females (oviparae) have mated. Aphids that interrupt parthenogenesis with sexual reproduction in this way are termed holocyclic, whether heteroecious or monoecious (Hardie 2018). Nonhost-alternating aphids remain either on the same host species or utilize closely related host species throughout the years; that is, they can sexually produce eggs on the same group of host plants that is fed on by all the parthenogenetic generations. Some aphid species reproduce continuously by parthenogenesis and never produce an egg stage (anholocyclic life cycle) (Hardie 2018). More importantly, they are such a potential harmful factor in plants since aphid populations increase in a very short period of time by means of many telescopic generations (Blackman and Eastop 2000). Also, aphids exhibit dimorphism with apterae and winged morphs, responsible for long-distance dispersal contributing to virus spread (Loxdale et al. 2018).

Aphids are considered one of the most important pests worldwide not only because of the direct damage they cause but also because of their alimentary habits that involve indirect damage. They excrete honeydew, fostering the development of sooty molds that eventually reduces the amount and quality of production. However, the most important indirect damage caused by aphids is the transmission of plant viruses (Field et al. 2018). Most plant viruses require insect vectors in order to move to a new healthy host plant. Hemipteran vector mouthparts are designed for the efficient extraction of plant sap, described as "piercing-sucking," but they are also well adapted and involved in the acquisition and inoculation of plant viruses. Within the proboscis, the mandibular and maxillary stylets form a fine elongated needlelike structure capable of penetrating plant cells walls without causing major damage to plant tissues.

Aphid vectors make their initial foraging decisions by integrating visual and odor cues, which convey information about plant presence, identity, and quality (Visser et al. 1996). Flying aphids are attracted to plant-reflected wavelengths (in the green-yellow part of the visual spectrum) during landing and by UV light when taking off from a plant and looking for a new host (Kennedy et al. 1961). After contacting the new host plant, vectors assess additional cues from leaf or stem surfaces, parenchyma, and vascular tissues through olfactory and gustatory sensory systems (Powell et al. 2006). The insect probing, feeding, and dispersal behaviors in response to plant cues directly determine the probability that virions will be acquired, retained, and transported (Fereres 2016; Mauck et al. 2018). Also, it has been proposed that viruses evolve traits that induce host phenotypes and effects on vectors that encourage virus spread. There is an increasing evidence of plant viruses manipulating vector behavior to favor their spread to new host plants by interacting with the vector tissues following acquisition from infected hosts (Carmo-Sousa et al. 2016; Mauck et al. 2018, 2019).

3 Virus-Aphid Interactions

3.1 Plant Virus Circulation Within the Vector: Non-circulative and Circulative Viruses

Vector-borne plant viruses are obligate intracellular parasites that depend on the host machinery to multiply and invade plant tissues, changing the physiology of their host plants. Plant viruses have been classified in two major categories depending on the site at which the virus is retained within the vector body: *circulative viruses (CV)* and *non-circulative (NC)* (Fereres and Raccah 2015) (Table 1). Within CV category, virus particles cross gut barriers of the vector and enter the circulatory system and ultimately reach the salivary glands, with few virus genera also able to propagate within the vector (Table 1). On the other hand, NC virus particles are attached to the aphid cuticle for a short period of time without any circulation within the vector's body (Blanc et al. 2014).

Table 1 Principal characteristics of the modes of virus transmission by insects

Feature	Non-circulative (NC)		Circulative (CV)	
	Nonpersistent (NP)	Semipersistent (SP)	Persistent propagative	Persistent nonpropagative
Acquisition time	Seconds to minutes	Seconds to hours	Minutes to hours	
Inoculation time	Seconds to minutes	Seconds to hours	Minutes to hours	
Retention time	Seconds to minutes/hours	Minutes to hours/ days	Days to months	
Latent period	Not required	Not required	Required	
Replication in the vector	No	No	Yes	No
Passage through molt	Negative	Negative	Positive	
Insect species specificity	Low	Intermediate	High	
Association with plants	Not restricted to phloem	Some restricted to the phloem	Restricted to the phloem	

NC viruses are also divided in two different categories depending on the duration of virus retention in the vector: NP and SP viruses (Fereres and Raccah 2015; Mauck et al. 2018) (Table 1). NP viruses persist briefly in the vector (few hours) and are acquired and inoculated during brief intracellular stylet punctures (seconds), without a latent period in the vector. However, SP viruses persist for longer time in the vector (several hours to days) and need longer periods of time (hours) for acquisition and inoculation, with also no latent period. In both cases, virus transmissibility is lost when the vector molts. CV viruses are also commonly referred to as P viruses since they persist for days/months in the vector (Fereres and Raccah 2015). These viruses also need longer periods of time for their transmission (many hours to days) and require a latent period in the vector (Table 1). It is important to note here that the time required for virus acquisition and inoculation is not a given fixed access period to the source or test plant, but is intimately dependent on specific aphid stylet activities that are mandatory for transmission. This will be illustrated in the following sections of the present chapter.

3.2 *Localization and Retention Sites of Plant Viruses in the Vector: Stylet-Borne, Foregut-Borne, and Salivary Gland-Borne*

The mode of transmission of NC viral particles has been classified according to the way they attach or bind to the vector in two different modes: the "CP" and "helper" strategies. Certain NC particles, i.e., *Cucumber mosaic virus* (CMV, *Cucumovirus*), bind to the aphid cuticle by simply interaction of proteins of the capsid protein (CP) and denominated as "CP strategy" (Fig. 1) (Normand and Pirone 1968; Gera et al. 1979; Chen and Francki 1990). However, some other NC viruses need the presence of an additional virus-encoded nonstructural protein (commonly named as "helper component") that mediates the virus-vector interaction (Pirone and Blanc 1996; Uzest and Blanc 2016). NC viruses transmitted in a NP manner are retained in

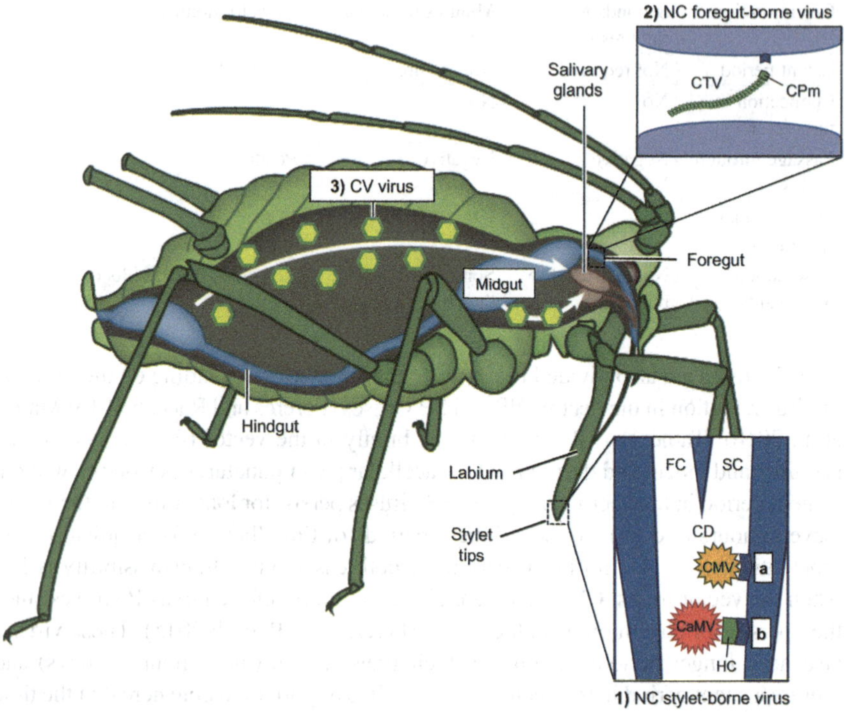

Fig. 1 General view of the localization of plant virus particles in the aphid vector. (1) Non-circulatively transmitted, stylet-borne viruses. Within this this type of virus transmission, virus particles use two different strategies to bind the aphid mouthparts: (**a**) capsid-strategy, with interaction of the capsid proteins directly with the protein of the aphid cuticle lining and (**b**) helper-strategy, with the presence of an additional helper protein (*HC* helper component) to bind the viral particle. Non-circulatively transmitted, stylet-borne aphids could be transmitted either in a nonpersistent (CMV) or semipersistent (CaMV) manner by aphids. (2) Non-circulative, foregut-borne viruses. These viruses are transmitted in a semipersistent manner and bind to the aphid cuticle directly through the interaction of the capsid protein (CTV; Killiny et al. 2016). (3) Circulatively transmitted viruses. Virus particles cross the aphid epithelium in the midgut and hindgut to be later transported through the hemolymph to finally reach the accessory salivary glands to be inoculated in the host plant together with the aphid saliva (e.g., *Barley yellow dwarf virus*; BYDV). *NC* non-circulative, *CV* circulative, *FC* food canal, *SC* salivary canal, *CD* common duct, *HC* helper component, *CMV Cucumber mosaic virus*, *CaMV Cauliflower mosaic virus*, *CTV Citrus tristeza virus*, *CPm* minor capsid protein. Figure adapted from *Localizing viruses in their aphid vectors*, authored by S. Blanc, M. Drucker & M. Uzest (2014). Annual Review of Phytopathology 52: 403–425. © 2014 Annual Reviews

the specific area at the stylet tips called common duct, where both the salivary and food canal fuse together (Martín et al. 1997). Therefore, these viruses are usually defined as stylet-borne viruses. However, NC viruses transmitted by aphids in a SP manner could be either retained in the common duct (stylet-borne) (e.g., CaMV) or in the foregut (foregut-borne) of the vector (e.g., CTV), but further research is needed to determine the precise location where NC viruses are retained (Fig. 1).

Specific retention sites of SP viruses within the vector mouthparts have been discovered for a couple of viruses. For instance, the semipersistently transmitted CaMV follows the HC strategy, but in this case, the retention to the aphid cuticle is mediated by two helper proteins (denominated P2 and P3) (Drucker et al. 2002; Palacios et al. 2002; Moreno et al. 2012). The P2 protein binds to a specific aphid mouthpart structure named acrostyle that is located in the common duct of the maxillary stylets where both food and salivary canals join together (Uzest et al. 2007). P2 is acquired primarily from epidermal and mesophyll cells during brief intracellular punctures, while the complex P3-CaMV is mainly acquired by aphids during the phloem phase (Palacios et al. 2002; Drucker et al. 2002). However, the SP and phloem-limited closterovirus CTV has been found to be retained in the foregut of its main vector *Toxoptera citricida* (Hemiptera: Aphididae) (Killiny et al. 2016) (Fig. 1). Also, the semipersistently transmitted virus *Anthriscus yellows virus* and *Parsnip yellow fleck virus* have been found to be retained in the foregut of its aphid vectors (Murant et al. 1976).

On the other hand, CV viruses are carried in the interior of the vector body after they cross the gut and salivary gland membranes (Gray and Gildow 2003; Hogenhout et al. 2008). During acquisition of a CV from the host plant, virus particles circulate through the vector body and pass the intestinal cells to the hemocoel, circulating towards the accessory salivary glands where they can be retained (Gildow 1999). In the case of CV luteoviruses, such as *Barley yellow dwarf virus* (BYDV, *Luteovirus*), 12–16 h are required for to be actively transported through the gut epithelial cell and released into the aphid hemocoel (latent period) (Garret et al. 1996). Virus particles survive in the hemolymph for several weeks and are passively transported to the accessory salivary glands (Gildow and Gray 1993; Gray and Gildow 2003) to be injected together with the saliva after aphid stylet intrusion into the phloem sieve elements of the host plant (Prado and Tjallingii 1994) (Fig. 1). But as explained later, transmission of CV may occur before aphids penetrate and salivate in the phloem sieve elements.

4 Plant Virus Localization in the Host Plant: Non-phloem and Phloem-Limited Viruses

Whereas all viruses transmitted in a NP manner by aphids are distributed in all plant tissues, most SP and P transmitted viruses are mainly limited to the phloem tissues of the host plant (Esau 1960). Viruses that are found in all tissues of the host plants are transmitted during short intracellular punctures by aphids in epidermal cells, so that this type of virus transmission is optimized with short probing periods of time (Powell et al. 1995). However, phloem-limited viruses replicate and circulate mainly in phloem cells of the host plant. Therefore, only aphid stylet activities associated with penetrations of phloem cells of the host plant are associated with the transmission of this type of viruses (Prado and Tjallingii 1994; Jiménez et al. 2018).

4.1 Phloem-Limited Viruses Transmitted by Aphids

Phloem-limited pathogens represent a significant research challenge because they are difficult to detect within plants and infected plants often exhibit variable symptoms that develop slowly. Moreover, these pathogens have complex infection cycles involving both plant hosts and insect vectors (Bendix and Lewis 2018). Phloem-limited viruses often require longer times of aphid feeding than non-phloem-limited viruses for their transmission and longer persistence of the virus particles in the vector for successful transmission. Consequently, phloem-limited viruses are only found within the semipersistently or persistently transmitted virus categories. Two of the most important cases of study have been viruses within the *Closterovirus* genus (family *Closteroviridae*) as well as *Luteovirus* (family *Luteoviridae*).

4.1.1 Closteroviruses

Closteroviruses comprise a unique case of study within virus transmitted in a SP manner by aphids, with exceptionally long filamentous virus particles (1350–2000 nm in length and ~12 nm in diameter) (Agranovsky et al. 1994; Peremyslov et al. 2004). The virion of closteroviruses present a flexuous filamentous shape composed of single-stranded RNA building a long "body" and short "tail" whose principal components are the major and minor capsid proteins, respectively (Peremyslov et al. 2004).

The most studied phloem-limited closterovirus have been *Beet yellows virus* (BYV) and *Citrus tristeza virus* (CTV) (Esau 1968; Price 1966). Whereas infection by CTV is really restricted to plant species within the Rutaceae family, BYV has a wide range of host plants, with *Beta vulgaris* as the most important commercial crop affected (Godfrey and Mauk 1993). Successful transmission of these viruses is achieved only when the aphid stylets reach phloem cells, thus taking longer time for their transmission, similar to circulative viruses. Whereas the optimum acquisition access period (AAP) and inoculation access period (IAP) have been well studied for closteroviruses (Sylvester 1956; Bennett 1960; Limburg et al. 1997; Campolo et al. 2014), the specific aphid stylet activities in phloem associated with their transmission remained unknown until recent studies were developed (Jiménez et al. 2018). The knowledge regarding the transmission mechanisms of closteroviruses has been mainly derived from studies using the model BYV and the vector *M. persicae* (Sylvester 1956; Bennett 1960; Jiménez et al. 2018).

BYV has a wide aphid vector range as it is transmitted by more than 20 aphid species. The main aphid vectors have been showed to be *Aphis fabae* and *M. persicae*, with the latter being the most efficient vector (Watson 1938; Heathcote and Cockbain 1964; Kirk et al. 1991; Limburg et al. 1997). The optimum AAP and the IAP of BYV have been accurately described for BYV. Whereas Sylvester's work (1956) showed 12 h as the optimum time to reach maximum of BYV acquisition (25–30%), Bennett's work (1960) reflected 6 h as being the one related to the highest acquisition rate (58%). However, the maximum inoculation efficiency was detected after 6–10 h (11%) and after 1 h (57%) of feeding in Sylvester and Bennett works, respectively. The higher BYV transmission efficiencies obtained by Bennett is explained by the larger number of insects used per test plant (three aphids/plant). However, the specific *M. persicae* stylet activities associated with successful inoculation or acquisition of BYV that could explain these differences between the IAP and the AAP remained unknown until recent studies (Jiménez et al. 2018). The latter work reflected an optimum AAP of 9.5 h and 3 h as an IAP for BYV based on the study of the aphid probing behavior in association with BYV acquisition and inoculation. However, the time frame required for virus acquisition or inoculation by an aphid will depend on when specific stylet activities will occur.

4.1.2 Luteoviruses

Barley yellow dwarf virus (BYDV, *Luteovirus*) is the type member of the *Luteoviridae* family and it has been thoroughly studied. BYDV is one of the most important virus diseases in cereal crops and grasses worldwide, infecting over 150 species in the Poaceae family (D'Arcy 1995). Moreover, many Poaceae species still need to be studied for their ability to be potential hosts of BYDV since new species are continuously identified as hosts (Ingwell et al. 2014; Ingwell and Bosque-Pérez 2015). BYDV infection primary causes stunted growth and yellow coloration of leaves due to a collapse and further necrosis of sieve element (SE) cells, with same cell damage also observed in companion cells (CCs) and phloem parenchyma cells (PPCs) (Esau 1957).

Twenty-five aphid species have been reported to transmit BYDV (Halbert and Voegtlin 1995), with the bird cherry-oat *Rhopalosiphum padi* (L.) (Hemiptera: Aphididae) and the English grain aphid *Sitobion avenae* (Fabricius) (Hemiptera: Aphididae) as the main vectors (Gildow 1990). The IAP has been accurately studied for BYDV, with aphids transmitting the virus at high efficiency after hours. *R. padi* was reported to transmit BYDV with an efficiency of 47.1 and 58.6% (for PAV and RPV isolates, respectively) after an IAP of 2 h (Power et al. 1991). Opposite to BYV, several works have focused on the transmission of BYDV by monitoring the aphid probing behavior of the two main aphid vector species: *R. padi* and *S. avenae* (Scheller and Shukle 1986; Prado and Tjallingii 1994).

5 The Electrical Penetration Graph (EPG) Technique: An Excellent Tool to Study Virus Transmission by Pierce-Sucking Insects

The electrical penetration graph (EPG) technique allows us to study the different aphid stylet activities produced by aphids in plant tissues (McLean and Kinsey 1964; Tjallingii 1978). Aphids are attached to a gold wire of an insect electrode and inserted into the input connector of an amplifier (EPG probe). EPG recording is performed inside a Faraday cage to prevent electrical noise. EPG data acquisition and analysis are conducted by specific software (Fig. 2). The different EPG waveforms produced by aphids (Fig. 3) have been previously described and associated to specific stylet tip positions in the plant tissues. The patterns so far described for aphids are:

- *Non-probing (np):* flat line associated to the non-probing activity of the aphid prior to stylet penetration into the leaf cuticle.
- *Waveform C:* waveform with extracellular potential. During this waveform, aphid stylets are located in the extracellular pathway (apoplast) before reaching phloem tissues. Two other waveforms are usually included in waveform C: (1) waveform A, representing the first intercellular stylet penetration of the aphid stylets in the epidermis and (2) waveform B, associated with the production of gelling saliva and the salivary sheath.

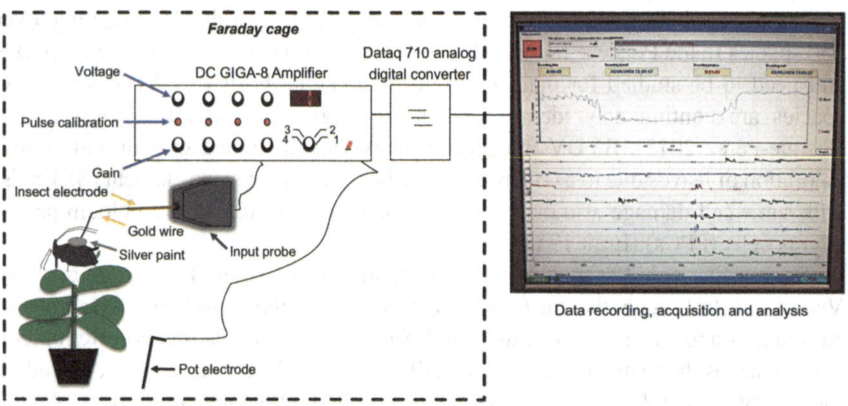

Fig. 2 General view of an electrical penetration graph (EPG) setup (Tjallingii 1978). Insects are attached to the gold wire using silver conductive paint. The end of the gold wire is also attached to a copper wire that is soldered to a brass nail using flux enhancement. This electrode nail is inserted into the input connector of the first-stage amplifier (EPG probe). The plant electrode is inserted into the potting soil of a plant. EPG recordings are performed inside a Faraday cage to prevent electrical noise. Figure reprinted from *Non-circulative virus transmission by aphids: new insights into the mechanisms of transmission and the interference for retention sites in the vector*. Ph. D. Thesis, authored by J. Jiménez (2019). Polytechnic University of Madrid, Spain

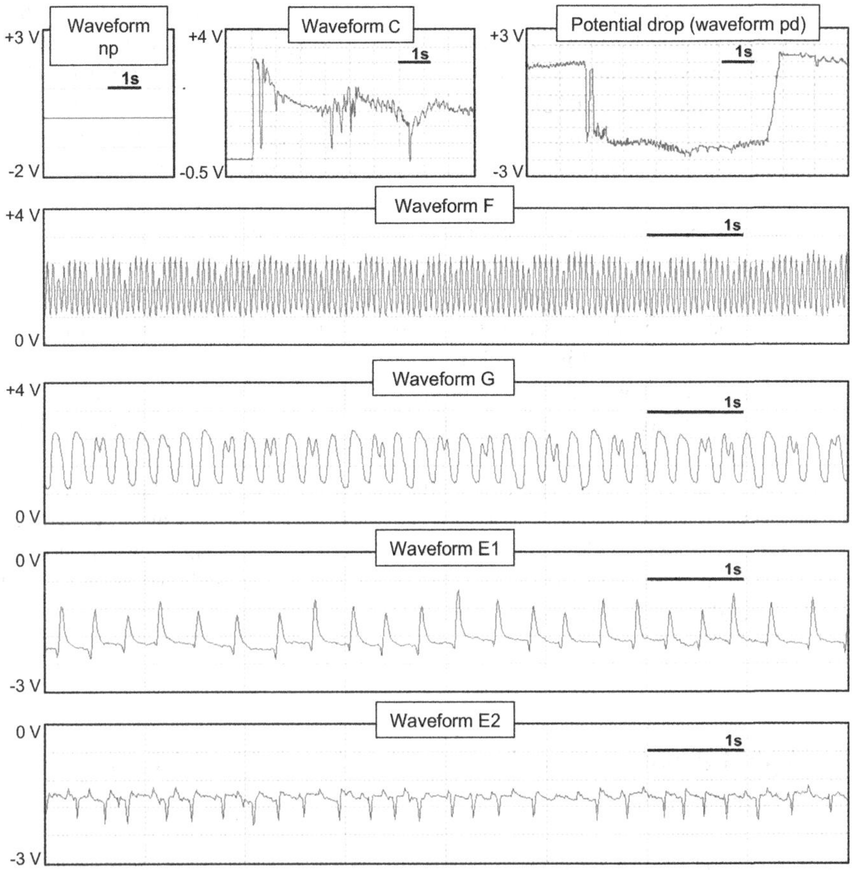

Fig. 3 Main EPG waveforms described for aphids (Tjallingii 1978). Y-axis: output voltage expressed in volts (V). X-axis: EPG recording time expressed in seconds. Figure reprinted from *Non-circulative virus transmission by aphids: new insights into the mechanisms of transmission and the interference for retention sites in the vector*. Ph. D. Thesis authored by J. Jiménez (2019). Polytechnic University of Madrid, Spain

– *Waveform pd (potential drop):* this waveform represents an intracellular stylet penetration. Three phases have been described within the pd: phase I, with a potential drop produced by aphid stylets insertion through the plasmalemma; phase II, representing the stylet tip presence inside the cell; and phase III, with stylet withdrawal from the cell. Phase II is also divided in three subphases (subphase II-1, II-2, and II-3). Subphase II-1 represents saliva injection into the cytoplasm and is responsible for the inoculation of viruses transmitted in a NP manner (Martín et al. 1997; Powell 2005). Subphase II-3 represents ingestion of cell contents and acquisition of viruses transmitted in a NP manner (Powell et al.

1995; Collar et al. 1997). Specific aphid activities during subphase II-2 are not well understood. It has been postulated that either aphid salivation or egestion of plant sap previously ingested by the aphid is occurring during the subphase II-2 of the pd (Moreno et al. 2012).

- *Waveform E:* this waveform represents an intracellular potential with aphid stylet tip presence in the sieve elements. Two different patterns are observed within that waveform: E1 and E2. Waveform E1 represents ejection of watery saliva into sieve elements, and it has been associated with the inoculation of persistently transmitted, phloem-limited viruses (Prado and Tjallingii 1994). Waveform E2 represents phloem sap ingestion with periodic ejection of watery saliva, and it has been associated with the acquisition of persistently transmitted, phloem-limited viruses (Prado and Tjallingii 1994).
- *Waveform F:* this waveform is observed when the aphid encounters mechanical difficulties during penetration of plant tissues, likely representing derailed stylet activities.
- *Waveform G:* waveform with extracellular potential representing active ingestion from xylem vessels. Even though this waveform does not play any role in virus transmission by aphids, active xylem ingestion is key in the transmission of xylem-limited bacteria such as *Xylella fastidiosa* by Cicadellidae (Almeida and Backus 2004) or by Aphrophoridae (Cornara et al. 2020).

6 Aphid Probing Behavior Associated with Plant Virus Transmission

Monitoring of the feeding behavior of aphids by the EPG technique to determine specific aphid stylet activities involved in successful inoculation and acquisition of plant viruses has been paramount in studies concerning aphid-vectored viruses (Scheller and Shukle 1986; Prado and Tjallingii 1994; Martín et al. 1997; Fereres and Moreno 2009; Moreno et al. 2012; Fereres 2016). These studies have provided novel information about the key aphid stylet activities involved in transmission of NP, SP, and P viruses by aphids, as well as elucidating the main aphid activities (salivation, ingestion, egestion) underlying the different EPG waveforms.

The different EPG waveforms involved in the transmission of non-phloem-limited viruses transmitted in a NP or SP manner by aphids have been well studied. Inoculation of NP viruses occurs during the pd subphase II-1, with acquisition process taking place during the subphase II-3 (Powell et al. 1995; Martín et al. 1997). Inoculation of some SP viruses, such as CaMV, is also associated with brief superficial stylet punctures or potential drops (pds), similarly to NP viruses. However, only the subphase II-2 plays a role in the inoculation of CaMV (Moreno et al. 2012), but the mechanisms involved in dislodging CaMV particles from the acrostyle during subphase II-2 remain unknown. Subphase II-2 could represent an additional salivation phase with possible egestion of cell content previously ingested (Moreno et al. 2012).

Several gaps existed concerning the mechanism of transmission of phloem-limited viruses transmitted in a SP or P manner by aphids. However, recent EPG studies have revealed the mechanisms of transmission of phloem-limited viruses. Regarding SP aphid-transmitted viruses, the newly characterized phloem-pd, a brief intracellular puncture produced in companion or sieve element cells, has been found as the key aphid pattern involved in the inoculation of the phloem-limited semipersistent virus BYV (Jiménez et al. 2018, 2020a). However, its acquisition is associated with long times of sap phloem ingestion (E2 waveform). In the case of phloem-limited viruses transmitted in P manner, the inoculation is associated with the production of two brief intracellular punctures (phloem-pd and short-E1) presumably produced by aphids in phloem cells. Transmission efficiency increases gradually and is proportional to the duration of salivation in phloem cells (Jiménez et al. 2020b).

7 Transmission of Phloem-Limited Viruses by Aphids

7.1 Characterization of the Phloem-pd and the Short-E1: Two Newly EPG Patterns Associated with the Inoculation of Phloem-Limited Viruses

A new type of potential drop (pd) has been recently identified and characterized as key aphid pattern in the inoculation of semipersistently transmitted, phloem-limited viruses (Jiménez et al. 2018). Preliminary observations of EPG signals from *M. persicae* on sugar beet plants in Prof. Alberto Fereres' team suggested a new type of pd (named as phloem-pd) that deviated from the regular or standard pd (Chen et al. 1997), the intracellular stylet puncture occurring during stylet pathway from the very beginning of a probe (Tjallingii 1978). The features of phloem-pd differed from the standard pd in showing lower number of intervals or frequency of the downward peaks in subphase II-2, as well as in a lower magnitude of the potential drop. Intervals were defined as the time between two peaks in subphase II-2. The most obvious feature of phloem-pds was its significantly lower potential drop magnitude (output voltage mean 2530.18 $\pm$ 76.24 mV) than the potential drop observed on standard-pds (output voltage mean 3002.54 $\pm$ 81.93 mV) (Fig. 4a, b). The magnitude of the potential drop of the phloem-pd was always lower than the one observed for the standard pd, representing 84.24 $\pm$ 1.02% of the mean value of the potential drop magnitude observed in the preceding standard pd. Interestingly, the potential drop magnitude of E1 phase (output voltage 2328.23 $\pm$ 76.12 mV) was similar to the potential drop magnitude observed for the phloem-pd but was significantly different to that of the potential drop of standard pds. In addition, the frequency of intervals between downward peaks in subphase II-2 was more than twice as lower in phloem-pd (2.65 $\pm$ 0.06 int./s) than in standard pd (5.51 $\pm$ 0.11 int./s) (Fig. 4b). In summary, phloem-pds were mainly distinguished from the standard-pds based on the

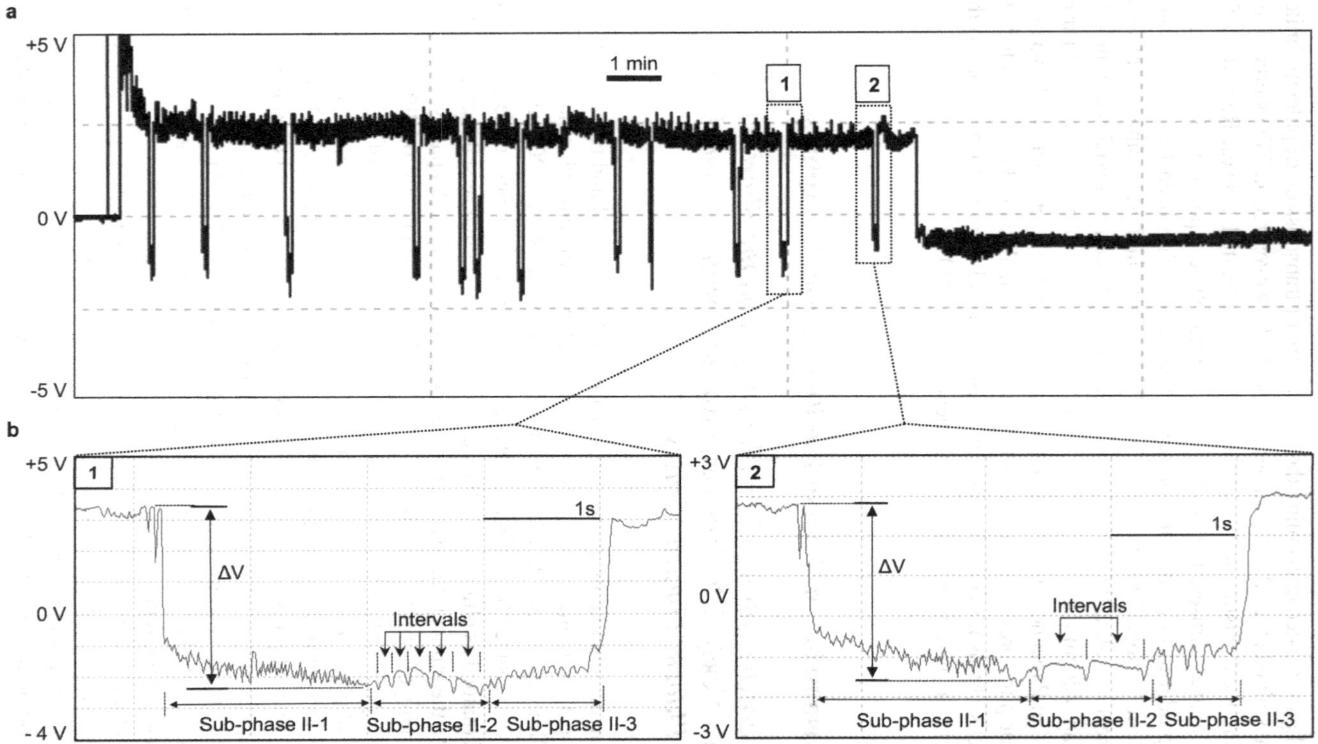

Fig. 4 (**a**) Overview of electrical penetration graphs produced by *Myzus persicae* on sugar beet. (**b**) Expanded view of the standard and phloem-pd: (1) standard-pds, intracellular punctures produced in nonvascular tissues; (2) phloem-pd, intracellular puncture produced in phloem cells. The main potential drop characteristics studied to distinguish the two potential drop waveforms are indicated: ΔV (potential drop magnitude [mV]), number of intervals/s in the

subphase II-2 (Int./s). An interval is defined by the distance between two downward peaks in the subphase II-2 of the pd. Y-axis: output voltage expressed in volts (V). X-axis: EPG recording time expressed in seconds. Figure adapted from *Newly distinguished cell punctures associated with transmission of the semipersistent phloem-limited Beet yellows virus*, authored by J. Jiménez, W.F. Tjallingii, A. Moreno & A. Fereres (2018). Journal of Virology, 92: e01076–18. © 2018 American Society for Microbiology

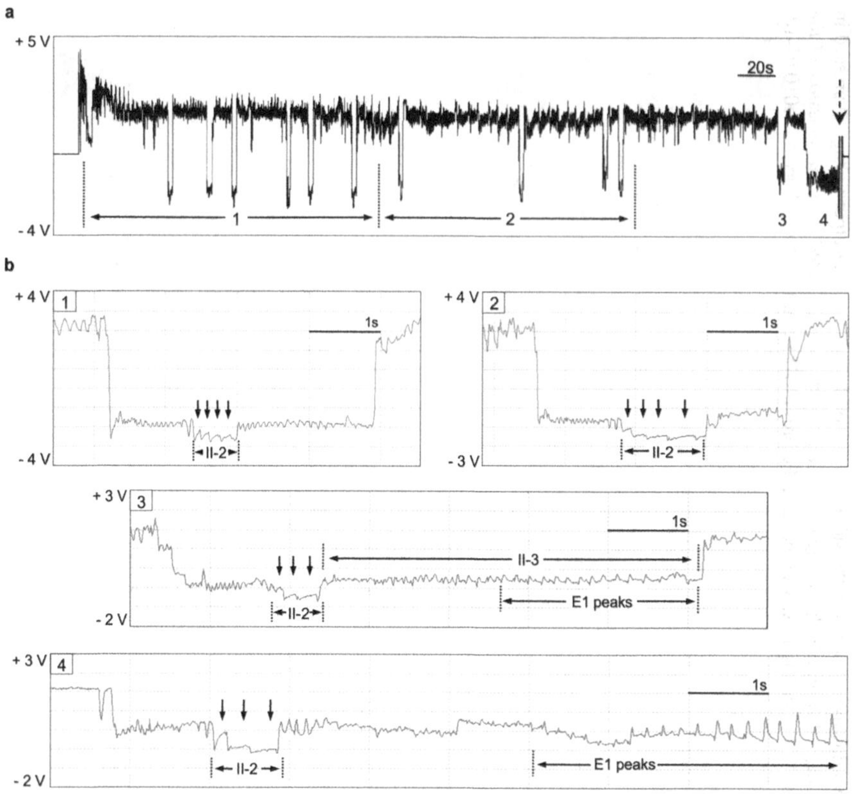

Fig. 5 (**a**) General view of a standard EPG recording produced by viruliferous *Rhopalosiphum padi* in *Hordeum vulgare* artificially ended after the observation of a single sieve element salivation phase (E1 waveform). Down arrow indicates the exact point where the recording was artificially terminated. Numbers along the recording indicate the types of potential drops produced by the aphid. (**b**) View of the different intracellular activities performed by *R. padi* in barley: (1) standard intracellular puncture (standard-pd), (2) intracellular puncture in phloem (phloem-pd), (3) short salivation phase into sieve elements (short-E1), and (4) sieve element continuous salivation phase (E1 waveform). The short-E1 is particularly distinguished from previous phloem-pds and subsequent E1 waveform due to the occurrence of E1 peaks in the subphase II-3. Down arrows refer to each interval (distance between two downward peaks) in the subphase II-2. Y-axis: output voltage expressed in volts (V). X-axis: EPG recording time expressed in seconds. Figure reprinted from *Barley yellow dwarf virus can be inoculated during brief intracellular punctures in phloem cells before the sieve element continuous salivation phase*, authored by J. Jiménez, M. Arias-Martin, A. Moreno, E. Garzo, A. Fereres (2020). Phytopathology Special Focus Issue, Vol. 110, Number 1, pp. 85–93. © 2020 The American Phytopathological Society

magnitude of the potential drop and the number of intervals between peaks within subphase II-2.

The lower potential drop magnitude of pds could be explained by the fact that phloem-pd is occurring in a different tissue from where standard-pds are produced. There are far less or even no plasmodesmata between phloem and mesophyll cells than between mesophyll cells and phloem cells, leading to a physiological isolated status of the phloem from the mesophyll, which may result in a different electrical status (van Bel and van Rijen 1993). The decreased pd magnitude seems to be caused by the higher electrical resistance rather than to real membrane potential difference. This higher resistance between the stylet tips (intracellular recording electrode) in sieve element/companion cell (SE-CC) complex and the extracellularly connected soil electrode than when inserted in a mesophyll cell (and possibly in the phloem parenchyma cells as well) may be due to the relative isolation of the SE-CC complex from the outer tissues in terms of plasmodesmata (van Bel and van Rijen 1993). This was checked for sugar beet when observing the few plasmodesmata connections between SE/CC complex with the surrounding cells and then suggesting an apoplastic loading into the sieve tube (Gamalei 1989).

Jiménez et al. (2020b) showed that the phloem-pd reported for *M. persicae* was also produced by *R. padi*, showing similar electrical properties to the intracellular stylet punctures in the sieve element/companion cell complex. However, *R. padi* also produced an extra intracellular pattern showing an elongated II-3 subphase with characteristic positive peaks similar to those produced during the E1 waveform. A general view of an EPG recording produced by *R. padi* in barley is indicated in Fig. 5a. The particular pattern observed during the II-3 subphase was not observed in the phloem-pd described for *M. persicae*. Therefore, this type of intracellular pattern was named as "short-E1" (Fig. 5b) because of its resemblance to the E1 waveform. The short-E1 was considered as a distinct type of long potential drop because it exhibited a distinct II-1 and II-2 subphase but with an extended II-3 subphase with positive peaks very similar to those of the E1 waveform (Fig. 5b).

The differences of potential drop voltage between standard-pds, phloem-pds, and short-E1 are more intriguing than those previously reported for *M. persicae*. In case of *R. padi* a gradual decrease in potential drop magnitude is observed from the standard-pds until the phloem phase is reached (Fig. 5). In fact, the paucity of plasmodesmata connections between the PPCs and the thin-walled SEs (and adjoining CCs) and thick-walled SEs supports the hypothesis of an apoplastic pathway (Evert et al. 1996). That could result in a relative isolation of both thin-walled SEs/CCs and thick-walled SEs complexes in barley, as observed for other plant species following an apoplastic phloem loading (van Bel and Van Rijen 1993).

7.2 *Inoculation of Closteroviruses*

The feeding behavior of several other vector species in association to SP phloem-restricted viruses was studied with some virus species such as *Criniviruses* and *Waikaviruses*. *Lettuce chlorosis virus* (LCV, *Closterovirus*) inoculation by the whitefly *Bemisia tabaci* (Hemiptera; Aleyrodidae) [formerly *Bemisia argentifolii*] was showed to occur primarily during the phloem salivation (E1) (Johnson et al. 2002) the same as for *Maize chlorotic dwarf virus* (MCDV, *Waikavirus*) transmitted by the black-faced leafhopper *Graminella nigrifrons* (Wayadande and Nault 1993). In those works, little inoculation was detected prior to the production of any E1 waveform by the insects, suggesting that likely some unknown brief penetrations could be playing a role in the transmission of phloem-limited viruses.

A novel recent study revealed that the inoculation of the phloem-limited closteroviruses is mainly associated with the production of the phloem-pd in the tissues of the plant (Jiménez et al. 2018). The highest transmission rate was obtained when the feeding process of the aphid was interrupted after the production of the phloem-pd (80%, with a single viruliferous aphid per test plant). No BYV inoculation was obtained during the extracellular pathway phase (C waveform) or when a single or more standard-pds occurred, which are associated with aphid activities in nonvascular tissues.

The lower magnitudes of the potential drop in the phloem-pds appeared to reflect that this feature is caused by a higher electrical resistance between the stylet tips when located in phloem cells as compared to non-phloem cells. The specific plant tissues in which a pd is occurring cannot be derived from the EPG as such but by microscopical studies in association with the EPG technique and cryofixation presented in the next section of this chapter.

7.3 Inoculation of Luteoviruses

Several works focused on the transmission of BYDV by monitoring the aphid probing behavior of the two main aphid vector species: *R. padi* and *S. avenae*. Early work by Scheller and Shukle (1986) using an AC-EPG device showed that the inoculation of BYDV was higher as the number of aphid stylets penetrations in phloem increased. Primary BYDV inoculation was obtained during the so-called X-waveform, an EPG pattern associated with the contact of aphid stylets with phloem cells before sap ingestion from the phloem (Ip). Some years later, aphid stylet activities involved in BYDV transmission were also studied for *R. padi* using a DC-EPG device, with virus transmission mainly associated with waveforms E1 (inoculation) and E2 (acquisition) (Prado and Tjallingii 1994). Besides authors observed that inoculation occurred, at a very low rate, during EPG recordings that were interrupted before the sieve element continuous salivation phase (E1 waveform) (Prado and Tjallingii 1994). However, these specific aphid activities involved in inoculation of BYDV prior to E1 waveform were unknown.

A recent study has showed that the phloem-pd already described for *M. persicae* and the short-E1 are involved in the inoculation of BYDV prior to E1 waveform (Jiménez et al. 2020b). BYDV inoculation occurred at a low rate after the production of a phloem-pd (7/47; 14.9%). Interestingly, BYDV inoculation efficiency was doubled when *R. padi* was allowed to produce a short-E1 before the occurrence of the E1 waveform (13/39; 33.3%). Aphids that performed a short-E1 (and also phloem-pds) prior to the E1 waveform doubled the BYDV inoculation rate (17/30;

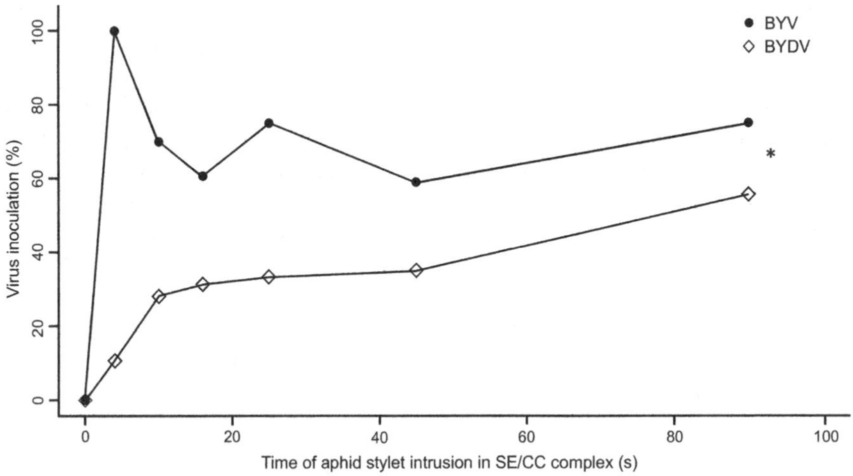

Fig. 6 Comparison of the relationship between total time of salivation in phloem tissues by *Rhopalosiphum padi* and *Myzus persicae* and successful inoculation of BYDV and BYV, respectively. Asterisk indicates significant differences between BYV and BYDV inoculation efficiencies in the period of time studied according to a linear model comparison by a likelihood ratio test ($p < 0.05$). Figure reprinted from *Barley yellow dwarf virus can be inoculated during brief intracellular punctures in phloem cells before the sieve element continuous salivation phase*, authored by J. Jiménez, M. Arias-Martin, A. Moreno, E. Garzo, A. Fereres (2020). Phytopathology Special Focus Issue, Vol. 110, Number 1, pp. 85–93. © 2020 The American Phytopathological Society

56.7%) in comparison to aphid that performed only a single phloem-pd but no short-E1 (5/21; 23.8%).

Opposite to BYV, a positive relationship between time of aphid salivation in phloem cells and increased efficiency of virus transmission was observed for BYDV. Whereas the inoculation efficiency of BYV reached the highest value after 4 s (a single phloem-pd) of aphid salivation into the SE/CC complex (6/6; 100%), the inoculation efficiency of BYDV was only 10.7% (3/28), reaching maximum inoculation efficiency after 45–90 s of aphid salivation into phloem cells (56.6%) (Fig. 6).

8 Visualization of Aphid Stylet Penetrations in Phloem Sieve Elements and Companion Cells During the Occurrence of the Phloem-pd

The phloem-pd was strongly associated with aphid stylet penetrations in either companion cells or sieve elements because phloem-limited BYV particles cannot move from the mesophyll into phloem tissues. The inability of the virus particles to move from cell-to-cell and reach the phloem keeps them in the nonvascular cells, thus avoiding systemic infection. Whether phloem-pds are produced by aphids in either SE cells or CC, or in both type of cells needed to be confirmed. They were tentatively named phloem-pds awaiting verification that they were specifically associated with intracellular punctures of phloem cells.

A combination of three different techniques was used in a recent study to localize the stylet tips in plant cells during the occurrence of the phloem-pd produced by *M. persicae* in sugar beet plants (Jiménez et al. 2020a). Authors combined EPG, cryofixation, and later confocal laser-scanning observation of aphid stylets in the plant. The feeding behavior of aphids was monitored, and stylets were fixed in situ in the sugar beet plant tissue by cryofixation with liquid nitrogen during the production of the novel phloem-pd. This was followed by observation of the stylets or salivary sheath tracks in the plant tissue by both confocal laser-scanning microscopy (CLSM). Studies of punctured cells showed that phloem-pd did represent intracellular penetrations in phloem companion and sieve element cells.

Control treatment with EPG recording terminated during them production of standard-pds by *M. persicae* revealed the stylet penetration of mesophyll cells or bundle sheath (Fig. 7a). Samples fixed during a phloem-pd showed the intrusion of aphid stylets into the phloem cells. Aphid stylet tips punctured indistinctly cells of the sieve elements (Fig. 7b) (four out of seven samples) or the sieve element/ companion cells (three out of seven samples) (Fig. 7c). Recordings terminated during the production the E1 waveform revealed the penetration of sieve element cells (Fig. 7d).

These results confirmed the initial hypothesis proposed by Jiménez et al. (2018), strongly suggesting that phloem-pd could represent brief stylet punctures into the phloem cells. Similarities in voltage drop magnitude of phloem-pds and E1 waveform, and the difference from standard-pds is supporting evidence that phloem-pds represent stylet punctures into the SE/CC complex, while standard-pds are not associated with penetration of the SE/CC complex. SEs and CCs are highly interconnected symplastically via numerous specialized plasmodesmata, called pore-plasmodesmata units (PPUs) (van Bel and Kempers 1997; van Bel 2003). This results in SEs and their associated CCs having the same electrical status (van Bel and van Rijen 1993), thus explaining the fact of same voltage drop in both phloem-pd and subsequent E1 waveform. However, in plants that load phloem apoplastically (such as sugar beet; Gamalei 1989), the SE/CC complex has very few symplastic connections with surrounding cells, thus having a different electrical status than the SE/CC complex (van Bel and van Rijen 1993; Hafke et al. 2005, 2013).

The ability to distinguish SE/CC pds from non-SE/CC pds may provide insights into the location of plant factors that affect successful feeding. The novel results of stylets penetrations of SE and CC phloem cells during the phloem-pd added to the current understanding of aphid stylet activities into phloem tissues, since first phloem contact did occur in a time slightly earlier than previously thought.

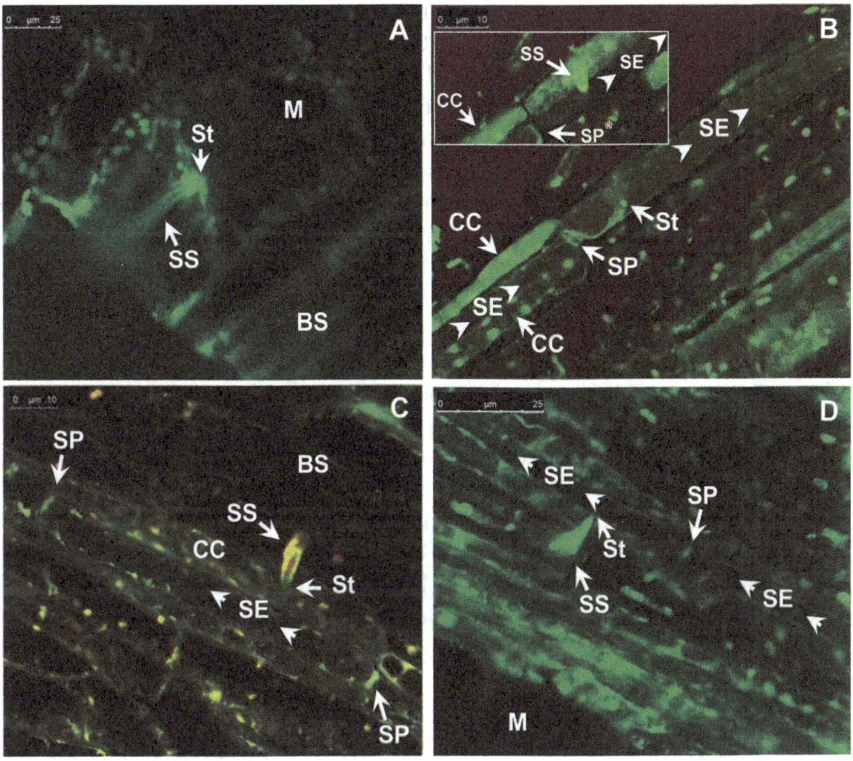

Fig. 7 Confocal laser-scanning microscope (CLSM) micrographs of sugar beet leaf tissues penetrated by *M. persicae* stylets. (**a**) Apex of the salivary sheath containing the stylets located in a mesophyll from a sample cryofixed during the tenth consecutive standard-pd in a probe. (**b**) Apex of the salivary sheath containing the stylets in contact with a sieve element from a sample fixed during a phloem-pd. Inset: image from a previous slice in the Z-stack showing the stylets encased in the salivary sheath; the SE, CC, and SP∗ in the inset are the same cells and sieve plate labeled SE, CC, and SP, respectively, in the main Figure **b**. (**c**) Apex of the salivary sheath containing the stylets in contact with a companion cell from a sample fixed during a phloem-pd. (**d**) Stylet tips inserted in a sieve element cell from a sample fixed during an E1 waveform. SS (aphid stylets encased in the salivary sheath), St (stylet tip), M (mesophyll), SE (sieve element), SP (sieve plate), CC (companion cell), BS (bundle sheath). Figure reprinted by permission from Springer Nature Customer Service Centre GmbH: Springer, Arthropod-Plant Interactions. *The phloem-pd: a distinctive brief sieve element stylet puncture prior to sieve element phase of aphid feeding behavior*. J. Jiménez, E. Garzo, J. Alba-Tercedor, A. Moreno, A. Fereres & G.P. Walker. © 2020 Springer

9 Conclusions and Further Research

The novel finding of the phloem-pd opens a new field of study regarding aphid interactions with the host plants as well as plant virus transmission. The phloem-pd seems not to be a unique feature of *M. persicae*, as it has also been observed for

R. padi and *Sitobion avenae* (Jiménez et al. unpublished data). Further microscopical or histological studies combined with EPGs and cryofixation are needed in order to study the type of phloem cell punctured by *R. padi* during the occurrence of the phloem-pd and the short-E1. Also, specific EPG-assisted transmission experiments need to be conducted to show if phloem-limited viruses are actually transmitted either in companion cells, sieve elements, or both.

The novel characterization of the phloem-pd produced by aphids in phloem tissues and likely involvements in other aphid-plant-virus interaction is a newly field of study. Every E1 waveform studied for both *M. persicae* (Jiménez et al. 2018, 2020a) and *R. padi* (Jiménez et al. 2020b) was preceded by at least one phloem-pd or phloem-pd and/or short-E1 in case of the latter. Therefore, these brief stylet tips penetrations would be playing an important role in aphid-plant interactions at the sieve element level, likely triggering wound responses before the continuous salivation in a sieve element (E1 waveform) (Medina-Ortega and Walker 2015; Garzo et al. 2018; Peng and Walker 2018). On the other hand, the EPG variable "time to first phloem sieve element phase" is sometimes interpreted as being affected entirely by plant qualities encountered by the aphid in non-sieve elements. However, the period preceding the first E1 waveform likely includes earlier penetrations of the SE/CC complex, and thus the time to first E1 waveform may be influenced by factors in sieve elements in addition to non-sieve element cells. The ability to distinguish SE/CC pds from non-SE/CC pds may provide insights into the location of plant factors that affect successful feeding.

The specific mechanism that could trigger the production of a different intracellular pattern by the aphid once a SE or CC is reached is intriguing. The phloem-pd represents an aphid activity well standardized and distinguished from the numerous standard intracellular punctures produce in the SE and CC. One could suggest that determined type of cells could be recognized by the aphids once a cell penetration is produced, as the change in the pattern within the phloem-pd represents an aphid-triggered change. However, these differences in the electrical signal during the subphase II-2 of the phloem-pd and the mechanisms developed by the aphid during this subphase and responsible for BYV inoculation during phloem-pds have not yet been identified. This question might be answered using the methodology that Martín et al. (1997) and Moreno et al. (2012) used to identify subphase II-1 and subphase II-2 as the subphases when nonpersistent viruses and CaMV, respectively, are inoculated. Additionally, the retention site of BYV in the aphid vector has not been determined, since attempts to identify the retention site of BYV in *M. persicae* using the same methodology of Chen et al. (2011) have so far been unsuccessful (unpublished data). The foregut, cibarium, precibarium, and stylet food canal are not in the path of saliva secretions and consequently egestion (ingestion-egestion theory – Harris and Harris 2001) would be the only feasible responsible for inoculation if any of those viruses are retained above the common duct of the maxillary stylets. The role of the newly described phloem-pd in other aphid-plant-pathogen interactions, such as plant defense responses before the E1 waveform, requires further investigations.

References

Agranovsky AA, Koonin EV, Boyko VP, Maiss E, Frötschl R, Lunina NA, Atabekov JG (1994) *Beet yellows closterovirus*: complete genome structure and identification of a leader papain-like thiol protease. Virology 198(1):311–324

Almeida RPP, Backus EA (2004) Stylet penetration behaviors of *Graphocephala atropunctata* (Signoret) (Hemiptera, Cicadellidae): EPG waveform characterization and quantification. Ann Entomol Soc Am 97(4):838–851

Bendix C, Lewis JD (2018) The enemy within: phloem-limited pathogens. Mol Plant Pathol 19 (1):238–254

Bennett CW (1960) Sugar beet yellows disease in the United States. In: USDA technical bulletin no. 1218. Agricultural Research Service. U.S. Department of Agriculture, Washington, DC, pp 1–63

Blackman RL, Eastop VF (2000) Aphids on the world's crops: an identification and information guide, 2nd edn. Wiley, New York

Blanc S, Drucker M, Uzest M (2014) Localizing viruses in their aphid vectors. Annu Rev Phytopathol 52:403–425

Campolo O, Chiera E, Malacrinò A, Laudani F, Fontana A, Albanese GR, Palmeri V (2014) Acquisition and transmission of selected CTV isolates by *Aphis gossypii*. J Asia Pac Entomol 17(3):493–498

Carmo-Sousa M, Moreno A, Plaza M, Garzo E, Fereres A (2016) *Cucurbit aphid-borne yellows virus* (CABYV) modifies the alighting, settling and probing behaviour of its vector *Aphis gossypii* favouring its own spread. Ann Appl Biol 169(2):284–297

Chen B, Francki RIB (1990) *Cucumovirus* transmission by the aphid *Myzus persicae* is determined solely by the viral coat protein. J Gen Virol 71:939–944

Chen JQ, Martín B, Rahbé Y, Fereres A (1997) Early intracellular punctures by two aphid species on near-isogenic melon lines with and without the virus aphid transmission (Vat) resistance gene. Eur J Plant Pathol 103(6):521–536

Chen AYS, Walker GP, Carter D, Ng JCK (2011) A virus capsid protein component mediates virion and transmission by its insect vector. Proc Natl Acad Sci U S A 108(40):16777–16782

Collar JL, Avilla C, Fereres A (1997) New correlations between aphid stylet paths and nonpersistent virus transmission. Environ Entomol 26(3):537–544

Cornara D, Marra M, Morente M, Garzo E, Moreno A, Saponari M, Fereres A (2020) Feeding behaviour in relation to spittlebug of Xylella fastidiosa. J Pest Sci. https://doi.org/10.1007/s10340-020-01236-4

D'Arcy CJ (1995) Symptomology and host range of barley yellow dwarf. In: D'Arcy DJ, Burnett PA (eds) Barley yellow dwarf: 40 years of progress. APS, St Paul, pp 9–28

Drucker M, Froissart B, Hebrard E, Uzest M, Ravallee M, Esperandieu P, Mani JC, Pugniere M, Roquet F, Fereres A, Blanc S (2002) Intracellular distribution of viral gene products regulates a complex mechanism of *Cauliflower mosaic virus* acquisition by its aphid vector. Proc Natl Acad Sci U S A 99(4):2422–2427

Esau K (1957) Phloem degeneration in Gramineae affected by the barley yellow-dwarf virus. Am J Bot 44:245–251

Esau K (1960) Cytologic and histologic symptoms of beet yellows. Virology 10(1):73–85

Esau K (1968) Viruses in plant hosts: form, distribution, and pathologic effects. University of Wisconsin Press, Madison

Evert RF, Russin WA, Botha CEJ (1996) Distribution and frequency of plasmodesmata in relation to photoassimilate pathway and phloem loading in the barley leaf. Planta 198:572–579

Fereres A (2016) Aphid behavior and the transmission of noncirculative viruses. In: Brown JK (ed) Vector-mediated transmission of plant pathogens. APS, St Paul, pp 31–45

Fereres A, Moreno A (2009) Behavioural aspects influencing plant virus transmission by homopteran insects. Virus Res 141(2):158–168

Fereres A, Raccah B (2015) Plant virus transmission by insects. In: eLS. Wiley, Chichester

Field LM, Bass C, Davies TG, Williamson MS, Zhou JJ (2018) Aphid genomics and its contribution to understanding aphids as crop pests. In: van Emden HF, Harrington R (eds) Aphids as crop pests, 2nd edn. CAB International, Wallingford, pp 37–49

Foster SP, Paul VL, Slater R, Warren A, Denholm I, Field LM, Williamson MS (2014) A mutation (L1014F) in the voltage-gated sodium channel of the grain aphid, *Sitobion avenae*, is associated with resistance to pyrethroid insecticides. Pest Manag Sci 70(8):1249–1253

Gamalei Y (1989) Structure and function of leaf minor and veins in trees and herbs. A taxonomic review. Trees 3(2):96–110

Garret A, Kerlan C, Thomas D (1996) Ultrastructural study of acquisition and retention of potato leafroll luteovirus in the alimentary canal of its aphid vector, *Myzus persicae* Sulz. Arch Virol 141:1279–1292

Garzo E, Fernández-Pascual M, Morcillo C, Fereres A, Gómez-Guillamón ML, Tjallingii WF (2018) Ultrastructure of compatible and incompatible interactions in phloem sieve elements during the stylet penetration by cotton aphids in melon. Int J Insect Sci 25(4):631–642

Gera A, Loebenstein G, Raccah B (1979) Protein coats of two strains of *Cucumber mosaic virus* affect transmission by *Aphis gossypii*. Phytopathology 69(4):396–399

Gildow FE (1990) Current status of barley yellow dwarf in the United States: a regional report. In: Burnet PA (ed) World perspectives on barley yellow dwarf. CIMMYT, Mexico, pp 11–20

Gildow FE (1999) Luteovirus transmission and mechanisms regulation vector specificity. In: Smith HG, Barker H (eds) The luteoviridae. CAB International, Oxon, pp 88–112

Gildow FE, Gray SM (1993) The aphid salivary gland basal lamina as a selective barrier associated with vector-specific transmission of *Barley yellow dwarf virus*. Phytopathology 83:1293–1302

Godfrey LD, Mauk PA (1993) Interactive effects of aphid injury and *Beet yellows virus* on sugar beet photosynthesis and yield (Abstr.). J Sugar Beet Res 30:95

Gray SM, Gildow FE (2003) Luteovirus-aphid interactions. Annu Rev Phytopathol 41:539–566

Hafke JB, van Amerongen JK, Kelling F, Furch ACU, Gaupels F, van Bel AJE (2005) Thermodynamic battle for photosynthate acquisition between sieve tubes and adjoining parenchyma in transport phloem. Plant Physiol 138:1527–1537

Hafke JB, Höll SB, Kühn C, van Bel AJE (2013) Electrophysiological approach to determine kinetic parameters of sucrose uptake by single sieve elements or phloem parenchyma cells in intact *Vicia faba* plants. Front Plant Sci 4:274

Halbert SC, Voegtlin D (1995) Biology and taxonomy of vectors of barley yellow dwarf viruses. In: D'Arcy CL, Burnett PA (eds) Barley yellow dwarf: 40 years of progress. APS, St Paul, pp 217–258

Hardie J (2018) Life cycles and polyphenism. In: van Emden HF, Harrington R (eds) Aphids as crop pests, 2nd edn. CABI, Wallingford, pp 81–97

Harris KF, Harris LJ (2001) Ingestion-egestion theory of cuticula-borne virus transmission. In: Harris KF, Smith OP, Duffus JE (eds) Virus-insect-plant interactions. Academic Press, New York, pp 111–132

Heathcote GD, Cockbain AJ (1964) Transmission of *Beet yellows virus* by alatae and apterous aphids. Ann Appl Biol 53(2):259–266

Hodges LR, McLean DL (1969) Correlation of transmission of bean yellow mosaic virus with salivation activity of Acrythosiphon pisum (Homoptera: Aphididae). Ann Entomol Soc Am 62:1398–1401

Hogenhout S, Ammar ED, Whitfield AE, Redinbaugh MG (2008) Insect vector interactions with persistently transmitted viruses. Annu Rev Phytopathol 46:327–359

Ingwell LL, Bosque-Pérez NA (2015) The invasive weed Ventenata dubia is a host of barley yellow dwarf virus with implications for an endangered grassland habitat. Weed Res 55:62–70

Ingwell LL, Zemetra R, Mallory-Smith C, Bosque-Pérez NA (2014) Arundo donax infection with barley yellow dwarf virus has implications for biofuel production and non-managed habitats. Biomass Bioenergy 66:426–433

Jiménez J (2019) Non-circulative virus transmission by aphids: new insights into the mechanisms of transmission and the interference for retention sites in the vector. PhD thesis, Polytechnic University of Madrid

Jiménez J, Tjallingii WF, Moreno A, Fereres A (2018) Newly distinguished cell punctures associated with transmission of the semipersistent phloem-limited *Beet yellows virus*. J Virol 92:e01076–e01018

Jiménez J, Arias-Martín M, Moreno A, Garzo E, Fereres A (2020a) *Barley yellow dwarf virus* can be inoculated during brief intracellular punctures in phloem cells before the sieve element continuous salivation phase. Phytopathology 110(1):85–93

Jiménez J, Garzo E, Alba-Tercedor J, Moreno A, Fereres A, Walker GP (2020b) The phloem-pd: a distinctive brief sieve element stylet puncture prior to sieve element phase of aphid feeding behavior. Arthropod Plant Interact 14:67–78

Johnson DD, Walker GP, Creamer R (2002) Stylet penetration behavior resulting in inoculation of a semipersistently transmitted closterovirus by the whitefly *Bemisia argentifolii*. Entomol Exp Appl 102(2):115–123

Kennedy JS, Booth CO, Kershaw WJS (1961) Host finding by aphids in the field. Ann Appl Biol 49:1–21

Killiny N, Harper SJ, Alfaress S, El Mohtar C, Dawson WO (2016) Minor coat and heat-shock proteins are involved in binding of *Citrus tristeza virus* to the foregut of its aphid vector, *Toxoptera citricida*. Appl Environ Microbiol 82(21):6294–6302

Kirk M, Temple SR, Summers CG, Wilson LT (1991) Transmission efficiencies of field-collected aphid (Homoptera: Aphididae) vectors of *Beet yellows virus*. J Econ Entomol 84(2):638–643

Limburg DD, Mauk PA, Godfrey LD (1997) Characteristics of *Beet Yellows Closterovirus* transmission to sugar beets by *Aphis fabae*. Phytopathology 87(7):766–771

Loxdale HD, Edwards O, Tagu D, Vorburger C (2018) Population genetic issues: new insights using conventional molecular markers. In: van Emden HF, Harrington R (eds) Aphids as crop pests, 2nd edn. CABI, Wallingford, pp 50–80

Martín B, Collar JL, Tjallingii WF, Fereres A (1997) Intracellular ingestion and salivation by aphids may cause the acquisition and inoculation of non-persistently transmitted plant viruses. J Gen Virol 78(10):2701–2705

Mauck KE, Chesnais Q, Shapiro LR (2018) Evolutionary determinants of host and vector manipulation by plant viruses. Adv Virus Res 101:189–250

Mauck KE, Kenney J, Chesnais Q (2019) Progress and challenges in identifying molecular mechanisms underlying host and vector manipulation by plant viruses. Curr Opin Insect Sci 33:7–18

McLean DK, Kinsey MG (1964) A technique for electronically recording aphid feeding and salivation. Nature 202:1358–1359

Medina-Ortega KJ, Walker G (2015) Faba bean forisomes can function in defence against generalist aphids. Plant Cell Environ 38(6):1167–1177

Moreno A, Tjallingii WF, Fernández-Mata G, Fereres A (2012) Differences in the mechanism of inoculation between a semi-persistent and non-persistent aphid-transmitted plant virus. J Gen Virol 93(3):662–667

Murant AF, Roberts IM, Elnagar S (1976) Association of virus-like particles with the foregut of the aphid *Cavariella aegopodii* transmitting the semi-persistent viruses Anthriscus yellows and Parsnip yellow fleck. J Gen Virol 31:47–57

Normand RA, Pirone TP (1968) Differential transmission of strains of *Cucumber mosaic virus* by aphids. Virology 36(4):538–544

Oerke EC (2006) Crop losses to pests. J Agric Sci 144(1):31–43

Palacios I, Drucker M, Blanc S, Leite S, Moreno A, Fereres A (2002) *Cauliflower mosaic virus* is preferentially acquired from the phloem by its aphid vectors. J Gen Virol 83:3163–3171

Peng HC, Walker GP (2018) Sieve element occlusion provides resistance against *Aphis gossypii* in TGR-1551 melons. Insect Sci 27(1):33–48

Peremyslov VV, Andreev IA, Prokhnevsky AI, Duncan GH, Taliansky ME, Dolja VV (2004) Complex molecular architecture of beet yellows virus particles. PNAS 101(4):5030–5035

Pirone TP, Blanc S (1996) Helper-dependent vector transmission of plant viruses. Annu Rev Phytopathol 34(1):227–247

Powell G (1991) Cell membrane punctures during epidermal penetrations by aphids: consequences for the transmission of two potyviruses. Ann Appl Biol 119(2):313–321

Powell G (2005) Intracellular salivation is the aphid activity associated with inoculation of non-persistently transmitted viruses. J Gen Virol 86(10):469–472

Powell G, Pirone T, Hardie J (1995) Aphid stylet activities during potyvirus acquisition from plants and an in vitro system that correlate with subsequent transmission. Eur J Plant Pathol 101 (4):411–420

Powell G, Tosh CR, Hardie J (2006) Host plant selection by aphids: Behavioral, evolutionary, and applied perspectives. Annu Rev Entomol 51(1):309–330

Power AG, Seaman AJ, Gray SM (1991) Aphid transmission of *Barley yellow dwarf virus*: inoculation access periods and epidemiological implications. Phytopathology 81(5):545–548

Prado E, Tjallingii WF (1994) Aphid activities during sieve elements punctures. Entomol Exp Appl 72(2):157–165

Price WC (1966) Flexuous rods in phloem cells of lime plants infected with *Citrus tristeza virus*. Virology 29(2):285–294

Scheller HV, Shukle RH (1986) Feeding behavior and transmission of *Barley yellow dwarf virus* by *Sitobion avenae* on oats. Entomol Exp Appl 40(2):189–195

Song Q, Zheng YJ, Xue Y, Sheng WG, Zhao MR (2017) An evolutionary deep neural network for predicting morbidity of gastrointestinal infections by food contamination. Neurocomputing 226:16–22

Strange RN, Scott PR (2005) Plant disease: a threat to global food security. Annu Rev Phytopathol 43(1):83–116

Sylvester ES (1956) *Beet yellows virus* transmission by the green peach aphid. J Econ Entomol 49 (6):789–800

Tjallingii WF (1978) Electronic recording of penetration behavior by aphids. Entomol Exp Appl 24 (3):721–730

Uzest M, Blanc S (2016) Molecular mechanisms involved in noncirculative virus–vector interactions. In: Brown JK (ed) Vector-mediated transmission of plant pathogens. APS, St Paul, pp 59–72

Uzest M, Gargani D, Drucker M, Hebrard E, Garzo E, Candresse T, Fereres A, Blanc S (2007) A protein key to plant virus transmission at the tip of the insect vector stylet. PNAS 104 (46):17959–17964

van Bel AJE (2003) The phloem, a miracle of ingenuity. Plant Cell Environ 26(1):125–149

van Bel AJE, Kempers R (1997) The pore/plasmodesm unit; key element in the interplay between sieve element and companion cell. Prog Bot 58:278–291

van Bel AJE, van Rijen HVM (1993) Microelectrode-recorded development of the symplastic autonomy of the sieve elements/companion cells complex in the stem phloem of *Lupinus luteus* L. Planta 192(2):165–175

van Hemden HF (2018) Host-plant resistance. In: van Emden HF, Harrington R (eds) Aphids as crop pests, 2nd edn. CAB International, Wallingford, pp 515–532

Visser JH, Piron PGM, Hardie J (1996) The aphid's peripheral perception of plant volatiles. Entomol Exp Appl 80(1):35–38

Watson MA (1938) Further studies on the relationship between Hyoscyamus virus 3 and the aphid *Myzus persicae* (Sulz.) with special reference to the effects of fasting. Proc R Soc Lond B 125:144–170

Wayadande AC, Nault LR (1993) Leafhopper probing behavior associated with chlorotic dwarf virus transmission. Phytopathology 83:522–526

Biotechnology for Biofuel Production

Bethanie Viele, Rebecca Ellingston, Dan Wang, Yerim Park, Riley Higgins, and Heather D. Coleman

Contents

Abstract The commercial production of lignocellulosic biofuels relies heavily on a reduction in production costs. These costs are high in part due to the challenge of biomass deconstruction and the complex nature of the secondary cell wall. The removal of lignin from the carbohydrates, and the subsequent or concurrent hydrolysis of the polysaccharides into monomers for fermentation continue to hamper large commercialization efforts. One solution for this is to tailor biomass for the production of fuels. Here we review work in the field of plant biotechnology, with a focus on dicots, to alter the secondary cell wall, produce plant made enzymes, increase biomass production, and create sterile lines. We conclude by laying out future directions for research to support the production of cost-effective lignocellulosic fuels.

B. Viele, R. Ellingston, D. Wang, Y. Park, R. Higgins, and H. D. Coleman (✉)
Biology Department, Syracuse University, Syracuse, NY, USA
e-mail: bmviele@syr.edu; rmelling@syr.edu; dwang36@syr.edu; ypark102@syr.edu; rhiggins@syr.edu; hcoleman@syr.edu

© Springer Nature Switzerland AG 2020
Progress in Botany (2021) 82: 383–404, https://doi.org/10.1007/124_2020_39,
Published online: 16 December 2020

1 Introduction

Global dependence on fossil fuels continues despite improved understanding of the climate implications of these emissions (Quéré et al. 2018). The ecological cost of these fuels as well as the high demand and concerns over energy security have led to the need for development of green energy sources such as biomass derived fuels (Rodionova et al. 2017). These biofuels are derived from the chemical conversion of organic material from organisms such as plants or algae as opposed to the nonrenewable sources from which fossil fuels arise. The two primary types of biomass derived fuels are first generation and second generation, or lignocellulosic, biofuels (Rodionova et al. 2017). First generation fuels utilize starch from potential feed and food sources such as corn kernels or wheat, whereas second generation fuels are produced from the structural carbohydrates in plant secondary cell walls (SCW; Baig et al. 2019). One of the primary benefits of lignocellulosic fuels is their independence from lands used for food production (Correa et al. 2019). The biomass used for the production of these second generation fuels can be sourced from agricultural waste, or from dedicated energy crops that may be grown on land that is less optimal for agriculture (Rocha-Meneses et al. 2017).

Lignocellulosic fuels are derived from the SCW, a thick structural cell wall layer located between the plasma membrane and the thinner primary cell wall (McCahill and Hazen 2019). The SCW is comprised of the polyphenol lignin, as well as cellulose and hemicellulose, both of which can be hydrolyzed into carbohydrate monomers, and fermented into the biomass derived fuel (Wang et al. 2016; Mahon and Mansfield 2019). Cellulose is the largest carbohydrate component of the SCW, making up approximately 40–50% of softwood and hardwood trees (Rocha-Meneses et al. 2017; Zhong et al. 2019). Cellulose is formed of D-glucose units with β-1,4 glycosidic bonds that can be broken into cellobiose and then cleaved into glucose molecules (Rodrigues Mota et al. 2018). This is typically done through enzymatic hydrolysis, using multiple cellulases to allow for a high glucose yield to be fermented into ethanol (Rodrigues Mota et al. 2018; Kumar et al. 2019; Liu et al. 2019; Chu et al. 2019). Glucose from cellulose is the primary desired product from hydrolysis of the SCW for use in biofuel synthesis; however, hemicellulose can also be utilized. Hemicellulose, contributing approximately 30% of the cell wall, is less abundant and more complex than cellulose, and varies compositionally across species. Hemicellulose is composed of carbohydrates such as xylose, arabinose, and mannose rather than singularly glucose (Wang et al. 2016; Rocha-Meneses et al. 2017). Hemicellulose within the SCW serves primarily to form a complex cross-linked structure with cellulose and lignin, providing strength and rigidity to the plants (Petridis and Smith 2018); however, these sugars can also be used for biofuel production (Dodd and Cann 2009). They are more readily broken down into monomer sugars than cellulose, though acetic acid is a major byproduct of hydrolysis due to acetylation (Liu et al. 2019). The acetyl groups, from xylan in particular as it is the most abundant polysaccharide in hemicellulose, become free and form acetic acid in the pretreatment process, leading to hemicellulose solubilization and inhibition of ethanol fermentation (Liu et al. 2019; Tian et al. 2019; Chu et al. 2019). While pretreatment increases the accessibility of cellulose, this process prevents hemicelluloses from additionally being easily used for biofuel production. The third major component of the SCW is lignin, which contributes about 20–30% of wood in hardwoods and softwoods but is not currently utilized for biofuels (Rocha-Meneses et al. 2017; Kim et al. 2019). Lignin provides rigidity and mechanical strength to the plant and is a complex phenolic polymer which crosslinks with xylan and other hemicelluloses (Xie et al. 2016; Madadi et al. 2017; Terrett and Dupree 2019). Due to this, lignin contributes to extreme recalcitrance within the plant, making the SCW difficult to break down into lignocellulosic fuel. While not utilized in second generation fermentable fuels, there is potential for lignin to be a valuable source for other types of fuels such as jet fuel (Shen et al. 2019).

Despite the potential value of biofuels from an economic and environmental perspective, there are numerous biological challenges related to lignocellulosic biofuel production, ranging from expensive resource requirements to natural barriers against SCW breakdown. Recalcitrance, broadly defined as the resistance of the SCW to the release of sugar for conversion into biofuel, is one of the primary challenges existing in the cost-effective production of lignocellulosic biofuels (Gilna et al. 2017). Crystalline cellulose microfibrils interact with hemicelluloses and lignin within the SCW, forming lignin–carbohydrate complexes which make access and extraction of celluloses and hemicelluloses expensive in terms of both cost and time (Liu et al. 2019). Biomass must undergo pretreatment to increase cellulose accessibility through decrystallization and chain disentanglement, as well as to separate lignin from the desired polysaccharides before fermentation (Seidl and Goulart 2016; Ghasemi et al. 2017). Various approaches to reduce the challenge of recalcitrance through genetic manipulation of the cell wall chemistry and structure have been examined, and recent work will be explored in this paper. In particular, research has focused on decreasing the lignin content of the plant cell wall, but often reduced lignin compromises plant growth through impairment of vascular system development (Coleman et al. 2008; Voelker et al. 2011; Pereira et al. 2018). Alternative approaches have involved modifying the crystallinity of cellulose or altering cellulose or hemicellulose levels, allowing for more susceptibility to enzymes in fermentation (Bali et al. 2016). One such study has had some success in manipulating putative "recalcitrance" genes in poplar to reduce xylan levels, finding no reduced growth traits but improved saccharification (Biswal et al. 2015).

In addition to the challenge of accessing the carbohydrates within the plant biomass, there is the challenge of producing large amounts of biomass rapidly. This can require a large input of fertilization. Numerous studies have assessed the impact of various nutrient fertilizers and have found a significant increase in biomass with their application in a variety of plant species including poplar and eucalyptus (Cooke et al. 2005; da Silva et al. 2016). The most commonly studied nutrient is nitrogen (N). Only 30–50% of nitrogen applied to crops is utilized by the plants, leaving a large percentage of N in the environment (Hodge et al. 2000; Masclaux-

Daubresse et al. 2010). Although N is abundant, N leaching acts as an environmental pollutant. The N runoff is transported from crops to rivers and other bodies of water which leads to eutrophication, or the overabundance of a nutrient. As a result, the mass addition of N to crops contributes to harmful algae blooms, anoxic waters, and other serious concerns (Le Moal et al. 2019). Improving plant uptake of N in order to reduce the cost of fertilization and the loss of N to the environment is an important challenge that must be addressed for the success of lignocellulosic fuels. In addition to these fertilization requirements, there may be more intense irrigation requirements, particularly as water becomes more and more depleted with changing climates (Kattel 2019). Agriculture accounts for approximately 70% of water use from freshwater systems, an estimate subject to change as biofuel production grows to replace fossil fuels (Pastor et al. 2019). With greater understanding of lignocellulosic biomass and improvements in nutrient and water use, biomass feedstock production will be less ecologically expensive.

Finally, these challenges of improved biomass quantity and quality are most likely to be addressed by biotechnological solutions. This requires that plants produced by these methods can be used in the field for the production of biomass. Current research is also addressing both methods to induce sterility and also assessing the environmental impact of these plants (Strauss et al. 2017). This work is integral to the success of lignocellulosic biofuel production.

Research into second generation biofuels utilizes multiple different species. *Arabidopsis thaliana* acts as a useful dicot model organism, while members of the *Populus* family, Eucalypts, and willow are also commonly studied, as trees which can produce large amounts of biomass on non-agricultural lands, making them potential biofuel feedstocks. Many other species, both monocots and dicots, also hold potential for use in the production of biomass for lignocellulose fuels; for the purpose of this review we focus on dicot species.

2 Modification of Secondary Cell Wall Biosynthesis

There are four major steps involved in the biochemical conversion of lignocellulosic biomass to ethanol: biomass production, pretreatment to disrupt the cell wall, enzymatic hydrolysis to release glucose monomers from structural carbohydrates, and the fermentation of monomeric carbohydrates into ethanol (Fig. 1; Huang et al. 2019). Higher biofuel yield or easier pretreatment can be achieved by using biomass that possesses lower cell wall recalcitrance and the use of suitable chemical pretreatment for the specific biomass being used (de Souza et al. 2019). Strategies aimed at changing the composition of the SCW have provided some success in reducing recalcitrance. Additional strategies for improving yield include increasing the carbohydrate content of the cell wall, reducing the crystallinity of the cellulose, or altering the complexity of the cell wall structure.

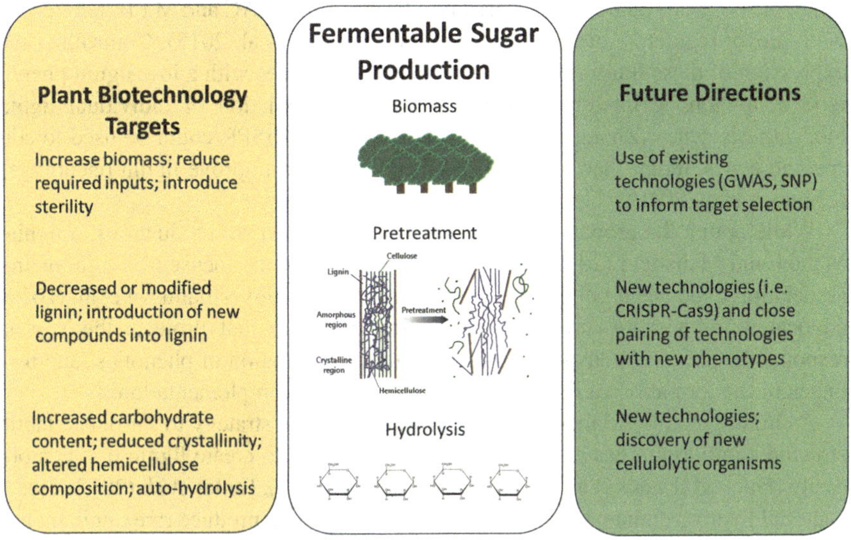

Fig. 1 Schematic of current targets of biotechnology and the expected future directions for continued advancement in the production of lignocellulosic biofuels

2.1 *Lignin*

Lignin is one of the major components of the SCW and is comprised of dimethoxylated (syringyl, S), monomethoxylated (guaiacyl, G), and non-methoxylated (p-hydroxyphenyl, H) phenylpropanoid units which are derived from *p*-hydroxycinnamyl alcohols (Martínez et al. 2005). It provides a rigid structure to the plant cell wall and, in processing, restricts the hydrolysis of cellulose and hemicelluloses (Isikgor and Becer 2015). Removal of lignin is therefore a necessary step in the production of lignocellulosic biofuel (Bušić et al. 2018).

The pretreatments required for the removal of lignin are very costly and potentially toxic to the environment (Hamelinck et al. 2005). The capability to produce biofuel feedstocks with reduced lignin would consequently have positive impacts in many capacities, and the modification of lignin gene expression has been well explored (reviewed in Chanoca et al. 2019). However, when there is a reduction in lignin content, there are other potentially negative impacts on the plant, including increased susceptibility to pests, decreased resistance to drought, and reduced growth (Moura et al. 2010). A number of strategies exist for reducing or altering lignin in plants; however, some recent reviews have emphasized that strictly reducing the amount of lignin throughout the plant is unlikely to be the solution to improving biomass for biofuel production (Mahon and Mansfield 2019; Muro-Villanueva et al. 2019). There are still other options addressing lignin recalcitrance, including manipulating expression of transcription factor genes controlling the lignin biosynthesis pathway. Transcription factors in the NAC and MYB families are well-known regulators of SCW biosynthesis (Nakano et al. 2015). Controlling the expression of these transcription factors can result in trees with a low-lignin phenotype in a more scripted way than the direct manipulation of individual lignin biosynthesis genes (Zhang et al. 2018). In addition, CRISPR could be used to edit specific genes, rather than reducing their expression, to improve lignin phenotypes (Chanoca et al. 2019).

While noting the growth penalties regularly associated with reductions in lignin, Mahon and Mansfield (2019) proposed research directions focused on minimizing association of adjacent polymers within the cell wall. Muro-Villanueva et al. (2019) advocated for a more directed modification of lignin that protects the vessels responsible for water transport and reduces the accumulation of phenolics, intermediates in the formation of lignin, using vessel specific complementation.

As highlighted by Mahon and Mansfield (2019), the strategy to introduce labile structures into lignin has proved valuable as a method to create lignin that is more easily removed (Eudes et al. 2012; Wilkerson et al. 2014). Eudes et al. (2012) used a bacterial hydroxycinnamoyl-CoA hydratase-lyase gene to produce rare lignin monomers they termed DO reducers (C_6C_1 monomer) to reduce the polymerization of lignin. Although this proved effective in altering the lignin structure, there was still a small growth penalty. Wilkerson et al. (2014) expressed a feruloyl–coenzyme A (CoA) monolignol transferase (FMT) gene from *Angelica sinensis* in poplar to drive the production of monolignol ferulate conjugates in lignin. These conjugates, which introduce readily digestible ester bonds into the lignin, allow for near doubling of glucose release following a mild alkaline pretreatment, with no growth deficit. Further analysis on these lines showed improved response to ionic liquid pretreatments (Kim et al. 2017).

As explored in Muro-Villanueva et al. (2019), another option is the reduction of lignin biosynthesis in fibers only, which would serve to reduce the negative effects of cell wall modification on plant growth and development. This could potentially be achieved through knockout of a gene, with complementation of the same gene using a vessel specific primer, or by targeting gene knockout to fibers. Both of these options have been explored. De Meester et al. (2018) used a vessel specific artificial promoter Secondary Wall NAC Binding Element of the xylem cysteine protease 1 to reintroduce cinnamoyl coenzyme A reductase 1(ccr1) into the *ccr1 Arabidopsis* mutant line. This resulted in recovery of biomass levels beyond that of wild type and a resultant glucose release per plant of nearly four times that of the wild type. Smith et al. (2017) demonstrated reduced lignification and increased saccharification in *Arabidopsis* by using a fiber specific promoter driving the expression of the same gene. Liang et al. (2019) used a fiber specific promoter (NST3/SND1) to drive Cas9 expression to knock out the expression of hydroxycinnamoyl-CoA shikimate/quinate hydroxycinnamoyl transferase (HCT). The CRISPR edited *HCT* lines were not different in size from wild type, but did display the expected increase in H lignin. Taken together, this work suggests that improvements in biomass quality can be made without detrimental impacts on plant growth if the modification is well targeted.

An additional interesting approach takes advantage of enzyme promiscuity to overload an enzyme in the lignin biosynthetic pathway with a non-traditional substrate in order to alter lignin production. Eudes et al. (2016) engineered *Arabidopsis* to produce protocatechuate, which they define as a HCT competitive inhibitor. By providing a nontraditional substrate to this lignin biosynthesis enzyme, the lignin content was reduced and saccharification increased.

2.2 *Cellulose*

Cellulose is the most abundant component of the SCW and is seen as an almost inexhaustible resource for greener fuels and products (Klemm et al. 2005). This, in addition to its pure composition of glucose monomers, makes cellulose the most desired portion of the SCW for the production of lignocellulosic biofuels. The synthesis of cellulose is facilitated by cellulose synthase (CesA) proteins which interact with one another to form a rosette-shaped plasma membrane-spanning CesA complex (CSC) (Holland et al. 2000; Li et al. 2016). Three isoforms of *CesA*, *CesA4*, *CesA7*, and *CesA8*, are essential to SCW biosynthesis of cellulose, and come together at the cell plasma membrane to link glucose chains into cellulose fibers (Li et al. 2016; Xi et al. 2017; Hu et al. 2018; Speicher et al. 2018). These CSCs synthesize cellulose microfibrils from β-1,4-linked glucose arranged in linear chains with cellobiose as the repeating unit (Rodrigues Mota et al. 2018; Turner and Kumar 2018). The linear glucan chains are arranged by the CSCs into a multilaminar structure where the chains interact with their neighbors through hydrogen bonds to form cellulose microfibrils (Zhang et al. 2015; Lindh et al. 2016; Xi et al. 2017). These cellulose microfibrils provide the framework for the cell wall and are the primary load-bearing component of the SCW (Mueller and Malcolm Brown 1980; Ruel et al. 2012). Studies have examined the potential in overexpressing SCW CesAs to improve cellulose content but have been unsuccessful. However, a study looking at the impact of increasing the expression of primary cell wall *CesAs*, including *CesA2*, *CesA5*, and *CesA6*, found improved cellulose synthesis and plant growth in *Arabidopsis* (Hu et al. 2018). Other research promoted the biosynthesis of cellulose and SCW thickening in *Arabidopsis* through hydrogen peroxide (H_2O_2) production under salt stress signaling CesA activity (Shafi et al. 2019).

The structure of cellulose provides a number of challenges for biofuel production. One of these is crystallinity index, associated with larger degrees of polymerization and microfibril orientation (Xi et al. 2017). Cellulose exists in two primary forms within the SCW depending on the organization and compactness of microfibrils (Sun et al. 2016). Crystalline cellulose represents the majority and is relatively resistant to hydrolysis processes aiming to break down the cellulose into glucose molecules. The second form of cellulose is amorphous and includes areas with low molecular order or weaker inter- and intramolecular hydrogen bonds and less organization (Karimi and Taherzadeh 2016; Hattori and Arai 2016). Amorphous regions are much more easily hydrolyzed due to their greater ability to adsorb water as opposed to crystalline cellulose. Due to the difficulties in hydrolyzing cellulose I, or native cellulose, even after the removal of lignin, studies have begun to examine the potential benefits of a synthetic allomorph of cellulose called cellulose II (Karimi and Taherzadeh 2016). Cellulose II has an antiparallel arrangement of its chains and a monoclinic lattice arrangement, contributing to decreased crystallinity, increased surface area, and greater hydrophilicity, all of which are ideal for biofuel production (Nagarajan et al. 2017).

2.3 Hemicellulose

Hemicellulose is the third major component of the plant SCW. In dicots, the majority of hemicellulose is glucuronoxylan (xylan), but there are also smaller amounts of glucomannan and galactoglucomannan (Kumar et al. 2016). Hemicellulose interacts closely with lignin and cellulose, which contributes to the complexity of the SCW. Pectin, an additional but minor component of the SCW, contributes to this complexity through forming a gel-like matrix in which hemicellulose and cellulose are embedded (Tomassetti et al. 2015).

Using targeted RNAi construct-driven downregulation of a gene encoding galacturonosyl transferase 12, *GAUT12* (*IRX8*), a gene involved in glucuronoxylan and pectin biosynthesis, in poplar led to increased biomass production and decreased recalcitrance (Biswal et al. 2015). In a later study, when GAUT12 was overexpressed, the resultant plants had decreased biomass and increased recalcitrance (Biswal et al. 2018b). This gene also impacts the formation of homogalacturonan (HG), a pectin, which is a minor part of the SCW. When overexpressed, the HG content in the plants increased and when silenced, the HG content decreased, following the same pattern as xylan levels. In other work by the same group, downregulation of *GAUT4*, a gene encoding enzyme responsible for pectin polysaccharides and glycan biosynthesis, resulted in increased biomass and improved saccharification (Biswal et al. 2018a), along with decreased pectins including HG and reduced crosslinking between carbohydrates and lignin (Li et al. 2019). Additionally, endopolygalacturonase (*EPG*), a well-known pectinase, has been examined as a potential additive to enzymatic hydrolysis in order to decrease the recalcitrance of pectin in the SCW. Experiments utilizing EPG found increased efficiency of hydrolysis and decreased recalcitrance (Latarullo et al. 2016).

The modification of hemicellulose results in some of the same growth challenges as the modification of lignin and cellulose. Recommendations include the use of wood-specific promoters or heat inducible enzymes or promoters (Donev et al. 2018).

3 Enzyme Production in Plants

Hydrolytic enzymes are essential for the breakdown of cellulose and hemicellulose into fermentable carbohydrate monomers. These enzymes include cellulases and hemicellulases (Niu et al. 2019). Both these enzyme classes work together to break down the SCW by hydrolyzing cellulose and hemicellulose respectively. They are of high importance in various industries beyond biofuels for the production of glucose and have been extensively studied (Srivastava et al. 2018). They are produced by many types of organisms, including fungi, plants, and animals. Cellulases from bacteria and fungi, including *Trichoderma reesei,* are often used in lignocellulosic biofuel production and are generally produced in microbes (Mojsov 2016; Singh et al. 2019). This microbial production of cellulases can be very expensive (Srivastava et al. 2018); an alternate production method may be producing them in the biofuels crop itself (Xiao et al. 2016).

There are three major families of cellulase enzymes: endoglucanase, exoglucanase, and β-glucosidase. These enzymes work together to hydrolyze the β-1,4 links in the cellulose chains (Mojsov 2016). Endoglucanase works within the cellulose chain where it breaks bonds in no particular pattern. Exoglucanase is much more specific as it hydrolyzes bonds near the end of the glucose chains. Exoglucanase releases β-cellobiose and other short fermentable oligosaccharides (Srivastava et al. 2018). β-glucosidase further breaks down the β-cellobiose into a simple sugar, β-glucose (Mojsov 2016).

Hemicellulases make up another family of enzymes that hydrolyze the various polysaccharides that compose hemicellulose. As hemicellulose is comprised of a complex subset of heteropolymers made up of multiple carbohydrates, the family of hemicellulase enzymes is also diverse and includes xylanases and mannanases. These enzymes work together to break down the polymer into monomeric sugars (Gupta et al. 2016). Though cellulase research is more prevalent, *in planta* expression of hemicellulases has also been shown to be a valuable tool for decreasing plant recalcitrance (Tsai et al. 2017). Hemicellulose is primarily made up of xylan, which is broken down into the monomer sugar xylose with the help of the hemicellulases endo-β-xylanase and β-xylosidase (Niu et al. 2019). Co-expression of a β-glucosidase with a xylose isomerase resulted in better xylose production efficiency in vitro since β-glucosidase is often inefficient and the rate-limiting step in xylose production (Niu et al. 2019).

Lignin modifying enzymes (LMEs) are another group of enzymes involved in plant cell wall break down. Laccases and peroxidases are the most common and best studied types of LMEs in plants (Bilal et al. 2018). These enzymes break down the complex matrix of lignin that envelops the SCW and limits the accessibility of the desired carbohydrates. Lignin degradation is a complicated process as there is no single method by which the enzymes work, but research in this area has shown that these enzymes are capable of reducing lignin in plants when their amino acid codons are altered by site-directed mutagenesis to create higher efficiency (Madzak et al.

2006). A bacterial-produced peroxidase, DypB, was expressed in tobacco and targeted specifically to either the endoplasmic reticulum (ER) or the cytosol (Ligaba-Osena et al. 2017). In both cases, the growth of the plants was not negatively impacted despite a 200% increase in saccharification when compared to the wild type tobacco plants likely due to a decrease in lignin in the transgenic plants. The reduction of lignin, which increased access to the cellulose, was attributed to its depolymerization by DypB (Ligaba-Osena et al. 2017). Ectopic expression of peroxidase genes can also alter cellulose accumulation and metabolic pooling which impacts growth and development.

Current research is primarily working with *in planta* expression of enzymes as an alternative or supplement to applying microbial derived enzymes. One of the potential concerns associated with overexpressing desired enzymes in the plants is stunted growth and poor development in the transgenic plants. Many recent studies have shown that it is possible to overcome this challenge by targeting the enzymes to various cellular components such as the apoplast and the ER as reviewed by Park et al. (2016). For example, when overexpressing an endoglucanase TrCel5A in tobacco, it was shown that targeting the enzymes to the ER did not result in detrimental effects on plant growth, contrary to a previous study where apoplast targeting of the same enzyme had a growth penalty (Klose et al. 2013, 2015).

Apoplast-targeted expression of a hyperthermophilic endoglucanase from the bacteria *Thermotoga neapolitana* in hybrid poplar resulted in plants with altered cell wall compositions and biomass that was easier to breakdown. One of the transgenic lines did not require pretreatment in order to reach the same level of glucose release as the pretreated wild type plants (Xiao et al. 2018). Similar results were seen with apoplast-targeted expression of hyperthermophilic endoglucanase and xylanase genes in *Arabidopsis* with no detrimental impact on height or fresh weight but an increase in glucose release by enzymatic hydrolysis (Mir et al. 2017).

Overexpression of cellulases is clearly a viable option for improving lignocellulosic biomass for biofuel production. In addition to avoiding harmful effects on plant growth, there is an increased rate of digestion of the SCW. Pretreatment and enzymatic hydrolysis, which together break down the SCW into fermentable sugars, are expensive and time-consuming steps in the biofuel production process, but overexpressing cellulases has shown to be as effective and more convenient, as reviewed in Akram et al. (2018).

4 Improving Biomass Production

4.1 *Nutrients*

Another major goal with regards to plant biotechnology for biofuels production is increasing biomass yield. In order to maximize growth, plants often require supplemental nutrients in the form of fertilizers. In addition to altering nutrient availability, it is possible to increase growth by manipulating the genes that regulate or directly participate in nutrient uptake and allocation.

Phosphorus and N are the most well-studied nutrients, as they are most often growth limiting. Phosphorus tends to be growth limiting in old soils due to a limited amount of mobile phosphorus (Wiersum 1962), whereas N is limiting in young soils in part due to rapid mineralization of available N (Wiersum 1962; Stanford and Smith 1972).

Supplying adequate levels of N typically results in increased plant growth and biomass production. Poplar trees subjected to deficient, control, and fertilized N levels were assessed for changes in biomass accumulation and concentrations of amino acids, hormones, and N across tissues. N deficient trees allocated their growth in opposing patterns to that of N normal and N fertilized, with N deficient trees expending resources to increase root biomass and N normal and N fertilized trees having increased above ground biomass (Luo et al. 2019). While a deficient N status resulted in decreased growth, there is a limit to the increase in growth that can be afforded by increased N supply. Providing 2 mM of N resulted in significantly increased leaf biomass compared to no additional N, and 50 mM of N also resulted in increased biomass, the plants developed syllepsis which persisted for the entire experiment (Cooke et al. 2005). Syllepsis, the growth of lateral branches in the same year as bud formation, has previously been shown to have a positive impact on total biomass production (Moreno-Cortés et al. 2012). The plasticity of growth in relation to available nutrients allows plants to respond to their immediate environment, but can also be useful to identify opportunities to modify the plant to improve uptake and use of nutrients.

Altering the expression of genes involved in N metabolism can also have significant impacts on plant growth. Overexpression of cystolic glutamine synthetase (GS1) has been shown to improve yield in a number of species, and in poplar has been shown to also alter the cell wall chemistry (Jing et al. 2004; Coleman et al. 2012). In addition, there is evidence that poplar overexpressing GS1 has increased nitrate uptake, making it ideal for phytoremediation in addition to having improved biomass production (Castro-Rodríguez et al. 2016). Overexpression of GS1 in *Betula pubescens* caused increased root growth and root complexity, as well as increases in amino acid concentrations. Glutamic acid, aspartic acid, and glutamine along with auxin were significantly increased in transgenic plant shoots compared to wild type controls (Lebedev et al. 2018). Improving the growth rate of plants has important implications for the production of fuels from biomass, and success using genetically modified plants with altered N metabolism (reviewed in Cánovas et al. 2018) may also reduce the need to increase fertilization.

Many studies also assessed phosphorus fertilization in tandem with N. In *Arabidopsis*, increased N availability resulted in increased concentrations of N, potassium, and magnesium in the leaves of *Arabidopsis*, while increased phosphorus availability resulted in increased concentrations of phosphorus and calcium in leaves with a reduction in carbon concentration (Yan et al. 2019a). This result implies that increasing phosphorus fertilization will increase the concentration of other required elements but may reduce carbon acquisition. Another study in *Arabidopsis* found that supplemental N increases growth of leaves and stems while supplemental phosphorus increases stems and fruiting bodies (Yan et al. 2019b). N and phosphorus are highly linked and should be researched in tandem to create the best possible increase in growth. This is due in part to genes encoding for phosphorus transporters such as PHOSPHATE2 (PHO2) being regulated by available nitrate and expressed under phosphorus starvation in correlation with nitrate transporters such as NITRATE TRANSPORTER 1.1 (NRT1.1). Crosstalk between NRT1.1 and PHO2 was observed, as each impacts the expression of the other in periods of phosphorus deficiency (Medici et al. 2019).

4.2 Hormones

In addition to the improvements in growth that can be induced by altering the plant metabolism of nutrients, altering hormones can also impact plant growth and biomass production. One study found that gibberellin, auxin, and brassinosteroids all contribute to growth of apple trees, and manipulation of brassinosteroid concentration caused an increase in internode length, stem biomass, and leaf biomass. In addition, increases in expression of cell growth related genes such as *MYB2*, *CESA*, and *CYCD1* were observed (Zheng et al. 2019). Consistent with this, overexpression of *CYP85A3*, a brassinosteroid biosynthesis gene, resulted in increased diameter, internode length, and height in transgenic poplar relative to wild type (Jin et al. 2017).

Abscisic acid is a classic plant hormone known not only for repressing leaf abscission and preventing bud break, but also for inducing drought tolerance. When overexpressed in poplar, abscisic acid was found to increase biomass production under drought conditions (Yu et al. 2019). Jeon et al. (2016) found that overexpressing *PdGA20ox1*, gibberellin 20 oxidase 1, increased biomass production, SCW thickening, glucose levels, and xylem differentiation in both plant species relative to wild type. Unlike previous studies, they also found decreased negative effects on root formation when expressing *PdGA20ox1* under the control of a xylem specific promoter as opposed to a constitutive promoter (Eriksson et al. 2000; Mauriat et al. 2014; Jeon et al. 2016). Manipulating genes involved in hormone biosynthesis may lead to new ways to increase biomass without requiring additional agricultural inputs.

5 Introducing Sterility

The use of genetically modified plants as biomass for fuel production is attractive, as the characteristics produced can be challenging to obtain through breeding programs, and for some species, the breeding cycle may be very long (Klocko et al. 2018). One of the most common concerns with genetically modified trees is their potential for wide dispersal of seed and pollen. Removing flowering capability thereby inducing plant sterility is a potential way to reduce concerns about the use of genetically modified species. The goal of sterility in genetically modified trees began in 1987 and has been progressing to prevent biosafety implications of releasing modified trees into the wild (Committee on Environmental Impacts Associated with Commercialization of Transgenic Plants, Board on Agriculture and Natural Resources NRC 2002). Since then, a condition for further commercialization of genetically modified trees is the development of genetic containment strategies, but this work remains challenging due to market and regulatory restrictions (Strauss et al. 2017).

There are many possible ways to reduce the potential of gene flow. Some methods are horticultural, such as harvesting before maturity, growing varieties that cannot interbreed, or creating sterile hybrids (Lu et al. 2019). Others require the identification of specific target genes with roles in flowering and take advantage of tissue specific promoters to target reproductive tissues and induce sterility or alter timing of flowering.

Engineered induction of male sterility has been engaged to prevent pollen mediated transgene flow. The anther is required for pollen development, including the tapetum cells which aid in pollen development. In some plant species, success in male sterility has been achieved using genetic ablation techniques and tapetum specific promoters. One example is the pea Endothecium 1 (PsEND1) promoter driving the expression of barnase-barstar in tobacco (Roque et al. 2007). This approach has been effective in all plant species that have been tested to produce full anther ablation with no mature pollen grains produced (Roque et al. 2019). Other similar approaches include the overexpression of a restriction enzyme under the control of a tomato pollen specific promoter in tobacco, which was also effective in producing male sterility (Millwood et al. 2016). A male cone specific promoter derived from *Pinus radiata* (PrMC2) was used to express *BARNASE* in pine and eucalyptus tree to successfully produce male sterility (Zhang et al. 2012).

Poplar is dioecious, with potential for long distance wind transport of pollen and seed. Recently, a large field trial of poplar engineered for sterility was assessed (Klocko et al. 2018). This trial included a diverse set of approaches (23 constructs) and two female and one male poplar clones. The goal of the research was to identify modes of bisexual sterility of poplar that did not produce off target effects. This trial built on previous work by the group that showed targeting of *LEAFY* and *AGAMOUS* successfully produced sterility in a female poplar clone and reduced pollen amount and size in apple trees respectively (Klocko et al. 2016a, b). In the field trial, these two genes in combination resulted in male sterility and female floral alterations, while *LEAFY* on its own resulted in female and male sterility (Klocko et al. 2018). Additionally from this work, *AGAMOUS* and *SEEDSTICK* were identified as strong modification targets to achieve gene containment (Lu et al. 2019).

Huang et al. (2016) found that creating a gene fusion of *SOLODANCERS* with *BARNASE* under the control of the SDS promoter resulted in ablation of both microspore and megaspore mother cells creating both male and female sterility. When the fusion was used, with the entire SDS coding region, there was no negative impact on growth and development in tobacco or *Arabidopsis* (Huang et al. 2016).

Another recent paper reported *Arabidopsis* sterility due to double mutations in the TFIIB-related factor (BRF) family, which plays important roles in RNA polymerase transcription. *BRF1* and *BRF2* are highly involved with the reproductive system, and the double mutation results in a high degree of aborted macrogametes and microgametes, and complete failure in zygote generation, inducing sterility (Zhang et al. 2019).

6 Conclusions

The challenges associated with the cost-effective production of biofuels are largely linked to the production of large amounts of high-quality biomass. Current fertilizer requirements, pretreatment methods, and enzymes for hydrolysis are costly to the environment and production process. As presented above, the use of plant biotechnology tools to produce improved plants, both in terms of increased biomass or improved biomass quality, continues to advance and is assisted by novel application of new technologies.

While significant advances have been made with existing technologies, the advent and rapid improvement of genomic and biotechnology tools such as genome editing (e.g., CRISPR) and the ability to introduce SNPs will allow for significant improvements over the status quo. Using the results of GWAS populations to identify the most relevant SNPs, the genome can be specifically tailored to produce the optimal combination for biomass production and quality (reviewed in Myburg et al. 2019). A final challenge remains in the regulation of transgenic trees and crops (Chang et al. 2018), and the high costs associated with bringing these crops to market. Improvement through plant biotechnology continues to advance the potential of lignocellulosic fuels; however, more work is needed to achieve cost competitive fuel production.

References

Akram F, ul Haq I, Imran W, Mukhtar H (2018) Insight perspectives of thermostable endoglucanases for bioethanol production: a review. Renew Energy 122:225–238

Baig KS, Wu J, Turcotte G (2019) Future prospects of delignification pretreatments for the lignocellulosic materials to produce second generation bioethanol. Int J Energy Res 43:1411–1427. https://doi.org/10.1002/er.4292

Bali G, Khunsupat R, Akinosho H et al (2016) Characterization of cellulose structure of *Populus* plants modified in candidate cellulose biosynthesis genes. Biomass Bioenergy 94:146–154. https://doi.org/10.1016/j.biombioe.2016.08.013

Bilal M, Iqbal HMN, Hu H et al (2018) Metabolic engineering and enzyme-mediated processing: a biotechnological venture towards biofuel production – a review. Renew Sustain Energy Rev 82:436–447

Biswal AK, Hao Z, Pattathil S et al (2015) Downregulation of *GAUT12* in *Populus deltoides* by RNA silencing results in reduced recalcitrance, increased growth and reduced xylan and pectin in a woody biofuel feedstock. Biotechnol Biofuels 8:41. https://doi.org/10.1186/s13068-015-0218-y

Biswal AK, Atmodjo MA, Li M et al (2018a) Sugar release and growth of biofuel crops are improved by downregulation of pectin biosynthesis. Nat Biotechnol 36:249–257. https://doi.org/10.1038/nbt.4067

Biswal AK, Atmodjo MA, Pattathil S et al (2018b) Working towards recalcitrance mechanisms: increased xylan and homogalacturonan production by overexpression of *GAlactUronosylTransferase12* (*GAUT12*) causes increased recalcitrance and decreased growth in *Populus*. Biotechnol Biofuels 11:1–26. https://doi.org/10.1186/s13068-017-1002-y

Bušić A, Marđetko N, Kundas S et al (2018) Bioethanol production from renewable raw materials and its separation and purification: a review. Food Technol Biotechnol 56:289–311. https://doi.org/10.17113/ftb.56.03.18.5546

Cánovas FM, Cañas RA, de la Torre FN et al (2018) Nitrogen metabolism and biomass production in forest trees. Front Plant Sci 9:1449. https://doi.org/10.3389/fpls.2018.01449

Castro-Rodríguez V, García-Gutiérrez A, Canales J et al (2016) Poplar trees for phytoremediation of high levels of nitrate and applications in bioenergy. Plant Biotechnol J 14:299–312. https://doi.org/10.1111/pbi.12384

Chang S, Mahon EL, MacKay HA et al (2018) Genetic engineering of trees: progress and new horizons. In Vitro Cell Dev Biol Plant 54:341–376. https://doi.org/10.1007/s11627-018-9914-1

Chanoca A, de Vries L, Boerjan W (2019) Lignin engineering in forest trees. Front Plant Sci 10:912. https://doi.org/10.3389/fpls.2019.00912

Chu Q, Song K, Hu J et al (2019) Integrated process for the coproduction of fermentable sugars and lignin adsorbents from hardwood. Bioresour Technol 289:121659. https://doi.org/10.1016/j.biortech.2019.121659

Coleman HD, Samuels AL, Guy RD, Mansfield SD (2008) Perturbed lignification impacts tree growth in hybrid poplar – a function of sink strength, vascular integrity, and photosynthetic assimilation. Plant Physiol 148:1229–1237. https://doi.org/10.1104/pp.108.125500

Coleman HD, Cánovas FM, Man H et al (2012) Enhanced expression of glutamine synthetase (*GS1a*) confers altered fibre and wood chemistry in field grown hybrid poplar (*Populus tremula* × *alba*) (717-1B4). Plant Biotechnol J 10:883–889. https://doi.org/10.1111/j.1467-7652.2012.00714.x

Committee on Environmental Impacts Associated with Commercialization of Transgenic Plants, Board on Agriculture and Natural Resources NRC (2002) Environmental effects of transgenic plants: the scope and adequacy of regulation. National Academies Press, Washington

Cooke JEK, Martin TA, Davis JM (2005) Short-term physiological and developmental responses to nitrogen availability in hybrid poplar. New Phytol 167:41–52. https://doi.org/10.1111/j.1469-8137.2005.01435.x

Correa DF, Beyer HL, Fargione JE et al (2019) Towards the implementation of sustainable biofuel production systems. Renew Sustain Energy Rev 107:250–263. https://doi.org/10.1016/j.rser.2019.03.005

da Silva RML, Hakamada RE, Bazani JH et al (2016) Fertilization response, light use, and growth efficiency in Eucalyptus plantations across soil and climate gradients in Brazil. Forests 7:117. https://doi.org/10.3390/f7060117

De Meester B, de Vries L, Özparpucu M et al (2018) Vessel-specific reintroduction of CINNAMOYL-COA REDUCTASE1 (CCR1) in dwarfed *ccr1* mutants restores vessel and xylary fiber integrity and increases biomass. Plant Physiol 176:611–633. https://doi.org/10.1104/pp.17.01462

de Souza WR, Pacheco TF, Duarte KE et al (2019) Silencing of a BAHD acyltransferase in sugarcane increases biomass digestibility. Biotechnol Biofuels 12:111. https://doi.org/10.1186/s13068-019-1450-7

Dodd D, Cann IKO (2009) Enzymatic deconstruction of xylan for biofuel production. GCB Bioenergy 1:2–17. https://doi.org/10.1111/j.1757-1707.2009.01004.x

Donev E, Gandla ML, Jönsson LJ, Mellerowicz EJ (2018) Engineering non-cellulosic polysaccharides of wood for the biorefinery. Front Plant Sci 9:1537. https://doi.org/10.3389/fpls.2018.01537

Eriksson ME, Israelsson M, Olsson O, Moritz T (2000) Increased gibberellin biosynthesis in transgenic trees promotes growth, biomass production and xylem fiber length. Nat Biotechnol 18:784–788. https://doi.org/10.1038/77355

Eudes A, George A, Mukerjee P et al (2012) Biosynthesis and incorporation of side-chain-truncated lignin monomers to reduce lignin polymerization and enhance saccharification. Plant Biotechnol J 10:609–620. https://doi.org/10.1111/j.1467-7652.2012.00692.x

Eudes A, Pereira JH, Yogiswara S et al (2016) Exploiting the substrate promiscuity of hydroxycinnamoyl-CoA: shikimate hydroxycinnamoyl transferase to reduce lignin. Plant Cell Physiol 57:568–579. https://doi.org/10.1093/pcp/pcw016

Ghasemi M, Alexandridis P, Tsianou M (2017) Cellulose dissolution: insights on the contributions of solvent-induced decrystallization and chain disentanglement. Cellulose 24:571–590. https://doi.org/10.1007/s10570-016-1145-1

Gilna P, Lynd LR, Mohnen D et al (2017) Progress in understanding and overcoming biomass recalcitrance: a BioEnergy Science Center (BESC) perspective. Biotechnol Biofuels 10:285. https://doi.org/10.1186/s13068-017-0971-1

Gupta VK, Kubicek CP, Berrin JG et al (2016) Fungal enzymes for bio-products from sustainable and waste biomass. Trends Biochem Sci 7:633–645. https://doi.org/10.1016/j.tibs.2016.04.006

Hamelinck CN, Van Hooijdonk G, Faaij APC (2005) Ethanol from lignocellulosic biomass: techno-economic performance in short-, middle- and long-term. Biomass Bioenergy 28:384–410. https://doi.org/10.1016/j.biombioe.2004.09.002

Hattori K, Arai A (2016) Preparation and hydrolysis of water-stable amorphous cellulose. ACS Sustain Chem Eng 4:1180–1186. https://doi.org/10.1021/acssuschemeng.5b01247

Hodge A, Robinson D, Fitter A (2000) Are microorganisms more effective than plants at competing for nitrogen? Trends Plant Sci 5:304–308. https://doi.org/10.1016/S1360-1385(00)01656-3

Holland N, Holland D, Helentjaris T et al (2000) A comparative analysis of the plant cellulose synthase (*CesA*) gene family. Plant Physiol 123:1313–1323. https://doi.org/10.1104/pp.123.4.1313

Hu H, Zhang R, Feng S et al (2018) Three AtCesA6-like members enhance biomass production by distinctively promoting cell growth in *Arabidopsis*. Plant Biotechnol J 16:976–988. https://doi.org/10.1111/pbi.12842

Huang J, Smith AR, Zhang T, Zhao D (2016) Creating completely both male and female sterile plants by specifically ablating microspore and megaspore mother cells. Front Plant Sci 7:30. https://doi.org/10.3389/fpls.2016.00030

Huang J, Xia T, Li G et al (2019) Overproduction of native endo-β-1,4-glucanases leads to largely enhanced biomass saccharification and bioethanol production by specific modification of cellulose features in transgenic rice. Biotechnol Biofuels 12:11. https://doi.org/10.1186/s13068-018-1351-1

Isikgor FH, Becer CR (2015) Lignocellulosic biomass: a sustainable platform for the production of bio-based chemicals and polymers. Polym Chem 6:4497–4559. https://doi.org/10.1039/C5PY00263J

Jeon HW, Cho JS, Park EJ et al (2016) Developing xylem-preferential expression of *PdGA20ox1*, a gibberellin 20-oxidase 1 from *Pinus densiflora*, improves woody biomass production in a hybrid poplar. Plant Biotechnol J 14:1161–1170. https://doi.org/10.1111/pbi.12484

Jin YL, Tang RJ, Wang HH et al (2017) Overexpression of *Populus trichocarpa CYP85A3* promotes growth and biomass production in transgenic trees. Plant Biotechnol J 15:1309–1321. https://doi.org/10.1111/pbi.12717

Jing ZP, Gallardo F, Pascual MB et al (2004) Improved growth in a field trial of transgenic hybrid poplar overexpressing glutamine synthetase. New Phytol 164:137–145. https://doi.org/10.1111/j.1469-8137.2004.01173.x

Karimi K, Taherzadeh MJ (2016) A critical review of analytical methods in pretreatment of lignocelluloses: composition, imaging, and crystallinity. Bioresour Technol 200:1008–1018

Kattel GR (2019) State of future water regimes in the world's river basins: balancing the water between society and nature. Crit Rev Environ Sci Technol 49:1107–1133. https://doi.org/10.1080/10643389.2019.1579621

Kim KH, Dutta T, Ralph J et al (2017) Impact of lignin polymer backbone esters on ionic liquid pretreatment of poplar. Biotechnol Biofuels 10:101. https://doi.org/10.1186/s13068-017-0784-2

Kim JY, Lee HW, Lee SM et al (2019) Overview of the recent advances in lignocellulose liquefaction for producing biofuels, bio-based materials and chemicals. Bioresour Technol 279:373–384

Klemm D, Heublein B, Fink HP, Bohn A (2005) Cellulose: fascinating biopolymer and sustainable raw material. Angew Chem Int Ed 44:3358–3393

Klocko AL, Borejsza-Wysocka E, Brunner AM et al (2016a) Transgenic suppression of AGAMOUS genes in apple reduces fertility and increases floral attractiveness. PLoS One 11: e0159421. https://doi.org/10.1371/journal.pone.0159421

Klocko AL, Brunner AM, Huang J et al (2016b) Containment of transgenic trees by suppression of LEAFY. Nat Biotechnol 34:918–922. https://doi.org/10.1038/nbt.3636

Klocko AL, Lu H, Magnuson A et al (2018) Phenotypic expression and stability in a large-scale field study of genetically engineered poplars containing sexual containment transgenes. Front Bioeng Biotechnol 6:100. https://doi.org/10.3389/fbioe.2018.00100

Klose H, Günl M, Usadel B et al (2013) Ethanol inducible expression of a mesophilic cellulase avoids adverse effects on plant development. Biotechnol Biofuels 6:53. https://doi.org/10.1186/1754-6834-6-53

Klose H, Günl M, Usadel B et al (2015) Cell wall modification in tobacco by differential targeting of recombinant endoglucanase from *Trichoderma reesei*. BMC Plant Biol 15:54. https://doi.org/10.1186/s12870-015-0443-3

Kumar M, Campbell L, Turner S (2016) Secondary cell walls: biosynthesis and manipulation. J Exp Bot 67:515–531. https://doi.org/10.1093/jxb/erv533

Kumar G, Dharmaraja J, Arvindnarayan S et al (2019) A comprehensive review on thermochemical, biological, biochemical and hybrid conversion methods of bio-derived lignocellulosic molecules into renewable fuels. Fuel 251:352–367. https://doi.org/10.1016/j.fuel.2019.04.049

Latarullo MBG, Tavares EQP, Maldonado GP et al (2016) Pectins, endopolygalacturonases, and bioenergy. Front Plant Sci 7:1401. https://doi.org/10.3389/fpls.2016.01401

Le Moal M, Gascuel-Odoux C, Ménesguen A et al (2019) Eutrophication: a new wine in an old bottle? Sci Total Environ 651:1–11. https://doi.org/10.1016/j.scitotenv.2018.09.139
Lebedev VG, Korobova AV, Shendel GV et al (2018) Effect of glutamine synthetase gene overexpression in birch (*Betula pubescens*) plants on auxin content and rooting in vitro. Dokl Biochem Biophys 480:143–145. https://doi.org/10.1134/S1607672918030043
Li S, Bashline L, Zheng Y et al (2016) Cellulose synthase complexes act in a concerted fashion to synthesize highly aggregated cellulose in secondary cell walls of plants. Proc Natl Acad Sci U S A 113:11348–11353. https://doi.org/10.1073/pnas.1613273113
Li M, Yoo CG, Pu Y et al (2019) Downregulation of pectin biosynthesis gene *GAUT4* leads to reduced ferulate and lignin-carbohydrate cross-linking in switchgrass. Commun Biol 2:1–11. https://doi.org/10.1038/s42003-018-0265-6
Liang Y, Eudes A, Yogiswara S et al (2019) A screening method to identify efficient sgRNAs in *Arabidopsis*, used in conjunction with cell-specific lignin reduction. Biotechnol Biofuels 12:130. https://doi.org/10.1186/s13068-019-1467-y
Ligaba-Osena A, Hankoua B, Dimarco K et al (2017) Reducing biomass recalcitrance by heterologous expression of a bacterial peroxidase in tobacco (*Nicotiana benthamiana*). Sci Rep 7:17104. https://doi.org/10.1038/s41598-017-16909-x
Lindh EL, Bergenstråhle-Wohlert M, Terenzi C et al (2016) Non-exchanging hydroxyl groups on the surface of cellulose fibrils: the role of interaction with water. Carbohydr Res 434:136–142. https://doi.org/10.1016/j.carres.2016.09.006
Liu CG, Xiao Y, Xia XX et al (2019) Cellulosic ethanol production: progress, challenges and strategies for solutions. Biotechnol Adv 37:491–504
Lu H, Klocko AL, Brunner AM et al (2019) RNA interference suppression of AGAMOUS and SEEDSTICK alters floral organ identity and impairs floral organ determinacy, ovule differentiation, and seed-hair development in *Populus*. New Phytol 222:923–937. https://doi.org/10.1111/nph.15648
Luo J, Zhou J-J, Masclaux-Daubresse C et al (2019) Morphological and physiological responses to contrasting nitrogen regimes in *Populus cathayana* is linked to resources allocation and carbon/nitrogen partition. Environ Exp Bot 162:247–255. https://doi.org/10.1016/j.envexpbot.2019.03.003
Madadi M, Penga C, Abbas A (2017) Advances in genetic manipulation of lignocellulose to reduce biomass recalcitrance and enhance biofuel production in bioenergy crops. J Plant Biochem Physiol 5:2. https://doi.org/10.4172/2329-9029.1000182
Madzak C, Mimmi MC, Caminade E et al (2006) Shifting the optimal pH of activity for a laccase from the fungus *Trametes versicolor* by structure-based mutagenesis. Protein Eng Des Sel 19:77–84. https://doi.org/10.1093/protein/gzj004
Mahon EL, Mansfield SD (2019) Tailor-made trees: engineering lignin for ease of processing and tomorrow's bioeconomy. Curr Opin Biotechnol 56:147–155. https://doi.org/10.1016/j.copbio.2018.10.014
Martínez ÁT, Speranza M, Ruiz-Dueñas FJ et al (2005) Biodegradation of lignocellulosics: microbial, chemical, and enzymatic aspects of the fungal attack of lignin. Int Microbiol 8:195–204. https://doi.org/10.2436/im.v8i3.9526
Masclaux-Daubresse C, Daniel-Vedele F, Dechorgnat J et al (2010) Nitrogen uptake, assimilation and remobilization in plants: challenges for sustainable and productive agriculture. Ann Bot 105:1141–1157. https://doi.org/10.1093/aob/mcq028
Mauriat M, Petterle A, Bellini C, Moritz T (2014) Gibberellins inhibit adventitious rooting in hybrid aspen and *Arabidopsis* by affecting auxin transport. Plant J 78:372–384. https://doi.org/10.1111/tpj.12478
McCahill IW, Hazen SP (2019) Regulation of cell wall thickening by a medley of mechanisms. Trends Plant Sci 24:853–866. https://doi.org/10.1016/j.tplants.2019.05.012

Medici A, Szponarski W, Dangeville P et al (2019) Identification of molecular integrators shows that nitrogen actively controls the phosphate starvation response in plants. Plant Cell 31:1171–1184. https://doi.org/10.1105/tpc.18.00656

Millwood RJ, Moon HS, Poovaiah CR et al (2016) Engineered selective plant male sterility through pollen-specific expression of the EcoRI restriction endonuclease. Plant Biotechnol J 14:1281–1290. https://doi.org/10.1111/pbi.12493

Mir BA, Myburg AA, Mizrachi E, Cowan DA (2017) *In planta* expression of hyperthermophilic enzymes as a strategy for accelerated lignocellulosic digestion. Sci Rep 7:11462. https://doi.org/10.1038/s41598-017-11026-1

Mojsov KD (2016) Aspergillus enzymes for food industries. In: New and future developments in microbial biotechnology and bioengineering. Elsevier, Amsterdam. https://doi.org/10.1016/C2014-0-00304-8

Moreno-Cortés A, Hernández-Verdeja T, Sánchez-Jiménez P et al (2012) CsRAV1 induces sylleptic branching in hybrid poplar. New Phytol 194:83–90. https://doi.org/10.1111/j.1469-8137.2011.04023.x

Moura JCMS, Bonine CAV, de Oliveira Fernandes Viana J et al (2010) Abiotic and biotic stresses and changes in the lignin content and composition in plants. J Integr Plant Biol 52:360–376. https://doi.org/10.1111/j.1744-7909.2010.00892.x

Mueller SC, Malcolm Brown R (1980) Evidence for an intramembrane component associated with a cellulose microfibril-synthesizing complex in higher plants. J Cell Biol 84:315–326. https://doi.org/10.1083/jcb.84.2.315

Muro-Villanueva F, Mao X, Chapple C (2019) Linking phenylpropanoid metabolism, lignin deposition, and plant growth inhibition. Curr Opin Biotechnol 56:202–208. https://doi.org/10.1016/J.COPBIO.2018.12.008

Myburg AA, Hussey SG, Wang JP et al (2019) Systems and synthetic biology of forest trees: a bioengineering paradigm for woody biomass feedstocks. Front Plant Sci 10:775. https://doi.org/10.3389/fpls.2019.00775

Nagarajan S, Skillen NC, Irvine JTS et al (2017) Cellulose II as bioethanol feedstock and its advantages over native cellulose. Renew Sustain Energy Rev 77:182–192

Nakano Y, Yamaguchi M, Endo H, Rejab NA (2015) NAC-MYB-based transcriptional regulation of secondary cell wall biosynthesis in land plants. Front Plant Sci 6:1–18. https://doi.org/10.3389/fpls.2015.00288

Niu Y, Wu L, Shen Y et al (2019) Coexpression of β-xylosidase and xylose isomerase in *Saccharomyces cerevisiae* improves the efficiency of saccharification and fermentation from xylo-oligosaccharides. Cellulose 26:7923–7937. https://doi.org/10.1007/s10570-019-02650-3

Park SH, Ong RG, Sticklen M (2016) Strategies for the production of cell wall-deconstructing enzymes in lignocellulosic biomass and their utilization for biofuel production. Plant Biotechnol J 14:1329–1344

Pastor AV, Palazzo A, Havlik P et al (2019) The global nexus of food–trade–water sustaining environmental flows by 2050. Nat Sustain 2:499–507. https://doi.org/10.1038/s41893-019-0287-1

Pereira L, Domingues-Junior AP, Jansen S et al (2018) Is embolism resistance in plant xylem associated with quantity and characteristics of lignin? Trees Struct Funct 32:349–358. https://doi.org/10.1007/s00468-017-1574-y

Petridis L, Smith JC (2018) Molecular-level driving forces in lignocellulosic biomass deconstruction for bioenergy. Nat Rev Chem 2:382–389. https://doi.org/10.1038/s41570-018-0050-6

Quéré C, Andrew R, Friedlingstein P et al (2018) Global carbon budget 2018. Earth Syst Sci Data 10:2141–2194. https://doi.org/10.5194/essd-10-2141-2018

Rocha-Meneses L, Raud M, Orupõld K, Kikas T (2017) Second-generation bioethanol production: a review of strategies for waste valorisation. Agron Res 15:830–847

Rodionova MV, Poudyal RS, Tiwari I et al (2017) Biofuel production: challenges and opportunities. Int J Hydrogen Energy 42:8450–8461. https://doi.org/10.1016/j.ijhydene.2016.11.125

Rodrigues Mota T, Matias de Oliveira D, Marchiosi R et al (2018) Plant cell wall composition and enzymatic deconstruction. AIMS Bioeng 5:63–77. https://doi.org/10.3934/bioeng.2018.1.63

Roque E, Gómez MD, Ellul P et al (2007) The PsEND1 promoter: a novel tool to produce genetically engineered male-sterile plants by early anther ablation. Plant Cell Rep 26:313–325. https://doi.org/10.1007/s00299-006-0237-z

Roque E, Gómez-Mena C, Hamza R et al (2019) Engineered male sterility by early anther ablation using the pea anther-specific promoter PsEND1. Front Plant Sci 10:819. https://doi.org/10.3389/fpls.2019.00819

Ruel K, Nishiyama Y, Joseleau JP (2012) Crystalline and amorphous cellulose in the secondary walls of *Arabidopsis*. Plant Sci 193–194:48–61. https://doi.org/10.1016/j.plantsci.2012.05.008

Seidl PR, Goulart AK (2016) Pretreatment processes for lignocellulosic biomass conversion to biofuels and bioproducts. Curr Opin Green Sustain Chem 2:48–53. https://doi.org/10.1016/j.cogsc.2016.09.003

Shafi A, Gill T, Zahoor I et al (2019) Ectopic expression of *SOD* and *APX* genes in *Arabidopsis* alters metabolic pools and genes related to secondary cell wall cellulose biosynthesis and improve salt tolerance. Mol Biol Rep 46:1985–2002. https://doi.org/10.1007/s11033-019-04648-3

Shen R, Tao L, Yang B (2019) Techno-economic analysis of jet-fuel production from biorefinery waste lignin. Biofuels Bioprod Biorefin 13:486–501. https://doi.org/10.1002/bbb.1952

Singh A, Rodríguez Jasso RM, Gonzalez-Gloria KD et al (2019) The enzyme biorefinery platform for advanced biofuels production. Bioresour Technol Rep 7:100257. https://doi.org/10.1016/j.biteb.2019.100257

Smith RA, Schuetz M, Karlen SD et al (2017) Defining the diverse cell populations contributing to lignification in *Arabidopsis* stems. Plant Physiol 174:1028–1036. https://doi.org/10.1104/pp.17.00434

Speicher TL, Li PZ, Wallace IS (2018) Phosphoregulation of the plant cellulose synthase complex and cellulose synthase-like proteins. Plants 7:52. https://doi.org/10.3390/plants7030052

Srivastava N, Srivastava M, Ramteke PW, Mishra PK (2018) Synthetic biology strategy for microbial cellulases. In: New and future developments in microbial biotechnology and bioengineering. Elsevier, Amsterdam. https://doi.org/10.1016/C2014-0-00304-8

Stanford G, Smith SJ (1972) Nitrogen mineralization potentials of soils. Soil Sci Soc Am J 36:465. https://doi.org/10.2136/sssaj1972.03615995003600030029x

Strauss SH, Jones KN, Lu H et al (2017) Reproductive modification in forest plantations: impacts on biodiversity and society. New Phytol 213:1000–1021. https://doi.org/10.1111/nph.14374

Sun S, Sun S, Cao X, Sun R (2016) The role of pretreatment in improving the enzymatic hydrolysis of lignocellulosic materials. Bioresour Technol 199:49–58. https://doi.org/10.1016/j.biortech.2015.08.061

Terrett OM, Dupree P (2019) Covalent interactions between lignin and hemicelluloses in plant secondary cell walls. Curr Opin Biotechnol 56:97–104. https://doi.org/10.1016/j.copbio.2018.10.010

Tian D, Shen F, Yang G et al (2019) Liquid hot water extraction followed by mechanical extrusion as a chemical-free pretreatment approach for cellulosic ethanol production from rigid hardwood. Fuel 252:589–597. https://doi.org/10.1016/j.fuel.2019.04.155

Tomassetti S, Pontiggia D, Verrascina I et al (2015) Controlled expression of pectic enzymes in *Arabidopsis thaliana* enhances biomass conversion without adverse effects on growth. Phytochemistry 112:221–230. https://doi.org/10.1016/j.phytochem.2014.08.026

Tsai AYL, Chan K, Ho CY et al (2017) Transgenic expression of fungal accessory hemicellulases in *Arabidopsis thaliana* triggers transcriptional patterns related to biotic stress and defense response. PLoS One 12:1–22. https://doi.org/10.1371/journal.pone.0173094

Turner S, Kumar M (2018) Cellulose synthase complex organization and cellulose microfibril structure. Philos Trans R Soc A Math Phys Eng Sci 376:20170048. https://doi.org/10.1098/rsta.2017.0048

Voelker SL, Lachenbruch B, Meinzer FC et al (2011) Transgenic poplars with reduced lignin show impaired xylem conductivity, growth efficiency and survival. Plant Cell Environ 34:655–668. https://doi.org/10.1111/j.1365-3040.2010.02270.x

Wang Y, Fan C, Hu H et al (2016) Genetic modification of plant cell walls to enhance biomass yield and biofuel production in bioenergy crops. Biotechnol Adv 34:997–1017. https://doi.org/10.1016/j.biotechadv.2016.06.001

Wiersum LK (1962) Uptake of nitrogen and phosphorus in relation to soil structure and nutrient mobility. Plant Soil 16:62–70. https://doi.org/10.1007/BF01378158

Wilkerson CG, Mansfield SD, Lu F et al (2014) Monolignol ferulate transferase introduces chemically labile linkages into the lignin backbone. Science 344:90–93. https://doi.org/10.1126/science.1250161

Xi W, Song D, Sun J et al (2017) Formation of wood secondary cell wall may involve two type cellulose synthase complexes in *Populus*. Plant Mol Biol 93:419–429. https://doi.org/10.1007/s11103-016-0570-8

Xiao Y, Poovaiah C, Coleman HD (2016) Expression of glycosyl hydrolases in lignocellulosic feedstock: an alternative for affordable cellulosic ethanol production. Bioenergy Res 9:1290–1304. https://doi.org/10.1007/s12155-016-9766-7

Xiao Y, He X, Ojeda-Lassalle Y et al (2018) Expression of a hyperthermophilic endoglucanase in hybrid poplar modifies the plant cell wall and enhances digestibility. Biotechnol Biofuels 11:225. https://doi.org/10.1186/s13068-018-1224-7

Xie S, Ragauskas AJ, Yuan JS (2016) Lignin conversion: opportunities and challenges for the integrated biorefinery. Ind Biotechnol 12:161–167. https://doi.org/10.1089/ind.2016.0007

Yan Z, Eziz A, Tian D et al (2019a) Biomass allocation in response to nitrogen and phosphorus availability: insight from experimental manipulations of *Arabidopsis thaliana*. Front Plant Sci 10:598. https://doi.org/10.3389/fpls.2019.00598

Yan Z, Hou X, Han W et al (2019b) Effects of nitrogen and phosphorus supply on stoichiometry of six elements in leaves of *Arabidopsis thaliana*. Ann Bot 123:441–450. https://doi.org/10.1093/aob/mcy169

Yu D, Wildhagen H, Tylewicz S et al (2019) Abscisic acid signalling mediates biomass trade-off and allocation in poplar. New Phytol 223:1192–1203. https://doi.org/10.1111/nph.15878

Zhang C, Norris-Caneda KH, Rottmann WH et al (2012) Control of pollen-mediated gene flow in transgenic trees. Plant Physiol 159:1319–1334. https://doi.org/10.1104/pp.112.197228

Zhang N, Li S, Xiong L et al (2015) Cellulose-hemicellulose interaction in wood secondary cell-wall. Model Simul Mater Sci Eng 23:085010. https://doi.org/10.1088/0965-0393/23/8/085010

Zhang J, Xie M, Tuskan GA et al (2018) Recent advances in the transcriptional regulation of secondary cell wall biosynthesis in the woody plants. Front Plant Sci 9:1535. https://doi.org/10.3389/fpls.2018.01535

Zhang K, Yang W, Yu H et al (2019) Double mutation of BRF1 and BRF2 leads to sterility in *Arabidopsis thaliana*. Biochem Biophys Res Commun 516:969–975. https://doi.org/10.1016/j.bbrc.2019.06.138

Zheng L, Gao C, Zhao C et al (2019) Effects of brassinosteroid associated with auxin and gibberellin on apple tree growth and gene expression patterns. Hortic Plant J 5:93–108. https://doi.org/10.1016/j.hpj.2019.04.006

Zhong R, Cui D, Ye Z (2019) Secondary cell wall biosynthesis. New Phytol 221:1703–1723. https://doi.org/10.1111/nph.15537

Modelling Urban Tree Growth and Ecosystem Services: Review and Perspectives

T. Rötzer, A. Moser-Reischl, M. A. Rahman, R. Grote, S. Pauleit, and H. Pretzsch

Contents

T. Rötzer (✉), A. Moser-Reischl, and H. Pretzsch
Chair for Forest Growth and Yield Science, Technical University Munich, Freising, Germany
e-mail: thomas.roetzer@tum.de

M. A. Rahman and S. Pauleit
Chair of Strategic Landscape Planning and Management, Technical University of Munich, Freising, Germany

R. Grote
Karlsruhe Institute of Technology, Institute of Meteorology and Climate Research, Atmospheric Environmental Research, Garmisch-Partenkirchen, Germany

© Springer Nature Switzerland AG 2020
Progress in Botany (2021) 82: 405–464, https://doi.org/10.1007/124_2020_46,
Published online: 28 November 2020

Abstract Climate change in combination with increasing urbanization is a major challenge for our cities. Ecosystem services from the urban green play a significant role in mitigating the negative effects. Urban tree growth models are appropriate tools for the quantification of ecosystem services in some cases in dependence of the plant growth dynamics and of the changing environment. We report about the state of the art in modelling urban tree growth and ecosystem services and describe the background of urban tree growth and the provision of ecosystem services. Furthermore, we present basic growth model principles and describe and compare existing urban tree growth models. Finally we discuss the use of urban tree growth models, uncover advantages and disadvantages of the single urban tree growth models and indicate current limitations and future venues in modelling.

Keywords Climate change, Tree species, Urban green, Urban planning, Urban tree growth models

1 Objectives of Growth Models for Urban Trees

Ecosystem services (ES) of the urban green can also mitigate climate change impacts of cities. Thus, green spaces and especially trees in cities are essential for sustainable and climate-resilient urban futures. They enhance the provision of ecosystem services, including carbon sequestration, reduction of the urban heat island effect, storm water management, improvement of air quality, cultural values and aesthetics, and thus support human health and well-being. It is a challenge, however, to quantify these ecosystem services in highly variable urban environments. Urban tree growth models which include modules for the estimation of ecosystem services are developed and applied to target these objectives. A comparative review of available models is still missing, though.

This review aims to present and assess the state of the art in modelling urban tree growth and ecosystem services. The objectives are (1) to introduce the relationships between urban tree growth in relation to environmental conditions and the provision of ecosystem services as a theoretical background for urban tree growth modelling, (2) to describe basic growth model principles, (3) to compare existing urban tree growth models and (4) to discuss the use of these models, particularly under present and future environmental conditions in cities. This way, advantages and disadvantages of the urban tree growth models are uncovered. Deficits in modelling are described in order to be overcome in future developments of modelling tools.

1.1 Tree Growth and Dimensional Changes

1.1.1 History and Background of Growth Studies

In forestry, growth studies about the size development of single trees and forest stands over time have a long history and are a fundamental tool for growth modelling (Bartelink 1996; Causton 1985; Jarvis and Leverenz 1983). Modelling of tree growth is usually based on measurement of diameter at breast height (dbh). This was stated with the pipe model theory (Chiba 1998; Shinozaki et al. 1964a, b) and the functional carbon balance theory (Mäkelä 1990). They mention that tree compartments such as tree height and crown volume can be derived based on other tree structures such as the dbh. Allometric relationships about the dimensional development of tree compartments like dbh, tree height and crown dimensions are necessary to predict tree growth. They are sufficiently available for forest tree species and forest stands (see, e.g. Cienciala et al. 2005; Pretzsch and Schütze 2005; Takashima et al. 2009; Watt and Kirschbaum 2011; Budhathoki et al. 2008). For urban trees, however, such studies are rare (Aguaron and McPherson 2012; McHale et al. 2009; Yoon et al. 2013).

In recent decades, studies of, e.g. Beatty and Heckman (1981), Dahlhausen et al. (2016), Moser et al. (2015) and Peper et al. (2014a) provided data on the structural development of urban tree species. In the past years, about 20 studies have been published providing equations about the structural development of trees in urban regions (Fig. 1a). A literature review was conducted using Internet search engines on topics such as 'urban tree growth', 'allometry urban trees' and 'growth relationships

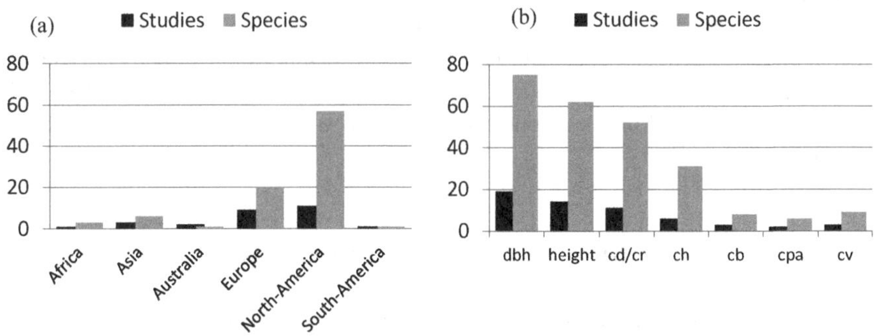

Fig. 1 Number of published studies on allometric relationships and studied tree species within an urban content by continents (**a**) and sorted by the analysed relationship with age (**b**). *dbh* diameter at breast height, *cd/cr* crown diameter and/or crown radius, *ch* crown height, *cb* crown base, *cpa* crown projection area, *cv* crown volume

of urban trees'. Most of these studies have been conducted in North America and Europe. Only a few studies are available for Asia, Africa and Latin America. In total, 74 different tree species have been studied, again mostly from North America and Europe. The studies were based on the development of tree compartments over age and considered in particular the stem diameter at breast height, dbh (Fig. 1b). Data on height, crown radius and crown diameter were also provided in many studies, while other crown parameters such as crown projection area (cpa) or crown height were rarely given.

In forestry, due to the similar conditions in forests of the same type and climate region, allometric equations can be easily transferred to another stand of a similar type within the same climate region. Such a transfer is much more challenging for urban tree equations due to the diverse environmental influences and growing sites of urban trees (McHale et al. 2009). While forest stands are mainly dependent on light and water availability as well as soil type, growing conditions of urban trees vary strongly across a city and depend to a great extent on the often changing growing conditions at a very small scale. Further, the growth response differences of urban trees compared to traditional forest trees may also originate from a lack of competition due to low density, poor soil environments with limited nutrients and soil moisture and the space in the planting pits (Yoon et al. 2013).

Urban trees can be found at very diverse growing sites ranging from park trees, street canyons at main traffic sites, over living areas, to industrial areas and trees growing at main plazas, at parking sites, as well as along railways providing very contrasting growing conditions. Trees growing at a typical street canyon face changed radiation and water conditions compared to trees at a public or private square or when growing in a park or garden (Kjelgren and Clark 1992; Moser et al. 2015). They often suffer from a shortage of water availability, space, oxygen and soil for roots due to small planting pits (Morgenroth and Buchan 2009; Nielsen et al. 2007; Rahman et al. 2013). Further, reduced light and space due to buildings, pruning, de-icing salts and other influences such as dog urine, pollutants and vandalism may further impact tree growth negatively (Beatty and Heckman 1981; Rodríguez Martín et al. 2015). Moreover, depending on the country and the city, urban trees often receive pruning and irrigation, which do not follow a typical management plan of a forest stand. For example, urban trees in Germany are usually irrigated only after planting and in extremely dry years. In other countries with warmer and drier climates, regular irrigation of urban trees during summer is common.

1.1.2 Deriving Tree Growth for Management Purposes

The need of growth equations for urban trees is getting stronger due to several reasons. In the past, urban tree species selection and planting was mostly based on aesthetic features of a species and on the principle of conformity. Today, urban trees are expected to fulfil multiple functions, also called ecosystem services (TEEB 2011), including regulation of the microclimate, carbon sequestration, water

Fig. 2 Growth expected over age

retention, provision of habitat for animals and more (MA 2005; TEEB 2011). Optimizing these functions requires strategically planned planting and management of trees that builds on comprehensive knowledge of their growth and predictive capacity in order to forecast their performance (e.g. Bodnaruk et al. (2017)). Techniques, such as GIS-based databases and online databases, mobile devices and apps with tree information for pedestrians, have facilitated development of tree inventories as well as access to tree data and growth information (e.g. My City's Trees (2019), TreePlotter (2019)). With knowledge about the expected growth of the most common urban tree species (Fig. 2) under various climate and site conditions, tree species selection in cities can be optimized specifically for the site in question, saving money for tree management and replacement.

Detailed growth equations can help to place the right tree species at a certain planting site. Rötzer and Pretzsch (2018), for example, provided data for four common urban tree species in Central Europe and the expected growth in different cities of varying climate and growth conditions. The study contains data sets for cities with dry climates and cities with high amounts of precipitation; expected growth data are given at street canyons, public squares and parks. Based on this study as well as on the study of Rötzer et al. (2019), information on the ecosystem services of these four tree species is provided. This way, it can be decided which tree species fits best for the specific climate of a growing site. In addition, the expected growth rate of the planted tree and its ecosystem services can be derived. Similar data, though not to such high detail regarding different growing sites within a city, are provided by McPherson et al. (2016) for several US cities. The data given by McPherson et al. (2016) are categorized based on the climate region of the cities, making an easy transfer of the growth equations to other cities of similar climate possible. For example, tree species like sycamore *Platanus x acerifolia* Münchh. or *Quercus nigra* L. can develop into very large trees, being able to reach crown diameters of up to 30 m. Moreover, they are categorized as very drought-tolerant species and can be planted in various climate conditions, including those characterized by pronounced periods of drought (Niinemets and Valladares 2006; Roloff 2013). Other species like horse chestnut *Aesculus hippocastanum* L. reach maximum crown diameters of only 18 m. Further, this species has a low drought tolerance and should only be planted in mild climate regions (Niinemets and Valladares 2006; Roloff 2013).

1.1.3 Tree Characteristics for the Calculation of Ecosystem Services

The structural developments of common urban tree species are of interest because of a direct relationship between tree compartments and their ecological benefits (Troxel et al. 2013). Urban green infrastructures, in particular urban trees, provide many ecosystem services for a pleasant microclimate for pedestrians (Bolund and Hunhammar 1999; Rahman et al. 2017). Urban trees can improve environmental conditions by reducing the surface temperature with their shade and by reducing the air temperature through transpiration (Dimoudi and Nikolopoulou 2003; Rahman et al. 2019b). Further improvements are the reduction of rainwater runoff (Xiao et al. 2000a), the air pollutant filtering (Grote et al. 2016), the carbon storage (Nowak et al. 2013) as well as wind and noise buffering (see Sect. 1.2 for more detail). The tree dimensions and their structural development over time are important dynamic input factors to calculate their ES, especially those services that are related to crown size and leaf area. In particular, the leaf area (as leaf area index LAI or leaf area density LAD) is related to the ecosystem services transpiration and shading. Leaf area is a key factor to investigate physiological processes, air pollutant filtering, rainfall interception (Peper and McPherson 1998), photosynthesis (Kramer and Kozlowski 1979), transpiration (McPherson and Peper 2012) and respiration (McPherson and Peper 2012). Several studies (see, e.g. Stoffberg et al. (2008), Rahman et al. (2019b) and Xiao et al. (2000a, b)) found a direct relationship of cooling with the canopy area and the leaf area of a tree.

Moreover, the carbon sequestration of trees – the reduction of atmospheric CO_2 – is closely associated with tree growth and tree biomass accumulation. The older a tree is, the more woody biomass can be produced to store carbon. Many studies, e.g. Maco et al. (2003), Soares et al. (2011) and Nowak and Crane (2002), have quantified the carbon storage of urban trees on a local and regional scale. McPherson (1998), for example, stated that urban forests store approximately half as much carbon as native forests. However, there are only few studies available providing detailed allometric equations for common urban tree species (Yoon et al. 2013). Only recently, studies of McHale et al. (2009), Yoon et al. (2013), McPherson (1998), Golubiewski (2006) and Nowak and Crane (2000) show the relationships of biomass and carbon storage, respectively, with tree structures like dbh. However, the available studies cover only a few species, age classes and climate regions. There are several approaches to close this gap by modelling tools such as i-Trees (Streets), computing the carbon storage of urban forests on a broad scale. These modelling tools, though, need to be based on accurate allometric equations for specific species, ages and climate regions as well, to provide exact estimations of the actual carbon storage of urban trees (McHale et al. 2009; McPherson and Simpson 2001; Pataki et al. 2006). The study of Rötzer et al. (2019) is a first approach to close the missing link between upscaling tools like i-Trees and specific allometric relationships by providing a growth model to estimate the carbon storage of different species for individual trees.

With ongoing climate change and in view of the heat island effects in cities, information on the provided ES is of high importance for a sustainable city planning and the human thermal comfort in cities (e.g. Konarska et al. 2016; Rahman et al. 2015, 2017).

1.2 *Ecosystem Services of Urban Trees*

Ecosystem services can be defined as the components of urban greening that are directly enjoyed, consumed or used to produce specific measureable human benefits (Escobedo et al. 2011) and can be divided into provision, regulation, support and cultural services (Kuyah et al. 2017). Here, we focus more on the regulating services, which can be directly measured under the tree in comparison to the areas without trees, for example, shade and cooling, interception of rainfall, reducing runoff and interception and assimilation of airborne particles. Interestingly all those benefits are also directly linked to tree growth and vitality (Rahman et al. 2014, 2015). For calculating the typical benefits of trees, either researchers have generally relied on small-scale surveys and experiments, or they have used mathematical models to estimate the potential benefits (Ennos et al. 2014).

1.2.1 Cooling Benefits

With the increasing urban heat island (UHI) effect, intensive research has been conducted to understand the impacts of urban trees on the urban microclimate (Bowler et al. 2010; Smithers et al. 2018). The cooling benefits of trees are mainly caused by two main factors. Firstly, the canopies of trees provide shade to the ground level and secondly, with evapotranspiration, trees reduce the amount of heat available to warm the air around them (Rahman et al. 2019b).

1.2.2 Shade Effect

Tree canopies actually reduce the input of short-wave radiation by about 25–90% (Konarska et al. 2016; Rahman et al. 2020). Consequently up to 40°C surface temperature differences between shaded asphalt surfaces under the dense canopies of trees have been reported (Armson et al. 2013b; Rahman et al. 2019b). Through this process, downward long-wave radiation fluxes are also diminished and can create 'refuge' areas for people on hot days. The effective temperature for people, as measured by thermal comfort indices such as the Physiologically Equivalent Temperature (PET) or globe temperatures, can be reduced by 7–15°C under the shade of trees (Matzarakis et al. 1999; Müller et al. 2014). However, surface cooling potential of trees is also dependent on two major variables: firstly, the investigated surfaces over which the surface cooling potential is investigated since the thermal properties of green spaces or different kinds of paved surfaces such as asphalt, concrete and brick pavers are contrastingly different (Rahman et al. 2018, 2019b) and secondly, tree structural characteristics such as tree shape, canopy size and density as well as the features of tree leaves (Georgi and Dimitriou 2010). Among them, canopy density showed the greatest effect in terms of surface temperature reduction (Armson et al. 2013b). Gillner et al. (2015) showed that every unit of leaf area density (LAD) decrease increases asphalt surface temperature by 4.6°C. More explicitly, Rahman et al. (2019b) showed the decrease of grass surface temperature by 3°C and asphalt by 6°C for every unit of increase in leaf area index (LAI).

1.2.3 Transpirational Cooling

Trees can also cool down the air temperature through the process of evapotranspiration (Rahman et al. 2017a; Shashua-Bar et al. 2009), which consumes energy. The direct impact of transpiration on reducing air temperature within or below the tree canopy varies between 1°C and 8°C (Georgi and Zafiriadis 2006; Rahman et al. 2017b; Taha et al. 1997). These consistent results from small-scale studies showed that the magnitude varies depending on vegetation type, size and climatic conditions. According to the laws of gas exchange, the faster the trees can photosynthesize, the faster is the carbon sequestration through their stomata. Hence, they will lose water quicker and provide more transpirational cooling (Ennos et al. 2014). Therefore, the simple measure of tree growth could eventually provide the estimate of the overall transpiration cooling. However, along with the tree structural measures of growth such as height, dbh or crown density (Rahman et al. 2015), species differences due to xylem anatomy or water use efficiency can also influence the regulation of their stomata. Hence, they can have an effect on air temperature amelioration (Moser et al. 2017; Stratopoulos et al. 2018).

Therefore, process-based simulations of tree growth and cooling effects depending on the site and environmental conditions are important. Researches such as Moser et al. (2015, 2016) uncovered the influence of age, species, water and nutrient supply and other site conditions on the allometric relationships while analysing the growth patterns of urban trees. Based on these data, Rötzer et al. (2019) developed the process-based growth model CityTree and calculated annual cooling potentials by transpiration between 21,675 and 51,649 kWh tree^{-1}, which correspond to 59 and 75 W m^{-2}. These values are in line with previous experimental studies of individual *Tilia cordata* Mill. and *Robinia pseudoacacia* L. trees in Munich (Moser-Reischl et al. 2019a; Rahman et al. 2017a).

1.2.4 Runoff Reduction

In urban environments, impervious built surfaces create an extensive area of hardscape and significantly transform the local hydrological cycle (Elliott et al. 2018). Individual street trees can intercept through their leaves and branches between 14 and 44% of rainfall (Livesley et al. 2014; Xiao et al. 2000c), channel water to the base of the trunk (stem flow) between 0.3 and 22.8% (Livesley et al. 2014; Schooling and Carlyle-Moses 2015) and infiltrate most of the remaining amount of the precipitation towards subsurface soil (Armson et al. 2013a; Elliott et al. 2018). Szota et al. (2019) found an average storm water retention by urban trees of on average 18.3% but with a maximum of 43.7%.

Undoubtedly, evapotranspiration is the key component of the urban hydrologic pathway with up to 80% of the annual precipitation (Cleugh et al. 2005; Rahman et al. 2019a). Accordingly, the main contribution of urban trees in terms of reduction of surface water runoff is highlighted as interception and evapotranspiration (Armson et al. 2013a; Zölch et al. 2017). A limited number of studies until now have directly measured the runoff reduction by urban trees (Rahman et al. 2019a). One experimental study in Manchester, UK, showed that a single small *Acer campestre* L. tree planted on the centre of a 9 m^2 asphalt plot could reduce surface runoff of the total catchment area by 62% through water infiltration into the tree pit (Armson et al. 2013a).

Tree growth and structure play a significant role in all mechanisms of storm water runoff reduction. Interception loss can be largely attributable to the tree size, leaf area and density, leaf and branch angle and smoothness (Berland et al. 2017). Transpiration water loss is highly dependent on leaf density and canopy extension (Armson et al. 2013b; Rahman et al. 2015, 2018), drought resistance and growth behaviour (Moser-Reischl et al. 2019a; Moser et al. 2017) and soil infiltration potential to above- and belowground growth rate (Bartens et al. 2009; Rahman et al. 2019a; Rötzer et al. 2017).

1.2.5 Air Pollution Removal

Plants usually remove air pollutants through deposition to their surfaces (foliage, bark) and through their stomatal uptake (Grote et al. 2016). Air pollution concentrations and aerodynamic boundary layer resistance as well as stomatal conductance play a significant role in determining the deposition rates (Wesely and Hicks 2000). Therefore, canopy structure and foliage characteristic (leaf shape, surface properties, physiology) are major determinants of the pollution removal potentials of single trees (Grote et al. 2016).

Particle interception by tree canopies especially at forested vegetative surfaces have been intensively studied and showed significant effect of trees on both gaseous and particle decomposition or interception loss. For instance, a reduction of total black carbon (fraction of PM $\leq$ 2.5 μm) by a single *Acer* and *Quercus* tree of about 12% of the outside concentration was reported (Brantley et al. 2014). However, pollution removal rate varied up to 20-fold based on species characteristics (Janhäll 2015) such as high leaf transpiration in the case of *Tilia* compared to *Platanus* (Dzierzanowski et al. 2011) or the presence of waxes, salts and ions (Altimir et al. 2006). Firstly, the notion of higher canopy density of increasing ecosystem services may not be necessarily true for all, at least for air pollution removal (Tiwary et al. 2016), and the structural variability at leaf level is more important. Secondly, concentrations of urban air pollutants, mostly from vehicular traffic-derived particulate matter, decrease with increasing height (Hofman et al. 2013). Therefore, taller trees with a higher growth rate might not be as effective as slow growing smaller to medium high tree species. At the same time, tree species with faster growth tend to show higher water use efficiency and indirectly reduce their stomatal conductance and finally the chance of stomatal decomposition of the gaseous pollutants.

In addition to the pollution removal, a higher concentration of pollutants might inhibit stomatal conductance and tree growth. On the other hand, trees actually can negatively affect air quality by emission of primary organic particles and biogenic volatile organic compounds (BVOCs) especially under stressed conditions. BVOCs play a role in the formation of O_3, secondary organic aerosols and PM in urban environments (Calfapietra et al. 2013). Consequently, pollen allergens, which are considered as the primary disservice of urban trees, along with urban pollutants might even be more harmful (Beck et al. 2013). Therefore, considering the growth pattern of urban trees in relation to their stress tolerance (especially against heat, drought, wind) is very important to optimize ES and minimize disservices.

1.3 Influence of Present and Future Climate Conditions and Changed Environments

The urban environment has been changed dramatically over the years when compared to the rural surroundings (Bridgeman et al. 1995). These environmental changes are a consequence of the replacement of vegetated surfaces by artificial surfaces, such as buildings, paved squares and roads, as well as the human activities that influence different land uses such as residential, commercial, industrial, recreational and transport. Therefore, the physical make-up of the city in a three-dimensional perspective and land use together are considered as key determinants of environmental conditions in urban areas (Larondelle et al. 2014; Pauleit and Breuste 2011).

In the following, we briefly present some of the main modifications of relevance for urban tree life that may need to be reflected in urban tree models if these should be able to accurately estimate tree growth in the urban environment. In all of this, it needs to be kept in mind that cities can be extremely heterogeneous at various scales. At a broader scale, they are a mosaic of different land uses with their respective configurations of built-up and open spaces such as different types of residential areas consisting of detached houses with private gardens, terraced houses, free-standing blocks of flats with semi-public green spaces or perimeter blocks which are a feature of centres in Central European cities (Pauleit and Breuste 2011). In contrast, in urban areas of other regions, other urban morphology types may prevail, including informal settlements (e.g. Woldegerima et al. (2016)). All of these types are distinct by physical parameters such as the amount of built cover, vegetated areas and single components of green spaces such as tree cover. Urban morphology types (e.g. Pauleit and Duhme (2000); Gill et al. (2008)), the 'HERCULES' classification grounded in the patch concept (Cadenasso et al. 2007) or 'Local Climate Zones' (Stewart and Oke 2012) are some of the approaches to represent this spatial heterogeneity for urban ecological or urban climate studies.

The urban population may also greatly vary in terms of its density, e.g. Schwarz (2010) showed how social and economic characteristics have implications on the use and management of open spaces. At a fine scale, open space types such as streets may differ in width, height of the surrounding buildings, orientation and hence the amount of solar radiation they receive, traffic intensity and more (Gebert et al. 2019). All of this can have an impact on growth conditions of the trees which makes growth modelling an extremely challenging task. The following can therefore only provide some pointers to general features of the urban environment, which, however, can vastly vary within urban areas at a small scale.

1.3.1 Local Climate Modifications

Air temperatures in city centres can increase by 2–3°C on annual average depending on city size but up to 12°C during the night of hot summer days with low wind speeds (EPA 2003). The consequence of this so-called urban heat island effect (Oke 1989, 2011) may be a prolongation of the growing periods in cities of the mid and high latitudes (Jochner et al. 2012; Rötzer et al. 2000) which means longer tree growth in the city. Pretzsch et al. (2017) confirmed an overall positive effect of elevated temperatures in urban areas for trees' growth when compared to their rural counterparts, particularly in the boreal climate zone. However, this effect cannot be seen in temperate and Mediterranean climates, likely because of more severely limited water supply in the urban situations.

Beyond these general features of urban climates, cities are characterized by climatic differences at a fine scale (Oke et al. 2017). Urban microclimates are a consequence of the modifications of the climatic energy balance within the built-up environment. These are changes of incoming and outgoing radiative fluxes due to urban morphology and modifications of airflows. The use of different materials with their respective capacities to reflect solar irradiation, absorb and store heat may further modify the urban climate, as well as the conversion of solar irradiation into latent heat via evapotranspiration. Changes of the amount of solar irradiation are of particular importance for urban tree growth. Built areas can be characterized by sharp contrasts between areas where sun can reach the ground and shady areas, e.g. in very narrow streets or when tall buildings block the sunlight. Information on the physiological impacts of variations in radiation with their effects on microclimatic conditions on urban trees is rare. A recent study, however, in a N-S street canyon in Melbourne showed that within-canyon variability had a greater effect on the photoperiod than the difference between urban and rural sites (Gebert et al. 2019). Peaks in short-wave and long-wave radiation loadings occurred in late morning to early afternoon on the west side but early to mid-afternoon on the east side. Maximum carbon assimilation on both sides of the street was observed for both species included in this study (*Eucalyptus olivacea*, *Olea europaea* L.) at the beginning of the respective photoperiods, but then gradually declined due to increasing water vapour deficits.

1.3.2 Air Pollution

In urban areas the concentration of substances that are considered as air pollutants due to their negative effects on human health is generally elevated and often exceeding the limits recommended by the World Health Organization (WHO 2018). Major pollutants are nitrogen oxides, sulphur dioxide, carbon oxide, ozone and particulate matters with aerodynamic diameter of less than 10 μm (PM10) (Pugh et al. 2012). Strong differences in the composition and concentration of these substances exist between cities depending on air quality standards, prevailing industries, car traffic, house heating systems and ventilation of the city. Green infrastructure, of which trees are an important component, can contribute to reduce air pollution levels, but this effect is highly context dependent. It can also reduce air quality, e.g. if trees block ventilation in highly trafficked streets or via the emission of biogenic volatile organic compounds (see respective section in this paper and Hewitt et al. (2019)).

Negative effects of air pollution on tree growth have been extensively researched (Däßler 1991). The impacts of air pollution on urban trees have been less studied and disentangled from other growth factors. Therefore, there is a general lack of quantitative information on air quality/urban tree health relations. However, the concentration of fine particles (PM10) explained 41% of the interannual variability of tree ring increment of one species (*Tipuana tipu* (Benth.) Kuntze) in São Paulo. Metals such as Al, Ba and Zn also showed a negative effect on the growth of this species (Locosselli et al. 2019). In a study from China, on the other hand, it was shown that ozone concentrations as encountered in urban areas of this country may significantly reduce the growth of common urban tree species (e.g. *Platanus orientalis L.*, *Robinia pseudoacacia L.*) (Gao et al. 2016), while an Indian study showed that the deposition of dust on leaf surfaces negatively affected their functionality. Tree species had a different sensitivity to dust deposition (Jeet Chaudhary and Rathore 2019). In contrast, higher concentrations of nitrogen oxides had no negative effects on growth of *Abies sachalinensis* (Mast.) in Sapporo, Japan, when compared with rural stands (Moser-Reischl et al. 2019b).

1.3.3 Urban Soils

Due to the long history of human settlement, sometimes extending over thousands of years, soils within urban areas have changed completely (Sauerwein 2011). Huge variations of soils in the urban environment make generalizations very difficult, though. The current surface can be several metres above the original level due to frequent destructions of buildings because of wars, fires, natural catastrophes, etc. Top soils have been removed for buildings and paved spaces such as squares and streets, while the composition of remaining soils has been modified, e.g. by adding fertilizer or heavy metals. Content of organic substances may be strongly reduced as is the life of organisms in the soil whether earthworms, bacteria or fungi. Soil structure is often characterized by a higher content of coarse materials, meaning a lower water holding capacity, while soil pH is often elevated due to alkaline leachates from building debris and waste, reducing nutrient availability (Jim 1998). Moreover, urban soils are often heavily compacted from construction activities (Randrup 1997), reducing soil aeration and moisture (Morgenroth and Buchan 2009).

Not least, toxic substances in soils can negatively affect tree growth. In northerly countries with cold winters, the use of de-icing salt (normally sodium chloride) on roads is a major problem for trees along streets (Dobson 1991). High salt concentrations may lead to crown dieback, and leaves, shoot and branches can be damaged very fast over a few months, while also the entire tree may die off, particularly when it is a young or a newly planted tree (Dobson 1991). However, probably the major change of soils in urban areas is surface sealing which means not only the complete removal of topsoil but also the inhibition of aeration and water infiltration (Scalenghe and Marsan 2009). Moreover, re-radiation of heat from sealed surface can lead to higher transpirational water demand and crown temperatures. Not least, temperatures of open spaces with a high degree of sealed soils are higher than for green open spaces. For instance, the comparison between two squares in the City of Munich showed an average temperature difference of 2.8°C during end July to end October, with elevated temperatures being one of the significant drivers for increased sap flux density in the studied narrow-leaved lime trees (*Tilia cordata* Mill.) (Rahman et al. 2017).

For the urban areas in the conterminous United States, average percentage of sealed soils was estimated at 17.5% with significant variation between the urban areas of US states (Nowak and Greenfield 2012). Comparative data on soil sealing for Europe's urban areas does not exist. For single cities it has been estimated that the average degree of soil sealing within the area of the city proper (i.e. excluding surrounding open spaces such as farmland within administrative boundaries) can be around 30–50% of the surface area (e.g. Artmann (2013) for Munich, Germany; Gill et al. (2008) for Greater Manchester, UK). Values may be above 80% in densely built city centres (Gill et al. 2008; Pauleit and Duhme 2000). However, for tree growth, the site scale and in particular the rooting conditions are of greater importance (Pauleit 2003; Rahman et al. 2011). Especially in streets, the growth of tree roots is often confined to small volumes of soils of few cubic metres which will translate into reduced growth.

Overall, negative effects of soil sealing and soil compaction on trees have been reported in several studies (e.g. Rahman et al. (2011); Watson et al. (2014)). Effects on tree growth have not always been conclusive, though. In a 4-year experiment, soil compaction treatments underneath pin oak (*Quercus palustris* Münchh.) trees did reduce oxygen diffusion rates in the upper soil levels; however, fine root growth was not significantly affected (Watson and Kelsey 2006). Similarly, Fini et al. (2017) did not find significant differences in root growth of newly planted trees (*Celtis australis* L. and *Fraxinus ornus* L.) underneath soil cover types (impervious, pervious, porous and control), while Volder et al. (2014), Viswanathan et al. (2011) and Weltecke and Gaertig (2012) established negative effects in their respective studies. Recently, Stratopoulos et al. (2019b) could show in a rainfall exclusion experiment with nursery trees that response of three species to induced drought was different and related to their investment into root growth as compared to shoot growth.

1.3.4 Hydrology

The replacement of natural soils by artificial surfaces of buildings, paved streets and squares strongly alters the natural hydrology of river catchments (Bridgeman et al. 1995). It increases the quantity of storm water runoff at the surface, from where it is mainly led to sewage systems (Illgen 2011). Consequently, water infiltration into the soils and groundwater replenishment are strongly reduced which can lead to less water being available for trees and the lowering of the water table. Such quantitative changes in hydrology are dependent on urban morphology and the local patterns of precipitation regarding annual distribution of rainfall amounts and intensities. For the City of Leipzig, it has been estimated that the surface runoff increased to 282% in 2003 when compared with the situation in 1870 (within today's administrative boundaries, still including a significant amount of farmland) (Haase 2009).

Irrigation can increase the amount of water available for the green areas. Irrigation of urban trees is a widespread measure in the hotter and more arid regions, but more rarely applied to the trees in temperate zones (except for establishment of trees after planting). Yet, even in the temperate regions, trees may experience lack of water when rooting space is limited. In an experiment in the Scandinavian city of Copenhagen street, trees still positively responded with increased growth even under the highest application of water indicating that the limits of optimum water supply had not yet been reached despite being applied in a period with above-average precipitation (Bühler et al. 2006). Maximum irrigation also extended the growth period of the trees by 7 days.

1.3.5 Physical Stress

Further factors may influence the growing environment for trees in urban areas. For instance, physical impacts that result from injuries, e.g. of the bark through cars that bump against the trunk when parking or root injuries caused by diggings, e.g. for utility maintenance in streets can have severe negative consequences for tree growth. However, studies that would quantify these effects for the stand of trees in a city are not known to the authors. Further on, tree pruning needs to be mentioned (Clark and Matheny 2010). Trees in streets and on squares need to be pruned to uplift the crown so that, e.g. lorries can pass underneath (approx. 4–4.5 m in height). Pruning may also be necessary when trees get too close to buildings or aboveground electricity lines. Not least, trees are pruned for aesthetic purposes, e.g. to give the crown the shape of a cube. All of these measures can have more or less strong impacts on the form of the crown and the amount of biomass and leaves in it and hence have a feedback on tree growth. For instance, removing the leader shoot showed marked effects on growth patterns (reducing stem diameter growth, among other) as well as increasing leaf area at the expense of leaf mass per area in an experiment with Norway maple (Fini et al. 2015).

In this section, it has been shown that the urban environment is a very special habitat for trees that differs largely from the natural habitats where most of the trees have evolved, i.e. woodlands. However, huge variation exists across the urban landscape. Streets and paved squares, rooftops and underground car parks are mostly characterized by strongly limiting growth conditions for trees. Bigger gardens and public green spaces such as parks and cemeteries may offer excellent tree habitats, on the other hand. Thus, it is in particular the site scale that determines growth conditions for urban trees (Pauleit 2003). Therefore, much energy is spent nowadays on finding suitable tree species for tough urban sites (e.g. Roloff et al. (2009); Saebo et al. (2003); Sjöman et al. (2012); Stratopoulos et al. (2019a)). Simultaneously, ways to improve site conditions, e.g. by adequate provision of soil volume, enhancing substrates, testing water pervious surfaces, combining tree trenches with storm water management and more, are explored (e.g. Alvem and Bennerscheidt (2009); Fini et al. (2015)).

1.3.6 Climate Change

Climate change will lead to further changes of growth conditions for urban trees. Regional variations of climate change make it difficult to make general statements; however, it seems to be safe to say that annual average air temperatures will continue to increase in the very majority of cities. The global analysis of urban trees in the study by Pretzsch et al. (2017) implies that recent climate change has accelerated tree growth both outside and inside urban areas but that positive effects have been more pronounced within urban areas. However, the study also clearly indicates that trees can only benefit from warming if water availability is not constrained. Therefore, it is likely that trees in already hot climate with extended periods of drought during the growing season such as in the Mediterranean will rather suffer from climate change as scenarios suggest even hotter and drier summers.

1.3.7 Tree Planting

In all of this, it should not be forgotten that planting styles may influence the availability of resources and hence tree growth. While trees in woodland habitats experience competition between the trees for light, nutrients and water, the situation is different for single tree plantings on squares or in gardens. The planting styles will certainly influence rate of tree growth, whether a tree invests more into its height or expands the crown laterally.

Summing up, trees planted in urban areas are growing in environments that are often very different from their native woodland habitats. Trees clearly respond to the urban environment concerning their growth. Therefore, measurements taken in natural habitats or managed forests cannot be applied to urban areas. Modelling of tree growth and ES provided by urban trees need to build on growth data obtained from urban trees. They also have to take into account urban environmental parameters. As has been shown in this section, the particular challenge lies in the huge variability of the urban environment that needs to be reflected in urban tree modelling to provide accurate estimates and plausible predictions of, e.g. development of tree size over time, diameter increment, storage of biomass and carbon or cooling via shading and transpiration.

2 Growth Model Principles and Allometric Relationships

Dale et al. (1985) reviewed growth models developed since the 1960s and made a distinction between (1) 'forest growth models' which employ empirical functions to simulate (potential) dimensional changes in dependence to site-specific resource availability and some kind of competition measure and (2) 'community models' that are based on assimilation and loss functions of carbon and allocate net gain into dimensional growth generally using some kind of allometric relationship.

According to Schenk (1996), the approaches can be differentiated into those:

(a) Using a sole empirical growth modification
(b) Using physiological process descriptions combined with empirical growth equations (hybrid approaches)
(c) Deriving also biomass and dimensional changes (process-based) from carbon budget calculations (mostly restricted to simple forest structures)

Further reviews presented specific approaches and suggest differentiations by temporal and spatial resolution of processes and the detail by which tree structural changes were described (e.g. Constable and Friend (2000); Le Roux et al. (2001); Mäkelä (2003)).

Urban tree growth models can principally belong to any of these approaches (empirical, hybrid, fully process-based). However, since so far biomass acquisition is generally considered less important in urban trees than in forests, mostly empirical approaches are employed to estimate growth development. Another reason for the relatively simple representation of growth in urban tree models is that these trees do not develop in homogeneous forests but grow under individual conditions, often without direct competition of other trees or in very heterogeneously structured, multi-species groups. An estimate of individual dimensional change in dependence on specific environmental conditions is therefore highly demanding regarding the initialization and provision of driving variables. Nevertheless, addressing shading (leaf area), cooling (transpiration) and air pollution removal (leaf area, stomata behaviour), properties and processes need to be considered that require to consider some relation to physiology. Thus, physiological processes that are already accounted for could potentially be linked to the dimensional development that is why we will give a short overview about how tree growth can be classified into and described with various approaches. However, in order to be responsive to a changing environment, a growth model needs to be to some degree mechanistic, i.e. based on physiological processes that interact or build upon each other and are related to temperature and other factors.

Tree growth can be differentiated into dimensional and mass changes. Dimensional changes include in particular height and crown size (i.e. crown width), which are important determinants for shading and air flow. They are further needed to determine if trees may get in conflict with obstacles such as buildings. Other dimensions that can be considered are rooting depth and the tree trunk diameter at 1.3 m height which is called diameter at breast height (dbh). This diameter is easy to measure and thus commonly used for building and evaluating models. It is also suitable to derive relations to other dimensional tree variables which are less accessible. These variables are the height of the tree (height-diameter ratio, hdr) and the width of the crown (crown-stem diameter ratio, cdr) and are depicted in Fig. 3. Together with tree height, dbh can also be used to calculate the volume of the tree trunk.

The consideration of biomass is particularly important for the determination of ecosystem services such as cooling by transpiration, air pollution removal and carbon sequestration. The latter is defined by wood accumulation in the stem, branches and coarse roots and is closely related to the stem volume growth. To derive woody biomass from volume, fixed wood density and fixed ratios between woody compartments are most commonly assumed although this would not be true for longer observation periods, in particular if young trees are involved. This is similarly true for foliage biomass, respectively, foliage area, which is of particular importance for physiological exchange processes such as transpiration or pollution removal. It can be calculated on a tree basis assuming empirical relationships to stem dimensions, i.e. dbh, as well as to crown size variables. Biomass compartments of trees can also be mechanistically derived (see below). In this case, other biomass

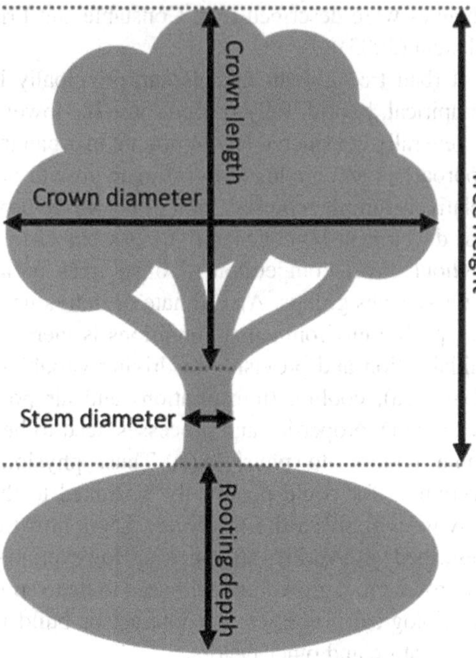

Fig. 3 Dimensional parameters of an individual to derive relationships and biomass

fractions have to be considered to close the carbon balance. Besides wood and foliage, the most common fractions are fine roots and a reserve compartment that can be used to supply respiration requirements or fruiting.

According to the classifications described above, approaches to represent the increase in dimension and biomass of urban trees in models can be distinguished into those that are (1) using empirical functions for dimensional changes and deriving biomass from the respective dimensions and (2) calculating biomass increases based on physiological processes and using these to derive respective dimensional changes.

Empirical growth functions and biomass relations are literally based on data that have been taken in the past, usually averaging over a range of environmental conditions. Simple estimations of annual growth are mostly using the average diameter and height increment determined from tree ring and main-stem node measurements (e.g. Nowak (1994)). Other models use polynomials, exponential or logarithmic relations to describe the age-diameter relation as well as other allometric ratios (e.g. Peper et al. (2014b); McPherson et al. (2016)), possibly modified with further empirically derived constraints (e.g. 'urbanity', Moser-Reischl et al. (2018)). It should be noted that relationships derived from forests seem to fail describing the allometry of urban trees (McHale et al. (2009); Timilsina et al. (2017)). The need for a specific 'urban' parameterization is only partly related to the smaller competition at urban than at forest sites. The larger part of the variability observed in urban tree allometry origins from specific management such as pruning (Fini et al. 2015) and from the complicated pattern of environmental conditions within the city. For example, light competition and temperature have been shown to clearly affect the relations between diameter and height or stem volume (e.g. Franceschini et al. (2016)). Also, the rate of growth depends on the available resources (e.g. Kjelgren and Clark 1992; Dahlhausen et al. 2017). Although an empirical model could capture these dependencies if a sufficient degree of stratification can be achieved, accounting for a future change in environmental conditions is not possible.

The necessity to enable a closer relation to environmental conditions suggests the development of process-based models for tree growth, which have been established for the same reason in order to represent forest development. These models have separate functions to calculate (1) carbon assimilation (or photosynthesis); (2) gaseous loss of carbon (or respiration); (3) distribution (allocation) of carbon within the plant, i.e. into leaves, fine roots and woody tissue (branches, stem and coarse roots) as well as reserves or fruits (not always considered); and (4) senescence of the different tissues. For each of these processes, various principles have been proposed; for more details, the readers are referred to respective reviews (e.g. Landsberg and Sands (2010); Fontes et al. (2010)).

All of the described physiological processes are – or can be described as – dependent on environmental conditions. This is most obvious for photosynthesis, which clearly depends on local light and temperature conditions but is also influenced by CO_2 concentration, water and nutrient availability (Bernacchi et al. 2013). Carbon gain in urban areas would thus be expected to be higher as at nearby rural sites because of higher CO_2 concentrations and temperatures, but might be restricted by shaded conditions, e.g. in the vicinity to high buildings. Furthermore, air pollution effects, heat stress and a shortage of nutrient and water could be considered as limiting factors for carbon assimilation. Losses by respiration generally increase exponentially with temperature and linearly with nutrient supply (Cannell and Thornley (2000), although this representation might be too simple because it neglects that the effects might be dampened by acclimation processes (Atkin et al. 2005). The distribution of assimilated carbon might be calculated without any direct environmental dependency, for example, if it is allocated by fixed fractions or following hierarchical principles (Franklin et al. 2012; Litton et al. 2007). However, the application of sink-driven approaches or the functional balance principle will relate biomass growth to weather and chemistry conditions. For example, leaf flushing occurs in dependence on cumulative temperature and thus happens earlier under climate warming (e.g. Chmielewski and Rötzer 2001; Gunderson et al. (2012), and allocation to roots might be preferred over aboveground compartments if soil resources are limited (and vice versa) (Rötzer et al. 2009; Stratopoulos et al. 2019a). Other possible impacts are that allocation into specific compartments might be prevented or enhanced under specific conditions (e.g. low temperatures or drought). Also, senescence could be represented by simple parameterizations of leaf, fine root or sapwood longevity. However, tissue turnover also depends on stress conditions which might need consideration, particularly if trees are exposed to air and soil pollution or heat and drought extremes (e.g. Kikuzawa et al. (2013)).

A physiologically based model provides the woody biomass that then determines stem height and diameter growth, as well as the dimension of crown and root systems. Therefore, it can be combined with empirical functions representing a so-called hybrid approach as has been mentioned above (Girardin et al. 2008; Landsberg 2003). This combination describes the pathway of dimensional relationships, while the biomass growth determines the speed of the development. However, a more mechanistic approach requires to also vary dimensional relations such as hdr and cdr in dependence on individual environmental conditions. An early approach to account for competition (and thus resource availability) was proposed by Bossel (1996), which only accounted for two states (i.e. crown coverage is larger or smaller than ground area). More recent developments suggest to apply continuous measures of competition or light availability, such as a 'light resource index' that is based on tree density and size within a 'domain of influence' (Seidl et al. 2012). A comprehensive approach that is reliably covering varying environmental conditions throughout the lifespan of a tree is, however, still not available.

Urban trees are already exposed to environmental conditions that are expected to develop with ongoing climate change in other areas, i.e. increasing CO_2 concentration and temperatures (Lahr et al. 2018). In fact, these conditions seem to be responsible for an increase in biomass (Gregg et al. 2003) as well as tree dimension (Pretzsch et al. 2017) in urban areas. Also, flushing has been shown to occur earlier, probably contributing to these developments (Jochner et al. 2012; Rötzer et al. 2000). However, city conditions can also be stressful, providing less nutrients and water resources but higher air and soil pollution levels. Both beneficial and stressful conditions are highly variable and are expected to change further with climate alterations, changes in emission regimes and the implementation of pollution mitigation strategies. This further increases the need of suitable urban tree growth models that account for these dynamic environments.

3 Modelling Tree Allometry

3.1 Tree Shape as a Result of Inner and Outer Determinants

The genetically determined shape of a tree under ideal conditions (genotype) can be modified by environmental conditions (Poorter 1999). Result of this modification is the phenotype. Latter is the result of inner and outer determinants of the genotype and phenotype (Fig. 4). In forests and parks, the shape of crown, root and stem is modified mainly by the competition with the neighbouring trees for resources. In urban areas, the tree shape is more determined by components of the inanimate environment like houses, walls, fences and pavement. Crowns can grow towards the light; roots can reach even further towards water and nutrients as they have lower mechanical restrictions than branches. The stem of a tree can grow slimmer, cylinder-shaped, but unstable in order to reach more light. With decreasing competition or resource restriction, the stem gets more conic and stable.

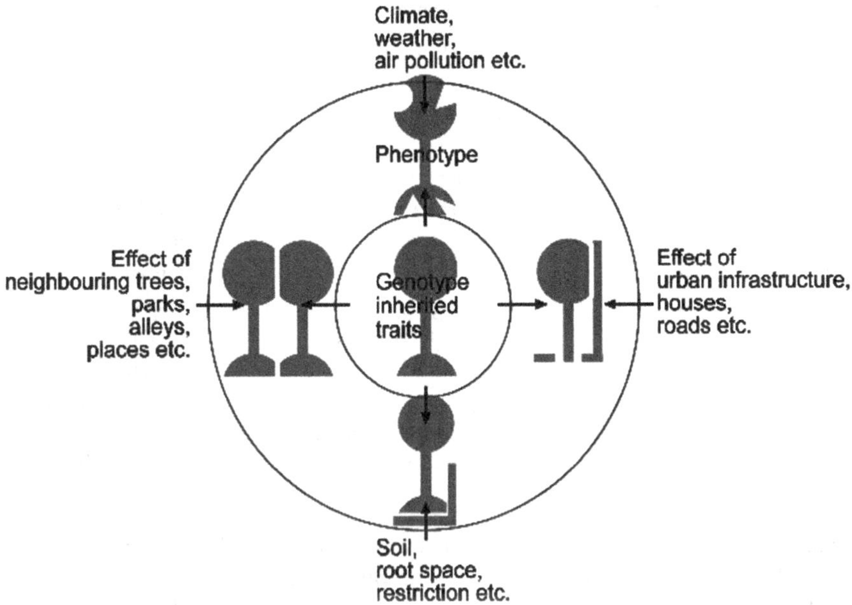

Fig. 4 Genotype and phenotype. Tree form and shape as a result of inner and outer determinants. Urban tree shape can be significantly determined by air pollution, infrastructures or root space restriction

Tree shape can be influenced by neighbouring trees and buildings. Wind can set trees in vibration in a way that they suffer mechanical abrasion of branches by neighbouring trees or buildings (crown shyness). The potential growth of the roots is often limited by stones, pavement or walls. If trees are pruned and trimmed, they can completely lose their natural shape; examples are sphere-shaped trees in parks or continuously cut trees under electric power lines. All this causes a broad variation of tree shapes and challenges for modelling and prognosticating it. However, fortunately, trees follow some basic allometric rules when increasing in size (Pretzsch 2010b).

3.2 The Allometric Concept

The shape development of organisms with their increasing size can be described and modelled by allometric rules (von Bertalanffy 1951). For the shape proportions between two dimensions x and y of an organism (e.g. between dbh and tree height), the allometric equation $y = a\ x^{\alpha}$ is a proven model (Huxley 1932; Teissier 1934; with $a =$ constant, $\alpha =$ exponent). It is often used in its double-logarithmic representation $\ln y = \ln a + \alpha \times \ln x$. However, the suitability for quantifying the relative growth velocity between two organs and the growth partitioning between

Table 1 Overview of primary allometric relationships for the ideal allometric plant and stand

Allometry	Exponent	Dependent variable	Independent variable
$h \propto d^{2/3}$	$\alpha_{h,\ d} = 2/3$	Tree height, h	Stem diameter, d
$v \propto d^{8/3}$	$\alpha_{v,\ d} = 8/3$	Stem volume, v	Stem diameter, d
$cr \propto d^{2/3}$	$\alpha_{cr,\ d} = 2/3$	Crown radius, cr	Stem diameter, d
$cpa \propto d^{4/3}$	$\alpha_{cpa,\ d} = 4/3$	Crown projection area, cpa	Stem diameter, d
$cv \propto v^{3/4}$	$\alpha_{cv,\ v} = 3/4$	Crown volume, cv	Stem volume, v
$la \propto d^{2}$	$\alpha_{la,\ d} = 2$	Leaf area, la	Stem diameter, d
$la \propto m^{3/4}$	$\alpha_{la,\ m} = 3/4$	Leaf area, la	Stem mass, m
$ms \propto mr$	$\alpha_{ms,\ mr} = 1$	Shoot mass, ms	Root mass, mr
$v_q \propto N^{-3/4}$	$\alpha_{v_q,N} = -3/4$	Volume mean stem, vq	Tree number, N
$N \propto d_q^{\ -2}$	$\alpha_{N,d_q} = -2$	Tree number, N	Mean stem diameter, dq

The relationship $h \propto d^{2/3}$ between tree height and stem diameter means that a diameter increase by 1% is linked with a height increase by 2/3 = 0.67%. An increase of tree mass by 1% is linked with an increase of leaf area by 3/4 = 0.75% (for explanation of the tree variables, see Fig. 7)

them is better reflected by its differentiated version: Between the growth rates *x*' (*dx*/*dt*) und *y*' (*dy*/*dt*) applies $dy/y/dx/x = \alpha$ (von Bertalanffy 1951).

The relative growth velocity between *x* and *y* is represented by the allometric exponent α; α is the exponent in $y = a\ x^{\alpha}$ and quantifies the shape expansion. It can be interpreted as partitioning or allocation ratio formula of the resources and growth between the organs *x* any *y* or the organ *x* and the body as a whole. It indicates that if *a* increases by 1%, y increases by α %. For instance, $\alpha_{\mathrm{h,d}} = 2/3$ means that height increases by 0.6% if the stem diameter increases by 1%.

Interestingly the shape development of plants and animals often follows overarching allometric rules and allometric exponents. Fundamental is, e.g. the relationship between the plant mass and its leaf area; the latter represents the plant's metabolism and biosynthesis. According to metabolic scaling, leaf area, la, and plant mass, m, follow generally the allometry $la \propto m^{3/4}$ (Enquist and Niklas 2001; Kleiber 1947; West et al. 1997). This means that leaf area increases just underproportionally with plant mass and causes the striking accumulation of standing stock in forest stands although their leaf area is limited. If plant mass increases by 1%, metabolism increases just by 3/4%. This 3/4 scaling assumes fractal scaling and inner packing of tubes and pipelines for matter transport in plants (West et al. 1997). Allometry further assumes that between diameter d and leaf area applies $la \propto d^2$ and between diameter and stem mass $m \propto d^{8/3}$ (see Table 1).

Former allometric models assumed geometric scaling, e.g. quadratic relations between leaf area and diameter, $la \propto d^2$, but also a cubic relationship between tree mass and diameter, $m \propto d^3$. They supposed according to Rubner (1931) that $la \propto m^{2/3}$. Figure 5 substantiates for trees the fractal 3/4 scaling for primary tree species in Europe. Based on 508 tree analyses, it shows the allometric exponent $\alpha_{\mathrm{la,\ m}} = 0.74 \pm 0.016$ for the relationship between leaf area and tree mass of Norway spruce (*Picea abies [L.] Karst.*), Scots pine (*Pinus sylvestris* L.), European beech (*Fagus sylvatica* L.) and common/sessile oak (*Quercus robur L., Quercus petraea*

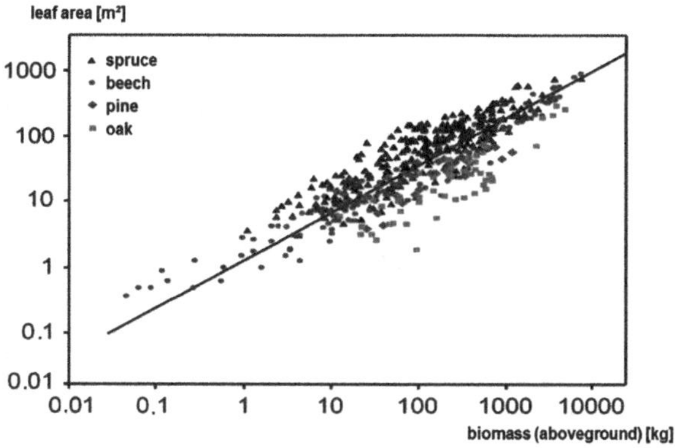

Fig. 5 Metabolic scaling of leaf area versus aboveground total plant mass mt $la \propto m^{\alpha_{la,m}}$ with overarching scaling exponent close to $\alpha_{la,\ m} = 3/4$. Pooled data of Norway spruce ($n = 280$), European beech ($n = 145$), Scots pine ($n = 31$) and sessile oak ($n = 52$). SMA regression analysis of leaf area versus total aboveground mass m yields $\alpha_{ml,\ mt} = 0.83 \pm 0.020$. Analogous estimation of scaling between leaf area, la, and plant mass, m, yielded $\ln(la) = -0.0176 + 0.7418 \times \ln(m)$ and an allometry exponent of $\alpha_{\mathrm{la,m}} = 0.7418 \pm 0.016$ (Pretzsch et al. 2012b)

(MATT.) LIEBL.). This means the empirical analysis indicates an exponent which is close to the ideal allometric plant according to the metabolic 3/4 scaling.

The allometric theory assumes general structural and functional relationships between the organs of organisms, e.g. that the relationship between leaf area and plant mass is $la \propto m^{3/4}$ (Enquist and Niklas 2001; Kleiber 1947; West et al. 1997) and the relationship between leaf area and dbh $la \propto d^2$ (Shinozaki et al. 1964a). So that the link between plant mass m and diameter results as $m \propto d^{8/3}$ (Niklas 1994; West et al. 1997, 2009). Table 1 gives an overview of some primary allometric relationships for the ideal allometric plant and stand that might be interesting for plant modelling (Pretzsch et al. 2002).

3.3 The Allometric Corridor

The concept of allometric partitioning (= allometric partitioning theory APT) assumes general and constant allometric relationships as shown in Table 1 for the allometric ideal plant (Enquist et al. 1998; Enquist and Niklas 2001; West et al. 1997). The concept of optimal resource and growth partitioning (optimal partitioning theory OPT) on the other hand assumes that those organs that are growth limiting are promoted in resource partitioning and size (McCarthy and Enquist 2007). The latter means that roots are promoted when water is limiting, whereas crown growth might be accelerated when light is limiting a plant's growth. Figure 6 shows that the two concepts can be combined (Pretzsch 2010a). Under optimal growing conditions, a plant may develop like an allometric ideal plant on the trajectories (black line in the middle) given by overarching exponents (Table 1). However, depending on its

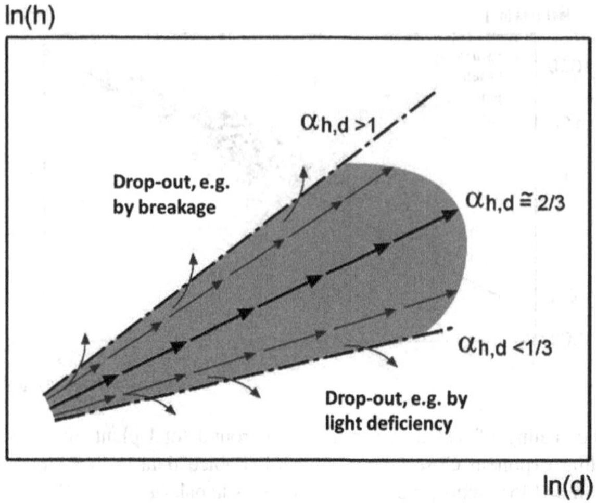

Fig. 6 Corridor of the allometric shape development by example of the height-stem diameter allometry. Theoretical trajectory of the allometric ideal plant under optimal growing conditions (black line with arrows in the middle). Realized h-d trajectories (grey lines with arrows) inside the allometric corridor (broken lines) as a result of adaptation to environmental conditions. Developments out of the allometric corridor (grey bended arrows) indicate drop out and failure

resource limitation and impediments, as shown in Fig. 4, it may deviate from the ideal trajectory according to its species-specific morphological variability (grey trajectories in the allometric corridor).

Optimal resource supply may enable, e.g. an ideal h-d development following $h \propto d^{2/3}$, $\alpha_{h,\,d} = 2/3$ as shown in Fig. 6 (black line with arrows in the middle). As a reaction on lateral restriction, the tree may promote the height growth on the expense of diameter growth ($\alpha_{h,\,d} > 2/3$) in order to keep access to light and remain in the play. In the understory, when in a sit-and-wait position, the height growth may be reduced, while the diameter still increases ($\alpha_{h,\,d} < 2/3$). If the height gets in relation to the diameter too large or small, the plant may fail due to stem breakage or shortage of light. This means, ideally, the plant follows the allometric trajectory; under suboptimal conditions, it deviates and adapts. If the growing conditions deviate too much from the required species-specific growing conditions, the plant may transgress the borders of the allometric corridor, fail and drop out of the population. The morphological plasticity allows the plant to adapt to spatially or temporally varying growing conditions. By the feedback loop resource availability and uptake → growth allocation → tree shape → resource availability and uptake, the plant can adapt even to unusual growing conditions in urban environments. However, the reference remains the allometric shape development reflected by overarching allometric rules.

3.4 Tree Allometry

3.4.1 Characteristics of a Tree's Crown Morphology, Crown Extension and Growing Area

The description and analysis of tree structure, growth and growth efficiency are mainly based on the tree variables shown in Fig. 7. To compare tree morphology, the following ratios between tree organ sizes are frequently used:

h/d = the h-d ratio using h (m) and dbh (cm) addresses tree stability.

cd/d = the cd-d ratio with crown diameter cd (m) and dbh (cm) addresses the crown extension.

dr/dbh = dr-dbh ratio with diameter of the main root, dr, and dbh quantifies the root-stem relationship. When based on increment cores from stem and main roots, this relationship provides insight into the allocation principle between stem and main root (Fig. 4).

cpa/sa = this ratio between crown projection area, cpa, and tree stand area, sa, is a measure for the interlocking of neighbouring crowns.

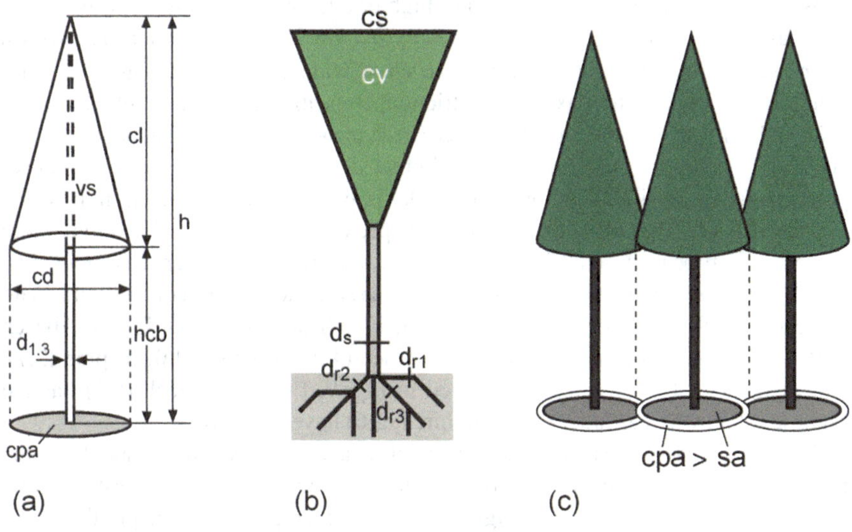

Fig. 7 Dendrometrical tree characteristics: (**a**) tree height, h; crown length, cl; height to crown base, hcb; crown diameter, cd; tree diameter at breast height, $d_{1.3}$; and stem volume, vs. Variable cpa represents the crown projection area. (**b**) Crown volume, cv; crown surface area, cs; diameter of the stem, dbh = ds = $d_{1.3}$; and diameter of the three tallest roots, dr1 ... dr3. (**c**) Tree stand area equivalent to growing area, sa, and crown projection area, cpa. Particularly in densely packed stands, cpa is often greater than sa

3.4.2 Crown Allometry

Of special relevance is the present and future lateral crown expansion characterized by the crown radius and crown projection area (Fig. 3) as it indicates any overlap with streets or buildings. The crown projection area cpa results from the mean crown radius of, e.g. 8 crown radius measurements ($\bar{r} = \sqrt{(r_1{}^2 + r_2{}^2 + \ldots. + r_8{}^2)/8}$) as follows: $cpa = \bar{r}^2 \times \pi$. It also can be used for calculating the crown volume and surface area which determine the absorption of light, fine particulates, habitat space and quality and other ecological services (Moser-Reischl et al. 2019a; Moser et al. 2015).

Figure 8 illustrates that the crown size can differ considerably between tree species. The data comes from crown measurements on long-term experimental plots in Germany and covers a broad range of tree ages and stand densities (Pretzsch 2014; Pretzsch et al. 2015).

Based on the 95% and 5% quantile of the cpa-d allometry (Fig. 8 upper and lower lines), the tree crown plasticity CPL, a relative measure, can be derived. Based on the lower quantile line ($cpa_{5\%} = a_{5\%} \times 25^{\alpha_{5\%}}$) and the upper one ($cpa_{95\%} = a_{95\%} \times 25^{\alpha_{95\%}}$), and using a reference diameter of 25 cm (where stand density in mono-layered stands is rather high or even at maximum), CPL is formulated as $CPL = cpa_{95\%,25}/cpa_{5\%,25} = (a_{95\%}/a_{5\%}) \times 25^{\alpha_{95\%}-\alpha_{5\%}}$. For a tree with a reference diameter of 25 cm, CPL indicates how wide a crown can range in solitary conditions in relation to maximum restriction. By setting the 95% in relation to the 5% width, any species-specific differences in shape and form (e.g. that beech crowns are a priori wider than spruces) are eliminated. CPL only indicates the relative potential for expansion which is of particular importance for competing in mixture.

Analyses of CPL values for various tree species in Europe revealed a maximum value for European beech of CPL = 5.1, i.e. the upper crown area is more than fivefold in size compared to its lower crown area under strong competition. With respect to CPL, the species represented in Fig. 8 are ranked as follows: beech (CPL = 5.1) > fir (4.7) > oak (4.5) > spruce (4.2) > maple (4.0) > pine (3.7). Out of the set of 14 species represented in Fig. 9, European beech (5.1) has the highest value, whereas alder (2.8) and birch (2.6) have the lowest.

In terms of a species' acclimation to intra- and interspecific competition, the maximum and minimum of its crown extension are of special interest as these indicate the range of plasticity. Figure 9a shows exemplarily for maple the upper (95%) and lower (5%) quantile of the cpa-d allometry in addition to the mean relationship. The ranking of species in terms of their 95% quantile is different. Species such as lime tree and silver birch lose when ranked in terms of their 95% quantile, whereas rather plastic species such as beech and silver fir win in this ranking. This emphasizes that deriving crown behaviour for solitary conditions from average behaviour in closed stands, especially from monocultures, is questionable. For example, beech, elm, maple and silver fir profit considerably in crown expansion when released from competition in monospecific or mixed stands (Fig. 9b, c).

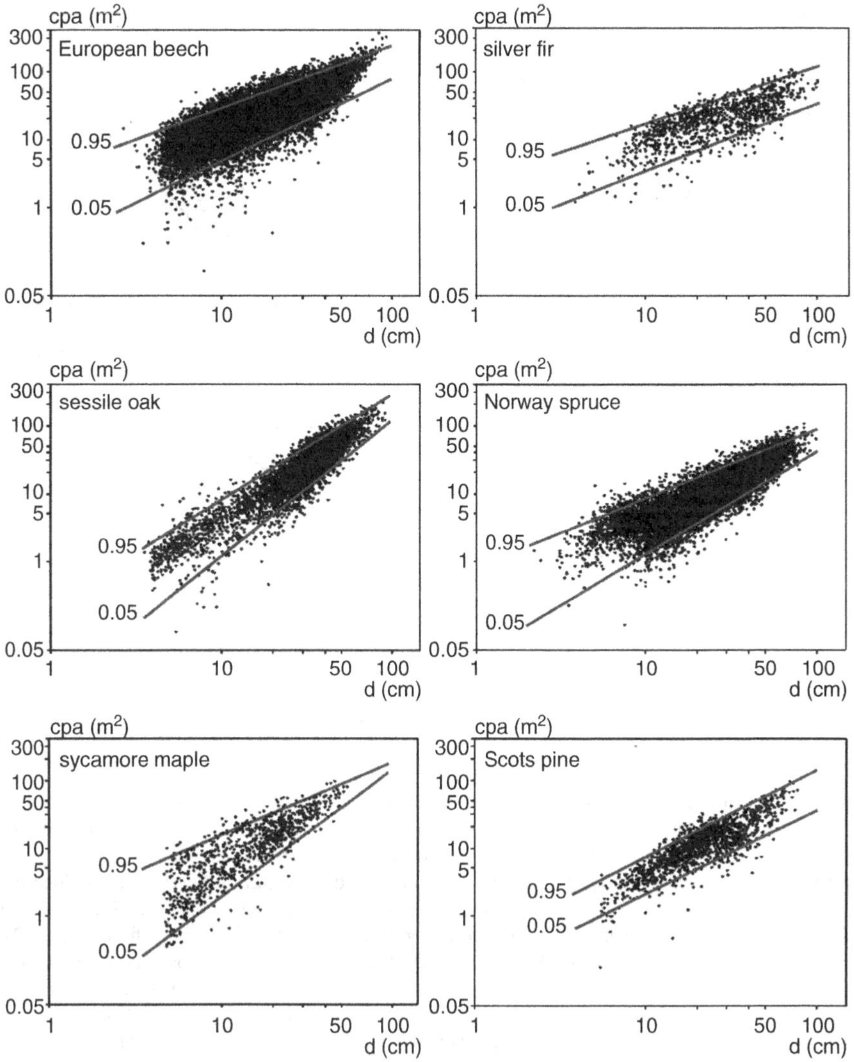

Fig. 8 Allometric relationships between stem diameter, d, and crown projection area, cpa, for European beech (*Fagus sylvatica* L.) ($n = 14{,}898$), silver fir (*Abies alba* Mill.) ($n = 1{,}079$), sessile/common oak (*Quercus petraea* (MATT.) LIEBL. and *Quercus robur* L.) ($n = 4{,}485$), Norway spruce (*Picea abies* [L.] Karst.) ($n = 10{,}724$), sycamore maple (*Acer pseudoplatanus* L.) ($n = 942$) and Scots pine (*Pinus sylvestris* L.) ($n = 1{,}609$) in even-aged and uneven-aged stands. Range of crown dimensions measured on long-term experimental plots which cover dense as well as very sparsely spaced stands. The upper and lower lines represent the 95% and 5% quantile regression $\ln(cpa) = a + \alpha \times \ln(d)$. The width of the scattering and the distance between the 95% and 5% quantile regression represents the crown plasticity. For the statistical characteristics of the quantile regressions, see Pretzsch (2014)

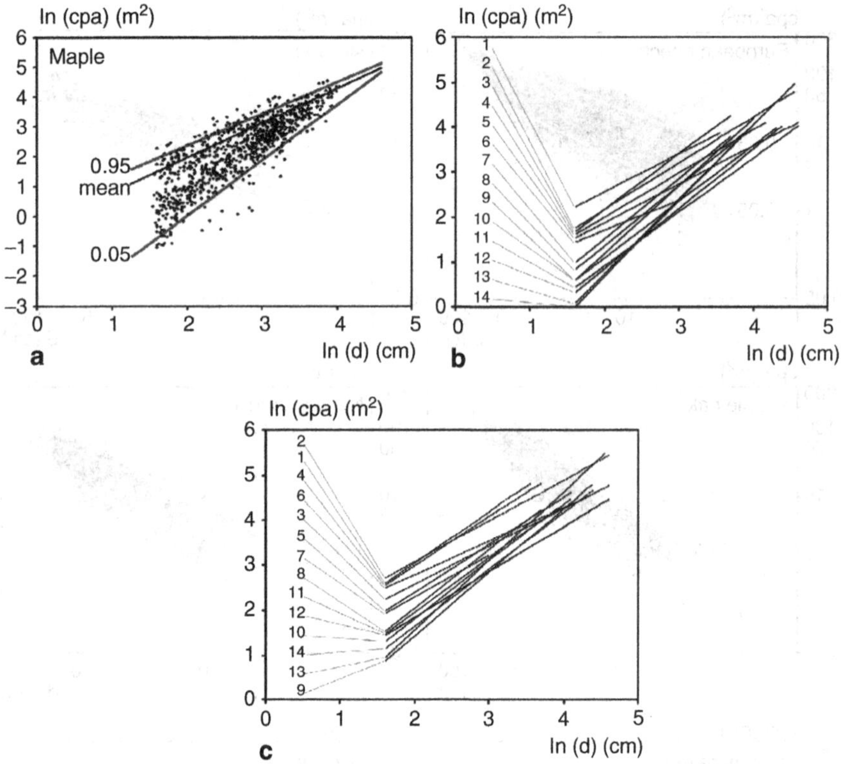

Fig. 9 Crown projection area to stem diameter relationship of various tree species. (**a**) Example derivation of the upper (95% quantile), mean and lower (5% quantile) ln(cpa)-ln(d) relationship for sycamore maple (*Acer pseudoplatanus* L.). (**b**) Interspecific variation in the mean $\ln(csa) - \ln(d)$ relationships. The model $\ln(cpa) = a + \alpha \times \ln(d)$ was fitted by OLS to the data and yielded the following parameters (a, α) for mountain elm (1.30, 0.67), European beech (0.18, 1.02), lime tree (0.08, 0.98), sycamore maple (−1.96, 1.55), hornbeam (−0.25, 1.24), silver fir (0.37, 0.82), rowanberry tree (1.13, 0.40), black alder (−0.85, 1.22), silver birch (−1.69, 1.48), Scots pine (−1.41, 1.27), Norway spruce (−1.43, 1.20), European larch (−1.69, 1.32), sessile oak/common oak (−2.48, 1.64) and common ash (−3.02, 1.79). (**c**) Interspecific variation in the 95% quantile of the $\ln(cpa) - \ln(d)$ relationships. The numbers denote different species: 1 mountain elm (*Ulmus glabra* Huds.), 2 European beech (*Fagus sylvatica* L.), 3 lime tree (*Tilia cordata* Mill.), 4 sycamore maple (*Acer pseudoplatanus* L.), 5 hornbeam (*Carpinus betulus* L.), 6 silver fir (*Abies alba* Mill.), 7 rowanberry tree (*Sorbus aucuparia* L.), 8 black alder (*Alnus glutinosa* (L.) Gaertn.), 9 silver birch (*Betula pendula* Roth), 10 Scots pine (*Pinus sylvestris* L.), 11 Norway spruce (*Picea abies* (L.) Karst.), 12 European larch (*Larix decidua* Mill.), 13 sessile oak/common oak (*Quercus petraea* (Matt.) Liebl./*Quercus robur* L.) and 14 common ash (*Fraxinus excelsior* L.)

The crown volume results as a product of crown projection area, crown length (tree height – height to the crown base) and the crown form factor f_c ($cv = cpa \times cl \times f_c$). For cylindric crown shapes (e.g. poplar, cypress), f_c is 1.0; for parabola-shaped crowns, f_c is 1/2 (lime tree, acacia); for cone-shaped crowns, f_c is 1/3 (spruce, fir); and for neiloid-shaped ones, f_c is 1/4 (ginko, larch).

For the crown surface area of a cylindric crown (top area + lateral area) applies $cs = cpa + 2 \times \pi \times \overline{r} \times cl$, with mean crown radius, $\overline{r}$, and crown length, cl (Pretzsch 2009). For the lateral area of cone-shaped crowns, the exact formula is $cs = \pi \times \overline{r} \times \sqrt{\overline{r}^2 + cl^2}$. The respective cs formulas for other crown shapes are more complicated and can be read off in textbooks of geometrics. As approximation for the surface area cs of various tree crown shapes, von Gadow (2003) suggested $cs = \frac{\pi \times \overline{r}}{6 \times cl^2} \times \left[\left(4 \times cl^2 \times \overline{r}^2\right)^{3/2} - \overline{r}^3 \right]$, with mean crown radius, $\overline{r}$, and crown length, cl.

3.4.3 Root Allometry

The percentage of roots in the total tree biomass ranges from 10 to 45% approximately depending on the growth conditions (Fogel 1983; Santantonio et al. 1977). The root mass is often estimated by expansion of the aboveground biomass by root expansion factor er with w(total) = w(above) × er (Rötzer et al. 2009). For the above range, it amounts to er = 1.11–1.81. Instead of the root factor, the literature also refers to the root/shoot ratio, which, for the above-mentioned values, lies between 10:90 and 45:55. The large variation in the root/shoot ratio can be explained by the optimal partitioning theory (McCarthy and Enquist 2007). The limitation of a resource leads to the promotion of growth of the plant organ responsible for supplying that critical resource (Comeau and Kimmins 1989; Keyes and Grier 1981).

In Fig. 10 four examples indicate the complexity of the root/shoot ratio. In the first example, where light, water and nutrient conditions are favourable (Fig. 10a), the tree shown develops a root/shoot ratio of 10: 90 (er = 1.11). Under adequate light conditions, but with a water or a nutrient deficiency (Fig. 10b), the tree invests more into root growth, especially fine roots. Thus the root/shoot ratio increases in favour of roots to 45:55, with er = 1.82. With an adequate water and nutrient supply, yet

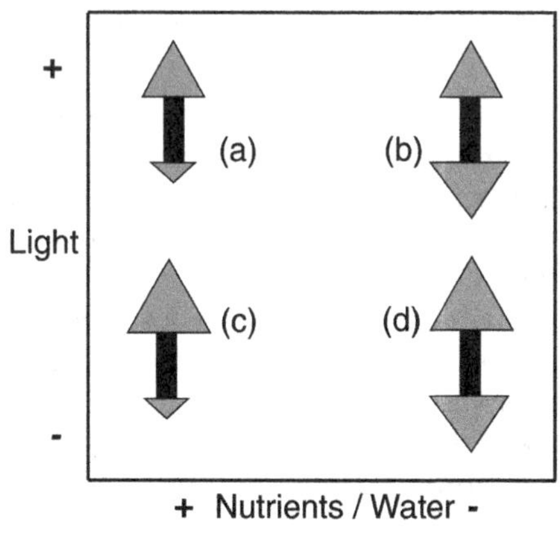

Fig. 10 Partitioning of total plant biomass on shoot and root organs in relation to supply of nutrients and water (x-axis) and supply of light (y-axis). Limitation of nutrient and water supply causes a partitioning in favour of roots. Limitation of energy supply raises the investment of biomass into shoots (by courtesy of Kimmins (1993), p. 13). The sizes of the triangles result in the ratios mentioned in the text

critical light conditions, e.g. on nutrient-rich soils or for understory trees (Fig. 10c), shoot growth is enhanced, so that the tree root/shoot ratio becomes 30:70, resulting in er = 1.43. In the case of rich soils, trees in the upperstory tend to allocate growth resources to extensive crown development, whereas trees in the understory often invest in height growth to escape the shade (Pretzsch 2014). If light, water and nutrient supply is limited (Fig. 10d), the root/shoot ratio may resemble the example of 10:90, where er = 1.11.

Aside from the large variation in the root/shoot ratio caused by different site conditions, Burschel et al. (1993) identified a size-dependent change in root/shoot ratios in several studies from 10:10 for small to 10:30 to 10:40 for large trees, corresponding to root factors of er = 2.0–1.25. In the absence of additional information, we recommend an approximate value for er of approximate 1.25.

When tree biomass is extrapolated from the stemwood, from merchantable volume or from the aboveground biomass without taking the site-specific root/shoot ratio into account, the root biomass is underestimated, especially on sites where growth is limited by the soil conditions.

3.5 Growing Area and Growing Space Requirement

The crown models, especially the allometries between stem diameter and crown radius and stem diameter and crown projection area, are also useful for planting and positioning trees in an urban environment. The cr and cpa values in Table 2 represent crown dimension of open grown trees. They might be used to assess the distance and growing space requirement in different states of the tree development. Conclusion by analogy might be used for estimation of crown measures for species not included in the list.

3.6 Link Between Tree Allometry and Growth

By TLidar technology, the primary variables such as height, diameter or crown diameter but also secondary variables such as crown volume, crown transparency or leaf area can be measured directly (Bayer et al. 2013, 2018). Compared to the forest, the measurement in the urban environment is often easier as the trees grow freely (Fig. 11).

By using TLidar, numerous structural parameters can be derived such as the percentage of wood and leaf area or branch angles or stem tapering. Further, several ecosystem services like tree shading can be derived. This information as well as the above shown allometric parameters and equations form an essential basis for (urban) tree growth models.

Table 2 Crown radius, cr, and crown projection area, cpa, of selected tree species

Stem diameter at 1.30		10 (cm)	20 (cm)	30 (cm)	40 (cm)	50 (cm)
Abies alba Mill.	cr	2.32	3.11	3.69	4.17	4.58
	cpa	16.84	30.30	42.74	54.54	65.90
Abies sachalinensis Mast.	cr	1.63	2.63	3.48	4.25	4.95
	cpa	8.37	21.78	38.09	56.63	77.03
Acer pseudoplatanus L.	cr	2.19	3.17	3.94	4.59	5.18
	cpa	15.09	31.64	48.78	66.32	84.17
Aesculus hippocastanum L.	cr	2.31	3.50	4.47	5.31	6.07
	cpa	16.72	38.47	62.64	88.53	115.77
Alnus glutinosa (L.) Gaertn.	cr	1.90	2.98	3.88	4.68	5.42
	cpa	11.34	27.96	47.39	68.91	92.13
Araucaria cunninghamii Aiton ex D. Don	cr	1.25	1.64	1.92	2.14	2.34
	cpa	4.90	8.41	11.54	14.45	17.19
Betula pendula Roth	cr	1.69	2.65	3.44	4.15	4.79
	cpa	8.96	22.01	37.23	54.06	72.20
Carpinus betulus L.	cr	2.96	4.32	5.40	6.32	7.14
	cpa	27.52	58.75	91.55	125.41	160.08
Fagus sylvatica L.	cr	2.99	4.15	5.02	5.75	6.38
	cpa	28.11	54.00	79.11	103.74	128.00
Fraxinus excelsior L.	cr	1.79	2.73	3.49	4.15	4.75
	cpa	10.11	23.41	38.23	54.16	70.95
Khaya senegalensis (Desr.) A.Juss.	cr	2.46	3.84	5.00	6.02	6.95
	cpa	18.96	46.43	78.41	113.72	151.72
Larix decidua Mill.	cr	1.42	2.33	3.11	3.81	4.47
	cpa	6.33	17.00	30.32	45.70	62.82
Picea abies [L.] Karst.	cr	1.65	2.34	2.88	3.33	3.73
	cpa	8.52	17.25	26.04	34.89	43.78
Picea glauca (Moench) Voss	cr	1.12	1.99	2.78	3.53	4.25
	cpa	3.95	12.44	24.35	39.20	56.71
Pinus sylvestris L.	cr	1.54	2.40	3.11	3.74	4.31
	cpa	7.50	18.15	30.46	43.97	58.45
Platanus x acerifolia Münchh.	cr	2.59	4.07	5.30	6.40	7.40
	cpa	21.01	51.95	88.24	128.49	171.98
Pseudotsuga menziesii (Mirbel) Franco	cr	1.57	2.64	3.59	4.45	5.27
	cpa	7.75	21.98	40.44	62.34	87.22
Quercus robur L.	cr	1.55	2.74	3.81	4.82	5.78
	cpa	7.59	23.52	45.58	72.88	104.88
Robinia pseudoacacia L.	cr	2.47	3.70	4.69	5.55	6.32
	cpa	19.24	43.12	69.14	96.65	125.32
Sorbus aucuparia L.	cr	2.04	2.80	3.38	3.86	4.28
	cpa	13.01	24.69	35.92	46.86	57.59
Tilia cordata Mill.	cr	2.67	3.97	5.01	5.90	6.71

(continued)

Table 2 (continued)

Stem diameter at 1.30		10 (cm)	20 (cm)	30 (cm)	40 (cm)	50 (cm)
	cpa	22.37	49.48	78.74	109.48	141.36
Ulmus glabra HUDS.	cr	3.07	4.57	5.77	6.82	7.75
	cpa	29.53	65.65	104.77	145.96	188.77

The cpa indicates the growing space requirement for open grown trees (Dahlhausen et al. 2016; Pretzsch et al. 2015)

Fig. 11 Towards 3D measurement of urban trees in their environment. Photograph (**a**) and TLidar scans (**b**) of a common oak (*Quercus robur L.*) on the campus Freising/Weihenstephan of the Technical University of Munich. Photo: Leonhard Steinacker. TLidar scan Martin Jacobs

4 Urban Tree Growth Models

In the future, information about growth, tree allometry, volume increment and dimensional changes of urban trees as well as their ES will be crucial for a sustainable and forward-looking planning of the green infrastructure in cities (Pauleit et al. 2017). Climate-sensitive growth models for urban trees are good tools to provide such data. However, in literature only a few growth models can be found which are able to more or less fulfil the above-mentioned criteria. We searched for models that include the growth of urban trees and estimate ES independent of spatial and temporal resolution. The most prominent ones that we found are UFORE (Nowak and Crane 2000; Nowak et al. 2008), i-Tree (i-Tree 2019; Nowak et al. 2008), CITYgreen (Longcore et al. 2004; Peng et al. 2008), CityTree (Rötzer et al. 2019) and UrbTree (Kramer and Oldengarm 2010). Hereafter, the structure, the design and the main processes of the models are introduced in more detail.

4.1 UFORE Model

The Urban Forest Effects (UFORE) model is able to simulate structures and functions of urban forests and was developed for North America by the US Forest Service (Saunders et al. 2011). For the simulations data regarding tree species and their dimensions as well as standard field-meteorological and pollution data are needed. As output information for entire populations and individual tree species are provided. These include information about the urban forest structure, the carbon storage and sequestration, volatile organic compounds, air pollution removal and transpiration, all of which are calculated within four modules (Fig. 12, Nowak and Crane (2000)). Further outputs are tree effects on building energy use and CO_2 emissions, insect and disease potential, pollen allergy rating and tree transpiration (Nowak et al. 2003, 2008).

Species composition, tree density and health as well as data on the foliage, i.e. structural information about the urban forest, are calculated in the anatomy module (=module A, Nowak et al. (2003)). Hereby, regression equations based on Nowak (1996) are used to estimate the leaf information of deciduous urban trees. Leaf information of conifers is a function of the mean LAI per height to width class of deciduous trees. Leaf biomass is derived by using species-specific leaf dry weights. In addition, species richness, population distribution, the Shannon-Wiener diversity index and the ground cover type distribution for the study area are quantified within the anatomy module.

The quantification of the volatile organic compounds (VOCs) that trees emit is estimated in dependence on tree species, leaf biomass and temperature (= module B). Standardized hourly values of emitted VOCs for a city are calculated by multiplying the leaf biomass of a species with emission factors of the genera. By

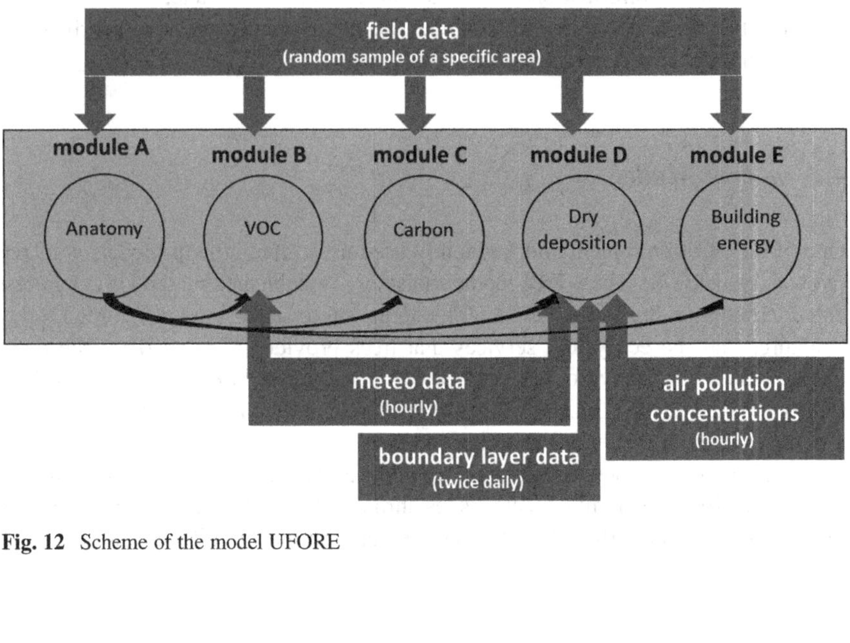

Fig. 12 Scheme of the model UFORE

using temperature, light correction factors and meteorological values of the city actual emission values are derived.

In module C the carbon storage of the urban green and the carbon sequestration are simulated on the basis of allometric equations from literature. Because most biomass equations are derived from forests, the biomass of urban trees, which is mostly smaller than the biomass of forest trees, is multiplied by 0.8. To calculate tree growth and thereby carbon sequestration, average diameter growth rates for trees in forest, in park-like and for open grown environments are used. Height growth is estimated according to Fleming (1988). At the end, the tree growth is adapted to the condition of the trees by using a multiplier for different condition classes (excellent, medium, etc.).

The dry deposition of the air pollution to the canopies of trees is simulated in module D. The annual removal of nitrogen and sulphur dioxide, of carbon monoxide as well as of ozone is calculated on an hourly basis, while the deposition of particulate matter is estimated on a daily basis (Saunders et al. 2011). They are functions of the deposition velocity and the pollutant concentration, which in turn depend on canopy resistances that are based on the photosynthetic active radiation, air temperature, humidity, wind speed and CO_2 concentration. As tree parameters the LAI, a bark area index (BAI), the distribution of deciduous and coniferous trees (d/c ratio) and phenological information must be available. While phenological data are input values, LAI (=6), BAI (= 1.7) and the d/c ratio (=9) are fixed values (Nowak and Crane 2000).

Module E describes the urban tree effects on building energy use and the avoiding of carbon from power plants (Nowak et al. 2008). For the latter a distinction is made between the avoidance of heating according to the energy effects of urban trees and the extent of cooling by urban trees. The values are estimated in dependence on a tree's shade and windbreak effects by using information on tree size and leaf type, climate region and information on the buildings and trees of a site.

The UFORE model is able to simulate the described parameters for urban and non-urban areas independent of size.

4.2 *i-Tree Model*

One of the most prominent and frequently used urban tree growth models is i-Tree (Nowak et al. 2008). The i-Tree model toolset was established by the USDA Forest and released in 2006. The different i-Tree tools can be used to quantify urban forest structures and the ecosystem services that trees provide (i-Tree 2019). The main i-Tree tools are Eco, Hydro, Canopy, Design, Landscape, Streets, Species and Vue (Fig. 13, i-Tree (2019): Lin et al. (2019)). This way, individual trees or trees of single regions or even of entire cities can be simulated.

The *Eco*-tool is a core module of i-Tree and was derived from the UFORE model. Using data from randomly located sites throughout a city, the module provides information on the urban forest structure, environmental effects and value to

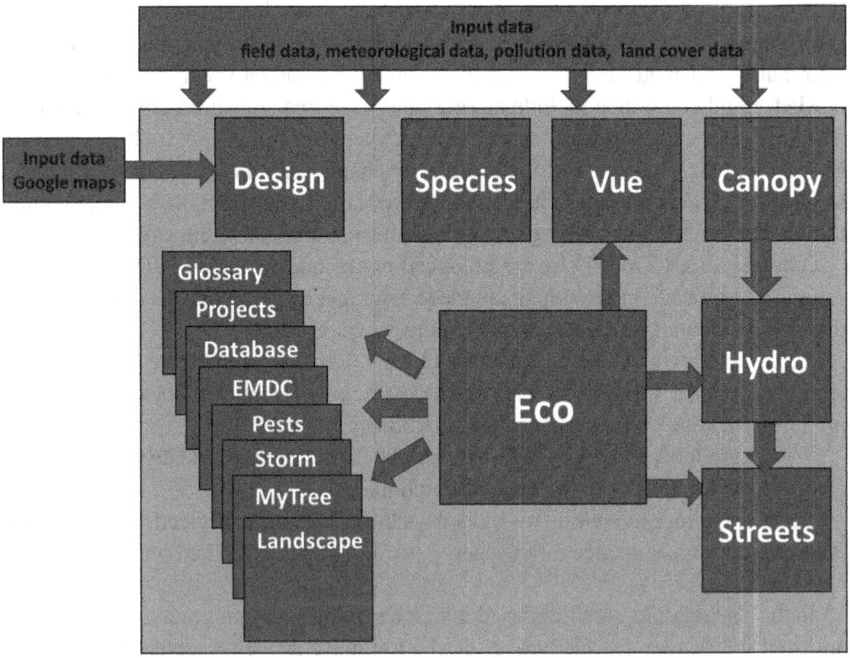

Fig. 13 Scheme of the i-Tree model toolset

communities. The calculations depend on average growth rates of trees as well as on biomass allometries and canopy structures of tree species altered by climate (Nowak et al. 2008). The tree species-specific average growth rates are based on urban street trees, park trees and forests standardized for Minnesota, USA. In addition, leaf area and biomass are estimated by using empirical regression equations. Input data for the Eco-module are field data and tree measurements as well as hourly local air pollution concentrations and weather data. On the basis of these information, the urban forest structure, the total carbon stored and the yearly net carbon sequestration are calculated. Pollution reduction, public health, energy effects and emissions are simulated according to the UFORE model. Further on, the annual avoided runoff of the tree species or strata, potential pest impacts on base on tree structure, host susceptibility and pest/disease range, and the compensatory and the economic values of the ecosystem services of the urban forest are simulated. A forecasting tool calculates tree and urban forest growth by taking tree structure simulations, environmental conditions and tree species characteristics into account. This way, also pollution removal, carbon storage and carbon sequestration can be projected on the basis of the estimated tree growth and leaf area (i-Tree 2019).

Another important basic tool within the i-Tree toolset is the *Hydro*-module. It analyses land cover scenarios and their hydrological impacts for different scales and calculates hydrological parameters in dependence on changes in the urban tree cover and on the impervious surfaces. While hourly stream flow alterations are estimated on the basis of the urban tree and impervious cover information, water quality changes are simulated by using data from the base program. For these simulations information on land cover, elevation, weather and model-specific parameters are needed. Land cover data include among others the tree and shrub cover percentage and the impervious surface. As a default weather data from 2005 to 2012 are used. Based on predefined soil data, which can be changed by the user, i-Tree calculates the hydrological parameters interception, evaporation, transpiration, snowmelt, infiltration, flow routing and storage as well as the stream flow at the gaging station.

Input data for i-Tree Hydro are produced in the module *Canopy*. The statistically estimated land cover types which are made available by aerial images can be used by urban forest managers to derive tree canopy cover estimates.

Ecosystem services and the structure of a municipality's street tree population are simulated by the module *Streets*. The ecosystem services simulated in the other modules are valued and transformed into monetary units.

Within the module *Species selector*, suitable tree species for an urban area can be chosen on the basis of environmental functions.

The *Design* tool (an online tool) displays the effects of tree selection and size and the tree locations of an area on the energy use and other ecosystem services by using Google Maps.

In the *Vue* module, satellite-based imageries of the National Land Cover Database are used to estimate the land cover of a community as well as selected ecosystem services of the urban forest. On the basis of planting scenarios, simulations for the future can be done.

Further modules of the i-Tree toolset are i-Tree Storm, i-Tree Landscape or i-Tree Pests (see Fig. 13, i-Trees (2019)).

4.3 CITYgreen Model

The model CITYgreen (Peng et al. 2008) was developed by American Forests in 1996 as a GIS extension tool to evaluate the ecosystem services of green space and translate them into economic values (American Forests 2002). It is a user-friendly and broadly accessible software. CITYgreen consists of modules to estimate air pollution removal, carbon sequestration and storage, storm water runoff reduction, energy conservation and wildlife and tree growth (Fig. 14).

Tree growth is modelled in dependence on height and the diameter at breast height of a tree. Based on these data, trees are classified as slow growing, medium growing or fast growing. For the classified trees, constant growth rates are available in the attribute database. These growth factors which are determined by American Forests are based on a great number of measurements of forest trees (Longcore et al. 2004). The module aims at predicting future tree growth and future ecosystems services of trees for a study site (Longcore et al. 2004).

The removal of carbon dioxide from the atmosphere through photosynthesis and the total amount of stored carbon is based on Nowak (1993) and by classifying the

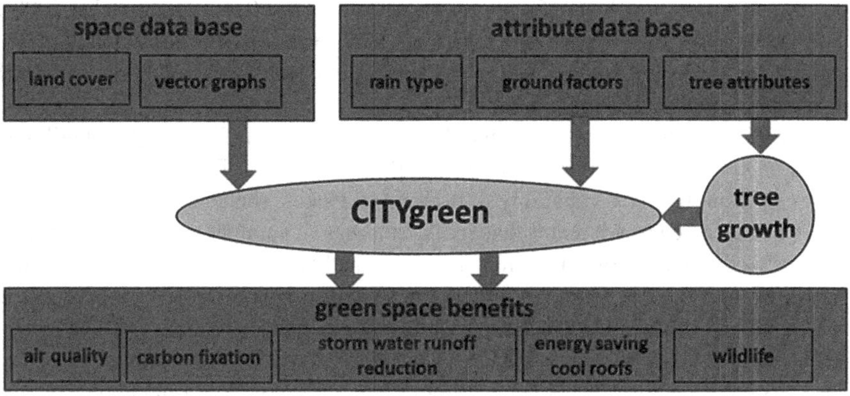

Fig. 14 Scheme of the model CITYgreen

trees of an area into young trees (type1), old trees (type2) or trees with a balanced age distribution (type3). For each type, specific multipliers are used to determine carbon storage and carbon sequestration (Peng et al. 2008). To estimate the amount of fixed carbon, the multipliers are combined with the overall size of the study site and its canopy coverage (Longcore et al. 2004). The classification is part of the attribute database (Fig. 14).

The air pollution removal module calculates air quality benefits by removing pollutants such as nitrogen dioxide (NO_2), carbon monoxide (CO), sulphur dioxide (SO_2), ozone (O_3) and particulate matter less than 10 μm (PM10). The module consists of an estimation of the pollution concentration in the analysed area and a subsequent calculation of the pollution removal rate and the amount of total pollution. The pollutant removal rate for an entire tree is determined based on the deposition velocity, the pollution concentration and a tree's leaf surface (Longcore et al. 2004) as well as on the basis of site-specific factors of the tree and/or forest coverage. An economic evaluation of the benefits from the pollutant removal is achieved by the estimation of the externality costs avoided to be paid for, e.g. health care.

The storm water runoff module is based on TR-55, a widely used urban hydrology model (Soil Conservation Service 1986). For the calculations, information is needed about the soil group, the slope percentage as well as the rainfall distribution and intensity, which falls during a typical 24 h extreme rainfall event within a 2-year period (Bruns and Fetcher 2008). The module estimates the runoff amount in dependence on the timing of the water movement through the area and the precipitation amount. These values as well as curves which are based on the water infiltration ability and on the land cover conditions for the main cover types determine which percentage of a given rainfall amount forms the direct runoff. Thus, the storm water runoff module calculates the effects of vegetation – especially trees – on the runoff amount and the peak flow in an urban area. It includes a comparison of current conditions to scenarios where all existing trees and their ecological benefits are lost in favour of suburban land use (Bruns and Fetcher 2008). The benefits of the urban green regarding the storm water runoff reduction are estimated by the potentially saved costs for building retention ponds or additional storm water management facilities.

The wildlife module values the wildlife of the study area. The rating of wildlife is based on secondary literature data for app. 100 tree species. The wildlife rating of a site results on the trees' capability to offer nesting sites, shelter and food.

The reduced energy amount that cools buildings in summer by direct shading of trees as well as the energy amount that reduces the heating energy in winter by wind break through trees is estimated in the energy conservation module. Trees are the most effective when situated near areas where the tree shade reduces cooling energy and with it the costs for energy. Thus, the energy conservation module calculates the energy saving for direct tree shading of buildings. It is, however, only applicable for one- or two-story buildings. The cool roof module assesses energy savings through roofs. Differently coloured roofs have different albedo values which can significantly change the energy balance. Again, this module is only suitable for one- or two-story buildings (Longcore et al. 2004).

4.4 CityTree Model

CityTree is a physiologically based growth model for single trees (Rötzer et al. 2019). The carbon and the water cycles of trees are described as biological, physical or chemical processes which are mainly based on environmental conditions. The tree development is primarily considered as matter change in the different compartments. Thus, the influence of a changing climate on the growth of urban trees can be simulated. Based on basic measurements, i.e. diameter at breast height and tree height, and a rough characterization of the soil, simulations can be done using CO_2 concentration of the atmosphere and climate data as driving forces of growth. The model is structured in eight modules (Fig. 15). Input values include all tree- and soil-specific information. Monthly data of temperature, radiation, air humidity, wind speed and precipitation form the base of the climate module (m1). In the module 'plant' (m2), the annual plant development is simulated.

The water balance module (m3) of CityTree model simulates the entire water balance cycle. Based on monthly precipitation amounts and an initial soil water content, the parameters interception, potential and actual evapotranspiration, surface runoff and percolation below the rooting zone are calculated. Hereby, potential evapotranspiration is calculated according to Penman (see DWA (2018)) which includes not only a temperature-humidity term and a ventilation term but also a radiation term. Water balance is linked with the growth and photosynthesis module via stomatal closure which in turn depends on the stomata conductivity.

In the photosynthesis module (m4), net assimilation is calculated according to Haxeltine and Prentice (1996) on basis of air temperature, photosynthetic active radiation, CO_2 concentration and day length. Using the LAI of the month and the tree

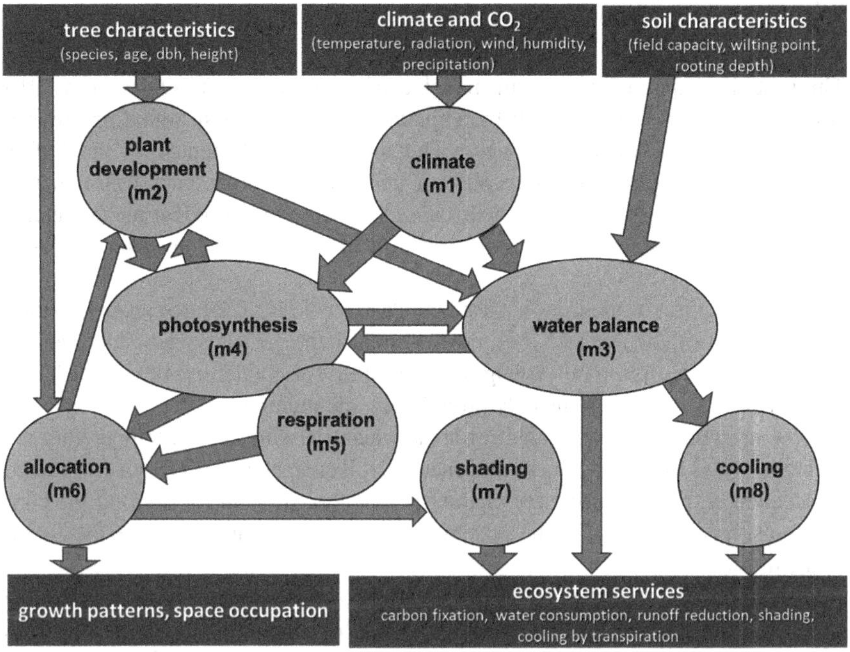

Fig. 15 Scheme of the model CityTree

crown projection area, a tree's monthly net assimilation rate can be estimated. After reducing net assimilation (= gross primary production) by a fixed share for growth and maintenance respiration (Pretzsch et al. 2012a), the resulting net primary production can be transformed into carbon or CO_2 units (module m5).

In the module allocation (m6), the net primary production, i.e. the gained carbon, is distributed within the tree, i.e. it is allocated into the stem, the branches, the foliage and the nonstructural carbon pool. Thus, the new tree structures such as stem diameter, tree height, stem and crown volume and crown projection area can be derived from the new biomasses of the tree compartments. Currently, a nutrient cycle is not yet included in the model.

In addition to carbon fixation, CityTree is able to simulate the ecosystem services shading (module m7) and cooling by transpiration (module m8). The constant energy that is needed for the transition of water from the liquid to the gaseous phase has to be multiplied with the transpiration sum of a tree to gain the cooling amount. The shading area is a function of crown diameter and tree height and is calculated for the longest day of the year. Along with the shaded area, the shade density of a tree is estimated using leaf area and crown volume.

4.5 UrbTree Model

The spatially explicit tree growth model UrbTree (Kramer and Oldengarm 2010) simulates individual tree growth by a spatial analysis of the surrounding environment on the basis of aerial images. The model is implemented in the Python programming language and depends on geographical processing techniques. It simulates differences in tree growth caused by the surface covering (vegetated, open, paved) by using techniques for spatial analysis. The update of spatial data is between 2 and 5 years.

The description of the trees is based on classifications of the tree shape (column, round or ellipse), the tree size (big, regular, small), the growth speed (slow, regular, fast) and the tree life phase (young, mature, old). The model uses existing data of individual trees, which are usually provided by the municipalities.

Tree growth can be simulated for areas where the climate, i.e. temperature and precipitation, is assumed to be homogeneous. Soil characteristics are not considered. Consequently, primarily the tree characteristics, their growing space and the competition situation and type of the surrounding surface covering determine the growth of the trees.

The tree database must comprise the location of every tree, its crown dimensions as well as the assigned classes, while the geodatabase includes the description of the environment (buildings, roads, grass, open surface, water) and the distances to objects which could influence growth and the surface structure (Fig. 16).

Through spatial analyses, the limiting factors for tree growth are estimated for all tree individuals. The limiting factors induced by the topography are based on the distance to nearby buildings and roads. A second limiting factor is based on the competition of neighbouring trees. It is represented as a decrease of horizontal crown and an increase of the vertical crown in the case when tree crowns overlap each other. A third limiting factor takes the surface perviousness in the direct neighbourhood of the tree crown into account. This way, the limiting factor is calculated in accordance to the percentage of open area within the tree crown area.

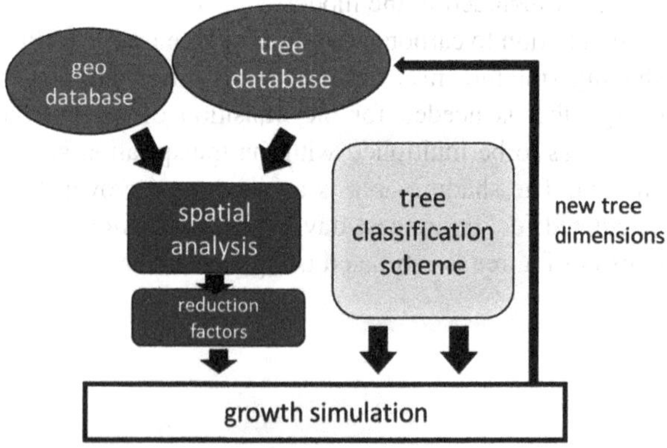

Fig. 16 Scheme of the model UrbTree

Using the reduction parameters and the growth parameters of the tree classification scheme, the new tree crown dimensions can be estimated which can be stored in the tree database.

These calculations are done iteratively on an annual base. Thus, the tree crown changes, which are calculated for every year, change the reduction factors annually.

4.6 *Further Urban Forest Models*

A main focus of the described urban tree growth models is on the description of tree or urban forest growth and the derivation of their ecosystem services. In several other urban forest models, the growth of trees or of urban forests or the vegetation development is regarded; however, their main focus is the microclimate or other physical effects of urban vegetation. Temporal, i.e. inter- and intra-annual, growth dynamics and structural changes of the vegetation are, however, not regarded in most of these models.

For example, CFD models, i.e. computational fluid dynamics models (e.g. Buccolieri et al. (2018)), link vegetation to thermal conditions and to air quality of urban areas. This way, along with the thermal properties of the vegetation (e.g. the transpirational cooling), also deposition and aerodynamic effects can be simulated.

An example is the ENVI-met model (Bruse and Fleer 1998) which allows to compute the micro-scale urban climate. It regards the urban environment by differentiating between buildings, impervious and natural surfaces and vegetation.

The green cluster thermal time constant (Green CTTC) model of Shashua-Bar and Hoffman (2002) simulates diurnal air temperature inside a 'wooded' site in a city based on vegetation effects. This way, the climate impact of trees and other green on urban areas can be studied.

On the other hand, the Town Energy Balance (TEB) urban canopy model is able to simulate climate change impacts at the city scale whereby street trees are regarded (Redon et al. 2017). On the basis of the mean height of trees, their LAI and albedo as well as on the basis of the cover fraction of trees, complex radiative estimations between the vegetation and built-up areas can be done.

Further models with a focus on the urban climate which are able to take vegetation into account are, for example, the vegetated urban canopy model (VUCM, Lee and Park (2008)), the building effect parameterization model with trees (BEPTree, Krayenhoff et al. (2014)), the vegetated urban canopy module (VUC) of the advanced regional prediction system (ARPS) model (Tavares et al. 2015) or the air-temperature response to green/blue-infrastructure tool (TARGET, Broadbent et al. (2019)).

Growth simulations can also be done based on dendrochronological data with subsequent statistical modelling (e.g. Canetti et al. (2017)). But – in a stricter sense – simulations based on such models are only applicable for the specific region, the time period and the tree species for which the analysis was done.

5 Model Comparisons

5.1 Model Aims and Base Characteristics

Models that simulate the growth of the urban green along with their ecosystem services such as water consumption, CO_2 fixation or cooling are scarcely available. We identified five urban tree growth models that are able to calculate tree growth as well as to provide information on their functions.

Most of the models were developed to assess tree structure, tree dimensions and/or ecosystem services (Table 3). Three out of the five models are capable of transforming the simulated ecosystem services into economic values. Except UFORE, all other models are single tree based, i.e. in a first step growth is calculated for a tree individual. Only the UrbTree model uses remotely sensed data for the growth simulations. Dynamic simulations over time can be done by the models CITYgreen, CityTree and UrbTree and to a limited extent by the model i-Tree.

Tree growth of three models is based on allometric equations, while the UrbTree model uses growth factors that are adapted to the fine-scale environment. Only the model CityTree simulates growth and ES process based, i.e. in dependence on biological, physical and chemical functions (Table 3). Hence, only two models are able to calculate growth in dependence of the climatic conditions and its changes. In the CityTree model, climate values are the direct drivers of growth, while modifiers are used in the i-Tree model to adapt growth to climatic conditions.

5.2 Demands on Input Data

The number of input parameters is closely related to the complexity of the growth and ES calculations. The process-based model CityTree, for example, needs detailed plant, climate and soil information to describe tree growth and the provision of ecosystem services (Table 4). Other models like UrbTree assume homogeneous climate and soil conditions.

If climate parameters are needed for the simulations in most cases, temperature, radiation, humidity and wind speed have to be made available. Only the model CityTree and CITYgreen use soil information for the simulations. As plant parameters tree species, diameter dbh and tree height must be known for all models. Information of the crown size is needed for the models UFORE, i-Tree, CityTree and UrbTree.

Table 3 Model aims and base characteristics of the tree growth models

Model	Main model aims	Simulation of tree growth	Simulation of ES	Economic valuation	Single tree based	Use of remotely sensed imageries	Temporal development	Growth calculations based on climate data
UFORE	Assessing urban tree population and ecosystem services	Allometric equations	Yes	Yes	No	No	No	No
i-Tree	Assessing urban forest structure, functions and values; evaluate urban vegetation-induced ecosystem services	Allometric equations	Yes	Yes	Yes	No	(Yes)	(Yes)
CITYgreen	Evaluate the ecosystem services of green space and translate them into economic values	Allometric equations	Yes	Yes	Yes	No	Yes	No
CityTree	Tree growth and ecosystem services based on climate change and change of environmental conditions and impacts of extreme events	Physiological functions	Yes	No	Yes	No	Yes	Yes
UrbTree	Tree growth and tree dimensions	Growth factors	No	No	Yes	Yes	Yes	No

Table 4 Main input data of the tree growth models

Model	Plant	Climate	Soil
UFORE	Species dbh Tree height crown dimensions	Temperature Radiation Humidity Wind speed CO_2 concentration	Not considered
i-Tree (Eco)	Species Tree position Crown dimensions	Temperature Radiation Humidity Wind speed CO_2 concentration	Not explicitly considered
CITYgreen	Species dbh Tree height	Not considered	Hydrological soil groups (classified by perviousness)
CityTree	Species dbh Tree height Crown dimensions	Temperature Radiation Precipitation Wind speed Humidity CO_2 concentration	Field capacity wilting point rooting depth
UrbTree	Species Tree position Crown dimensions	Considered as homogenous	Considered as homogenous

5.3 *Calculations and Simulation Outputs*

Based on the main publications of the single models, the methods of the growth and ES calculations as well as the key characteristics were derived (Table 5). As already mentioned before, most of the models are based on allometric equations (Table 3). The model type is empirical, i.e. the equations are statistically based. One model (CityTree) is physiologically based; one model (i-Tree) can be classified as hybrid, i.e. it can be described as a mixture between a pure empirical and a pure physiological model.

Only two models (CityTree and UrbTree) provide data for different time scales, while the other three models are static. On the other hand, the spatial scale of the simulations ranges from individual trees (CityTree, UrbTree) to regions (CITYgreen). Water balance parameters are estimated in the i-Tree, CITYgreen and CityTree model.

Tree growth calculations of the CITYgreen model depend exclusively on tree characteristics. The UFORE, i-Tree and UrbTree models need further information on the surface cover, and for the CityTree model, additional data of the climate conditions, of the tree phenology and of the soil must be available. This results in a broad spectrum of simulation outputs of this model which include not only information about the standing biomass but also values for the annual increments and for the tree size.

Table 5 Main model characteristics

Parameter	UFORE	i-Tree	CITYgreen	CityTree	UrbTree
Source	Nowak and Crane (2000), Nowak et al. (2008)	Nowak et al. (2008), i-Tree (2019)	Longcore et al. (2004), Peng et al. (2008)	Rötzer et al. (2019)	Kramer and Oldengarm (2010)
Model type	Empirical	Hybrid, semi-physiological	Empirical	Physiological	Empirical
Built on	Field data	Field data	Imagery data	Tree measurements	Imagery data
Input data					
Tree characteristics	X	X	X	X	X
Geodata	–	(X)	X	–	X
Climate data	Hourly (pollution module)	Hourly (pollution module)	Extreme prec. events	Monthly	–
Soil information	–	–	X	X	–
General information					
Temporal scale	–	–	–	Month	2–5 years
Spatial scale	Area	Area	Regional	Tree	Tree
Water balance simulation	–	X	X	X	–
Growth simulations are based on					
Tree characteristics	X	X[a]	X	X	X
Phenology				X	
Climate and weather				X	
Soil conditions				X	
Surface cover	X	X[a]		X	X
Other					Neighbouring areas
Simulation of					
Tree size				X	X
Biomass increment	X	X[a]		X	
Above- and belowground increment	X	X[a]		X	

(continued)

Table 5 (continued)

Parameter	UFORE	i-Tree	CITYgreen	CityTree	UrbTree
Derivation of ecosystem services					
Carbon storage	X	X[a]	X	X	
Shading		X	X	X	
Water consumption				X	
Runoff		X	X	X	
Wind buffering		X			
Cooling by transpiration				X	
Air quality	X	X	X		
Biodiversity		X	X		
Others	Building energy effects	Building energy effects			

[a]Identical to UFORE model

The number of simulated ecosystem services clearly differs between the models. Only tree size is calculated by the UrbTree model. All other models are able to simulate the carbon storage of the trees. Shading and runoff estimations can be given by the models i-Tree, CITYgreen and CityTree. Because the model CityTree includes a full water balance module, also the water consumption (transpiration) and the transpirational cooling of individual trees can be determined. Information on air quality aspects is given by the models UFORE, i-Tree and CITYgreen, while the i-Tree and CITYgreen model also provide information on biodiversity. A broad range of further ecosystem services like wind buffering or building energy effects are simulated by the i-Tree model.

6 Discussion and Conclusions

This review illustrates that studies and especially models for the development of urban trees are necessary and much-needed tools to derive accurate estimations of tree growth and provided ES such as cooling, shading, runoff, carbon sequestration and air pollution filtering. The amount of studies is increasing over time; especially in recent years, several publications about urban tree development have been published (e.g. Peper et al. 2014a; Moser et al. 2015; Dahlhausen et al. 2016; Moser-Reischl et al. 2018). However, the number of urban tree models is still fairly low and some of them are relatively old, especially when compared to the forest sector.

The development of accurate modelling tools for the urban green is very challenging because of the huge amount of different tree species occurring in cities worldwide. Only for Germany, GALK eV (2019) names more than 170 tree species. Further challenges are the different climate regions and the vast number of different growth conditions in cities as, for example, water shortage, compacted soils, small amount of rooting space, management such as pruning and irrigation, above- and belowground limitations or pollutant inputs. Moreover, the spatial scale of the growth models can vary, from single tree modelling to tree clusters or even whole urban forest stands. Therefore it is almost impossible to select a growth model for most of the common urban tree species worldwide, which is able to simulate their growth and ecosystem services for various climate conditions including all the mentioned growth conditions in cities.

When looking at the published models described above, a solid need of some missing but often important influencing parameters in urban tree growth models is obvious. Some of them are described below.

6.1 *Mortality*

In most models, the mortality of a tree is completely neglected despite its importance for planning and management. This aspect is of particular importance, since the longevity and maximum age of urban trees are mostly lower than those of their rural or forest conspecifics (Nowak et al. 2004; Roman and Scatena 2011). The age and the mortality of a tree are closely linked to its physiology, and to its growth. Thus, mortality could be considered as a process depending on its environmental. Especially when modelling a tree row or a tree stand, the specific mortality of a single tree and its removal from the stand should be included to accurately estimate the growth and services of a stand.

6.2 *Nutrient Cycles and Nutrient Supply*

In all models, the nutrient cycles and the nutrient supply of trees are not included. As described above for mortality, the nutrient supply for urban trees is different compared to forest trees. In urban areas, the input of pollutants is fairly higher due to anthropogenic pollutants (vehicle pollutants, industry pollutants) and other influences like dog urines and de-icing salts in winter. However, especially at highly sealed sites such as street canyons and public squares, the nutrient status of urban trees is often below the actual need (Meineke et al. 2013). Reduced nutrient inputs could lead to decreased biomass and leaf growth as well as to smaller ES provision (Fisher et al. 2012; Gessler et al. 2017). Besides the importance of nutrients for metabolic processes and growth, the absorption of nutrients by a removal from storm water also poses an ES (Denman et al. 2016), which could be included in urban tree growth model.

6.3 Fructification

Fructification is a further aspect which is still not realized in urban tree growth models, albeit it is also seldom included in most forest growth models. However, not only bud break and leaf fall are important cornerstones of the annual phenological cycle of a tree, but also flowering and fructification. The building of leaves, flowers and fruits contribute to biomass production and influence the water and nutrient cycle, but are also important when considering tree disservices such as waste production. Several tree species also show alternations after some years of fructification or extensive fructification (mast year) and do not have fruit production every year (e.g. fruit tree species, beech, oak, fir), which also has to be included in 'fructification modules' for urban tree growth models.

6.4 Other Influences

As described above, urban trees face several growth conditions that are different to forest trees, which, however, influence the growth behaviour and ES provision markedly, like soil sealing, increased temperatures or limited space. Most of these aspects are hard to include in tree growth models. The model CityTree, for example, provides an approach to include surface sealing as a growth diminishing factor. Soil sealing greatly affects the water balance of a tree, as the model CityTree illustrates. A higher amount of soil sealing was negatively affecting growth and ES provision of all tested tree species (Rötzer et al. 2019). Thus, the grade of soil sealing should also be integrated in modelling tools for urban trees.

Management practices such as cutting and reduced above- and belowground growing space greatly affect tree growth. Urban trees facing these effects often show reduced growth (Moser et al. 2015); therefore, the incorporation of such effects is absolutely necessary to accurately predict growth and ES (McPherson and Peper 2012).

6.5 Selection of an Appropriate Model

Along with the differing depth of the representation of the single processes, and along with missing parameters in the urban tree growth models, the models often provide static values for ES, because the ES of a tree is not synchronized with growth dynamics. For example, the UrbTree model calculates tree growth and ES without taking the climate into account. Thus, the implications of different climate conditions are not represented. The CityTree model, on the other hand, has a climate module, enabling the model to compute the effects of climate scenarios on the water balance, on the tree growth and on the ES. With certain limitations since the temporal development of trees is only partially given, this option is also provided by the i-Tree model.

Another important aspect is obvious when discussing tree growth and ES modelling of urban trees. Most models compute carbon storage (with exception of UrbTree), shading (with exception of UFORE and UrbTree), runoff (with exception of UFORE and UrbTree) and air quality (with exception of CityTree and UrbTree). ES such as water consumption (by CityTree), wind buffering (by i-Tree), cooling by transpiration (by CityTree) and biodiversity (i-Tree and CITYgreen) are only included in some models. Other services like the recreational effect or filtering of particulate matter are up to date not included in most of the models.

The selection of an urban tree growth model should always depend on the aims of the simulations and whether data of all input parameters are available. Also the choice which ecosystem services are most important depends on the aim of the study and whether the growth model is able to simulate the ES. If, for example, the mitigation of climate change or of the urban heat island effect is the study objective, then especially cooling by transpiration and shading are likely the most important ES.

When discussing the ES provision of urban trees, it might also be relevant to include disservices of urban trees in urban tree growth models. To name here are the emission of biogenic volatile compounds (BVOCs), release of allergens, litter fall and costs due to management or the invasive potential of uncontrolled spread (e.g. *Ailanthus altissima* (Mill.) Swingle). BVOCs are released by trees when experiencing stress, e.g. due to drought or insect attack or for attraction of pollinators (Grote and Haas 2013; Nowak and Dwyer 2007). However, the release of BVOCs is also depending on the type of BVOC and air temperature, though the relationship with tree stress is difficult to implement in a model. Further, the allergenic potential of urban tree species might be modelled through phenological simulations by a flowering module based on the temperature sum (Chmielewski and Rötzer 2001; Rötzer et al. 2010).

A model for urban tree development can be based on different scales (single tree, patch, city level). It can contain different modules and simulate different outputs for tree growth and tree dimensions as well as diverse ecosystem services (e.g. carbon storage, air pollutant filtering, transpiration). The five discussed models vary in all these categories; therefore, model selection should be based on the specific aim of the simulations. With all the presented models, urban tree development and different ES can be simulated, though the depth, accuracy and scale can vary to a great extent.

UrbTree and CITYgreen are at the one end of the spectrum, requesting relatively little input. At the other end of the spectrum is the CityTree model which is based on physiological equations which generally allows simulations of tree growth and derives ecosystem services for individual trees on a monthly basis. Not surprisingly, this model has the highest demands on data input, e.g. climate data and soil information. The other models reviewed, UFORE and i-Tree, are in between. Clearly, there is a trade-off between data required and potential of the models reviewed here to simulate tree growth accurately and derive ES that needs to be carefully considered in model selection for a specific purpose.

Overall, the reviewed models are able to improve the understanding of urban tree growth and their functions and values in the urban environment. Given the different nature of the models, a direct comparison of the models concerning their outputs seems hardly feasible. But nevertheless, rigorous testing of the validity and reliability of the chosen model is recommended before applying it.

Acknowledgements The authors thank the Bavarian State Ministry of the Environment and Consumer Protection for funding the projects TUF01UF-64971 'Urban trees under climate change: growth, functions & services, and perspectives' and TKP01KPB-71924 'Climate experience Würzburg: Influence of trees on the micro climate of the city of Würzburg'. We would also like to thank the German Research Foundation (DFG) for funding the project 'Impact of trees on the urban microclimate under climate change: Mechanisms and eco-system services of urban tree species in temperate, Mediterranean and arid major cities' (PR 292/21-1 and PA 2626/3-1). Our thanks also go the municipal authorities of Munich, Würzburg, Hof, Kempten, Bayreuth and Nuremberg, who supported the study and permitted tree sampling.

References

Aguaron E, McPherson EG (2012) Comparison of methods for estimating carbon dioxide storage by Sacramento's urban forest. In: Lal R, Augustin B (eds) Carbon sequestration in urban ecosystems. Springer, New York, pp 43–71

Altimir N, Kolari P, Tuovinen JP, Vesala T, Bäck J, Suni T, Kulmala M, Hari P (2006) Foliage surface ozone deposition: a role for surface moisture? Biogeosciences 3:209–228

Alvem B-M, Bennerscheidt C (2009) Baumstandortoptimierung und Regenwasserbewirtschaftung – Chancen für ein gemeinsames Vorgehen. In: Dujesiefken D (ed) Jahrbuch der Baumpflege 2009. Taspo Fachbuchservice, Braunschweig, pp 70–78

American Forests (2002) Regional ecosystem analysis Roswell. Georgia, Washington, p 8

Armson D, Stringer P, Ennos AR (2013a) The effect of street trees and amenity grass on urban surface water runoff in Manchester, UK. Urban For Urban Green 12:282–286

Armson D, Rahman MA, Ennos AR (2013b) A comparison of the shading effectiveness of five different street tree species in Manchester, UK. Arboricult Urban Forests 39:157–164

Artmann M (2013) Spatial dimensions of soil sealing management in growing and shrinking cities – a systemic multi-scale analysis in Germany. Erdkunde 67:249–264

Atkin OK, Bruhn D, Hurry VM, Tjoelker MG (2005) Evans review no. 2: the hot and the cold: unravelling the variable response of plant respiration to temperature. Funct Plant Biol 32:87–105

Bartelink HH (1996) Allometric relationships on biomass and needle area of Douglas-fir. For Ecol Manage 86:193–203

Bartens J, Day SD, Harris JR, Wynn TM, Dove JE (2009) Transpiration and root development of urban trees in structural soil Stormwater reservoirs. Environ Manag 44:646–657

Bayer D, Seifert S, Pretzsch H (2013) Structural crown properties of Norway spruce (*Picea abies* [L.] Karst.) and European beech (*Fagus sylvatica* [L.]) in mixed versus pure stands revealed by terrestrial laser scanning. Trees 27:1035–1047

Bayer D, Reischl A, Rötzer T, Pretzsch H (2018) Structural response of black locust (Robinia pseudoacacia L.) and small leaved lime (Tilia cordata mill.) to varying urban environments analyzed by terrestrial laser scanning: implications for ecological functions and services. Urban For Urban Green 35:129–138

Beatty RA, Heckman CT (1981) Survey of urban tree programs in the United States. Urban Ecol 5:81–102

Beck I, Jochner S, Gilles S, McIntyre M, Buters JTM, Schmidt-Weber C, Behrendt H, Ring J, Menzel A, Traidl-Hoffmann C (2013) High environmental ozone levels lead to enhanced allergenicity of birch pollen. PLoS One 8
Berland A, Shiflett SA, Shuster WD, Garmestani AS, Goddard HC, Herrmann DL, Hopton ME (2017) The role of trees in urban stormwater management. Landsc Urban Plan 162:167–177
Bernacchi CJ, Bagley JE, Serbin SP, Ruiz-Vera UM, Rosenthal DM, Vanloocke A (2013) Modelling C3 photosynthesis from the chloroplast to the ecosystem. Plant Cell Environ 36:1641–1657
Bodnaruk EW, Kroll CN, Yang Y, Hirabayashi S, Nowak DJ, Endreny TA (2017) Where to plant urban trees? A spatially explicit methodology to explore ecosystem service tradeoffs. Landsc Urban Plan 15:457–467
Bolund P, Hunhammar S (1999) Ecosystem services in urban areas. Ecol Econ 29:293–301
Bossel H (1996) TREEDYN3 forest simulation model. Ecol Model 90:187–227
Bowler DE, Buyung-Ali L, Knight TM, Pullin AS (2010) Urban greening to cool towns and cities: a systematic review of the empirical evidence. Landsc Urban Plan 97:147–155
Brantley HL, Hagler GSW, Deshmukh PJ, Baldauf RW (2014) Field assessment of the effects of roadside vegetation on near-road black carbon and particulate matter. Sci Total Environ 468-469:120–129
Bridgeman H, Warner R, Dodson J (1995) Urban biophysical environments. Oxford University Press, Oxford
Broadbent AM, Coutts AM, Nice KA, Demuzere M, Krayenhoff ES, Tapper NJ, Wouters H (2019) The air-temperature response to green/blue-infrastructure evaluation tool (TARGET v1. 0): an efficient and user-friendly model of city cooling. Geosci Model Dev 12:785–803
Bruns DA, Fetcher N (2008) CITYgreen watershed analysis of Toby Creek: an American Heritage River tributary. J Contemp Water Res Educ 138:29–37
Bruse M, Fleer H (1998) Simulating surface-plant-air interactions inside urban environments with a three dimensional numerical model. Environ Model Software 13:373–384
Buccolieri R, Santiago JL, Rivas E, Sanchez B (2018) Review on urban tree modelling in CFD simulations: aerodynamic, deposition and thermal effects. Urban For Urban Green 31:212–220
Budhathoki CB, Lynch TB, Guldin JM (2008) A mixed-effects model for the dbh–height relationship of shortleaf pine (*Pinus echinata* mill.). South J Appl For 32:5–11
Bühler O, Nielsen CN, Kristoffersen P (2006) Growth and phenology of established Tilia cordata street trees in response to different irrigation regimes. Arboric Urban For 32:3–9
Burschel P, Kürsten E, Larson BC (1993) Die Rolle von Wald und Forstwirtschaft im Kohlenstoffhaushalt. Forstl Forschungsber München 126:135
Cadenasso ML, Pickett STA, Schwarz K (2007) Spatial heterogeneity in urban ecosystems: reconceptualizing land cover and a framework for classification. Front Ecol Environ 5:80–88
Calfapietra C, Fares S, Manes F, Morani A, Sgrigna G, Loreto F (2013) Role of biogenic volatile organic compounds (BVOC) emitted by urban trees on ozone concentration in cities: a review. Environ Pollut 183:71–80
Canetti A, de Mattos PP, Braz EM, Netto SP (2017) Life pattern of urban trees: a growth-modelling approach. Urban Ecosyst 20:1057–1068
Cannell MGR, Thornley JHM (2000) Modelling the components of plant respiration: some guiding principles. Ann Bot 85:45–54
Causton DR (1985) Biometrical, structural and physiological relationships among tree parts. In: Cannel MGR, Jackson JE (eds) Attributes of trees as crop plants. Institute of Terrestrial Ecology, Huntingdon, pp 137–159
Chiba Y (1998) Architectural analysis of relationship between biomass and basal area based on pipe model theory. Ecol Model 108:219–225
Chmielewski FM, Rötzer T (2001) Response of tree phenology to climate change across Europe. Agric For Meteorol 108:101–112
Cienciala E, Černý M, Apltauer J, Exnerová Z (2005) Biomass functions applicable to European beech. J For Sci 51:147–154

Clark JR, Matheny N (2010) The research foundation to tree pruning: a review of the literature. Arboricult Urban For 36:110–120

Cleugh, H.A., Bui, E., Simon, D., Xu, J., Mitchell, V.G., 2005. The impact of suburban design on water use and microclimate

Comeau PG, Kimmins JP (1989) Above- and below-ground biomass and production of Lodgepole pine on sites with differing soil moisture regimes. Can J For Res 19:447–454

Constable JVH, Friend AL (2000) Suitability of process-based tree-growth models for addressing tree response to climate change. Environ Pollut 110:47–59

Dahlhausen J, Biber P, Rötzer T, Uhl E, Pretzsch H (2016) Tree species and their space requirements in six urban environments worldwide. Forests 7:19

Dahlhausen J, Rötzer T, Biber P, Uhl E, Pretzsch H (2017) Urban climate modifies tree growth in Berlin. Int J Biometeorol 62:795–808

Dale VH, Doyle TW, Shugart HH (1985) A comparison of tree growth models. Ecol Model 29:145–169

Däßler H-G (1991) Einfluß von Luftverunreinigungen auf die Vegetation. Ursachen – Wirkungen – Gegenmaßnahmen. Fischer Verlag, Jena

Denman L, May PB, Moore GM (2016) The potential role of urban forests in removing nutrients from Stormwater. J Environ Qual 45:207–214

Dimoudi A, Nikolopoulou M (2003) Vegetation in the urban environment: microclimatic analysis and benefits. Energ Buildings 35:69–76

Dobson MC (1991) De-icing salt damage to trees and shrubs. For Comm Bull 101:64

DWA (2018) Ermittlung der Verdunstung von Land- undWasserflächen. Merkblatt DWAM 504–1. Deutsche Vereinigung für Wasserwirtschaft, Abwasser und Abfall, Hennef, p 142

Dzierzanowski K, Popek R, Gawrońska H, Saebø A, Gawroński SW (2011) Deposition of particulate matter of different size fractions on leaf surfaces and in waxes of urban forest species. Int J Phytoremediation 13:1,037–1,046

Elliott RM, Adkins ER, Culligan PJ, Palmer MI (2018) Stormwater infiltration capacity of street tree pits: Quantifying the influence of different design and management strategies in New York City. Ecol Eng 111:157–166

Ennos AR, Armson A, Rahman MA (2014) How useful are urban trees: the lessons of the Manchester research project, trees, people and the built environment II. Institute of Chartered Foresters, pp 62–70

Enquist BJ, Niklas KJ (2001) Invariant scaling relations across tree-dominated communities. Nature 410:655–660

Enquist BJ, Brown JH, West GB (1998) Allometric scaling of plant energetics and population density. Nature 395:163–165

EPA UEPA (2003) Cooling summertime temperatures. Strategies to reduce urban heat islands. www.epa.gov/heatisland/resources/pdf/BasicsCompendium.pdf

Escobedo FJ, Kroeger T, Wagner JE (2011) Urban forests and pollution mitigation: analyzing ecosystem services and disservices. Environ Pollut 159:2078–2087

Fini A, Frangi P, Faoro M, Piatti R, Amoroso G, Ferrini F (2015) Effects of different pruning methods on an urban tree species: a four-year-experiment scaling down from the whole tree to the chloroplasts. Urban For Urban Green 14:664–674

Fini A, Frangi P, Moria J, Donzellic D, Ferrinia F (2017) Nature based solutions to mitigate soil sealing in urban areas: results from a 4-year study comparing permeable, porous, and impermeable pavements. Environ Res 156:443–454

Fisher JB, Badgley G, Blyth E (2012) Global nutrient limitation in terrestrial vegetation. Global Biogeochem Cycles 26:GB3007

Fleming LE (1988) Growth estimates of street trees in Central New Jersey. Rutgers University, New Brunswick, p 143

Fogel R (1983) Root turnover and productivity of coniferous forests. Plant and Soil 71:75–85

Fontes L, Bontemps J-D, Bugmann H, Van Oijen M, Gracia C, Kramer K, Lindner M, Rötzer T, Skovsgaard JP (2010) Models for supporting forest management in a changing environment. For Syst 19:8–29
Franceschini T, Martin-Ducup O, Schneider R (2016) Allometric exponents as a tool to study the influence of climate on the trade-off between primary and secondary growth in major north-eastern American tree species. Ann Bot 117:551–563
Franklin O, Johansson J, Dewar RC, Dieckmann U, McMurtrie RE, Brännström Å, Dybzinski R (2012) Modeling carbon allocation in trees: a search for principles. Tree Physiol 32:648–666
GALK eV (2019) In: Gartenamtsleiterkonferenz D (ed) GALK-Straßenbaumliste
Gao F, Calatayud V, García-Breijo F, Reig-Armiñana J, Feng Z (2016) Effects of elevated ozone on physiological, anatomical and ultrastructural characteristics of four common urban tree species in China. Ecol Indic 67:367–379
Gebert LL, Coutts AM, Tapper NJ (2019) The influence of urban canyon microclimate and contrasting photoperiod on the physiological response of street trees and the potential benefits of water sensitive urban design. Urban For Urban Greening 40:152–164
Georgi JN, Dimitriou D (2010) The contribution of urban green spaces to the improvement of environment in cities: Case study of Chania, Greece. Build Environ 45:1,401–1,414
Georgi NJ, Zafiriadis K (2006) The impact of park trees on microclimate in urban areas. Urban Ecosyst 9:195–209
Gessler A, Schaub M, McDowell NG (2017) The role of nutrients in drought-induced tree mortality and recovery. New Phytol 214:513–520
Gill S, Handley J, Pauleit S, Ennos R, Theuray N, Lindley S (2008) Characterising the urban environment of UK cities and towns: a template for landscape planning in a changing climate. Landsc Urban Plan 87:210–222
Gillner S, Vogt J, Tharang A, Dettmann S, Roloff A (2015) Role of street trees in mitigating effects of heat and drought at highly sealed urban sites. Landsc Urban Plan 143:33–42
Girardin MP, Raulier F, Bernier PY, Tardif JC (2008) Response of tree growth to a changing climate in boreal central Canada: A comparison of empirical, process-based, and hybrid modelling approaches. Ecol Model 213:209–228
Golubiewski NE (2006) Urbanization increases grassland carbon pools: effects of landscaping in Colorado's front range. Ecol Appl 16:555–571
Gregg JW, Jones CG, Dawson TE (2003) Urbanization effects on tree growth in the vicinity of New York City. Nature 424:183–187
Grote R, Haas E (2013) Modelling potential impacts of land-use change on BVOC-emissions by bioenergy production in Germany, impacts world 2013. Potsdam Institute for Climate Impact Research, Potsdam, pp 425–432
Grote R, Samson R, Alonso R, Amorim JH, Cariñanos P, Churkina G, Fares S, Thiec DL, Niinemets Ü, Mikkelsen TN, Paoletti E, Tiwary A, Calfapietra C (2016) Functional traits of urban trees: air pollution mitigation potential. Front Ecol Environ 14:543–550
Gunderson CA, Edwards NT, Walker AV, O'Hara KH, Campion CM, Hanson PJ (2012) Forest phenology and a warmer climate – growing season extension in relation to climatic provenance. Glob Chang Biol 18:2008–2025
Haase D (2009) Effects of urbanisation on the water balance – a long-term trajectory. Environ Impact Assess Rev 29:211–219
Haxeltine A, Prentice IC (1996) A general model for the light-use efficiency of primary production. Funct Ecol 10:551–561
Hewitt CN, Ashworth K, MacKenzie AR (2019) Using green infrastructure to improve urban air quality (GI4AQ). Ambio First Online
Hofman J, Stokkaer I, Snauwaert L, Samson R (2013) Spatial distribution assessment of particulate matter in an urban street canyon using biomagnetic leaf monitoring of tree crown deposited particles. Environ Pollut 183:123–132
Huxley JS (1932) Problems of relative growth. Lincoln Mac Veagh, Dial Press, New York

Illgen M (2011) Hydrology of urban environments. In: Niemelä J, Breuste JH, Elmqvist T, Guntenspergen G, James P, McIntyre M (eds) Urban ecology: patterns, processes, and applications. Oxford University Press, Oxford, pp 59–70

i-Tree (2019). www.itreetools.org

Janhäll S (2015) Review on urban vegetation and particle air pollution – deposition and dispersion. Atmos Environ 105:130–137

Jarvis PG, Leverenz JW (1983) Productivity of temperate, deciduous and evergreen forests. In: Lange OL, Nobel PS, Osmond CB, Ziegler H (eds) Physiological plant ecology IV. Encyclopedia of Plant Physiology. Springer-Verlag, Berlin, pp 233–280

Jeet Chaudhary I, Rathore D (2019) Dust pollution: its removal and effect on foliage physiology of urban trees. Sustain Cities Soc 51

Jim CY (1998) Urban soil characteristics and limitations for landscape planting in Hong Kong. Landsc Urban Plan 40:235–249

Jochner SC, Sparks TH, Estrella N, Menzel A (2012) The influence of altitude and urbanisation on trends and mean dates in phenology (1980–2009). J Biometeorol 56:387–394

Keyes MR, Grier CC (1981) Above-and below-ground net production in 40-years-old Douglas-fir stands on low and high productivity sites. Can J For Res 11:599–605

Kikuzawa K, Seiwa K, Lechowicz MJ (2013) Leaf longevity as a normalization constant in allometric predictions of plant production. PLoS One 8

Kimmins JP (1993) Scientific foundations for the simulation of ecosystem function and management in FORCYTE-11. Forestry Canada, Northern Forestry Centre, Edmonton, p 88

Kjelgren RK, Clark JR (1992) Microclimates and tree growth in tree urban spaces. J Environ Hortic 10:139–145

Kleiber M (1947) Body size and metabolic rate. Physiol Rev 27:511–541

Konarska J, Uddling J, Holmer B, Lutz M, Lindberg F, Pleijel H, Thorsson S (2016) Transpiration of urban trees and its cooling effect in a high latitude city. Int J Biometeorol 60:159–172

Kramer JK, Kozlowski TT (1979) Physiology of woody plants. Academic Press, London

Kramer H, Oldengarm J (2010) UrbTree: a tree growth model for the urban environment. Int Arch Photogramm Remote Sens Spat Inf Sci 38:4

Krayenhoff ES, Christen A, Martilli A, Oke TR (2014) A multi-layer radiation model for urban neighbourhoods with trees. Bound-Lay Meteorol 151:139–178

Kuyah S, Öborn I, Jonsson M (2017) Regulating ecosystem services delivered in agroforestry systems. In: Dagar JC, Tewari VP (eds) Agroforestry: anecdotal to modern science. Springer, Singapore, pp 797–815

Lahr EC, Dunn RR, Frank SD (2018) Getting ahead of the curve: cities as surrogates for global change. Proc R Soc B Biol Sci 285:1–9

Landsberg J (2003) Physiology in forest models: history and the future. For Biometry Modell Inf Sci 1:49–63

Landsberg J, Sands P (2010) Physiological ecology of forest production. Terrestrial ecology, 1st edn. Academic Press

Larondelle N, Hamstead ZA, Kremer P, Haase D, McPhearson T (2014) Applying a novel urban structure classification to compare the relationships of urban structure and surface temperature in Berlin and New York City. Appl Geogr 53:427–437

Le Roux X, Lacointe A, Escobar-Gutierrez A, Le Dizes S (2001) Carbon-based models of individual tree growth: a critical appraisal. Ann For Sci 58:469–506

Lee S-H, Park S-U (2008) A vegetated urban canopy model for meteorological and environmental modelling. Bound-Lay Meteorol 126:73–102

Lin J, Kroll CN, Nowak DJ, Greenfield EJ (2019) A review of urban forest modeling: implications for management and future research. Urban For Urban Greening 46:126–366

Litton CM, Raich JW, Ryan MG (2007) Carbon allocation in forest ecosystems. Glob Chang Biol 13:2089–2119

Livesley SJ, Baudinette B, Glover D (2014) Rainfall interception and stem flow by eucalypt street trees – The impacts of canopy density and bark type. Urban For Urban Green 13:192–197

Locosselli GM, de Camargo EP, Moreira TCL, Todesco E, de Fátima Andrade M, de André CDS, de André PA, Singer JM, Ferreira LS, Saldiva PHN, Buckeridge MS (2019) The role of air pollution and climate on the growth of urban trees. Sci Total Environ 666:652–661
Longcore T, Li C, Wilson JP (2004) Applicability of citygreen urban ecosystem analysis software to a densely built urban neighborhood. Urban Geogr 25:173–186
MA (2005) Millennium ecosystem assessment, ecosystems and human wellbeing. Synthesis, Washington
Maco SE, McPherson EG, Simpson JR, Peper PJ, Xiao Q (2003) City of San Francisco, California street tree resource analysis. In: CUFR-3, C.f.U.F.R., Pacific Southwest Research Station (ed) United States Forest Service, Davies
Mäkelä A (1990) Modelling structural-functional relationships in whole-tree growth; resource allocation. In: Dixon RK, Meldahl RS, Ruark GA, Warren WG (eds) Process modelling of forest growth responses to environmental stress. Timber Press, Portland, pp 81–95
Mäkelä A (2003) Process-based modelling of tree and stand growth: towards a hierarchical treatment of multiscale processes. Can J For Res 33:398–409
Matzarakis A, Mayer H, Iziomon MG (1999) Applications of a universal thermal index: physiological equivalent temperature. Int J Biometeorol 43:76–84
McCarthy MC, Enquist JB (2007) Consistency between an allometric approach and optimal partitioning theory in global patterns of plant biomass allocation. Funct Ecol 21:713–720
McHale MR, Burke IC, Lefsky MA, Peper PJ, McPherson EG (2009) Urban forest biomass estimates: is it important to use allometric relationships developed specifically for urban trees? Urban Ecosyst 12:95–113
McPherson EG (1998) Atmospheric carbon dioxide reduction by Sacramento's urban forest. J Arboricul 24:215–223
McPherson EG, Peper PJ (2012) Urban tree growth modeling. Arboricult Urban For 38:172–180
McPherson EG, Simpson JR (2001) Potential energy savings in buildings by an urban tree planting programme in California. Urban For Urban Green 2:73–86
McPherson EG, van Doorn NS, Peper PJ (2016) Urban tree database and allometric equations. US Department of Agriculture, Forest Service, Pacific Southwest Research Station, Albany, p 86
Meineke EK, Dunn RR, Sexton JO, Frank SD (2013) Urban warming drives insect pest abundance on street trees. PLoS One 8:e59687
Morgenroth J, Buchan GD (2009) Soil moisture and aeration beneath pervious and impervious pavements. Arboric Urban For 35:135–141
Moser A, Roetzer T, Pauleit S, Pretzsch H (2015) Structure and ecosystem services of small-leaved lime (Tilia cordata Mill.) and black locust (Robinia pseudoacacia L.) in urban environments. Urban For Urban Green 14:1110–1121
Moser A, Rotzer T, Pauleit S, Pretzsch H (2016) The urban environment can modify drought stress of small-leaved lime (Tilia cordata Mill.) and Black Locust (Robinia pseudoacacia L.). Forests 7
Moser A, Rahman MA, Pretzsch H, Pauleit S, Rotzer T (2017) Inter- and intraannual growth patterns of urban small-leaved lime (Tilia cordata mill.) at two public squares with contrasting microclimatic conditions. Int J Biometeorol 61:1095–1107
Moser-Reischl A, Uhl E, Rötzer T, Biber P, van Con T, Tan NT, Pretzsch H (2018) Effects of the urban heat island and climate change on the growth of Khaya senegalensis in Hanoi, Vietnam. For Ecosyst 5:37
Moser-Reischl A, Rahman MA, Pauleit S, Pretzsch H, Rötzer T (2019a) Growth patterns and effects of urban micro-climate on two physiologically contrasting urban tree species. Landsc Urban Plan 183:88–99
Moser-Reischl A, Rötzer T, Biber P, Ulbricht M, Uhl E, Qu L, Koike T, Pretzsch H (2019b) Growth of *Abies sachalinensis* along an urban gradient affected by environmental pollution in Sapporo, Japan. Forests 10:707
Müller N, Kuttler W, Barlag A-B (2014) Counteracting urban climate change: adaptation measures and their effect on thermal comfort. Theor Appl Climatol 115:243–257
My City's Trees (2019)

Nielsen CN, Bühler O, Kristoffersen P (2007) Soil water dynamics and growth of street and park trees. Arboric Urban For 33:231–245
Niinemets Ü, Valladares F (2006) Tolerance to shade, drought, and waterlogging of temperate northern hemisphere trees and shrubs. Ecol Monogr 76:521–547
Niklas KJ (1994) Plant allometry. University of Chicago Press, The scaling of form and process
Nowak DJ (1993) Atmospheric carbon reduction by urban trees. J Environ Manage 37:207–217
Nowak DJ (1994) Air pollution removal by Chicago's urban forest. In: McPherson EG, Nowak DJ, Rowntree RA (eds) Chicago's urban forest ecosystem: results of the Chicago urban forest climate project. USDA Forest Service, Radnor, pp 63–82
Nowak DJ (1996) Estimating leaf area and leaf biomass of open-grown deciduous urban trees. For Sci 42:504–507
Nowak DJ, Crane DE (2000) The Urban Forest Effects (UFORE) model: quantifying urban forest structure and functions. In: Hansen M, Burk T (eds) Integrated tools for natural resources inventories in the twenty-first century. US Department of Agriculture, Forest Service, North Central Forest Experiment Station, St. Paul, pp 714–720
Nowak DJ, Crane DE (2002) Carbon storage and sequestration by urban trees in the USA. Environ Pollut 116:381–389
Nowak DJ, Dwyer JF (2007) Understanding the benefits and costs of urban forest ecosystems. In: Kuser JE (ed) Urban and community forestry in the northeast. Springer, Berlin
Nowak DJ, Greenfield EJ (2012) Tree and impervious cover change in U.S. cities. Urban For Urban Green 11:21–30
Nowak DJ, Crane DE, Stevens JC, Hoehn RE (2003) The urban forest effects (UFORE) model: Field data collection manual. US Department of Agriculture Forest Service, Northeastern Research Station, Syracuse, pp 4–11
Nowak DJ, Kuroda M, Crane DE (2004) Tree mortality rates and tree population projections in Baltimore, Maryland, USA. Urban For Urban Green 2(3):139–147
Nowak DJ, Crane DE, Stevens JC, Hoehn RE, Walton JT, Bond J (2008) A ground-based method of assessing urban forest structure and ecosystem services. Arboric Urban For 36:347–358
Nowak DJ, Greenfield EJ, Hoehn RE, Lapoint E (2013) Carbon storage and sequestration by trees in urban and community areas of the United States. Environ Pollut 178:229–236
Oke TR (1989) The micrometeorology of the urban forest. Philos Trans R Soc Lond Ser B Biol Sci 324:335–349
Oke TR (2011) Urban heat islands. In: Douglas I, Goode D, Houck M, Wang R (eds) The Routledge handbook of urban ecology. Routledge, London, pp 120–131
Oke TR, Mills G, Christen A, Voogt JA (2017) Urban climates. Cambridge University Press, Cambridge
Pataki DE, Bowling DR, Ehleringer JR, Zobitz JM (2006) High resolution atmospheric monitoring of urban carbon dioxide sources. Geophys Res Lett 33:1–5
Pauleit S (2003) Towards successful urban street tree plantings: identifying the key requirements. Municipal Eng 156:43–56
Pauleit S, Breuste JH (2011) Land use and surface cover as urban ecological indicators. In: Niemelä J, Breuste JH, Elmqvist T, Guntenspergen G, James P, McIntyre M (eds) Urban ecology: patterns, processes, and applications
Pauleit S, Duhme F (2000) GIS assessment of Munich's urban forest structure for urban planning. J Arboricult 23:133–141
Pauleit S, Zölch T, Hansen R, Randrup TB, Konijnendijk van den Bosch CC (2017) Nature-based solutions and climate change – four shades of green. In: Kabisch S, Korn H, Stadler J, Bonn A (eds) Theory and practice of urban sustainability transitions. Nature-based solutions to climate change adaptation in urban areas. Linkages between science, policy and practice. Springer, Berlin
Peng L, Chen S, Liu Y, Wang J (2008) Application of CITYgreen model in benefit assessment of Nanjing urban green space in carbon fixation and runoff reduction. Front For China 3:177–182
Peper PJ, McPherson EG (1998) Comparison of five methods for estimating leaf area index of open grown deciduous trees. J Arboricult 24:98–111

Peper PJ, Alzate CP, McNeil JW, Hashemi J (2014a) Allometric equations for urban ash trees (Fraxinus spp.) in Oakville, Southern Ontario, Canada. Urban For Urban Green 13:175–183
Peper PJ, Alzate CP, McNeil JW, Hashemi J (2014b) Allometric equations for urban ash trees (Fraxinus spp.) in Oakville, Southern Ontario, Canada. Urban For Urban Green 13:175–183
Poorter L (1999) Growth responses of 15 rain-forest tree species to a light gradient: the relative importance of morphological and physiological traits. Funct Ecol 13:396–410
Pretzsch H (2009) Forest dynamics, growth and yield. Springer Verlag, Berlin
Pretzsch H (2010a) Forest dynamics, growth and yield: from measurement to model. Springer, Berlin
Pretzsch H (2010b) Re-evaluation of allometry: state-of-the-art and perspective regarding individuals and stands of woody plants. In: Lüttge U, Beyschlag W, Büdel B, Francis D (eds) Progress in botany, pp 339–369
Pretzsch H (2014) Canopy space filling and tree crown morphology in mixed-species stands compared with monocultures. For Ecol Manage 327:251–264
Pretzsch H, Schütze G (2005) Crown allometry and growing space efficiency of norway spruce (*Picea abies* [L.] Karst.) and European Beech (*Fagus sylvatica* L.) in pure and mixed stands. Plant Biol 7:628–639
Pretzsch H, Biber P, Ďurský J (2002) The single tree-based stand simulator SILVA: construction, application and evaluation. For Ecol Manage 162:3–21
Pretzsch H, Dieler J, Seifert T, Rötzer T (2012a) Climate effects on productivity and resource-use efficiency of Norway spruce (Picea abies [L.] Karst.) and European beech (Fagus sylvatica [L.]) in stands with different spatial mixing patterns. Trees Struct Funct 24
Pretzsch H, Matthew C, Dieler J (2012b) Allometry of tree crown structure. Relevance for space occupation at the individual plant level and for self-thinning at the stand level. In: Matyssek REA (ed) Growth and defence in plants. Springer, Berlin
Pretzsch H, Biber P, Uhl E, Dahlhausen J, Rötzer T, Caldentey J, Koike T, van Con T, Chavanne A, Seifert T, du Toit B, Farnden C, Pauleit S (2015) Crown size and growing space requirement of common tree species in urban centres, parks, and forests. Urban For Urban Green 14:466–479
Pretzsch H, Biber P, Uhl E, Dahlhausen J, Schütze G, Perkins D, Rötzer T, Caldentey J, Koike T, van Con T, Chavanne A, du Toit B, Foster K, Lefer B (2017) Climate change accelerates growth of urban trees in metropolises worldwide. Sci Rep 7:10
Pugh TAM, MacKenzie AR, Whyatt JD, Hewitt CN (2012) Effectiveness of green infrastructure for improvement of air quality in urban street Canyons. Environ Sci Technol 46:4692–7699
Rahman MA, Smith JG, Stringer P, Ennos AR (2011) Effect of rooting conditions on the growth and cooling ability of *Pyrus calleryana*. Urban For Urban Green 10:185–192
Rahman MA, Stringer P, Ennos AR (2013) Effect of pit design and soil composition on performance of *Pyrus calleryana* street trees in the establishment period. Arboric Urban For 39:256–266
Rahman MA, Armson D, Ennos AR (2014) Effect of urbanization and climate change in the rooting zone on the growth and physiology of Pyrus calleryana. Urban For Urban Green 13:325–335
Rahman MA, Armson D, Ennos AR (2015) A comparison of the growth and cooling effectiveness of five commonly planted urban tree species. Urban Ecosyst 18:371–389
Rahman MA, Moser A, Rötzer T, Pauleit S (2017) Below-canopy surface and air cooling effect of two contrasting tree species in urban street conditions. Landsc Urban Plan In Begutachtung
Rahman MA, Moser A, Rötzer T, Pauleit S (2017a) Microclimatic differences and their influence on transpirational cooling of *Tilia cordata* in two contrasting street canyons in Munich, Germany. Agric For Meteorol 232:443–456
Rahman MA, Moser A, Rötzer T, Pauleit S (2017b) Within canopy temperature differences and cooling ability of *Tilia cordata* trees grown in urban conditions. Build Environ 114:118–128
Rahman MA, Moser A, Gold A, Rötzer T, Pauleit S (2018) Vertical air temperature gradients under the shade of two contrasting urban tree species during different types of summer days. Sci Total Environ 633:100–111

Rahman MA, Moser A, Rötzer T, Pauleit S (2019a) Comparing the transpirational and shading effects of two contrasting urban tree species. Urban Ecosyst

Rahman MA, Moser A, Anderson M, Zhang C, Rötzer T, Pauleit S (2019b) Comparing the infiltration potentials of soils beneath the canopies of two contrasting urban tree species. Urban For Urban Green 38:22–32

Rahman MA, Hartmann C, Moser-Reischl A, von Strachwitz MF, Paeth H, Pretzsch H, Pauleit S, Rötzer T (2020) Tree cooling effects and human thermal comfort under contrasting species and sites. Agric For Meteorol 287:107947

Randrup TC (1997) Soil compaction on construction sites. J Arboricult 23:207–210

Redon EC, Lemonsu A, Masson V, Morille B, Musy M (2017) Implementation of street trees within the solar radiative exchange parameterization of TEB in SURFEX v8.0. Geosci Model Dev 10:385–411

Rodríguez Martín JA, De Arana C, Ramos-Miras JJ, Gil C, Boluda R (2015) Impact of 70 years urban growth associated with heavy metal pollution. Environ Pollut 196:156–163

Roloff A (2013) Bäume in der Stadt. Besonderheiten – Funktion – Nutzen – Arten – Risiken. Ulmer, Stuttgart

Roloff A, Korn S, Gillner S (2009) The climate-species-matrix to select tree species for urban habitats considering climate change. Urban For Urban Green 8:295–308

Roman LA, Scatena FN (2011) Street tree survival rates: meta-analysis of previous studies and application to a field survey in Philadelphia, PA, USA. Urban For Urban Green 10(4):269–274

Rötzer T, Pretzsch H (2018) Stadtbäume im Klimawandel II. Wuchsverhalten, Umweltleistungen und Perspektiven. TU München, Chair of Forest Growth and Yield Science, Freising

Rötzer T, Wittenzeller M, Häckel H, Nekovar J (2000) Phenology in central Europe – differences and trends of spring phenophases in urban and rural areas. Int J Biometeorol 44:60–66

Rötzer T, Seifert T, Pretzsch H (2009) Modelling above and below ground carbon dynamics in a mixed beech and spruce stand influenced by climate. Eur J Forest Res 128:171–182

Rötzer T, Dieler J, Mette T, Moshammer R, Pretzsch H (2010) Productivity and carbon dynamics in managed Central-European Forests depending on site conditions and thinning regimes. Forestry 83:483–496

Rötzer T, Häberle KH, Kallenbach C, Matyssek R, Schütze G, Pretzsch H (2017) Tree species and size drive water consumption of beech/spruce forests – a simulation study highlighting growth under water limitation. Plant and Soil 418:337–356

Rötzer T, Rahman MA, Moser-Reischl A, Pauleit S, Pretzsch H (2019) Process based simulation of tree growth and ecosystem services of urban trees under present and future climate conditions. Sci Total Environ 676:651–664

Rubner M (1931) Die Gesetze des Energieverbrauchs bei der Ernährung. Wien, Berlin

Saebo A, Benedikz T, Randrup TB (2003) Selection of trees for urban forestry in the Nordic countries. Urban For Urban Green 2:101–114

Santantonio D, Hermann RK, Overton WS (1977) Root biomass studies in forest ecosystems. Pedobiologia 17:1–31

Sauerwein M (2011) Urban soils – characterization, pollution, and relevance in urban ecosystems. In: Niemelä J, Breuste JH, Elmqvist T, Guntenspergen G, James P, McIntyre M (eds) Urban ecology: patterns, processes, and applications. Oxford University Press, Oxford

Saunders SM, Dade E, Niel K (2011) An urban forest effects (UFORE) model study of the integrated effects of vegetation on local air pollution in the Western Suburbs of Perth, WA. In: 19th international congress on modelling and simulation (MODSIM2011). Modelling and Simulation Society of Australia and New Zealand, Perth, Australia

Scalenghe R, Marsan FA (2009) The anthropogenic sealing of soils in urban areas. Landsc Urban Plan 90:1–10

Schenk HJ (1996) Modeling the effects of temperature on growth and persistence of tree species: a critical review of tree population models. Ecol Model 92:1–32

Schooling JT, Carlyle-Moses DE (2015) The influence of rainfall depth class and deciduous tree traits on stemflow production in an urban park. Urban Ecosyst 18:1261–1284

Schwarz N (2010) Urban form revisited – selecting indicators for characterising European cities. Landsc Urban Plan 96:29–47
Seidl R, Rammer W, Scheller RM, Spies TA (2012) An individual-based process model to simulate landscape-scale forest ecosystem dynamics. Ecol Model 231:87–100
Shashua-Bar L, Hoffman ME (2002) The Green CTTC model for predicting the air temperature in small urban wooded sites. Build Environ 37:1279–1288
Shashua-Bar L, Pearlmutter D, Erell E (2009) The cooling efficiency of urban landscape strategies in a hot dry climate. Landsc Urban Plan 92:179–186
Shinozaki KK, Nozumi YK, Kira T (1964a) A quantitative analysis of plant form-the pipe model theory. I. Basic analysis. Jpn J Ecol 14:97–105
Shinozaki KK, Yoda K, Hozumi K, Kira T (1964b) A quantitative analysis of plant form-Pipe model theory. II. Further evidences of the theory and its application in forest ecology. Jpn J Ecol 14:133–139
Sjöman H, Gunnarsson A, Pauleit S, von Bothmer R (2012) Selection approach of urban trees for inner-city environments: learning from nature. Arboric Urban For 38:194–204
Smithers RJ, Doick KJ, Burton A, Sibille R, Steinbach D, Harris R, Groves L, Blicharska M (2018) Comparing the relative abilities of tree species to cool the urban environment. Urban Ecosyst
Soares AL, Rego FC, McPherson EG, Simpson JR, Peper PJ, Xiao Q (2011) Benefits and costs of street 980 trees in Lisbon, Portugal. Urban For Urban Greening 10:69–78
Soil Conservation Service (1986) Urban hydrology for small watersheds. TR-55. United States Department of Agriculture
Stewart DI, Oke TR (2012) Local climate zones for urban temperature studies. Bull Am Meteorol Soc 93:1879–1900
Stoffberg GH, van Rooyen MW, van der Linde MJ, Groeneveld HT (2008) Predicting the growth in tree height and crown size of three street tree species in the City of Tshwane, South Africa. Urban For Urban Green 7:259–264
Stratopoulos LMF, Duthweiler S, Haberle KH, Pauleit S (2018) Effect of native habitat on the cooling ability of six nursery-grown tree species and cultivars for future roadside plantings. Urban For Urban Green 30:37–45
Stratopoulos LMF, Zhang C, Duthweiler S, Häberle KH, Rötzer T, Xu C, Pauleit S (2019a) Tree species from two contrasting habitats for use in harsh urban environments respond differently to extreme drought. Int J Biometeorol
Stratopoulos LMF, Zhang C, Häberle K-H, Pauleit S, Duthweiler S, Pretzsch H, Rötzer T (2019b) Effects of drought on the phenology, growth, and morphological development of three urban tree species and cultivars. Sustainability 11:5117
Szota C, Coutts AM, Thom JK, Virahsawmy HK, Fletcher TD, Livesley SJ (2019) Street tree stormwater control measures can reduce runoff but may not benefit established trees. Landsc Urban Plan 182:144–155
Taha H, Douglas S, Haney J (1997) Mesoscale meteorological and air quality impacts of increased urban albedo and vegetation. Energ Buildings 25:169–177
Takashima A, Kume A, Yoshida S, Murakami T, Kajisa T, Mizoue N (2009) Discontinuous DBH–height relationship of Cryptomeria japonica on Yakushima Island: effect of frequent typhoons on the maximum height. Ecol Res 24:1003–1011
Tavares R, Calmet I, Dupont S (2015) Modelling the impact of green infrastructures on local microclimate within an idealized homogeneous urban canopy. In: 9th international conference on urban climate (ICUC9), Toulouse (France), pp 1–6
TEEB (2011) The economics of ecosystems and biodiversity. TEEB manual for cities: ecosystem services in urban management
Teissier G (1934) Dysharmonies et discontinuités dans la Croissance. Hermann, Paris
Timilsina N, Beck J, Eames MS, Hauer R, Werner L (2017) A comparison of local and general models of leaf area and biomass of urban trees in USA. Urban For Urban Greening 24:157–163
Tiwary A, Williams ID, Heidrich O, Namdeo A, Bandaru V, Calfapietra C (2016) Development of multi-functional streetscape green infrastructure using a performance index approach. Environ Pollut 208:209–220
TreePlotter (2019)

Troxel B, Piana M, Ashton MS, Murphy-Dunning C (2013) Relationships between bole and crown size for young urban trees in the northeastern USA. Urban For Urban Greening 12:144–153

Viswanathan B, Volder A, Watson WT, Aitkenhead-Peterson JA (2011) Impervious and pervious pavements increase soil CO_2 concentrations and reduce root production of American sweetgum (*Liquidambar styraciflua*). Urban For Urban Greening 10:133–139

Volder A, Viswanathan B, Watson WT (2014) Pervious and impervious pavements reduce root production and decrease lifespan of fine roots of mature sweetgum trees. Urban Ecosyst 17:445–453

von Bertalanffy L (1951) Theoretische Biologie: II. Band, Stoffwechsel, Wachstum. A Francke AG, Bern

von Gadow K (2003) Waldstruktur und Wachstum. Universitätsdrucke Göttingen, University Press, Göttingen

Watson GW, Kelsey P (2006) The impact of soil compaction on soil aeration and fine root density of Quercus palustris. Urban For Urban Greening 4:69–74

Watson GW, Hewitt AM, Custic M, Lo M (2014) The management of tree root systems in urban and suburban settings II: a review of soil influence on root growth. Arboric Urban For 40:249–271

Watt MS, Kirschbaum MUF (2011) Moving beyond simple linear allometric relationships between tree height and diameter. Ecol Model 222:3910–3916

Weltecke K, Gaertig T (2012) Influence of soil aeration on rooting and growth of the Beuys-trees in Kassel, Germany. Urban For Urban Greening 11:329–338

Wesely ML, Hicks BB (2000) A review of the current status of knowledge on dry deposition. Atmos Environ 34:2261–2282

West GB, Brown JH, Enquist BJ (1997) A general model for the origin of allometric scaling laws in biology. Science 276:122–126

West GB, Enquist BJ, Brown JH (2009) A general quantitative theory of forest structure and dynamics. Proc Natl Acad Sci 106:7040–7045

WHO (2018) WHO ambient (outdoor) air quality database Summary results, update 2018. https://www.who.int/airpollution/data/AAP_database_summary_results_2018_final2.pdf?ua=1

Woldegerima T, Yeshitela K, Lindley S (2016) Characterizing the urban environment through urban morphology types (UMTs) mapping and land surface cover analysis: the case of Addis Ababa, Ethiopia. Urban Ecosystems

Xiao Q, McPherson EG, Ustin SL, Grismer ME, Simpson JR (2000a) A new approach to modeling tree rainfall interception. J Geophys Res 105:29–173

Xiao Q, McPherson EG, Ustin SL, Grismer ME, Simpson JR (2000b) Winter rainfall interception by two mature open-grown trees in Davis, California. Hydrol Process 14:763–784

Xiao QF, McPherson EG, Ustin SL, Grismer ME (2000c) A new approach to modeling tree rainfall interception. J Geophys Res Atmos 105:29173–29188

Yoon TK, Park C-W, Lee SJ, Ko S, Kim KN, Son Y, Lee K-H, Oh S, Lee W-K, Son Y (2013) Allometric equations for estimating the aboveground volume of five common urban street tree species in Daegu, Korea. Urban For Urban Green 12:344–349

Zölch T, Henze L, Keilholz P, Pauleit S (2017) Regulating urban surface runoff through nature-based solutions – an assessment at the micro-scale. Environ Res 157:135–144

9783030686192